THE WILEY BICENTENNIAL—KNOWLEDGE FOR GENERATIONS

*E*ach generation has its unique needs and aspirations. When Charles Wiley first opened his small printing shop in lower Manhattan in 1807, it was a generation of boundless potential searching for an identity. And we were there, helping to define a new American literary tradition. Over half a century later, in the midst of the Second Industrial Revolution, it was a generation focused on building the future. Once again, we were there, supplying the critical scientific, technical, and engineering knowledge that helped frame the world. Throughout the 20th Century, and into the new millennium, nations began to reach out beyond their own borders and a new international community was born. Wiley was there, expanding its operations around the world to enable a global exchange of ideas, opinions, and know-how.

For 200 years, Wiley has been an integral part of each generation's journey, enabling the flow of information and understanding necessary to meet their needs and fulfill their aspirations. Today, bold new technologies are changing the way we live and learn. Wiley will be there, providing you the must-have knowledge you need to imagine new worlds, new possibilities, and new opportunities.

Generations come and go, but you can always count on Wiley to provide you the knowledge you need, when and where you need it!

WILLIAM J. PESCE
PRESIDENT AND CHIEF EXECUTIVE OFFICER

PETER BOOTH WILEY
CHAIRMAN OF THE BOARD

Visualizing

GEOLOGY

Barbara W. Murck, PhD
University of Toronto

Brian J. Skinner, PhD
Yale University

Dana Mackenzie, PhD

BICENTENNIAL
BICENTENNIAL
1807
WILEY
2007
BICENTENNIAL
BICENTENNIAL

In collaboration with
THE NATIONAL GEOGRAPHIC SOCIETY

CREDITS

EXECUTIVE PUBLISHER Jay O'Callaghan
MANAGING DIRECTOR Helen McInnis
EXECUTIVE EDITOR Ryan Flahive
EXECUTIVE MARKETING MANAGER Jeffrey Rucker
DIRECTOR OF DEVELOPMENT Barbara Heaney
SENIOR DEVELOPMENT EDITOR Nancy Perry
MARKETING MANAGER Emily Streutker
PRODUCTION MANAGER Kelly Tavares
PRODUCTION ASSISTANT Courtney Leshko
ASSISTANT EDITOR Laura Kelleher
PROGRAM ASSISTANT Christine Moore
CREATIVE DIRECTOR Harry Nolan
COVER DESIGNER Harry Nolan
INTERIOR DESIGN Vertigo Design
PHOTO RESEARCHERS Tara Sanford/Mary Ann Price/
Stacy Gold, National Geographic Society
ILLUSTRATION COORDINATOR Anna Melhorn

Cover credits: Main image: Melissa Farlow/NG Image Collection; Inset images (from left to right): Richard Olsenius/NG Image Collection; Michael Melford/NG Image Collection; David McLain/NG Image Collection; Frans Lanting/NG Image Collection; Michael Melford/NG Image Collection.

This book was set in Times New Roman by Pre-Press Company, Inc., printed and bound by Quebecor World. The cover was printed by Phoenix Color.

To order books or for customer service please, call 1-800-CALL WILEY (225-5945).

ISBN 978-0-471-74727-7

Printed in the United States of America

10 9 8 7 6 5 4 3 2

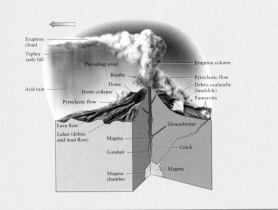

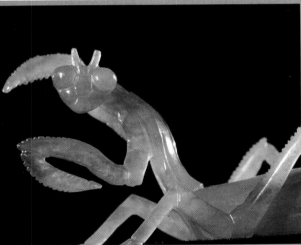

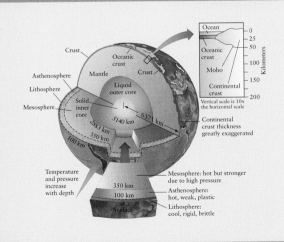

Visualizing Geology is designed to help your students learn effectively. Created in collaboration with the National Geographic Society and our Wiley Visualizing Consulting Editor, Professor Jan Plass of New York University, *Visualizing Geology* integrates rich visuals and media with text to direct students' attention to important information. This approach represents complex processes, organizes related pieces of information, and integrates information into clear representations. Beautifully illustrated, *Visualizing Geology* shows your students what the discipline is all about—its main concepts and applications—while also instilling an appreciation and excitement about the richness of the subject.

Visuals, as used throughout this text, are instructional components that display facts, concepts, processes, or principles. They create the foundation for the text and do more than simply support the written or spoken word. The visuals include diagrams, graphs, maps, photographs, illustrations, schematics, animations, and videos.

Why should a textbook based on visuals be effective? Research shows that we learn better from integrated text and visuals than from either medium separately. Beginners in a subject benefit most from reading about the topic, attending class, and studying well-designed and integrated visuals. A visual, with good accompanying discussion, really can be worth a thousand words!

Well-designed visuals can also improve the efficiency with which information is processed by a learner. The more effectively we process information, the more likely it is that we will learn. This processing of information takes place in our working memory. As we learn, we integrate new information in our working memory with existing knowledge in our long-term memory.

Have you ever read a paragraph or a page in a book, stopped, and said to yourself: "I don't remember one thing I just read?" This may happen when your working memory has been overloaded, and the text you read was not successfully integrated into long-term memory. Visuals don't automatically solve the problem of overload, but well-designed visuals can reduce the number of elements that working memory must process, thus aiding learning.

You, as the instructor, facilitate your students' learning. Well-designed visuals, used in class, can help you in that effort. Here are six methods for using the visuals in the *Visualizing Geology* in classroom instruction.

1. **Assign students to study visuals in addition to reading the text.**

   Instead of assigning only one medium of presentation, it is important to make sure your students know that the visuals are just as essential as the text.

2. **Use visuals during class discussions or presentations.**

   By pointing out important information as the students look at the visuals during class discussions, you can help focus students' attention on key elements of the visuals and help them begin to organize the information and develop an integrated model of understanding. The verbal explanation of important information combined with the visual representation can be highly effective.

3. **Use visuals to review content knowledge.**

   Students can review key concepts, principles, processes, vocabulary, and relationships displayed visually. Better understanding results when new information in working memory is linked to prior knowledge.

4. **Use visuals for assignments or when assessing learning.**

   Visuals can be used for comprehension activities or assessments. For example, students could be asked to identify examples of concepts portrayed in visuals. Higher-level thinking activities that require critical thinking, deductive and inductive reasoning, and prediction can also be based on visuals. Visuals can be very useful for drawing inferences, for predicting, and for problem solving.

5. **Use visuals to situate learning in authentic contexts.**

   Learning is made more meaningful when a learner can apply facts, concepts, and principles to realistic situations or examples. Visuals can provide that realistic context.

6. Use visuals to encourage collaboration.

Collaborative groups often are required to practice interactive processes such as giving explanations, asking questions, clarifying ideas, and argumentation. These interactive, face-to-face processes provide the information needed to build a verbal mental model. Learners also benefit from collaboration in many instances such as decision making or problem solving.

Visualizing Geology not only aids student learning with extraordinary use of visuals, but it also offers an array of remarkable photos, media, and film from the National Geographic Society collections. Students using *Visualizing Geology* also benefit from the long history and rich, fascinating resources of National Geographic.

National Geographic has also performed an invaluable service in fact-checking *Visualizing Geology:* they have verified every fact in the book with two outside sources, ensuring the accuracy and currency of the text.

Given all of its strengths and resources, *Visualizing Geology* will immerse your students in the discipline, and its main concepts and applications, while also instilling an appreciation and excitement about the richness of the subject area.

Additional information on learning and instructional design is provided in a special guide to using this book, *Learning from Visuals: How and Why Visuals Can Help Students Learn*, prepared by Matthew Leavitt of Arizona State University. This article is available at the Wiley web site: www.wiley.com/college/visualizing. The online *Instructor's Manual* also provides guidelines and suggestions on using the text and visuals most effectively.

Visualizing Geology also offers a rich selection of visuals in the supplementary materials that accompany the book. To complete this robust package the following materials are available: Test Bank with visuals used in assessment, PowerPoints, Image Gallery to provide you with the same visuals used in the text, web-based learning materials for homework and assessment including images, video, and media resources from National Geographic.

PREFACE

The goal of *Visualizing Geology* is to introduce students to Earth system science through the distinctive mode of visual learning that is the hallmark of Wiley Visualizing. Students will learn that the geological features we see and experience on the landscapes around us result from the interaction of many separate phenomena, which extend from the Earth's core to the fringes of outer space. *Visualizing Geology* organizes the interactions into three grand cycles: the tectonic cycle, the rock cycle, and the water cycle. We place special emphasis on plate tectonics because it is an organizing principle that unites many disparate observations into a coherent pattern of geologic activity on our planet. Case studies throughout the book and a final chapter devoted to Earth's resources will bring the science of geology into contact with students' everyday lives. Current events that students learn about in the newspapers or on TV news, ranging from global warming and landslides in Southern California, to the catastrophic tsunami in the Indian Ocean in December 2004, will fit into a larger picture of how Earth works and why such things happen. Students will also learn to appreciate the way that human actions affect Earth systems and vice versa.

The unique format of Wiley Visualizing allows us to reinforce the textual content with arresting images that are, in many cases, "the next best thing to being there." Geology, perhaps more than any other science, invites us to travel outside our familiar environment to distant parts of the world. Such travel is beyond the means of the ordinary college student (and perhaps some of their professors, too). Who better to take us on an armchair visit to these places than the photographers of National Geographic, who have been traveling the world and recording it visually for more than a century? The authors of *Visualizing Geology* were given exclusive access to National Geographic's vast photo archive. It was a particular treat to burrow through this archive and occasionally turn up the true photographic gem: a solitary wolf footprint protruding *above* the surrounding snow in Siberia; a fairy-tale oasis in the Sahara, showing both the majesty of the desert sands and the transforming touch of water. With such photos, and with features such as "Amazing Places" at the end of every chapter, we unabashedly seek to instill what words sometimes cannot: a sense of wonder about the planet we call home.

Visualizing Geology is organized as follows. In Chapters 1 through 4, we outline Earth system science as a whole, describe the various kinds of rocks and minerals, explain the ways in which geologists learn about Earth's changes over time, and present the unifying theory of plate tectonics.

In Chapters 5 and 6, we discuss the hazards of earthquakes and volcanoes, and explain how they relate to the tectonic cycle. Chapters 7 through 10 describe the major processes of the rock cycle—weathering, erosion, and metamorphism. In addition, the student will learn some of the basics of geologic maps and the terminology of structural geology. In Chapters 11 through 13, we turn our attention to the water cycle and explain the ubiquitous effects of water on Earth's surface, underground, and in the atmosphere. We devote a full chapter to deserts and glaciers, the two extreme environments that in recent years have become bellwethers of climate change. Finally, Chapters 14 and 15 re-integrate the various parts of the Earth system to draw conclusions about two topics of great interest to the students and to society as a whole: the history of life on Earth, and the future of the natural resources on which the modern population now depends.

A new feature of this book, we hope, is a freshness of tone and point of view that anticipates the questions and difficulties an introductory student is likely to have. To this end, Barbara Murck and Brian Skinner, both geologists and university professors, have enlisted a third author, Dana Mackenzie, a science journalist who is *not* a professional geologist but is experienced at explaining science to the general public. This opportunity for scientists to collaborate with science communicators has been a new experience and a delight for all three of us.

Visualizing Geology is intended as a textbook for an introductory college-level course in geology. Because our emphasis is on the processes of physical geology, it could be used as well for an introductory physical geology course. We do not expect that most of the students who read this book will go on to become geologists, but we hope that all will come to have a better understanding of, and appreciation for, their home planet as a result. For those students who do wish to take further courses in the field—and we hope there are many—we have provided a solid, sufficient and challenging background to do so with confidence.

ILLUSTRATED BOOK TOUR

A number of pedagogical features using visuals have been developed specifically for *Visualizing Geology*. Presenting the highly varied and often technical concepts woven throughout environmental science raises challenges for reader and instructor alike. The **Illustrated Book Tour** on the following pages provides a guide to the diverse features contributing to *Visualizing Geology's* pedagogical plan.

CHAPTER INTRODUCTIONS illustrate certain concepts in the chapter with concise stories about some of today's most pressing environmental issues. These narratives are featured alongside striking accompanying photographs. The chapter openers also include illustrated **CHAPTER OUTLINES** that use thumbnails of illustrations from the chapter to refer visually to the content.

VISUALIZING features are specially designed multi-part visual spreads that focus on a key concept or topic in the chapter, exploring it in detail or in broader context using a combination of photos and figures.

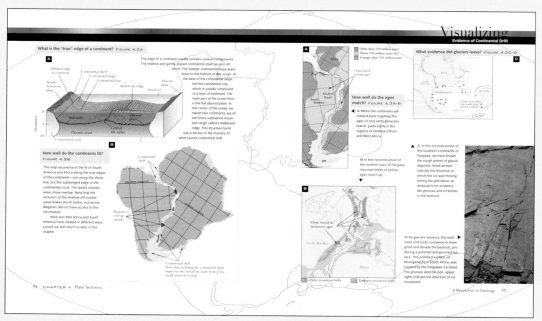

WHAT A GEOLOGIST SEES features highlight a concept or phenomenon, using photos and figures that would stand out to a professional in the field, and helping students to develop observational skills.

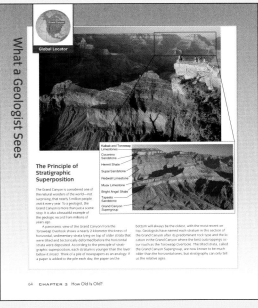

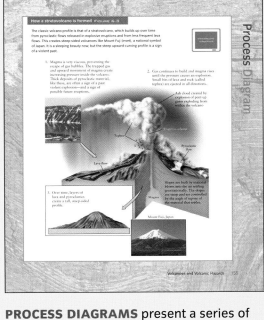

PROCESS DIAGRAMS present a series of figures or a combination of figures and photos that describe and depict a complex process, helping students to observe, follow, and understand the process.

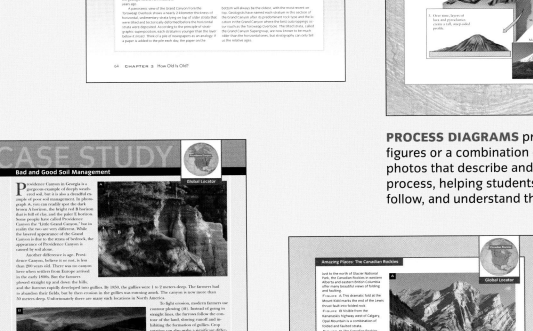

The illustrated **CASE STUDIES** that cap off the text sections of each chapter offer a wide variety of in-depth examinations that address important issues in the field of geology.

The **AMAZING PLACES** section takes the student to a unique location in the world that provides a vivid illustration of a theme in the chapter. Students could easily visit most of the Amazing Places someday and so continue their geologic education after they finish this book.

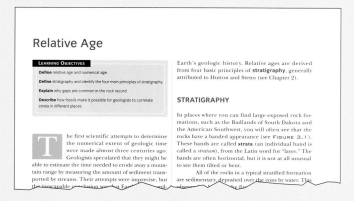

Relative Age

LEARNING OBJECTIVES

Define relative age and numerical age.

Define stratigraphy and identify the four main principles of stratigraphy.

Explain why gaps are common in the rock record.

Describe how fossils make it possible for geologists to correlate strata in different places.

The first scientific attempts to determine the numerical extent of geologic time were made almost three centuries ago. Geologists speculated that they might be able to estimate the time needed to erode away a mountain range by measuring the amount of sediment transported by streams. Their attempts were imprecise, but the inescapable conclusion was that Earth...

Earth's geologic history. Relative ages are derived from four basic principles of **stratigraphy**, generally attributed to Hutton and Steno (see Chapter 2).

STRATIGRAPHY

In places where you can find large exposed rock formations, such as the Badlands of South Dakota and the American Southwest, you will often see that the rocks have a banded appearance (see FIGURE 3.1). These bands are called **strata** (an individual band is called a *stratum*), from the Latin word for "layer." The bands are often horizontal, but it is not at all unusual to see them tilted or bent.

All of the rocks in a typical stratified formation are sedimentary, deposited over the eons by water. This observation led to the first...

Earth, and we will have much more to say about it in Chapter 14. Aside from their biologic interest, fossils also have great practical value to geologists. About the same time that James Hutton was working in Scotland, a young surveyor named William Smith was laying out canal routes in southern England. As the canals were excavated, Smith noticed that each group of strata contained a specific assemblage of fossils. In time, he could look at a specimen of rock from any sedimentary layer in southern England and name the stratum and its position in the sequence of strata. This skill enabled him to predict what kind of rock the canal excavators would encounter and how long it would take to dig through it. It also earned him the nickname "Strata."

The stratigraphic ordering of fossil assemblages is known as the principle of *faunal and floral succession*. Faunal means animals, floral means plants; succession means that new species succeed earlier ones as they evolve. William Smith's practical discovery turned out to be of great scientific importance. Geologists soon demonstrated that the faunal succession in

northern France is the same as that found by Smith in southern England. By the middle of the nineteenth century, it had become clear that faunal succession is essentially the same everywhere. Thus Smith's practical observations led to a means of worldwide **correlation**, which is the most important way of filling the gaps in the geologic record (see FIGURE 3.6).

■ **correlation**
A method of equating the ages of strata that come from two or more different places.

CONCEPT CHECK STOP

What are the four principles on which stratigraphy is based?

How can the principles of stratigraphy be used to determine the relative ages of strata?

What is the difference between conformable and unconformable strata?

How can fossils provide information about the relative ages of rocks?

LEARNING OBJECTIVES at the beginning of each section head indicate in behavioral terms what the student must be able to do to demonstrate mastery of the material in the chapter.

CONCEPT CHECK questions at the end of each section give students the opportunity to test their comprehension of the learning objectives.

GLOBAL LOCATOR MAPS, prepared specifically for this book by National Geographic, accompany figures addressing issues encountered in a particular geographic region. These feature insets of hemispheric locator maps help students visualize where the area discussed is situated on a continent.

NATIONAL GEOGRAPHIC SOCIETY MAPS are featured throughout the book.

■ **numerical age**
The age of a rock or geological feature in years before the present.

■ **unconformity**
A substantial gap in a stratigraphic sequence that marks the absence of part of the rock record.

MARGINAL GLOSSARY TERMS (in green boldface) introduce each chapter's most important terms, often reinforced with a thumbnail photograph. The second most important terms appear in **black boldface** and are defined in the text; the third level of important terms appear in italics.

ILLUSTRATIONS AND PHOTOS support concepts covered in the text, elaborate on relevant issues, and add visual detail. Many of the photos originate from National Geographic's rich sources.

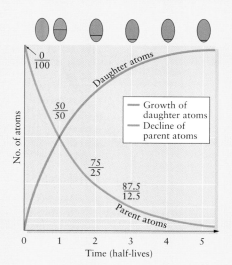

TABLES AND GRAPHS, with data sources cited at the end of the text, summarize and organize important information.

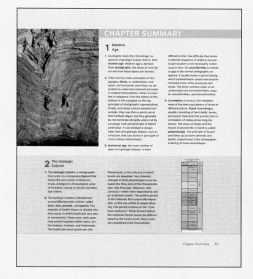

The **CHAPTER SUMMARY** revisits each learning objective and redefines each marginal glossary term, featured in boldface here, and included in a list of **KEY TERMS**. Students are thus able to study vocabulary words in the context of related concepts. Each portion of the Chapter Summary is illustrated with a relevant photo or line art from its respective chapter section.

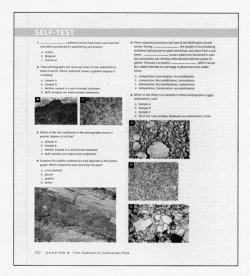

SELF-TESTS at the end of each chapter provide a series of multiple-choice questions, many of them incorporating visuals from the chapter, that review the major concepts.

WHAT IS HAPPENING IN THIS PICTURE? is an end-of-chapter feature that presents students with a photograph relevant to chapter topics but that illustrates a situation students are not likely to have encountered previously. The photograph is paired with questions designed to stimulate creative thinking.

MEDIA AND SUPPLEMENTS

Visualizing Geology is accompanied by a rich array of media and supplements that incorporate the visuals from the textbook extensively to form a pedagogically cohesive package. For example, a Process Diagram from the book appears in the Instructor's Manual with suggestions on using it as a PowerPoint in the classroom; it may be the subject of a short video or an online animation; and it may also appear with questions in the Test Bank, as part of the chapter review, homework assignment, assessment questions, and other online features.

Media Advisory Board:
Special thanks to the following people who contributed to creation and organization of the media and supplements for *Visualizing Geology*:

Michael Bradley
Eastern Michigan University

Mark Francek
Eastern Michigan University

Christopher DiLeonardo
Foothill College

Arthur Lee
Roane State Community College

INSTRUCTOR SUPPLEMENTS

VIDEOS

A rich collection of videos, many of them from the award-winning National Geographic Film Collection, have been selected to accompany and enrich the text. Each chapter includes two to three video clips, available online as digitized streaming video, that illustrate and expand on a concept or topic to aid student understanding.

This logo in the text indicates that video from the archives of the National Geographic Society is available.

 Accompanying each of the videos is contextualized commentary and questions that can further develop student understanding.

The videos are available on the main website: (www.wiley.com/college/murck).

POWERPOINT PRESENTATIONS AND IMAGE GALLERY

A complete set of highly visual PowerPoint presentations by Christopher DiLeonardo of Foothill College is available online to enhance classroom presentations. Tailored to the text's topical coverage and learning objectives, these presentations are designed to convey key text concepts, illustrated by embedded text art.

All photographs, figures, maps, and other visuals from the text are online and can be used as you wish in the classroom. These online electronic files allow you to easily incorporate them into your PowerPoint presentations as you choose, or to create your own overhead transparencies and handouts.

As noted previously, by using visuals in class discussions, you can help focus students' attention on key elements of the visuals and help them begin to organize the information and develop greater understanding. The classroom explanation of important information combined with the visual representation can be highly effective.

TEST BANK (AVAILABLE IN WILEYPLUS AND ELECTRONIC FORMAT)

The visuals from the textbook are also included in the Test Bank by Christopher DiLeonardo of Foothill College, Jeffery Richardson of Columbus State Community College, and David Schwimmer of Columbus State University. The test items include multiple choice and essay questions testing a variety of comprehension levels. The test bank is available in two formats: online in MS Word files and a Computerized Test Bank, a multi-platform CR-ROM. The easy-to-use test-generation program fully supports graphics, print tests, student answer sheets, and answer keys. The software's advanced features allow you to create an exam to your exact specifications.

INSTRUCTOR'S MANUAL (AVAILABLE IN ELECTRONIC FORMAT)

The Manual begins with a special introduction on *Using Visuals in the Classroom*, prepared by Matthew Leavitt of the Arizona State University, in which he provides guidelines and suggestions on how to use the visuals in teaching the course. For each chapter, materials by Christopher DiLeonardo of Foothill College include suggestions and directions for using web-based learning modules in the classroom and for homework assignments.

WEB-BASED LEARNING MODULES

A robust suite of multi-media learning resources have been designed for *Visualizing Geology*, again focusing on and using the visuals from the book. Delivered via the web, the content is organized in the following categories:

Tutorial Animations: Animations visually support the learning of a difficult concept, process, or theory, many of them built around a specific feature such as a Process Diagram, Visualizing piece, or key visual in the chapter. The animations go beyond the content and visuals presented in the book, providing additional visual examples and descriptive narration.

BOOK COMPANION SITE (WWW.WILEY.COM/COLLEGE/MURCK)

Instructor Resources on the book companion site include the Test Bank, Instructor's Manual, all illustrations and photos in the textbook in jpeg format, as well as select flash animations for use in classroom presentations.

WILEYPLUS

Visualizing Geology is available with **WileyPlus**, a powerful online tool that provides instructors and students with an integrated suite of teaching and learning resources in one easy-to-use Web site.

Wiley PLUS Helping Teachers Teach and Students Learn

www.wiley.com/college/murck

This title is available with *Wiley PLUS*, a powerful online tool that provides instructors and students with an integrated suite of teaching and learning resources in one easy-to-use Web site. *Wiley PLUS* is organized around the activities you and your students perform in class.

For Instructors

Prepare & Present: Create class presentations using a wealth of Wiley-provided resources—such as an online version of the textbook, PowerPoint slides, animations, overviews, and visuals from the Wiley Image Gallery—making your preparation time more efficient. You may easily adapt, customize, and add to this content to meet the needs of your course.

Create Assignments: Automate the assigning and grading of homework or quizzes by using Wiley-provided question banks, or by writing your own. Student results will be automatically graded and recorded in your gradebook. *Wiley PLUS* can link the pre-lecture quizzes and test bank questions to the relevant section of the online text.

Track Student Progress: Keep track of your students' progress via an instructor's gradebook, which allows you to analyze individual and overall class results to determine students' progress and level of understanding.

Administer Your Course: *Wiley PLUS* can easily be integrated with another course management system, gradebook, or other resources you are using in your class, providing you with the flexibility to build your course, your way.

For Students

Wiley PLUS provides immediate feedback on student assignments and a wealth of support materials. This powerful study tool will help your students develop their conceptual understanding of the class material and increase their ability to answer questions.

A "**Study and Practice**" area links directly to text content, allowing students to review the text while they study and answer. Resources include National Geographic videos, concept animations and tutorials, visual learning interactive exercises, and links to websites that offer opportunities for further exploration.

An "**Assignment**" area keeps all the work you want your students to complete in one location, making it easy for them to stay "on task." Students will have access to a variety of interactive self- assessment tools, as well as other resources for building their confidence and understanding. In addition, all of the pre-lecture quizzes contain a link to the relevant section of the multimedia book, providing students with context-sensitive help that allows them to conquer problem-solving obstacles as they arise.

A **Personal Gradebook** for each student will allow students to view their results from past assignments at any time.

Please view our online demo at **www.wiley.com/college/wileyplus.** Here you will find additional information about the features and benefits of *Wiley PLUS*, how to request a "test drive" of *Wiley PLUS* for this title, and how to adopt it for class use.

ACKNOWLEDGMENTS

STUDENT FEEDBACK/CLASS TESTING

In order to make certain that *Visualizing Geology* met the needs of current students, we asked several instructors to class test a chapter. The feedback that we received from students and instructors confirmed our belief that the visualizing approach taken in this book is highly effective in helping students to learn.

We wish to thank the following instructors and their students who provided us with helpful feedback and suggestions:

Laura Sue Allen-Long
Indiana University-Purdue University Indianapolis

Cathy Connor
University of Alaska Southeast

Chu-Yen Chen
University of Illinois at Urbana-Champaign

Robert Eves
Southern Utah University

Lynn Fielding
El Camino College

Pamela Gore
Georgia Perimeter College

Paul Grogger
University of Colorado, Colorado Springs

Erich Guy
Ohio University

Verner Johnson
Mesa State College

Amanda Julson
Blinn College

Arthur Lee
Roane State Community College

Ntungwa Maasha
Coastal Georgia Community College

Anthony Martorana
Chandler Gilbert Community College

Ryan Mathur
Juniata College

Katherine Miller
Florida A&M University

Keith Montgomery
University of Wisconsin, Marathon

Pamela Nelson
Glendale Community College

Donald Thieme
Georgia Perimeter College

REVIEWERS

Our sincere appreciation to the following professionals who offered their comments and constructive criticism as we developed this book:

Laura Sue Allen-Long
Indiana University-Purdue University Indianapolis

Laurie Anderson
Louisiana State University

Jake Armour
University of North Carolina Charlotte

Jerry Bartholomew
University of Memphis

Jay D. Bass
University of Illinois at Urbana-Champaign

Barbara Bekken
Virginia Polytechnic Institute and State University

Ross A. Black
University of Kansas

Theodore J. Bornhorst
Michigan Technological University

Michael Bradley
Eastern Michigan University

Michael Canestaro
Sinclair Community College

Richard L. Carlson
Texas A&M University

Victor V. Cavaroc
North Carolina State University

Cathy Connor
University of Alaska Southeast

Michael Dalman
Blinn College

John Dassinger
Chandler-Gilbert Community College

W. Crawford Elliott
Georgia State University

Robert Eves
Southern Utah University

Mike Farabee
Estrella Mountain Community College

David A. Foster
University of Florida

Carol D. Frost
University of Wyoming

Tracy Furutani
North Seattle Community College

William Garcia
University of North Carolina at Charlotte

Daniel Habib
Queens College

Frederika Harmsen
California State University, Fresno

Michael J. Harrison
Tennessee Technological University

Mary Anne Holmes
University of Nebraska-Lincoln

William Hoyt
University of Northern Colorado

Richard Josephs
University of North Dakota

Amanda Julson
Blinn College

Arthur Lee
Roane State Community College

Steven Lower
Ohio State University

Ronald Martino
Marshall University

Brendan McNulty
California State University, Dominguez Hills

Joseph Meert
University of Florida

Ken Miller
Rutgers University

Kenneth Rasmussen
Northern Virginia Community College

Bruce Simonson
Oberlin College

Debra Stakes
Cuesta College

Donald Thieme
Georgia Perimeter College

Carol Thompson
Tarleton State University

Harry Williams
University of North Texas

FOCUS GROUPS AND TELESESSION PARTICIPANTS

A number of professors and students participated in focus groups and telesessions, providing feedback on the text, visuals, and pedagogy. Our thanks to the following participants for their helpful comments and suggestions:

Sylvester Allred
Northern Arizona University

David Bastedo
San Bernardino Valley College

Ann Brandt-Williams
Glendale Community College

Natalie Bursztyn
Bakersfield College

Stan Celestian
Glendale Community College

O. Pauline Chow
Harrisburg Area Community College

Diane Clemens-Knott
California State University, Fullerton

Mitchell Colgan
College of Charleston

Linda Crow
Montgomery College

Smruti Desai
Cy-Fair College

Charles Dick
Pasco-Hernando Community College

Donald Glassman
Des Moines Area Community College

Mark Grobner
California State University, Stanislaus

Michael Hackett
Westchester Community College

Gale Haigh
McNeese State University

Roger Hangarter
Indiana University

Michael Harman
North Harris College

Terry Harrison
Arapahoe Community College

Javier Hasbun
University of West Georgia

Stephen Hasiotis
University of Kansas

Adam Hayashi
Central Florida Community College

Laura Hubbard
University of California, Berkeley

James Hutcheon
Georgia Southern University

Scott Jeffrey
Community College of Baltimore County, Catonsville Campus

Matther Kapell
Wayne State University

Arnold Karpoff
University of Louisville

Dale Lambert
Tarrant County College NE

Arthur Lee
Roane State Community College

Harvey Liftin
Broward Community College

Walter Little
University at Albany, SUNY

Mary Meiners
San Diego Miramar College

Scott Miller
Penn State University

Jane Murphy
Virginia College Online

Bethany Myers
Wichita State University

Terri Oltman
Westwood College

Keith Prufer
Wichita State University

Ann Somers
University of North Carolina, Greensboro

Donald Thieme
Georgia Perimeter College

Kip Thompson
Ozarks Technical Community College

Judy Voelker
Northern Kentucky University

Arthur Washington
Florida A&M University

Stephen Williams
Glendale Community College

Feranda Williamson
Capella University

Thanks for Participating in a Special Project

Visualizing Geology has been a special experience for the authors. It has been a bonding experience for us and, we hope, for the wonderful people at Wiley, National Geographic, and Pre-Press, plus all the skilled consultants who helped bring the book to completion. We set out to create a different, more engaging text, and to do so within the bounds of today's needs in these most challenging times for educators and students. At every step along the way, and especially those steps where we stumbled, a group of thoughtful, sensitive, and very professional experts steered us and cheered us on, and led us in the right direction.

The staff at John Wiley and Sons merits special thanks. Anne Smith and Helen McInnis had the bright idea to team up Murck and Skinner, full-time geologists and long-time collaborators, with Dana Mackenzie, a science writer with a mathematics background. It's hard for us to imagine a mix working better together. Once the collaboration started we were in almost daily contact with Charity Robey who set up 4-way conference calls and, and with her voice-of-reason, got us working smoothly together. Once under way, and with deadlines tightening, Nancy Perry stepped in and carried the beacon. If ever an editor should be acknowledged as an essential part of a team, it is Nancy. Behind the scenes were members of the skilled team that make a project work; our Executive Editor Ryan Flahive, Jay O'Callaghan the Publisher, Barbara Heaney, Director of Product and Market Development, and Kelly Tavares the Production Manager. Members of Wiley's photo and illustration staff, Hilary Newman, Tara Sanford, and Anna Melhorn, and media editor Tom Kulesa were ever ready to provide help and advice. Stacey Gold, Research Editor and Account Executive for the National Geographic Society, was an invaluable aid in leading us through the voluminous photo files of the Society. Mary Ann Price was at all times prompt in her responses and swift in her searches for essential images, often at critical times in the production cycle.

The book was prepared by the Pre-Press Company, Inc. and we especially thank Abigail Greshik, Project Manager, who worked on the entire project and in the end became a real member of the team. At the beginning of the project it was Gordon Laws of Pre-Press who made sure we started off on the right footing.

We say special thanks to all of those professional colleagues who read, thought about, and offered advice on our proposal, plans, writing and production. They are thanked and named elsewhere, but we authors want to acknowledge the essential part they played in the production of the book.

CONTENTS *in Brief*

Foreword v
Preface viii

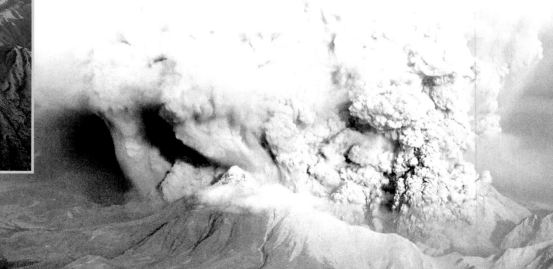

CONTENTS

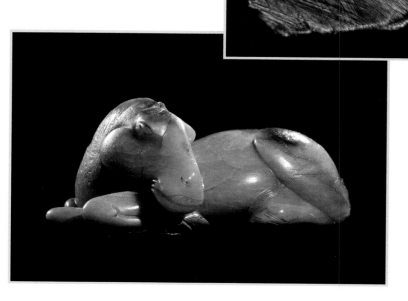

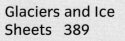

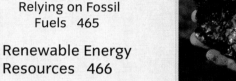

VISUALIZING FEATURES

Multi-part visual presentations that focus on a key concept or topic in the chapter.

xxx   Contents

PROCESS DIAGRAMS

A series or combination of figures and photos that describe and depict a complex process.

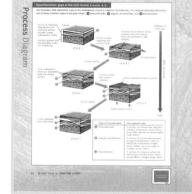

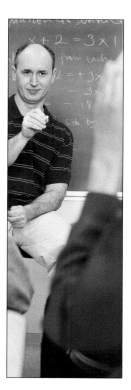

The first person to invent a car that runs on water...

... may be sitting right in your classroom! Every one of your students has the potential to make a difference. And realizing that potential starts right here, in your course.

When students succeed in your course—when they stay on-task and make the breakthrough that turns confusion into confidence—they are empowered to realize the possibilities for greatness that lie within each of them. We know your goal is to create an environment where students reach their full potential and experience the exhilaration of academic success that will last them a lifetime. *WileyPLUS* can help you reach that goal.

Wiley**PLUS** is an online suite of resources—including the complete text—that will help your students:

- come to class better prepared for your lectures
- get immediate feedback and context-sensitive help on assignments and quizzes
- track their progress throughout the course

"I just wanted to say how much this program helped me in studying... I was able to actually see my mistakes and correct them. ... I really think that other students should have the chance to use *WileyPLUS*."

Ashlee Krisko, *Oakland University*

www.wiley.com/college/wileyplus

80% of students surveyed said it improved their understanding of the material.

FOR INSTRUCTORS

WileyPLUS is built around the activities you perform in your class each day. With WileyPLUS you can:

Prepare & Present
Create outstanding class presentations using a wealth of resources such as PowerPoint™ slides, image galleries, interactive simulations videos, and more. You can even add materials you have created yourself.

Create Assignments
Automate the assigning and grading of homework or quizzes by using the provided question banks, or by writing your own.

Track Student Progress
Keep track of your students' progress and analyze individual and overall class results.

Now Available with WebCT and Blackboard!

"It has been a great help, and I believe it has helped me to achieve a better grade."

Michael Morris,
Columbia Basin College

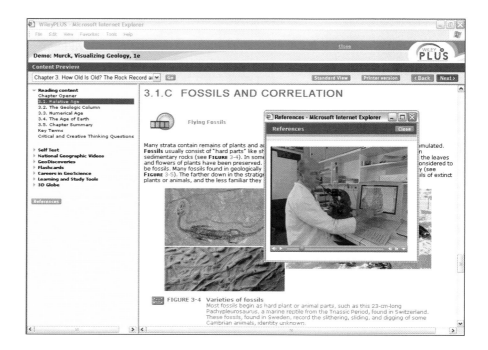

FOR STUDENTS

You have the potential to make a difference!
WileyPLUS is a powerful online system packed with features to help you make the most of your potential and get the best grade you can!

With WileyPLUS you get:

- A complete online version of your text and other study resources.
- Problem-solving help, instant grading, and feedback on your homework and quizzes.
- The ability to track your progress and grades throughout the term.

For more information on what *WileyPLUS* can do to help you and your students reach their potential, please visit www.wiley.com/college/*wileyplus*.

76% of students surveyed said it made them better prepared for tests. *

Visualizing
GEOLOGY

In collaboration with
THE NATIONAL GEOGRAPHIC SOCIETY

Earth as a Planet

Not far from Alice Springs in central Australia, a ring of hills called Gosses Bluff (right) juts up from the endlessly flat landscape. This curious formation is the relic of a great meteorite that crashed to Earth 142 million years ago. The impact and explosion blasted out a crater 24 kilometers in diameter, surrounded by a ring of debris. What we see today is the eroded remnant of the center of the impact site. You can see a ghostly, highly eroded remnant of the outer rim in a photograph from space (inset).

Our planet was formed about 4.56 billion years ago out of innumerable rocks very much like the Gosses Bluff meteorite. Though impacts of large meteorites are rare in the present era, scars from past impacts remind us that we are not isolated in space but, rather, are part of an active solar system. We share energy and matter with the rest of that system.

The successes of the Space Age have made us more conscious of Earth's place in the larger solar system. We have learned that Earth is a system, too. Its components—rocks, water, the atmosphere, living creatures—depend vitally on one another. Instead of studying various aspects of the planet in isolation, as we once did, geologists now try to look at the Earth system as an integrated whole. In *Visualizing Geology*, you will learn how Earth system science is clarifying our understanding of our Earth. You will also learn how Earth has changed and is continuing to change today.

Outer rim of crater

Gosses Bluff

What Is Geology?

LEARNING OBJECTIVES

Describe several of the many branches of geology.

Explain what it means to take a systems approach to geology.

Identify three types of systems.

Identify four major subsystems of the Earth system.

Explain how these subsystems interact, using the concept of cycles.

The word **geology** comes from two Greek roots: *geo-*, meaning "Earth," and *-logis*, meaning "study" or "science." The science called geology encompasses the study of our planet: how it formed; the nature of its interior; the materials of which it is composed; its water, glaciers, mountains, and deserts; its earthquakes and volcanoes; its resources; and its history—physical, chemical, and biological. Scientists who make a career of geology are *geologists*. Geology, like all sciences, is based on factual observations, testable hypotheses, reproducible procedures, and open communication of information.

geology The scientific study of Earth.

The study of geology is traditionally divided into two broad subject areas: **physical geology** and **historical geology**. Physical geology is concerned with understanding the *processes* that operate at or beneath the surface of Earth and the *materials* on which those processes operate. Some examples of geologic processes are mountain building, volcanic eruptions, earthquakes, river flooding, and the formation of ore deposits. Some examples of materials are minerals, soils, rocks, air, and water.

Historical geology, on the other hand, is concerned with the sequence of geologic *events* that have occurred in the past. These events can be inferred from the rock record, that is, evidence left in Earth's rocks (**FIGURE 1.1**). Through the findings of historical

Digging up the past FIGURE 1.1

The light layer of clay seen in this rock from Clear Creek, Colorado, provided geologists with crucial evidence concerning one of the great mysteries of the past: Why did the dinosaurs suddenly die out 65 million years ago? Layers of clay like this are found at many places in rocks that are 65 million years old. As you will learn in Chapter 14, many geologists believe that the clay represents the debris from an enormous meteorite impact—like the impact that created the hills seen in the introduction to this chapter, only much larger. Above the clay layer lies a layer that contains fragments of "shocked" or fractured quartz, also indicative of some very violent event, which may have contributed to, or even caused the extinction of the dionosaurs.

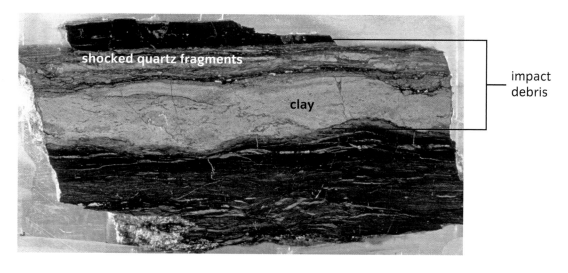

shocked quartz fragments

clay

impact debris

Geologists are privileged to work in some of the most exotic places on Earth—and beyond.

A Harrison (Jack) Schmidt, a planetary geologist, is the only scientist (so far) to walk on the Moon. Schmidt flew on the *Apollo 17* mission in 1972. Here, he is collecting a lunar sample to take back to Earth.

B Volcanologists get uncomfortably close to the 2002 eruption of Mount Etna in Sicily, Italy, to record the sounds of the eruption.

C A paleontologist dives into the waters off the Bahama Islands to study stromatolites, a living algal formation reminiscent of Earth's oldest fossils.

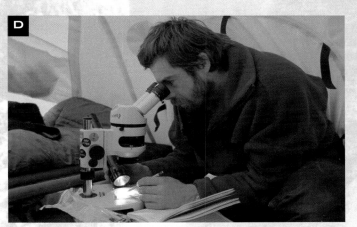

D Sheltered by a tent from the cold outside, this geologist is looking through a microscope at volcanic ash specimens collected in Antarctica.

geology, scientists seek to resolve questions such as when did the oceans form, why did the dinosaurs die out, when did the Rocky Mountains rise, and when and where did the first trees appear. Historical geology gives us a perspective on the past. It also establishes a context for thinking about present-day changes in our natural environment. *Visualizing Geology* is concerned mainly with physical geology, but it also deals with many lessons we can learn from historical geology.

Within the traditional domains of physical and historical geology there are many specialized disciplines, some of which are illustrated in FIGURE 1.2. Economic geology, for example, is concerned with the formation and occurrence of and the search for valuable mineral deposits. Environmental geology focuses on how materials and processes in the natural geologic environment affect—and are affected by—human activities. Volcanologists study volcanoes and eruptions, past and present; seismologists study earthquakes; mineralogists undertake the microscopic study of minerals and crystals; paleontologists study fossils and the history of life on Earth; structural geologists study how rocks break and bend. These specialties are needed because geology encompasses such a broad range of topics.

To a certain extent we are all geologists, even though only a few of us make a career out of geology. Everyone living on this planet relies on geologic resources: water, soil, building stones, metals, fossil fuels, gemstones, plastics (from petroleum), ceramics (from clay minerals), glass (from silica sand), salt (a mineral called halite), and many others. Geologic processes affect us every day. We also influence the geologic environment through our daily activities, whether we are drinking water that came from an aquifer or planting trees to control soil erosion. This book will help you to become better informed and more mindful of these interactions. As a result, you will be better equipped to make decisions about Earth materials and processes that affect your life.

EARTH SYSTEM SCIENCE

Traditionally, scientists have studied Earth by focusing on separate units—the atmosphere, the oceans, or a single mountain range—in isolation from the other units. However, the first photographs of Earth taken from space in the 1960s caused a dramatic rethinking of this traditional view (**FIGURE 1.3A**). For the first time, it was possible to see the whole planet in one sweeping view. We could see everything at a glance—the clouds, the oceans, polar ice caps, and the continents—all at the same time, and in their proper scale. The astronauts, like the rest of us, marveled at Earth's "overwhelming beauty . . . the stark contrast between bright colorful home and stark black

Earth from orbit FIGURE 1.3

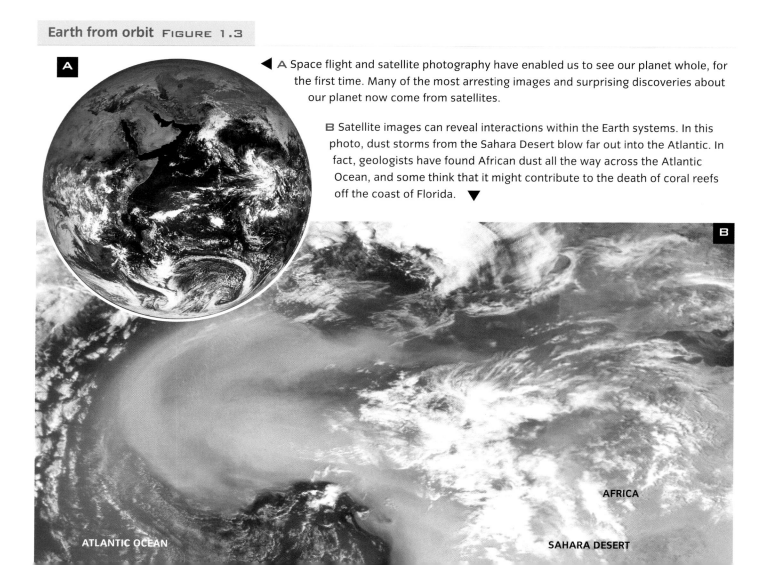

◀ **A** Space flight and satellite photography have enabled us to see our planet whole, for the first time. Many of the most arresting images and surprising discoveries about our planet now come from satellites.

B Satellite images can reveal interactions within the Earth systems. In this photo, dust storms from the Sahara Desert blow far out into the Atlantic. In fact, geologists have found African dust all the way across the Atlantic Ocean, and some think that it might contribute to the death of coral reefs off the coast of Florida. ▼

AFRICA

ATLANTIC OCEAN

SAHARA DESERT

The system concept FIGURE 1.4

This figure shows a variety of systems. The entire diagram—mountains, river, lake—is one kind of system known as a *watershed*. The individual pieces enclosed by boxes, such as the river, are also systems. Even a small volume of water or lake sediment (foreground boxes) can be considered a system.

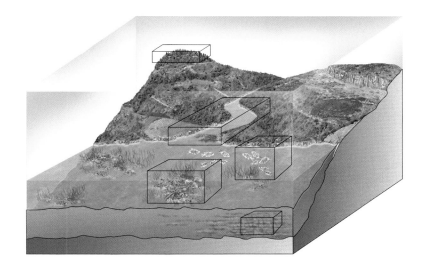

infinity" (Rusty Schweikart, *Apollo 9*). Yet from space it was also clear how small Earth is—just a dust speck compared to the vastness of the solar system and the universe. On such a small planet, it no longer made sense to study all the pieces separately. There was only one geology that mattered, not the geology of America or the Atlantic Ocean but the geology of the whole Earth.

Instruments carried by satellites in space have also given us new ways to study the relationships of the parts on a global scale, as we never could before (FIGURE 1.3B). This new, more all-inclusive view of geology is called *Earth system science*.

The system concept

A **systems** approach is a helpful way to break down a large, complex problem into smaller pieces that are easier to study without losing sight of the connections between those pieces. A system may be large or small, simple or complex (FIGURE 1.4). It could be the contents of the beaker in a laboratory experiment or the contents of an ocean. A leaf is a system, but it is also part of a larger system (a tree), which is part of a still larger system (a forest).

The fact that we distinguish a system from the rest of the universe for specific study does not mean that we ignore its surroundings. In fact, the nature of a system's

> **system** A portion of the universe that can be separated from the rest of the universe for the purpose of observing changes that happen in it.

boundaries is one of its most important defining characteristics. FIGURE 1.5 illustrates the three basic kinds of systems. The easiest to understand is an *isolated system*. In an isolated system, the boundaries prevent the system from exchanging either matter or energy with its surroundings. However, no isolated systems actually exist in the real world. It is possible to have boundaries that do a pretty good job of preventing the passage of matter, but no boundary is so perfectly insulating that it prevents energy from entering or escaping.

A second type of system, and the nearest thing to an isolated system in the real world, is a *closed system*. Such a system has boundaries that permit the exchange of

Three kinds of systems FIGURE 1.5

Scientists distinguish three kinds of systems: isolated, closed, and open. Most systems in geology are open.

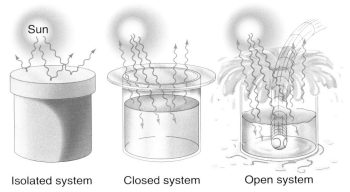

Sun

Isolated system   Closed system   Open system

Island or Open System?

The island of Bora Bora (above) is an exotic, though isolated, tourist destination. But how isolated is it, really?

Some islands may seem isolated to a tourist or to a shipwrecked sailor, but from the point of view of an Earth systems scientist, every island is an open system. (Remember from Figure 1.5 that an open system allows both matter and energy to cross its boundaries.)

Energy (in the form of sunlight) and matter (in the form of precipitation) reach the island from outside sources. The energy leaves the island as heat. The water either evaporates or drains into the sea. In the modern era, humans may also bring material into and out of the system, by importing and exporting resources.

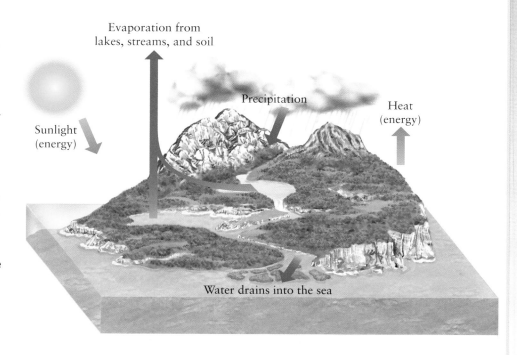

energy, but not matter, with its surroundings. An example of a closed system would be a perfectly sealed oven, which would allow the material inside to be heated but would not allow any of that material to escape. (Note that in real life, ovens do allow some vapor to escape, so they are not perfect examples of closed systems.)

The third kind of system, an *open system*, can exchange *both* matter and energy across its boundaries. An island offers a simple example (see What a Geologist Sees). The system concept can also be applied to artificial environments. For example, urban geographers and land use planners sometimes use a systems approach in the study of cities. Enormous flows of energy and materials occur across city borders.

The Earth system
Earth itself is a very close approximation to a closed system. Energy enters the Earth system as solar radiation. The energy is used in various biologic and geologic processes and then departs in the form of heat. Very little matter crosses the boundaries of the Earth system. We do lose some hydrogen and helium atoms from the outer atmosphere, and we gain some material in the form of meteorites. However, for most purposes, especially over the short term, we can treat Earth as a closed system. Over the long term, geologically speaking, we cannot

ignore the effects of meteorites, though, as illustrated by the introduction to this chapter.

The fact that Earth is a closed system (on the scale of the time humans have existed) has extremely important implications. Any change in one part of a closed system will eventually affect other parts of the system. For instance, when we divert a river to provide drinking water for a city, we may deplete the water resources somewhere else (**Figure 1.6**). The amount of matter in a closed system is fixed and finite. Therefore, the resources on this planet are *all we have* and, for the foreseeable future, *all we will ever have.* This means we must treat Earth resources with respect and use them wisely and cautiously. When we dispose of waste materials, we must remember that they will remain within the boundaries of the Earth system. As environmentalists sometimes say, "There is no *away* to throw things to."

The Earth system can be divided into four very large subsystems, which you can think of as Earth's principal reservoirs of materials and energy. These are the **lithosphere**, **biosphere**,

lithosphere
Earth's rocky, outermost layer.

biosphere The system consisting of all living and recently dead organisms on Earth.

A The Colorado River and its tributaries provide drinking water to 25 million people in California, Nevada, Arizona, Utah, New Mexico, Colorado, and Wyoming. It also irrigates 3.5 million acres of fields. Because of the massive diversion of water away from the river in those states, little water reaches the river's historic terminus, the Gulf of California (or Sea of Cortez) in Mexico.

B All that remains of the river delta at the Gulf of California are the vast mud flats seen in this photo. The branched "stream" you see here is *not* the Colorado River; it is an inlet carved by the tides in the soft mud of the delta.

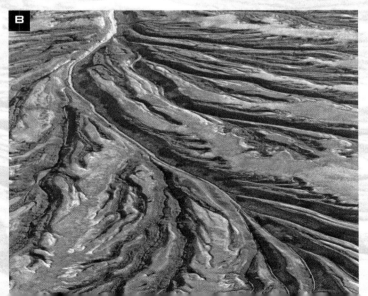

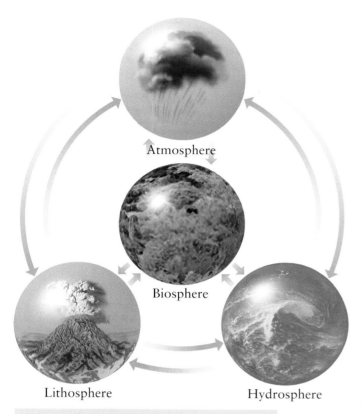

Earth's four "spheres" FIGURE 1.7

This figure illustrates Earth's four principal subsystems: lithosphere, biosphere, atmosphere, and hydrosphere. Materials and energy cycle among these subsystems, as shown by the arrows, making them open systems.

atmosphere, and **hydrosphere** (FIGURE 1.7). Each of these is an open system. They can be further subdivided into the many subsystems of interest to geologists; for example, the hydrosphere can be divided into oceans, glacial ice, streams, lakes, groundwater, and so on. The lithosphere may be what comes to mind first if you think of geology (incorrectly) as only the study of rocks—but in fact all four spheres play important roles in geology. Plants draw nutrients from the lithosphere and incorporate them into the

> **atmosphere** The envelope of gases that surrounds Earth.

> **hydrosphere** The system comprising all of Earth's bodies of water and ice, both on the surface and underground.

biosphere. When the plants die and decompose, some of the material they contain may enter the atmosphere, while other parts may fossilize and reenter the lithosphere. Rocks erode, and the minerals they contain become salts in the hydrosphere. Lakes evaporate and return their salt to the lithosphere. The exchanges of materials between spheres never stop. Some scientists refer to the collective interacting "whole" of these four systems as the *geosphere*.

CYCLES AND INTERACTIONS

Figure 1.7 represents the interactions between Earth's reservoirs in a simplistic fashion, but it is possible to be a good deal more precise about the nature of the flows by focusing on the way materials move, or cycle, among the reservoirs. In this book, we will discuss the Earth system as a series of three interrelated cycles that facilitate the movement of materials and energy among the reservoirs. These are the **water cycle** or **hydrologic cycle**, the **rock cycle**, and the **tectonic cycle**. They are sketched in FIGURE 1.8. It is not necessary for you to understand the details of this diagram yet; we will return to this figure repeatedly in later chapters and label each of the processes that are illustrated with icons and arrows here. The most important points to understand now are that the interactions form *cycles*—that is, processes without beginning or end—and that they are closely interconnected.

Because this is a book about physical geology, we will focus primarily on the lithosphere—that is, the outermost rocky portion of the solid Earth. However, the systems

> **hydrologic cycle** A model that describes the movement of water through the reservoirs of the Earth system; the water cycle.

> **rock cycle** The set of crustal processes that form new rock, modify it, transport it, and break it down.

> **tectonic cycle** Movements and interactions in the lithosphere and the internal Earth processes that drive them.

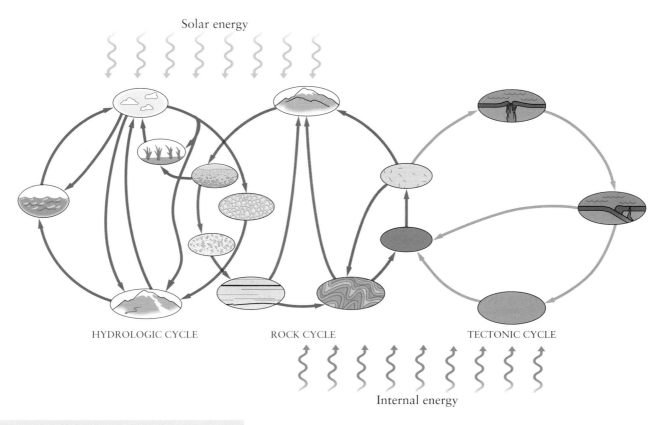

Solar energy

Internal energy

HYDROLOGIC CYCLE   ROCK CYCLE   TECTONIC CYCLE

Interconnected cycles FIGURE 1.8

The hydrologic cycle (left) circulates water through various reservoirs: the ocean, the atmosphere, the lithosphere (where it forms surface water and groundwater), and the biosphere (where it is incorporated into plants and animals). The cycle is completed when the water returns to the ocean. The rock cycle (center) describes crustal processes through which rocks are uplifted into mountains, then eroded and weathered (often by water), and the debris modified, transformed, or reformed into rock underground. The cycle is completed when the rocks are thrust up into mountains again. The tectonic cycle (right) explains where igneous rock comes from and how new crust is formed and recycled by large-scale motions of Earth's surface and interior. The Sun provides the energy that powers the hydrologic cycle. Heat from Earth's interior powers the tectonic cycle, and the rock cycle draws energy from both sources.

approach tells us that it is unreasonable—even impossible—to consider one part of the Earth system in isolation from the rest. As Figure 1.8 shows, we cannot fully understand where rocks come from without also understanding the hydrologic cycle and the tectonic cycle. Nor can we understand the lithosphere without learning something about the hydrosphere, atmosphere, and biosphere, as well as Earth's deep interior (which is distinct from the lithosphere). Thus, in this course, you will study not only geology, but also a little bit of oceanography, hydrology, meteorology, physics, chemistry, biology, and astronomy.

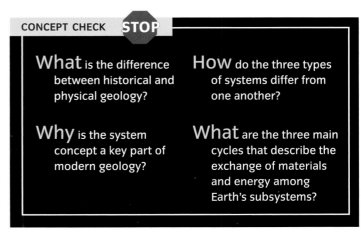

CONCEPT CHECK   STOP

What is the difference between historical and physical geology?

Why is the system concept a key part of modern geology?

How do the three types of systems differ from one another?

What are the three main cycles that describe the exchange of materials and energy among Earth's subsystems?

Earth in Space

A s geologists, we mainly study the processes that occur on Earth, our home planet, in isolation from the rest of the solar system. But at the outset, we must broaden our perspective and ask how Earth originated and how it came to be a unique and special place. The characteristics of the planet we live on today depend very much on how it was formed and what has happened to it over the past 4.56 billion years.

THE SOLAR SYSTEM

Earth is one of eight large objects traditionally called **planets** in our **solar system**, which consists of the Sun and the group of objects in orbit around it. In addition to the Sun and the planets, the solar system includes 140 known moons, a vast number of asteroids, millions of comets, and innumerable fragments of rock and dust called *meteoroids*. All the objects in our solar system move through space in smooth, regular orbits, held in place by gravitational attraction. The planets, asteroids, comets, and meteoroids orbit the Sun, and the moons orbit the planets.

We can separate the planets into two groups on the basis of their physical characteristics and distances from the Sun (**FIGURE 1.9**). The innermost planets—Mercury, Venus, Earth, and Mars—are small, rocky, and relatively dense. They are similar in size and chemical

Family portrait of the solar system FIGURE 1.9

Our solar system's eight recognized planets, shown to scale against the Sun. (Note, however, that the distances between planets are much greater than shown, and the planets never line up neatly like this.) Between the terrestrial and jovian planets lies the asteroid belt, consisting of more than 100,000 small pieces of rock that never coalesced into a planet. Beyond Neptune lies Pluto, formerly considered a ninth planet, and a large number of other icy objects, mostly like Pluto—some larger but mostly smaller—which collectively comprise the Kuiper Belt.

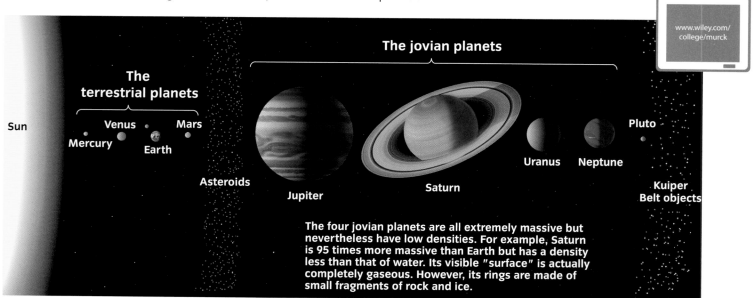

www.wiley.com/college/murck

The four jovian planets are all extremely massive but nevertheless have low densities. For example, Saturn is 95 times more massive than Earth but has a density less than that of water. Its visible "surface" is actually completely gaseous. However, its rings are made of small fragments of rock and ice.

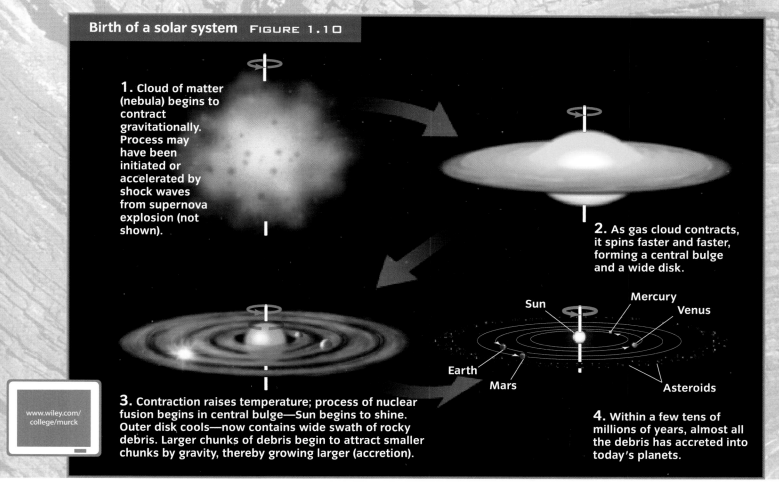

Birth of a solar system FIGURE 1.10

Process Diagram

1. **Cloud of matter (nebula) begins to contract gravitationally. Process may have been initiated or accelerated by shock waves from supernova explosion (not shown).**

2. **As gas cloud contracts, it spins faster and faster, forming a central bulge and a wide disk.**

3. **Contraction raises temperature; process of nuclear fusion begins in central bulge—Sun begins to shine. Outer disk cools—now contains wide swath of rocky debris. Larger chunks of debris begin to attract smaller chunks by gravity, thereby growing larger (accretion).**

4. **Within a few tens of millions of years, almost all the debris has accreted into today's planets.**

Sun — Mercury — Venus — Earth — Mars — Asteroids

www.wiley.com/college/murck

composition and are called *terrestrial planets* because they resemble Earth (*Terra* in Latin). The outer planets are much larger and more massive than the terrestrial planets, yet much less dense. These *jovian planets*—Jupiter, Saturn, Uranus, and Neptune—take their name from *Jove*, the name for Jupiter in Roman mythology. (Jupiter was the king of the gods, and the god of light and weather, among other things. In size, Jupiter is certainly the king of the planets.) The jovian planets probably have small solid centers that may resemble terrestrial planets, but much of their planetary mass is contained in thick atmospheres of hydrogen, helium, and other gases. The atmospheres are what we actually see when we observe these planets.

Pluto, until recently considered to be a ninth planet, doesn't fit into either of these planetary groups. It is much smaller and denser than the jovian planets but much less dense than the terrestrial planets. Recent discoveries suggest that Pluto is actually the nearest of a sizeable population of Pluto-like bodies beyond the orbit of Neptune called Kuiper Belt objects, some of which are actually larger than Pluto. Kuiper Belt objects are currently the focus of much research.

THE ORIGIN OF THE SOLAR SYSTEM

How did the solar system form? The answer to this ancient question is still incomplete, and it is an excellent example of a carefully researched scientific hypothesis. The *nebular theory*, originally formulated by the German philosopher Immanuel Kant in 1755 and now widely accepted as the best description of planetary formation, proposes that the solar system coalesced out of a swirling cloud of interstellar dust and gas called a *nebula* (FIGURE 1.10).

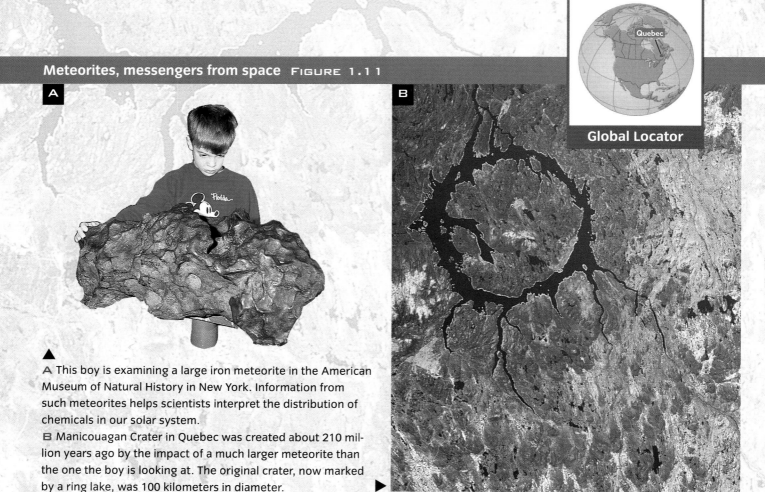

Global Locator

A This boy is examining a large iron meteorite in the American Museum of Natural History in New York. Information from such meteorites helps scientists interpret the distribution of chemicals in our solar system.

B Manicouagan Crater in Quebec was created about 210 million years ago by the impact of a much larger meteorite than the one the boy is looking at. The original crater, now marked by a ring lake, was 100 kilometers in diameter.

The nebular theory explains very well why the inner planets are rocky, while the outer planets contain a higher proportion of ice and gas. The temperature was higher in the portion of the nebula that was to become the innermost part of the solar system. Elements with a high melting point (iron, silicon, and so on—the chemicals that make up rock) would have been early to condense from the cloud. Meanwhile, a strong solar wind stripped much of the lighter gases, such as helium and hydrogen, from the inner planets. But the solar wind was not strong enough to do the same to the outer planets, which grew into gas giants. Volatile compounds, such as water and methane, condense only at lower temperatures. Therefore water ice, methane ice, and other ices are abundant on the moons and smaller bodies of the outer solar system, where the primordial nebula was cooler.

meteorite A fragment of extraterrestrial material that falls to Earth.

Planetary accretion, a twentieth-century supplement to the nebular theory, accounts for the existence of meteoroids and asteroids. According to the accretion hypothesis, all of the planets assembled themselves from rocky, metallic, and icy debris 4.56 billion years ago, shortly after the Sun itself was formed. Today's meteoroids are the debris that never managed to be swept up by any planet; sometimes pieces of this debris happen to fall to Earth as **meteorites**. As such, meteorites are fascinating relics of the early days of the solar system (FIGURE 1.11A).

A major consequence of the planetary accretion hypothesis, which scientists have grasped only since the Moon landings in 1969–1972, is the great importance of violent collisions in the history of the solar system. Every crater on our Moon, as far as we know, was formed by a meteorite impact. (The astronauts looked for volcanic craters, too, but never found a single one.) Scientists recognize impact craters on Earth, too, although they are harder to find because erosion and other geologic processes cover them up or erase them (**FIGURE 1.11 B**). A meteorite impact may have been responsible for the catastrophe that killed the dinosaurs more than 65 million years ago.

Even larger impacts than this were commonplace in the early solar system. Most planetary scientists now believe that Earth collided with another planetary body, roughly the size of Mars, around 4.5 billion years ago. The impact tilted Earth's axis of rotation at an angle to the plane of its orbit around the Sun, and that is why we have seasons. The impact must also have melted most of Earth's surface due to the tremendous amount of energy released. (At the hyper-speeds typical of cosmic impacts, every ton of the impactor strikes Earth with an energy equivalent to 100 tons of dynamite.) The collision completely destroyed the other planet, and blasted so much debris into orbit that for a little while Earth had rings much denser than Saturn's. Eventually the ring of debris coalesced into the most familiar of astronomical objects—Earth's Moon (**FIGURE 1.12**). This hypothesis explains the existence of a magma ocean early in lunar history (shown by rocks re-

trieved from the Moon). It also explains our Moon's relatively large size in contrast to other moons, which are many times smaller than their parent bodies, and accounts for certain chemical discrepancies and similarities between Earth and its Moon.

Such giant collisions were the inevitable final stage of planetary accretion, when most of the debris has been swept up and only larger objects remain. Signs of giant impacts abound in the solar system. One such impact probably caused Uranus's axis to tip over on its side. Pluto's moon, Charon, was probably created by an impact, because it is unusually large compared to Pluto. Perhaps Venus experienced a giant impact, too. Although it lacks a moon, it is the only planet that rotates in the *retrograde* (east-to-west) direction, an effect that could have been produced by a large glancing blow, essentially tipping the planet upside down.

Why is the accretion history of the planets important to geologists? It means that every rocky planet probably started out hot enough to melt either partially or completely. During the period of partial melting, terrestrial planets separated into layers of differing chemical composition, a process called **differentiation**. Earth

Formation of the Moon FIGURE 1.12

Impact

Some 4550 million years ago, the still-forming Earth (unrecognizable because it probably did not yet have oceans) runs into another growing planet, which scientists have named Theia.

Impact + 8 hours

Theia is obliterated, and its remnants—along with a good chunk of Earth's mantle—are blasted into orbit around Earth. The off-center impact has knocked Earth's axis of rotation askew.

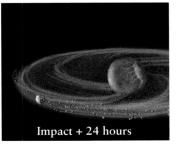

Impact + 24 hours

The debris spreads itself into a ring and begins to clump together.

Impact + 1 year

The largest clump starts to attract other fragments and is well on its way toward becoming the Moon. Moon's surface is initially molten. Earth has recovered its shape, leaving no trace of the most violent event in its history.

Process Diagram

differentiated into three layers: a relatively thin, low-density, rocky **crust**; a rocky, intermediate-density **mantle**; and a metallic, high-density **core**. Similar layers are present in Mercury, Venus, Mars, and the Moon, although they have different proportional sizes and compositions.

There are other important similarities among the terrestrial planets. All of them have experienced volcanic activity, which means they have, or had, an internal heat source. (We will explain where the heat originates in Chapter 3.) The volcanism is dominated by the formation of a volcanic rock called *basalt* (**FIGURE 1.13**). All of the planets have also been through intense cratering processes, although the signs are well hidden on Venus and Earth. Finally, all have lost their primordial atmospheres. The three that ended up with atmospheres (Earth, Mars, and Venus) evolved them over time, from material that leaked from their interiors via volcanoes.

- **crust** The outermost compositional layer of the solid Earth.

- **mantle** The middle compositional layer of Earth, between the core and the crust.

- **core** Earth's innermost compositional layer.

Basalt: the most common rock in the solar system FIGURE 1.13

Basaltic lava erupts from Mauna Loa volcano in Hawaii. When it cools, it will form the same kind of rock, called basalt, that forms most of the surface of Venus, Mars, and the "seas" (the dark-colored spots) of Earth's Moon. Though taken only a few years ago, this picture could represent what the surface of any of these planets looked like shortly after their formation.

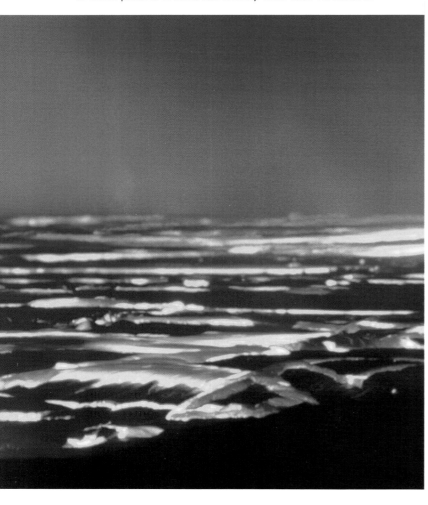

WHAT MAKES EARTH UNIQUE?

Venus and Mars, Earth's nearest neighbors, are in some ways very similar to our planet. In terms of size, Venus is nearly Earth's twin. Yet there is no chance of mistaking either of them for Earth (**TABLE 1.1**). Earth's blues, whites, and greens attest that it has three things neither Venus nor Mars, nor any other planet in our solar system, possesses: an oxygen-rich *atmosphere;* a *hydrosphere* that contains water as a solid, liquid, and vapor; and a *biosphere* full of living organisms.

The nature of Earth's solid surface is another special characteristic. Earth is covered by an irregular blanket of loose debris formed as a result of *weathering*—the chemical alteration and mechanical breakdown of rock caused by exposure to water, air, and living organisms. This layer is called *regolith* (from the Greek words for "blanket" and "stone"). It includes soil, river mud, desert sand, rock fragments, and all other unconsolidated debris. Other planets and moons with rocky surfaces have regolith too, but in those cases the regolith has formed primarily from endless pounding by impacting meteorites. Earth's regolith is also unique because it teems with life. When material from the biosphere becomes incorporated with rock material, the result is a special type of regolith unique to Earth, *soil.*

In spite of their similar origins, Venus, Earth, and Mars have profound geologic differences that have made Earth the only one that is hospitable for life.

	Venus	Earth	Mars
Atmosphere	White in photos, 97% carbon dioxide; temperature averages a blistering 480°C (hot enough to melt lead)	78% nitrogen, 21% oxygen; average temperature 14.6°C	Thin and insufficient to retain much heat; average temperature −63°C; temperatures usually too low to melt water ice (0°C)
Hydrosphere	Exists only as vapor in the atmosphere due to high temperatures	Contains water as solid, liquid, and vapor	Water cannot exist in liquid form on the surface due to low temperatures and pressure
Biosphere	None	Only known biosphere	None known

plate tectonics The movement and interactions of large fragments of Earth's lithosphere, called plates.

continental crust The older, thicker, and less dense part of Earth's crust; the bulk of Earth's land masses.

(You may read in some places about "lunar soil" and "Martian soil," but such usage of the word "soil" is geologically incorrect. The correct terms are "lunar regolith" and "Martian regolith.")

Another unique property of Earth is the nature and extent of its tectonic activity. **Plate tectonics** have shaped Earth's continents and oceans and govern, to a large extent, the location of Earth's volcanoes and the occurrence of earthquakes. Tectonic activity has given Earth two different kinds of crust—the thin, basaltic **oceanic crust** and the thick, granitic **continental crust**. The latter seems to be unique to Earth; at least, Mars and Venus do not have it. Because the location of continents affects oceanic and atmospheric currents, plate tectonics exert a powerful influence on Earth's climate, which in turn

affects the evolution of life. We will have much more to say about tectonics in Chapter 4 and later chapters.

oceanic crust The thinner, denser, and younger part of Earth's crust, underlying the ocean basins.

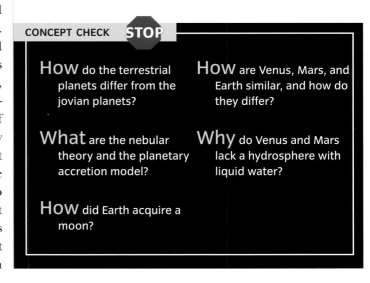

CONCEPT CHECK **STOP**

How do the terrestrial planets differ from the jovian planets?

What are the nebular theory and the planetary accretion model?

How did Earth acquire a moon?

How are Venus, Mars, and Earth similar, and how do they differ?

Why do Venus and Mars lack a hydrosphere with liquid water?

The Ever-Changing Earth

LEARNING OBJECTIVES

Explain Hutton's principle of uniformitarianism.

Describe why the Earth system can be so dynamic and yet appear so stable.

Identify several benefits of studying geology.

UNIFORMITARIANISM

Since material is constantly being transferred from one open system on Earth to another, you may wonder why these systems seem so stable. Why doesn't the sea become saltier or fresher? Why doesn't all the water in the world flow into the sea and stay there? Why should the chemical composition of the atmosphere be mainly nitrogen and oxygen (as it has been for millions of years)? How can rock that is two billion years old have the same composition as rock that is being formed today? If mountains are constantly being worn down by erosion, why are there still high mountains? The answers to these questions are the same: Materials cycle from one reservoir to another, but the reservoirs themselves don't change noticeably because the different parts of the cycle balance each other—the amounts added approximately equal the amounts removed. While a mountain is worn down in one part of the cycle, a new mountain is being built up in another part. This cycling of materials has been going on since Earth was formed, and it continues today.

A fundamental principle of geology, attributed to an eighteenth-century Scottish geologist named James Hutton, is based on this idea. This principle is called **uniformitarianism**, and one way to express it is to state that "the present is the key to the past." We can examine any rock, however old, and compare its characteristics with those of similar rocks forming today. We can then infer that the ancient rock likely formed in a similar environment through similar processes and on a comparable time scale. For example, in many deserts today we can see gigantic dunes formed from sand grains transported by the wind. Because of the way they form, dunes have a distinctive internal structure (**FIGURE 1.14**). Using the principle of uniformitarianism, we can infer that a rock composed of cemented grains of sand and having the same distinctive internal structure as modern dunes is the remains of an ancient dune.

TIME AND CHANGE

Hutton's principle of uniformitarianism provides a first step toward understanding Earth's history. Geologists have used this principle to explain Earth's features in a logical manner. In so doing, they have discovered that Earth is incredibly old. An enormously long time is needed to erode a mountain range or for huge quantities of sand and mud to be transported by streams, deposited in the ocean, and cemented into rocks, and for the rocks to be uplifted to form a mountain. Yet the cycle of erosion, formation of new rock, uplift, and more erosion has been repeated many times during Earth's long history.

One ongoing process that Hutton did not know about is the slow motion of Earth's plates. In a nutshell, plate tectonics involves the motion of about a half dozen large, curved fragments of Earth's outermost rocky layer, the lithosphere. Like so much else about Earth, we can now observe this motion from space: global positioning satellites can measure the shifting of plates in centimeters per year. By the principle of uniformitarianism, this process must also have operated in the past. When we extrapolate this imperceptibly slow motion over millions of years, we discover a stunning result, which is supported by many decades of scientific observation: Earth's continents were in very different positions in the past (**FIGURE 1.15** on pages 20–21). This leads us to a more sophisticated understanding of Hutton's principle. The physical *processes* that occur on

uniformitarianism The concept that the processes governing the Earth system today have operated in a similar manner throughout geologic time.

A A distinctive pattern of wind-deposited sand grains can be seen in a trench dug in this modern sand dune near Yuma, Arizona.

B A similar pattern can be seen in sandstone rocks millions of years old, in the Vermilion Cliffs Wilderness Area in Arizona. We can infer from the similar patterns that these ancient rocks were once sand dunes.

Earth have not changed over time, but the physical *conditions* of Earth have changed dramatically. Sea levels drop and rise; the chemical composition of Earth's atmosphere fluctuates, albeit ever so slowly. The cycles maintain a balance, but in doing so the sizes of the reservoirs may change and the speed of cycles and processes may increase or decrease. This is an especially important lesson today, when it appears our planet has entered a period of climatic change.

Throughout this book we will explore the process of plate tectonics in much greater depth. We will see how virtually every aspect of the Earth system owes its essential character to the existence of plate tectonics. Because plate tectonics has become the overarching theme in geology, it is viewed as a *unifying theory*. Practically every aspect of geologic science, including the histories of the atmosphere, hydrosphere, and biosphere, is connected to the motion of lithospheric plates.

Visualizing

Earth's changing face FIGURE 1.15

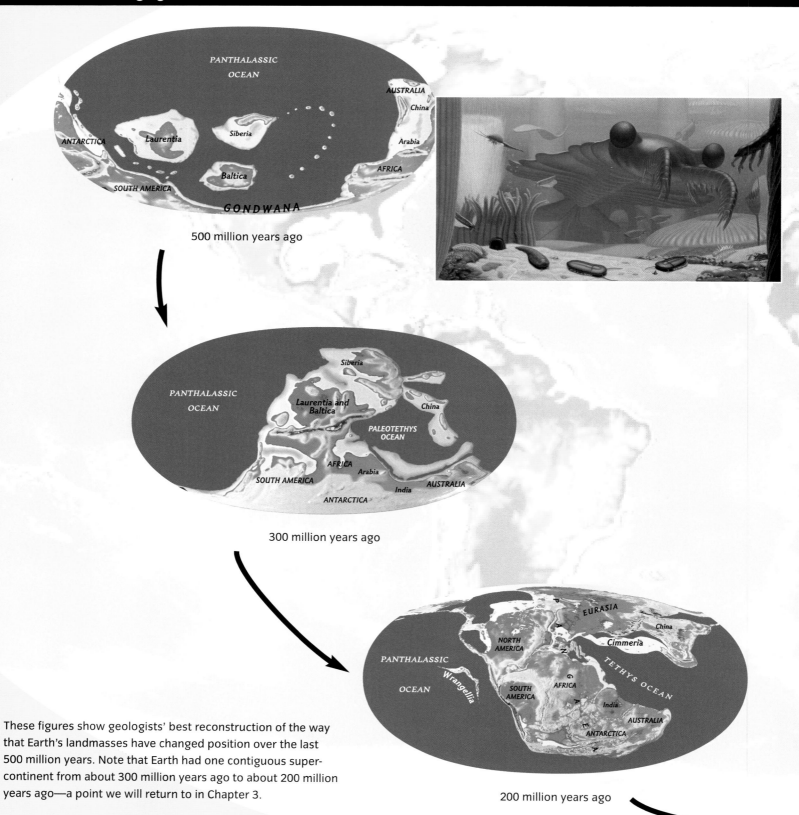

500 million years ago

300 million years ago

200 million years ago

These figures show geologists' best reconstruction of the way that Earth's landmasses have changed position over the last 500 million years. Note that Earth had one contiguous super-continent from about 300 million years ago to about 200 million years ago—a point we will return to in Chapter 3.

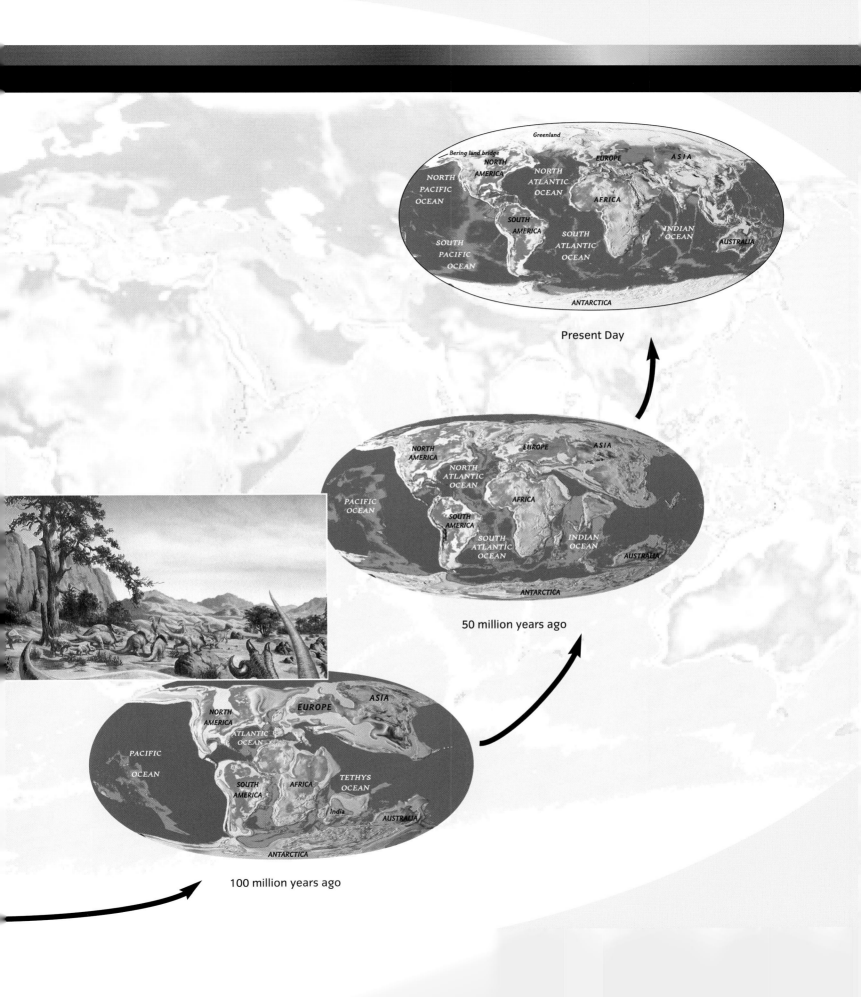

Present Day

50 million years ago

100 million years ago

WHY STUDY GEOLOGY?

With this brief introduction to geology and the Earth system, you have probably deduced some of the reasons why it is important to study geology. We need to understand Earth materials because we depend on them for all of our material resources—the minerals, rocks, and metals with which we construct our built environment; the energy with which we run it; the soil that supports agriculture and other plant life; and the air and water that sustain life itself. Many Earth resources are limited and require knowledgeable and thoughtful management. The materials of Earth also have physical and chemical properties that affect us, such as their tendency to flow or fail during a landslide, their capacity to hold or transmit fluids such as water or oil, or their ability to absorb waste or prevent it from migrating.

We have learned that Earth is essentially a closed system, which means that all materials remain within the system. Therefore, it is important to understand how materials move from one reservoir to another. It is also important to understand the time scales that govern these processes in order to gain some perspective on the changes that we see occurring in the natural environment. Some Earth processes are hazardous—that is, damaging to human interests. These *geologic hazards* include earthquakes, volcanic eruptions, landslides, floods, and even meteorite impacts. The

Lethal eruption FIGURE 1.16

Mount Pinatubo in the Philippines erupted in 1991. Volcanologists predicted the eruption, making it possible to evacuate residents from the area and prevent thousands of deaths. When it erupted, the volcano sent this lethal cloud of searing, dust-laden gases rolling down its flanks, to spread rapidly across the surrounding plains. This particular car and driver escaped, but many houses, trees, and fields were smothered with volcanic ash.

Global Locator

more we know about these hazardous processes, the more successful we will be in protecting ourselves from future natural disasters (FIGURE 1.16).

Finally, Earth is our home planet. The features that make Earth unique and the powerful geologic processes that characterize the Earth system are a constant source of awe and fascination to those who study them. It makes sense to deepen and refine our understanding of the planet we live on.

From its beginnings a couple of centuries ago, geology has been an interdisciplinary science, because Earth operates through the interactions of biologic, physical, and chemical processes. Yet we are discovering that the interactions are more complex and dynamic than we would have believed only decades ago. We are still learning about the complexities and interrelationships of subsystems such as climate, ocean currents, and shifting continents. We now appreciate more profoundly our own role in geologic change as well as the need to study the Earth system as a whole rather than in separate fragments.

Visualizing Geology starts your study of Earth. If you are planning to become a geologist, this book will be an introduction to some of the many fascinating possibilities that await you in your career. If you are taking this course out of personal interest or to fulfill a degree requirement, you will emerge more aware of the geologic nature of our planet and better prepared to make informed decisions about the natural processes that affect your life on a daily basis.

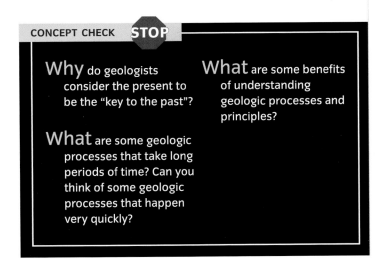

CONCEPT CHECK STOP

Why do geologists consider the present to be the "key to the past"?

What are some benefits of understanding geologic processes and principles?

What are some geologic processes that take long periods of time? Can you think of some geologic processes that happen very quickly?

The most amazing place of all, however, is Earth itself (pages 24–25), the only world in the universe where we *know* that life exists.

This is a computer rendering, compiled from many satellite images, of Earth at night. (In reality, the entire Earth is never dark at the same time.) Note how the lights are brightest in places with a high population density. Also, note how human settlement concentrates on coastlines. Why? What geologic features explain the dark areas? Can you spot where you live or identify any major cities?

NATIONAL GEOGRAPHIC

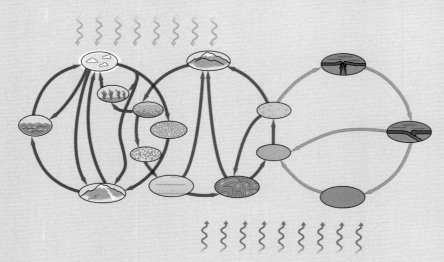

1 What Is Geology?

1. **Geology** is the scientific study of Earth, including its formation and internal structure; the materials it is composed of and the properties of those materials; its chemical and physical processes; and its physical, chemical, and biologic history. Geology is divided into two broad categories, **physical geology** and **historical geology**. Physical geology focuses on the *materials* and *processes* of the Earth system. Historical geology seeks to establish the chronology of geologic *events* in Earth's history. Today geology is a diverse science, with dozens of subspecialties, including environmental geology, volcanology, seismology, mineralogy, paleontology, tectonics, and structural geology.

2. Geologists study Earth today using *Earth system science*. This concept comes from the discovery that Earth is an integrated **system** of interconnected and interdependent parts. Individual systems within the larger Earth system can be big or small, and can vary greatly in complexity, but regardless of size, each system operates within an identifiable boundary. There are three kinds of systems, *isolated*, *open*, and *closed*; the properties of the boundary determine the kind of system. In an isolated system, boundaries prevent the system from taking in or releasing any energy or matter. Because there is no perfect boundary against the passage of energy, isolated systems do not exist in the real world. A closed system has a boundary that permits the passage of energy, but not of matter, in and out of the system. The third kind of system, an open system, permits the exchange of both matter and energy across its boundary. Most environmental and geologic systems in the natural world are open systems.

3. Earth is considered a closed system, though some small amounts of matter, such as meteorites, do cross its boundary. The Earth system consists of four principal open subsystems, including the **atmosphere**, the envelope of gas that surrounds Earth; the **hydrosphere**, comprising all of Earth's water; all of Earth's living organisms, known as the **biosphere**; and the **lithosphere**, Earth's rocky outer layer. Materials and energy are stored for varying lengths of time in each of these systems or reservoirs and can move among them via innumerable pathways and processes. Each of the four Earth subsystems can be further broken down into a vast number of still smaller subsystems, all of which are open.

4. An important component of Earth system science is the monitoring and study of the movement of materials among the subsystems. Because materials can move in both directions between subsystems, the movement of materials and energy among the reservoirs of the Earth system are called cycles. The **hydrologic cycle** describes the movement of all of Earth's water, above and below ground, through the reservoirs of the Earth system. The **rock cycle** describes the crustal processes that create, modify, transport, and break down rock. The **tectonic cycle** includes the movements and interactions of Earth's tectonic plates along with the internal processes that drive those movements.

2 Earth in Space

1. Earth is one of eight bodies in the **solar system** recognized as planets. In addition to the Sun and **planets**, the solar system includes a vast number of moons, asteroids, comets, and fragments of rock and dust called *meteoroids*. The four inner planets, or *terrestrial planets*, Mercury, Venus, Earth, and Mars, are similar in many ways. They are all small, rocky, and relatively dense, and they have similar sizes and chemical composition. The four outer planets, or *jovian planets*, in contrast, consist of huge gaseous atmospheres with small solid cores, giving them very low densities overall. The jovian planets are Jupiter, Saturn, Uranus, and Neptune. Pluto, until recently considered to be a ninth planet, is not a jovian planet because it is icy but not gassy. Instead, it is considered to be a "dwarf" planet, and part of the Kuiper Belt, a region at the outer edge of the solar system that contains a large number of small, icy objects.

2. The *nebular theory* is the generally accepted model that the solar system formed from the coalescence and condensation of a nebula, a cloud of interstellar gas and dust. This theory is supplemented by hypothesis known as *planetary accretion*. The planetary accretion hypothesis states that all of the planets assembled themselves by the accretion of debris within the nebula, and that today's meteoroids and asteroids are fragments that were never swept into the forming planets. The commonality of origin explains why all the terrestrial planets have experienced volcanic activity and have undergone extensive impact cratering. It is likely that an impact between early Earth and a small planet is responsible for the creation of our Moon.

3. The hypothesis of planetary accretion and intense early heating is important to geologists because it helps explain Earth's beginnings. Early in its history, Earth undervent **differentiation** into a dense, metallic **core**, a rocky **mantle**, and a brittle, rocky outer **crust**.

4. Earth is unique in the solar system in that it possesses an oxygen-rich *atmosphere*. Earth is also the only planet in the solar system with a *hydrosphere* in which water exists near the surface in solid, liquid, and gaseous forms, and a *biosphere* with living organisms. Finally, Earth is the only planet where true *soil* is formed from *regolith* by interactions among physical, chemical, and biologic processes, and where life as we know it could exist.

5. **Plate tectonics** is a unifying theory in geology. It describes the motion and interaction of large segments of the lithosphere. It is because of plate tectonics that Earth has two fundamentally different types of crust: the relatively thin, dense **oceanic crust** of basaltic composition, and the thicker, less dense **continental crust** of granitic composition.

3 The Ever-Changing Earth

1. The principle of **uniformitarianism** is one of the fundamental concepts of geology. It is based on the idea that the processes we see operating in the Earth system today have operated in a similar manner throughout much of geologic time. Uniformitarianism is based on the idea that "the present is the key to the past." That is, by studying processes that are occurring today, we can deduce much about past occurrences. More than two centuries of work by geologists has demonstrated the correctness of uniformitarianism.

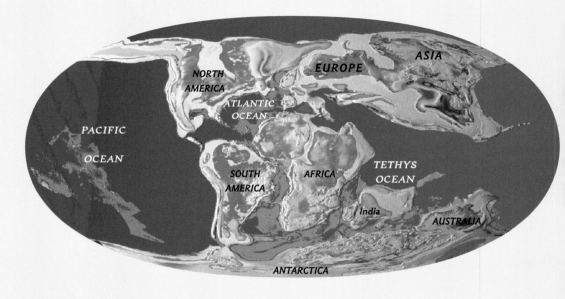

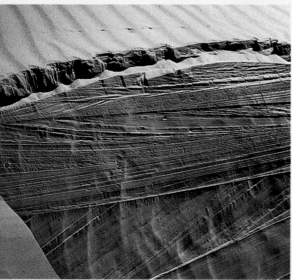

2. From a human standpoint, most geologic processes are incredibly slow. Geologists have discovered that Earth is roughly 4.56 billion years old, and that the rock cycle has been continuous throughout Earth's long history. Though the processes that occur on Earth have not changed, the rates of the different cycles, such as the rock cycle and plate tectonics, have differed over time. The *physical conditions* on Earth—such as the temperature and composition of the atmosphere, the level of the oceans, and the location of the continents—have also been dramatically different at times in the past.

3. The study of geology is important to human society for many reasons. Earth materials and processes affect our lives through our dependence on Earth resources; through *geologic hazards* such as volcanic eruptions, floods and earthquakes; and through the physical properties of the natural environment.

KEY TERMS

- geology p. 4
- system p. 7
- lithosphere p. 9
- biosphere p. 9
- atmosphere p. 10
- hydrosphere p. 10

- hydrologic cycle p. 10
- rock cycle p. 10
- tectonic cycle p. 10
- meteorite p. 14
- crust p. 16
- mantle p. 16

- core p. 16
- oceanic crust p. 17
- continental crust p. 17
- plate tectonics p. 17
- uniformitarianism p. 18

CRITICAL AND CREATIVE THINKING QUESTIONS

1. Do you think there may be life on a planet outside our solar system? What would the atmosphere of that planet be like? Must it have a hydrosphere? Why or why not?

2. Why is the systems approach so useful in studying both natural and artificial processes? Can you think of examples of artificial (that is, human-built) systems other than those given in the text? Are they open systems or closed systems? (Think about the materials and energy in them.)

3. How do you think the principle of uniformitarianism accounts for occasional catastrophic events such as meteorite impacts, huge volcanic eruptions, or great earthquakes?

4. In this chapter we have suggested that Earth is a close approximation of a natural closed system, and we have hinted at some of the ways that living in a closed system affects each of us. Can you think of some other ways?

5. In what ways do geologic processes affect your daily life?

What is happening in this picture ?

This rock, photographed in Saudi Arabia's Rhub al Khali (Empty Quarter), was discovered in 1965. It is believed to be the largest fragment of a meteorite that fell to Earth sometime before 1863 (when the first piece was discovered). How do you think these scientists can tell it is a meteorite? Why did it break up into pieces? Why is the desert a good place to look for meteorites? (Hint: Think about what would have happened to this rock if it had fallen in a jungle or a mountain range.)

1. _____ is fundamentally concerned with understanding the processes that operate at or beneath the surface of Earth and the materials on which those processes operate.

   a. Economic geology
   b. Physical geology
   c. Historical geology
   d. Environmental geology
   e. Planetary science

2. On this illustration, label each of the following systems:

   isolated system   closed system
   open system

3. The island depicted in the figure acts as a(n)

   a. intermittent system.
   b. closed system.
   c. solar system.
   d. open system.
   e. isolated system.

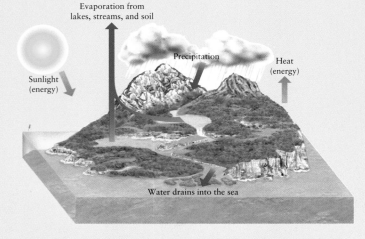

4. On the time scale of a human lifetime, Earth acts as a(n)

   a. intermittent system.
   b. closed system.
   c. solar system.
   d. open system.
   e. isolated system.

5. The _____ is a subset of the Earth system that comprises all of its bodies of water and ice, both on the surface and underground.

   a. atmosphere
   b. hydrosphere
   c. lithosphere
   d. ionosphere
   e. biosphere

6. The _____ explains how new crust is formed and recycled by large-scale motions of Earth's surface and interior.

   a. tectonic cycle
   b. rock cycle
   c. water (hydrologic) cycle
   d. nebular theory

7. In the illustration, in which stage of solar system evolution did the Sun first begin to shine? _____

   1.   2.   3.   4.

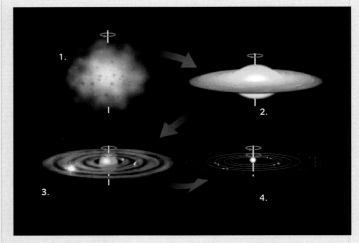

8. According to the planetary accretion model,

   a. planets evolved from molten material condensing from an early solar nebula.
   b. planets assembled themselves from meteorite-like debris.
   c. most of today's meteorites are the debris that never managed to be swept up by any planet.
   d. Both b. and c. are true.
   e. All of the above statements are true.

9. The inner planets of the solar system are rocky whereas the outer planets contain a higher proportion of ice and gas. This differentiation occurred in the early solar system because

   a. the rocky and metallic components, which have higher melting points, would have condensed at an early stage, and in the innermost region of the solar system.

   b. the solar wind was strong enough in the inner solar system to push the lighter gases, such as helium and hydrogen, to the outer solar system.

   c. volatile compounds, such as water and methane, condense at lower temperatures common to the outer solar system, where the primordial nebula was cooler.

   d. All of the above statements are true.

10. The photograph is of a basalt flow on the island of Hawaii. Which one of the following statements is true?

   a. Basaltic lava flows have only occurred on Earth.

   b. These types of flows have been common only on Earth and in the early history of the Moon.

   c. Basalt is the most common volcanic rock known in our solar system.

   d. Basaltic lava flows would have been common in the early history of Earth, but in modern times they have largely ceased.

   e. None of the above statements is true.

11. Earth, Mars, and Venus all have

   a. an oxygen- and nitrogen-rich atmosphere.

   b. a core, mantle, and crust.

   c. a hydrosphere with liquid water.

   d. a biosphere.

   e. All of the above statements are true.

12. These two photographs compare distinctive patterns in wind-blown sand in a modern dune (A) with similar structures inside an ancient sandstone (B). Our ability to infer that the structures inside the sandstone were formed by the same processes that formed them in a modern sand dune is an application of

   a. the accretionary hypothesis.

   b. plate tectonic theory.

   c. Hutton's principle of uniformitarianism.

   d. the nebular theory.

   e. planetary differentiation.

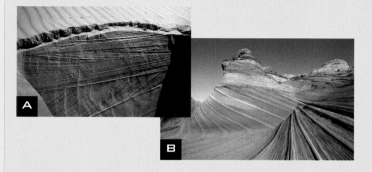

13. Hutton's principle of uniformitarianism

   a. states that the physical processes that act on Earth have not changed over time, even though the physical conditions of Earth have changed dramatically.

   b. states that the physical processes that act on Earth and the physical conditions of Earth have not changed over time.

   c. cannot be used to explain rapid fluctuations in the climate system.

   d. has been rendered obsolete by the modern unifying concept of plate tectonics.

14. The rates of some processes involved in the cycles of the Earth system, such as the rock cycle and tectonic cycle,

   a. have been constant over time.

   b. have varied over time .

   c. have been steadily increasing with time.

   d. have been steadily decreasing with time.

15. The study of geology is important because

   a. it helps us understand the processes that govern the Earth system.

   b. it is important to understand the time scales that govern Earth processes.

   c. it helps us understand and mitigate the potential threats of geologic hazards, such as earthquakes, volcanic eruptions, landslides, floods, and even meteorite impact.

   d. All of the above statements are true.

Earth Materials

SIRIU

This diamond comes from Point Lake in the Northwest Territories, Canada, where geologists Charles Fipke and Stewart Blusson discovered a rich diamond deposit in 1991. Fipke and Blusson theorized that occasional diamonds found in Wisconsin and elsewhere had been pushed all the way from northern Canada by glaciers during the last Ice Age. In spite of the skepticism of many geologists, they were proved right. Since 1998, when the first diamond mine opened at Point Lake, Canada has become the third-largest exporter of diamonds in the world, with two active mines and three more soon to open. Every diamond produced in Canada is marked by a minuscule engraving of a polar bear that attests to its origin in the Northwest Territories (inset on diamond).

Even apart from its rarity and value, diamond is a remarkable mineral. It is chemically simple: pure elemental carbon, the same chemical found in graphite and charcoal. But, unlike graphite and charcoal, natural diamond forms only at extraordinarily high pressures and temperatures, deep under Earth's surface. It is the hardest mineral known and an excellent conductor of heat. Its properties have made it a valuable material for industrial purposes—for abrasives, cutting tools, and high-pressure cells. If production costs can be brought down, artificial diamonds could one day replace silicon in our computers. A computer with diamond chips could run at much higher temperatures, because diamond has such desirable thermal properties.

In this chapter you will learn about rocks and minerals and why minerals of identical composition, such as diamond and graphite, have different properties. Some minerals are beautiful, some are important to the modern economy (but not beautiful), and some are vital to us as nutrients. Other minerals, such as certain varieties of asbestos, are potentially hazardous to human health. By studying and understanding rocks and minerals, we can learn to balance their positive and negative effects in our lives.

Point Lake

Global Locator

A2-031Z

Elements and Compounds

LEARNING OBJECTIVES

Define element, atom, compound, molecule, and ion.

Explain the difference between an atom and a molecule.

Describe the internal structure of an atom.

Identify four kinds of chemical bonding.

Explain how the kinds of bond in a material affect its physical properties.

All matter on and in Earth, including the page you are reading and the eyes you are reading it with, consists of one or more chemical elements. All of the chemical reactions that control our lives and make life on Earth possible depend on the ways in which these elements interact.

ELEMENTS, ATOMS, AND IONS

Chemical **elements** are the most fundamental substances into which matter can be separated and analyzed by ordinary chemical methods. Ninety-two naturally occurring elements are known and a number more have been synthesized by atomic scientists. Each element is identified by abbreviated symbols, such as H for hydrogen and Si for silicon. Some of the symbols come from the names of the elements in other languages, such as Fe for iron, from the Latin *ferrum*, and Na for sodium, from the Latin *natrium*. Others are named in honor of famous scientists, such as element 99, Es, einsteinium. The periodic table of the elements is shown in Appendix B.

Even the tiniest grain of dust is made up of innumerable particles, called **atoms**, which are much too small for the eye to see (**FIGURE 2.1**). In a pure sample of an element, every atom would be exactly the same kind. Atoms are the building blocks of chemistry, which account for all of an element's chemical properties and many of its physical ones, such as density and color. They are so tiny, about one billionth of a millimeter, that they cannot be seen at all with a conventional optical microscope. Specially designed microscopes have succeeded in imaging atoms, but it would be more accurate to say that they "feel" the atoms rather than see them.

Chemical reactions also produce the rocks, minerals, liquids, and gases that geologists study. For this reason, a quick review of chemical elements and atoms is a good place to begin our study of Earth materials. Atoms themselves are composed of even smaller particles, which have no independent chemical properties. The *nucleus* (plural *nuclei*) of an atom contains *protons*, with positive electric charges, and *neutrons*, which are electrically neutral. The number of protons in an atom— its *atomic number*— determines which element the atom is and gives the atom its chemical characteristics. Atomic numbers range from 1 for the lightest element, hydrogen, up to 92 for the heaviest naturally occurring element, uranium. Every element from atomic number 1 to 92 has either been synthesized in the laboratory or found in nature, so there are no new elements to be discovered in that range. However, some laboratories are still working on synthesizing heavier elements, and they have gotten up to element 112 (ununbium).

The number of protons plus the number of neutrons in the nucleus of an atom is the *mass number*. Atoms of a given element always have the same atomic number, but they can have different mass numbers. For example, there are three naturally occurring **isotopes** of carbon: carbon-12, carbon-13, and carbon-14. Each of the isotopes of carbon has 6 protons and thus an atomic number of 6. However, the three isotopes contain different numbers of neutrons: 6, 7, and 8 per atom, respectively (thus, different mass numbers: 12, 13, and 14).

element The most fundamental substance into which matter can be separated by chemical means.

atom The smallest individual particle that retains the distinctive chemical properties of an element.

isotopes Atoms with the same atomic number but different mass numbers.

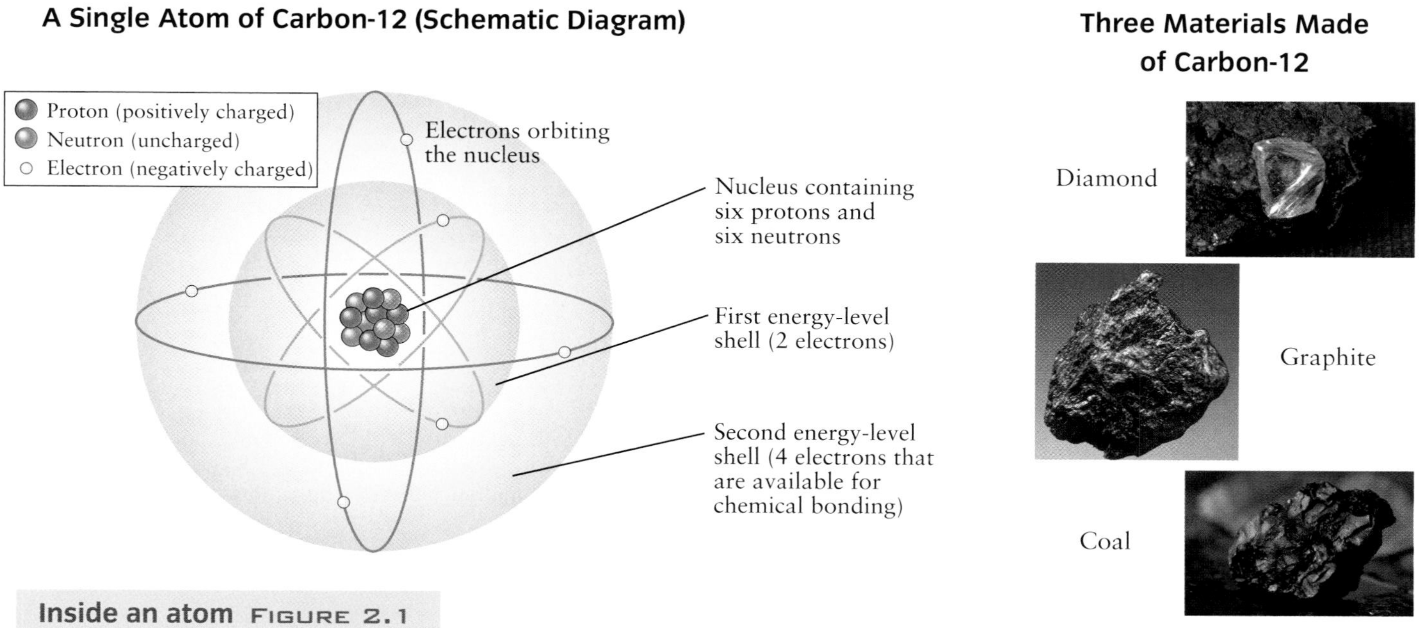

A Single Atom of Carbon-12 (Schematic Diagram)

Proton (positively charged)
Neutron (uncharged)
Electron (negatively charged)

Electrons orbiting the nucleus

Nucleus containing six protons and six neutrons

First energy-level shell (2 electrons)

Second energy-level shell (4 electrons that are available for chemical bonding)

Inside an atom FIGURE 2.1

Three Materials Made of Carbon-12

Diamond

Graphite

Coal

As shown in the schematic diagram of carbon-12 (left), six electrons orbit the nucleus in two complex paths called orbitals, rendered here (unrealistically) as circles. The orbitals arrange themselves in energy-level shells, which are more stable when completely filled. Besides being the sole component of diamond, graphite, and coal, carbon-12 is the most important atom in living beings.

The third small component of an atom is called an *electron*. Electron interactions help determine the makeup of ions and compounds. Electrons orbit the nucleus in complex patterns, shown schematically in Figure 2.1 as circles of different sizes. Electrons have a negative charge that is equal in magnitude but opposite in sign to the positive charge of the proton. In its ideal state, an atom has an equal number of protons and electrons and thus is electrically neutral. Under certain circumstances, however, an atom may gain or lose an electron during a chemical reaction. Although atoms can exchange electrons, they never exchange protons or neutrons as a result of chemical processes.

An atom that has lost or gained one or more electrons will have a net electric charge, and is called an **ion**. If the charge is positive, meaning that the atom has lost one or more electrons, the ion is called a *cation*. If the charge is negative, meaning that the atom has gained one or more electrons, the ion is called an *anion*. A convenient way to indicate ionic charges is to record them as superscripts. For example, Na^+ is the symbol for an atom of sodium that has given up an electron; Cl^- is the symbol for an atom of chlorine that has accepted an electron; and Fe^{2+} is the symbol for an atom of iron that has given up two electrons.

COMPOUNDS, MOLECULES, AND BONDING

Chemical **compounds** form when atoms of one or more elements combine with atoms of another element in a specific ratio. For example, sodium and chlorine combine to form sodium chloride (a mineral called halite, also known as table salt), which is written NaCl. For every Na atom in this compound, there is one Cl atom. The element that tends to form cations is written first; the element that tends to form anions is written second; and the relative numbers of atoms are indicated by subscripts. For example, water forms when hydrogen (a cation, H^+) combines with oxygen (an anion, O^{2-}) in the ratio of two atoms of hydrogen to one atom of oxygen.

compound
A combination of atoms of one or more elements in a specific ratio.

How ions and compounds form FIGURE 2.2

● Protons   ● Neutrons   ○ Electrons

Movement of electron

Li⁺   Ionic bond   F⁻

Lithium (Li) (element) has one electron in outer orbital and donates an electron.

+

Fluorine (F) (element) is one electron short of complete in outer orbital and borrows an electron.

Lithium fluoride (LiF) (compound) contains positively charged lithium and negatively charged fluorine, creating an ionic bond.

Thus, for water, we write H_2O. FIGURE 2.2 shows how lithium and fluorine combine to form an ionic compound, lithium fluoride, LiF.

Clearly, the properties of compounds are not the same as the properties of their constituent elements. For example, hydrogen and oxygen are both gases at surface temperature and pressure, whereas water is a liquid. Similarly, elemental sodium and elemental chlorine are both highly toxic, whereas their compound, salt, is essential for life.

molecule The smallest chemical unit that has all the properties of a particular compound.

bond The force that holds the atoms together in a chemical compound.

The smallest unit that has the properties of a given compound is a **molecule**. Do not confuse a molecule and an atom; the definitions are similar, but a molecular compound always consists of two or more atoms. Molecules are held together by electromagnetic forces known as **bonds**. Bonding involves the transfer of electrons from one atom to another or in some cases the sharing of electrons. There are four principal kinds of bonds, which are illustrated in FIGURE 2.3. You will see that different bonding explains the difference in properties of diamond and graphite.

Why have we spent all this time learning about elements, compounds, and bonding? Because minerals—the main building blocks of the solid Earth—occur in the form of chemical compounds (or sometimes as elements). The properties of these minerals depend very much on what chemicals they are made of and how those chemicals are put together. Now let's look more closely at the characteristics that define minerals and help us to identify them.

CONCEPT CHECK ⬛ STOP

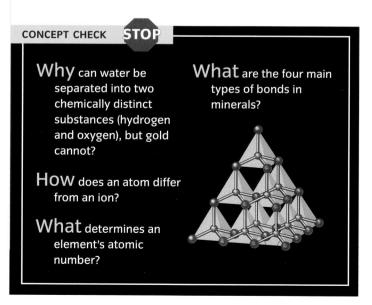

Why can water be separated into two chemically distinct substances (hydrogen and oxygen), but gold cannot?

How does an atom differ from an ion?

What determines an element's atomic number?

What are the four main types of bonds in minerals?

Ionic bonding: When one atom transfers an electron to another, as illustrated in Figure 2.2, an attractive force is set up that creates an ionic bond. In a crystal such as table salt (NaCl), the sodium ions (red) are attracted to all of the neighboring chlorine ions (gray), not just to one of them. Thus the ionic bonds form a cubic lattice. Compounds with ionic bonds tend to have moderate strength and hardness.

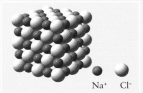

In table salt (sodium chloride, NaCl), each sodium cation is surrounded by chlorine anions.

Crystals of sodium chloride are rectangular, with straight edges.

Salt is a moderately hard solid that dissolves easily in water.

Covalent bonding: when electrons from different atoms "pair up," the force of this sharing is called a covalent bond. Note that electron sharing does not produce ions. These are the strongest chemical bonds, and elements and compounds with covalent bonds (such as diamond) tend to be strong and hard.

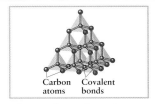

Diamond consists of carbon atoms connected in a network of covalent bonds. Each atom is connected to four others.

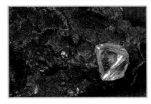

Diamond crystals appear in a rock called kimberlite. Covalent compounds are often strong and hard; diamond is one of the hardest substances known.

Cut and polished diamonds are prized gems. Tiny diamonds are used in industry for cutting and grinding instruments.

Metallic bonding: In metals, atoms are so tightly packed that electrons can be shared among several atoms. In fact, the outermost electrons are so loosely held that they can readily drift from one atom to another. This mobility of electrons explains why metals are good at conducting electricity and heat.

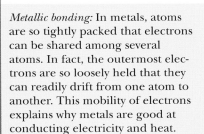

Atoms of gold are packed in the densest possible manner. Each atom is surrounded by, and in contact with, 12 other gold atoms

This nugget of gold was once embedded in rock, but weathering and erosion have removed most of the rock.

Gold is durable as well as malleable; it has been used as currency since ancient times.

Van der Waals bonding: A weak attraction can occur between electrically neutral molecules that have an asymetrical charge distribution. The positive end of one molecule will be attracted to the negative end of another molecule. For example, the carbon atoms in graphite form sheets in which each carbon atom has strong covalent bonds with three neighbors. The bonds between sheets are weak. This is why graphite feels slippery when you rub it between your fingers.

In graphite, carbon atoms form layers connected by covalent bonds. The layers are weakly held together by Van der Waals bonds.

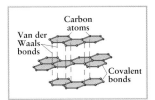

Graphite is not a strong material and can be easily crumbled into small particles.

The "lead" in pencils is really graphite. When you write, the pressure of your hand breaks off a trail of carbon particles.

www.wiley.com/college/murck

What Is a Mineral?

LEARNING OBJECTIVES

Identify four requirements for a solid material to be classified as a mineral.

Explain the principle of atomic substitution.

Explain why crystals tend to have flat faces with specific angles between them.

Identify at least six methods that geologists use to tell minerals apart.

Explain why color is one of the least dependable ways to identify a mineral.

T he word **mineral** has a specific connotation in science. To be classified as a mineral, a substance must meet certain requirements. A substance must be a *naturally occurring solid*, formed by *inorganic processes*, with a *characteristic crystal structure* and a *specific chemical composition*. Each of the items on this checklist is essential (**FIGURE 2.4**).

COMPOSITION OF MINERALS

An apparent exception to the rule that a mineral must have a specific chemical composition is a phenomenon called *atomic substitution*. In some cases, two elements can be similar enough in size and in bonding properties that they can substitute for each other in a mineral. For example,

mineral A naturally formed, solid, inorganic substance with a characteristic crystal structure and a specific chemical composition.

Mineral or not mineral? FIGURE 2.4

Ice is a mineral, though you may not usually think of it that way. It occurs in nature in the form of hexagonal crystals, and has a specific chemical formula (H_2O).

Water is not a mineral, because it is not a solid. This criterion also means that such naturally occurring substances such as oil and natural gas cannot be considered minerals.

Bones are a tricky case. They do contain the same chemical compound found in a common mineral called apatite, but they are not minerals because they form by organic processes. Thus the bone in this modern crocodile skull is not a mineral.

However, this fossilized crocodile skull, from the Kenyan National Museum, is composed of minerals. Why? During fossilization, the original materials were replaced in an inorganic process called *mineralization*.

www.wiley.com/college/murck

magnesium and iron ions (Mg^{2+} and Fe^{2+}) are so similar in size and electrical charge that one often takes the place of the other. The mineral olivine (an important component of Earth's mantle) can occur as pure Fe_2SiO_4 or pure Mg_2SiO_4 or an intermediate mixture in which some of the Fe^{2+} cations are replaced by Mg^{2+} cations. We show atomic substitution in the chemical formula by putting parentheses around the substituting elements and a comma between them. The formula for olivine, therefore, becomes $(Mg, Fe)_2SiO_4$, indicating that Mg and Fe can substitute for one another in this mineral.

The composition requirements for minerals specifically rule out materials whose composition varies so much that it cannot be expressed by an exact chemical formula. An example of such a material is glass, which is a mixture of many elements and can have a wide range of compositions.

Glass—even naturally formed volcanic glass—also fails the test of having a characteristic **crystal structure**. In a crystal, the atoms are arranged in regular, highly repetitive geometric patterns, as shown in FIGURE 2.5A on the following page. By contrast, the atoms in a liquid or in an *amorphous* solid like glass are mixed up or randomly jumbled. Sometimes amorphous solids are referred to as *mineraloids*; an example of a mineraloid is opal, a familiar stone that is often used in jewelry.

All specimens of a given mineral have an identical crystal structure. Extremely sensitive scanning tunneling microscopes enable scientists to look at the crystal structures of minerals and actually see the orderly arrangement of atoms in the mineral. As you can see in FIGURE 2.5B, the atoms in a crystalline material resemble the regular, orderly rows in an egg carton.

> ■ **crystal structure**
> An arrangement of atoms or molecules into a regular geometric lattice. Materials that possess a crystal structure are said to be crystalline.

Coal fails the second of the four tests for a mineral because it is derived from the remains of plant material and was formed as a result of organic processes.

Steel (being made in the background) fails the first of the four tests for a mineral because it does not occur naturally. It is formed by extensive human processing of naturally occurring ores, which are minerals.

Quartz is an easily recognizable mineral. Its chemical formula is SiO_2. Note that some minerals have very complex formulas. For example, phlogopite, a form of mica, is $KMg_3AlSi_3O_{10}(OH)_2$. The important thing is that the elements combine in specific ratios.

Although **opals** are typically included in books about minerals, they are not true minerals because they do not have a specific composition and lack a crystalline structure. Opals are *mineraloids*.

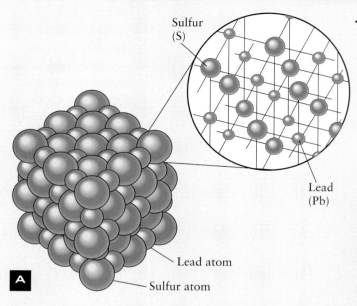

Sulfur
(S)

Lead
(Pb)

A Lead atom

Sulfur atom

A The atoms of all crystalline materials are arranged in orderly lattices, like the cubical lattice illustrated here for a mineral called galena (PbS), the main source of lead. The atoms are so small that a cube of galena 1 cm on an edge would contain 10^{22} atoms (that's 1 followed by 22 zeros). The inset shows an "exploded view" of the packing arrangement of atoms in a galena crystal. The atoms are shown pulled apart along the black lines to demonstrate how they fit together. Compare the arrangement of atoms in NaCl (Figure 2.3); it is the same as the arrangement of lead and sulfur atoms in PbS.

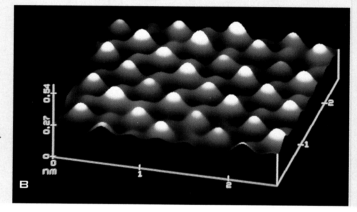

B

B Atoms are too small to see with an optical microscope, but a scanning tunneling microscope can show the atoms in a crystal. This is an image of a galena crystal; the sulfur atoms look like large bumps and the lead atoms like small ones.

TELLING MINERALS APART

The compositions and crystal structures of minerals influence their physical properties and characteristics. If we have an unidentified mineral sample (such as FIGURE 2.6), we can apply a few simple tests to determine what mineral it is, without taking it to a laboratory or using expensive equipment. The properties most often used to identify minerals are the quality and intensity of light reflected from the mineral, crystal form and habit of the mineral, hardness, tendency to break in preferred directions, color, and specific gravity or density. Color, perhaps the most obvious characteristic, is often the least reliable identifier. Let's look at the properties that are used to identify different minerals.

Luster   Suppose you collected a mineral sample like the one in Figure 2.6. What steps would you go through to identify it? One of the first things you would notice is how shiny it is, or what geologists call its **luster**. Different minerals can reflect light in dif-

A "mystery mineral" FIGURE 2.6

Geologists have several effective, low-tech methods to identify minerals. We will use them to identify the mineral pictured here. First, note its metallic luster and cubic habit. (The answer is on page 45.)

luster The quality and intensity of light that reflects from a mineral.

ferent ways as well as with different intensities. The luster of the mystery sample in Figure 2.6 is *metallic*, meaning that it looks like a polished metal surface. Some other kinds of luster you might encounter are *vitreous*, like that of glass; *resinous*, like that of resin; *pearly*, like that of pearl (**FIGURE 2.7**); or *greasy*, as if the surface were covered by a film of oil. Two minerals with almost identical color can have quite different lusters.

Crystal form and habit

The ancient Greeks were fascinated by ice. They were intrigued by the fact that needles of ice are six-sided and have smooth, planar surfaces. The Greeks called ice *krystallos*. Eventually, the word *crystal* came to be applied to any solid body that

has grown with flat or planar surfaces. The planar surfaces that bound a crystal are called *crystal faces*.

During the seventeenth century, scientists discovered that crystal form could be used to identify minerals. But certain properties seemed to vary widely from one sample to another. Under some circumstances, a mineral species may grow a thin crystal; under others, the same mineral species may grow a fat crystal, as **FIGURE 2.8** shows. It is apparent from the figure that the overall crystal size and the relative sizes of crystal faces are not the same for these two crystals of quartz. In fact, crystal size and the relative sizes of crystal faces are not definitive for any mineral.

In 1669 a Danish physician, Nils Stensen (better known by his Latin name, Nicolaus Steno) unraveled the mystery of crystal form. Steno demonstrated that the key property that identifies a given mineral is not the size of its faces, but rather the angles between them. According to Steno's Law, the angle between any corresponding pairs of crystal faces of a given mineral species is constant no matter what the overall shape or size of the crystal might be (see Figure 2.8).

Steno and other early scientists suspected that a mineral must have some kind of internal order that

Mineral luster FIGURE 2.7

Three examples of luster.

Vitreous
Quartz (SiO_2) has a glassy luster.

Resinous
Sphalerite (ZnS, a source of zinc) has a resinous luster, like dried tree resin.

Pearly
Talc ($MgSi_4O_{10}$ $(OH)_2$) has a pearly luster.

Crystal faces and angles FIGURE 2.8

Crystals of the same mineral may differ widely in shape. However, the angles between faces will remain the same in all specimens. In the two quartz crystals pictured, numbers identify equivalent faces. According to Steno's Law, the angle between faces 1 and 3 (for example) is the same in both specimens.

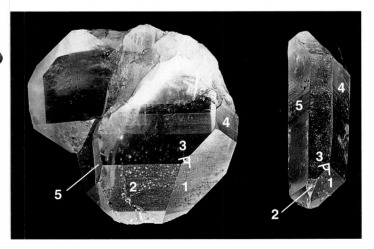

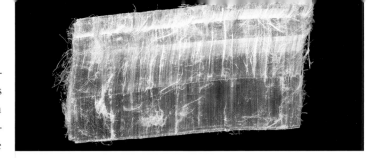

predisposes it to form crystals with constant interfacial angles. Proof that crystal form reflects a crystal's internal atomic structure finally arrived in 1912, when German scientist Max von Lauë beamed X-rays at crystals and showed that the diffracted rays matched the patterns a geometric array would create. More recently, scanning tunneling microscopes have given more direct proof of atomic lattice structure, as shown in Figure 2.5B.

Crystals with planar faces form most easily when mineral grains can grow freely in an open space. Because most mineral grains do not form in open, unobstructed spaces, nicely formed crystals are uncommon in nature. Usually other mineral grains get in the way. As a result, most mineral grains have an irregular shape. However, in both a crystal and an irregularly shaped grain of the same mineral, all the atoms present are packed in the same strict geometric pattern. This is why we use the term *crystal structure*, rather than crystal, in the definition of a mineral.

Some minerals grow in such distinctive ways that their shape—called the mineral's **habit**—can be used as an identification tool. Our mystery sample in Figure 2.6 clearly has a *cubic* habit, because it looks like a collection of interlocked cubes. A very different example is the mineral chrysotile, shown in **FIGURE 2.9**, which takes the

Fibers of asbestos FIGURE 2.9

Some minerals have distinctive growth habits, even though they do not develop well-formed crystal faces. The mineral chrysotile ($Mg_3Si_2O_5(OH)_4$) sometimes grows as fine, cotton-like threads that can be separated and woven into fireproof fabric. When the mineral occurs like this, it is said to have an *asbestiform* habit. Many different minerals can grow with *asbestiform* habits, and several are mined and commercially sold as asbestos.

form of fine fibers or threads. This *fibrous* habit is characteristic of asbestos minerals.

■ **habit** The distinctive shape of a particular mineral.

Hardness Hardness, like habit and crystal form, is governed by crystal structure and by the strength of the bonds

■ **hardness** A mineral's resistance to scratching.

The Mohs Scale* of relative hardness of minerals TABLE 2.1

		Relative Hardness Number	Reference Mineral	Hardness of Common Objects
	Softest	1	Talc	
		2	Gypsum	
		3	Calcite	Fingernail Copper penny
		4	Fluorite	
		5	Apatite	Pocketknife; glass
		6	Potassium feldspar	
		7	Quartz	
		8	Topaz	
		9	Corundum	
	Hardest	10	Diamond	

The ten minerals of the Mohs scale, starting with the softest, talc, in the upper left-hand corner, and proceeding across two rows to diamond, the hardest mineral in the lower right-hand corner.

*Named for Friedrich Mohs, a German mineralogist, who chose the 10 minerals of the scale.

"Weak bonds" between sheets

Muscovite (a potassium alumino-silicate mineral), a form of mica, cleaves so easily in one direction that it can be split by hand into flakes. The flakes suggest the leaves of a book, and geologists will often refer to a "mica book."

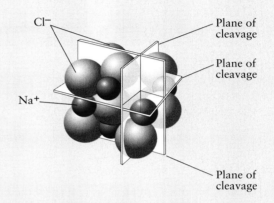

Cl⁻

Plane of cleavage

Plane of cleavage

Na⁺

Plane of cleavage

Halite, NaCl or table salt, has three distinct cleavage directions. No matter how small the bits you break it into, they will always have perpendicular faces.

between atoms. The stronger the bonds, the harder the mineral. Relative hardness values can be assigned by determining whether one mineral will scratch another, using the *Mohs relative hardness scale*. The scale is divided into 10 steps, each marked by a common mineral listed in **TABLE 2.1**. Talc, the basic ingredient of most body ("talcum") powders, is the softest mineral known and therefore is assigned a value of 1 on the relative hardness scale. Diamond, the hardest mineral, has a value of 10. The 10 steps of the hardness scale do not represent equal intervals of hardness; the important thing is that any mineral on the scale will scratch all minerals below it. For convenience, we often test relative hardness by using a common object such as a penny or a penknife as the scratching instrument, or glass as the object to be scratched. The hardness values of these objects are also shown in Table 2.1.

The mystery mineral in Figure 2.6 is relatively soft, and it scratches easily with a copper penny or a piece of fluorite. However, it is barely scratched by a fingernail. This puts its hardness somewhere around 2.5 on the Mohs hardness scale.

Cleavage If you break a mineral with a hammer or drop it on the floor so that it shatters, some of the broken fragments will be bounded by surfaces that are smooth and planar, called **cleavage** surfaces. In **FIGURE 2.10**, the muscovite "book" cleaves easily into thin sheets (the "pages" of the book) but does not cleave at all in any other direction. In certain cases, such as the halite (NaCl) fragments shown in Figure 2.10, several cleavage directions are present, and all of the fragments are bounded by smooth, planar surfaces. Don't confuse crystal faces and cleavage surfaces, even though the two often look alike. A cleavage surface is a breakage surface, whereas a crystal face is a growth surface. Also, note the difference between hardness and cleavage: a mineral might be quite

These cut gems are all synthetic sapphires, the mineral corundum. They are all the same type of material, but slight differences in chemical composition give them very different colors.

Uncut natural specimens of the same material, corundum (Al_2O_3), also have very different colors. The red crystals, rubies, are from Tanzania, and the blue crystals, sapphires, come from Newton, New Jersey. The ruby in the upper right-hand corner is about 2.5 cm (1 inch) across.

hard—that is, resistant to scratching—but it may still cleave easily in one or more directions. Diamond, the hardest mineral, has four directions of cleavage.

The directions in which cleavage occurs are governed by the crystal structure. Cleavage takes place along planes where the bonds between atoms are relatively weak, as in the case of muscovite, or where there are fewer bonds per unit area, as in the case of diamond. Because cleavage directions are directly related to crystal structure, the angles between equivalent pairs of cleavage directions are the same for all grains of a given mineral. Thus, analogously to Steno's Law for crystals, the angles between cleavage planes are constant. A small hand lens is usually enough for a geologist to spot these distinctive angles. As we remarked earlier, crystals and crystal faces are rare; but almost every mineral grain you see in a rock shows one or more breakage surfaces. That is why cleavage is such a useful aid in the identification of minerals.

The mystery mineral in Figure 2.6 has three directions of cleavage at right angles to each other.

Color and streak The color of a mineral, though often striking, is not a reliable means of identification (FIGURE 2.11). A mineral's color is determined by several factors, but the main determinant is chemical composition. Some elements can create strong color effects, even when they are present only as trace impurities. For example, the mineral corundum (Al_2O_3) is commonly white or grayish, but when small amounts of chromium are present as a result of atomic substitution of Cr^{3+} for Al^{3+}, corundum is blood red and is given the gem name *ruby*. Similarly, when small amounts of iron and titanium are present, the corundum is deep blue, producing another gem, *sapphire*. Many other colors such as green, gray, white, black, and pink are not useful in identifying minerals, because there are so many minerals that occur in these colors.

Color can be particularly confusing in opaque minerals that have metallic luster. This is because the color is partly a property of the size of the mineral grains. One way to reduce error is to prepare a **streak**, by rubbing a specimen of a metallic luster mineral on an

Mineral streak FIGURE 2.12

Although hematite (Fe_2O_3) is shiny, black or gray, and metallic looking, it makes a distinctive red-brown smear when rubbed on a porcelain streak plate.

> **streak** A thin layer of powdered mineral made by rubbing a specimen on an unglazed fragment of porcelain.

unglazed fragment of porcelain called a *streak plate*. The color of a streak is reliable because all the grains in the streak are very small and the effect of grain size is reduced. For example, hematite (Fe_2O_3), an important iron-bearing mineral, produces a reddish-brown streak, even though a specimen may look black and metallic (**FIGURE 2.12**). The streak of our mystery mineral from Figure 2.6 is gray.

Density Another important physical property of a mineral is how light or heavy it feels. We are all familiar with the fact that two equal-sized baskets have different weights when one is filled with feathers and the other with rocks. This means the rocks have greater **density** than the feathers. Minerals that have a high density, such as gold, have closely packed atoms. Minerals with a low density, such as ice, have less closely packed atoms.

It is not easy to measure the density of a mineral in a laboratory, because it requires dividing the mass by the volume, and the volume of an irregularly shaped grain is difficult to determine. This was the same problem that faced Archimedes more than 2000 years ago when he was asked to determine whether the gold in a crown was pure or alloyed with a less dense material. Geologists use a modern version of Archimedes' method. First, they measure the weight of the grain in air (W_A). Then they submerge it in water and measure the weight again (W_W). The weight in water is less than that in air, and the difference ($W_A - W_W$) is the weight of water displaced by the mineral grain. The ratio of W_A to ($W_A - W_W$) gives the *specific gravity* of the mineral. Because water has a density of 1.0 grams per cubic centimeter, the specific gravity is numerically equal to the density.

Many common minerals, such as quartz, have specific gravities in the range of 2.5 to 3.0. Fortunately, fancy equipment for measuring specific gravity is usually not necessary. If a mineral is substantially lighter than 2.5 g/cm^3, or substantially heavier than 3.0 g/cm^3, a geologist can immediately tell this by hefting it in his or her hand. Metallic minerals generally feel heavy, whereas minerals with vitreous luster tend to feel light. The mystery mineral in Figure 2.6 would feel quite a bit heavier than a normal rock of the same size, and if we took it back to the lab, we would find that its specific gravity is 7.5.

The mystery mineral revealed What is the identity of our mystery mineral in Figure 2.6? We have revealed several clues: It has metallic luster, a cubic habit, a hardness around 2.5, gray color and streak, three perpendicular cleavage directions, and an unusually high specific gravity of 7.5. A quick look in a table of minerals (such as the one in Appendix C) would confirm a geologist's suspicions: The mystery mineral is galena (PbS), a compound of lead and sulfur. This is the same mineral illustrated in Figure 2.5.

CONCEPT CHECK **STOP**

What requirements must be satisfied if a substance is to be called a mineral?

What is the difference between luster, color, and streak, and how are they related?

How can hardness of a mineral be used for identification?

What is the difference between a cleavage surface and a crystal face?

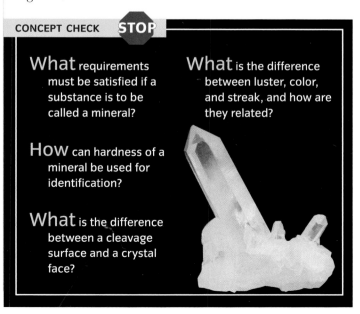

Mineral Families

G eologists have identified approximately 4000 mineral species. This number may seem large, but it is tiny compared with the number of synthetic materials such as ceramics, concrete, drugs like aspirin, and solid chemical reagents. The reason for the disparity between the number of minerals and the millions of solids that have been synthesized in laboratories becomes clear when we consider the relative abundances of the chemical elements in nature. Out of every kilogram of material in Earth's continental crust, only 12 elements are present in quantities greater than one gram: oxygen, silicon, aluminum, iron, calcium, magnesium, sodium, potassium, titanium, hydrogen, manganese, and phosphorus (see **FIGURE 2.13**). The abundant 12 account for 992.3 of the 1000 grams; all common minerals have compositions based on one or more of these abundant elements. The remaining 80 elements, combined, make up less than 1 percent of

Elements of the continental crust FIGURE 2.13

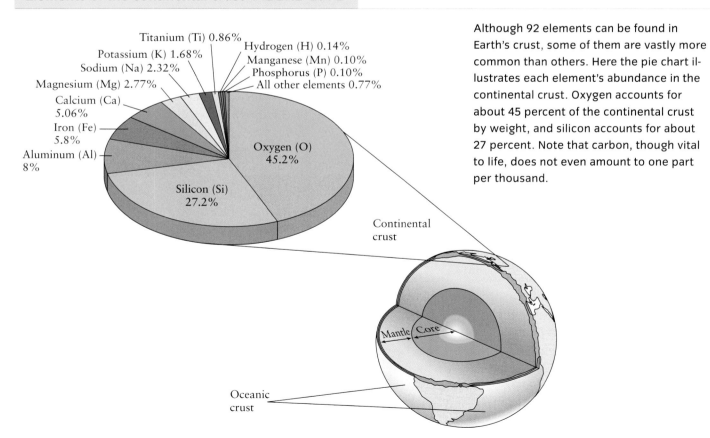

Titanium (Ti) 0.86%
Hydrogen (H) 0.14%
Manganese (Mn) 0.10%
Phosphorus (P) 0.10%
All other elements 0.77%
Potassium (K) 1.68%
Sodium (Na) 2.32%
Magnesium (Mg) 2.77%
Calcium (Ca) 5.06%
Iron (Fe) 5.8%
Aluminum (Al) 8%
Oxygen (O) 45.2%
Silicon (Si) 27.2%
Continental crust
Oceanic crust
Mantle Core

Although 92 elements can be found in Earth's crust, some of them are vastly more common than others. Here the pie chart illustrates each element's abundance in the continental crust. Oxygen accounts for about 45 percent of the continental crust by weight, and silicon accounts for about 27 percent. Note that carbon, though vital to life, does not even amount to one part per thousand.

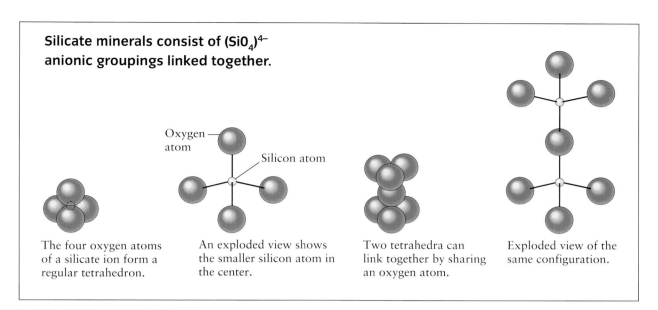

Silicate minerals consist of $(SiO_4)^{4-}$ anionic groupings linked together.

Oxygen atom

Silicon atom

The four oxygen atoms of a silicate ion form a regular tetrahedron.

An exploded view shows the smaller silicon atom in the center.

Two tetrahedra can link together by sharing an oxygen atom.

Exploded view of the same configuration.

Silicate links FIGURE 2.14

the crust by weight and less than 2 percent by volume. Minerals made of the scarcer elements occur only in small amounts, and ore deposits of scarce elements such as gold, uranium, and tin are rare and hard to find. However, these scarce elements can be extracted and used to synthesize a wide range of materials in the laboratory and in manufacturing for everyday use.

MINERALS OF EARTH'S CRUST

Mineral families are groups of minerals that are similar to one another in terms of chemistry or atomic structure, or (more commonly) both. Two elements alone—oxygen and silicon—make up more than 80 percent of the atoms in Earth's crust and comprise more than 70 percent of its mass. Thus it should be no surprise that the great majority of minerals contain one or both of these elements. The most common family of minerals, the **silicate minerals**, contain a strongly bonded complex anion called the *silica* anion that contains both silicon and oxygen, $(SiO_4)^{4-}$. The bonding in most silicate minerals is a mixture of ionic and covalent; as a result, silicates tend to be hard, tough minerals (**FIGURE 2.14**). The next most abundant family is the **oxide minerals**, which contain the simple oxide anion O^{2-}.

Other mineral groups are based on different anions, some complex and some simple. For example, *carbonates* are based on the complex $(CO_3)^{2-}$ anion, *sulfates* are based on the complex $(SO_4)^{2-}$ anion, *phosphates* are based on the complex $(PO_4)^{3-}$ anion, and *sulfides* are based on the simple S^{2-} anion. A few silicate minerals and a few oxide minerals, together with calcium sulfate and calcium carbonate, comprise the bulk of Earth's crust—an estimated 99 percent by volume. These common minerals, of which there are about 30, are called the **rock-forming minerals**. Rock-forming minerals are everywhere—not only in rocks, but also in soils and sediments and even in the dust that we breathe. Rock-forming minerals are common and inexpensive but nevertheless economically vital; you rely on them every day when you drive on a paved road or enter a concrete building.

Less common minerals are called **accessory minerals**. These are widely present in common rocks but in such small amounts that they do not determine the properties of the rocks. Though they are less abundant than the silicates and oxides, many are important economically. Some are ore minerals. As mentioned, galena is the main source of lead, and chalcopyrite ($CuFeS_2$) is the principal source of copper. Others have significant biological properties. The phosphate mineral apatite $[Ca_5(PO_4)_3(F,OH)]$ supplies the phosphorus for agricultural fertilizers and is also the fundamental compound

Polymerization and the common silicate minerals FIGURE 2.15

Silicate structure	Mineral/formula	Cleavage	Example of a specimen
 Single tetrahedron	Olivine Mg_2SiO_4	None	
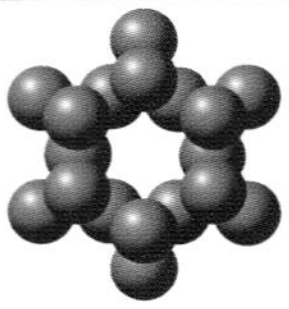 Hexagonal ring	Beryl (Gem form is emerald) $Be_3Al_2Si_6O_{18}$	One direction	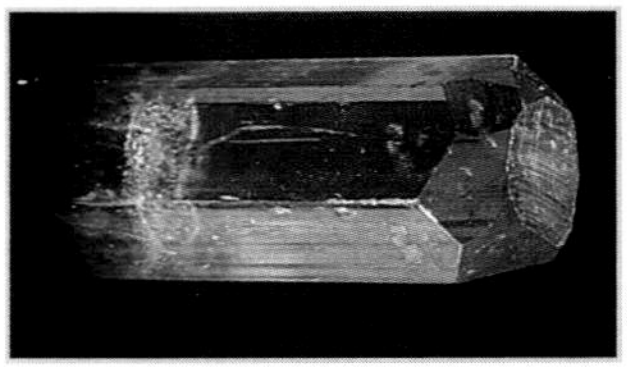
 Single chain	Pyroxene group $CaMg(SiO_3)_2$ (variety: diopside)	Two directions at 90°	
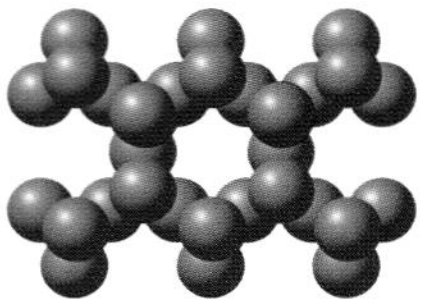 Double chain	Amphibole group $Ca_2Mg_5(Si_4O_{11})_2(OH)_2$ (variety: tremolite)	Two directions at 120°	
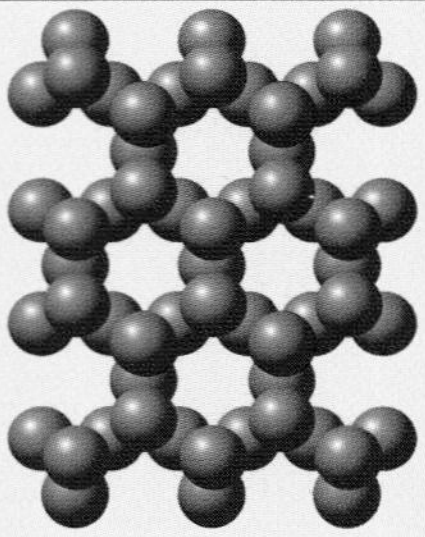 Sheet	Mica $KAl_2(AlSi_3O_{10})(OH)_2$ (variety: muscovite) $K(Mg,Fe)_3(AlSi_3O_{10})(OH)_2$ (variety: biotite)	One direction	
Too complex to show here, see FIGURE 2.16	Feldspar $KAlSi_3O_8$ (variety: orthoclase)	Two directions at 90°	
	Quartz SiO_2	None	

This chart summarizes the ways in which silica ions can polymerize to form minerals. Typical examples of each type are shown in the photographs. There are many other silicate materials in each category, but all of the principal categories that occur in nature are illustrated here.

Quartz FIGURE 2.16

One of the most abundant minerals, quartz is made up of silica ions that are tightly linked in a three-dimensional lattice. It has no cleavage planes and is very resistant to wear. For the purposes of illustration, the oygen atoms (blue) are shown as being smaller than the silicon atoms (red).

in our bones and teeth. What a Geologist Sees (page 50) shows other everyday applications of minerals.

Sometimes two minerals can have the same chemical formula, but different crystal structures. A very common example is calcium carbonate ($CaCO_3$), which forms two different minerals called calcite (the principal ingredient in limestone and marble) and aragonite (found in the shells of many organisms, such as mollusks). These different forms are called *polymorphs*. Likewise, graphite and diamond are both polymorphs of carbon, as explained in Figure 2.3. Remarkably, water ice has 14 different polymorphs that have been discovered so far, only one of which occurs naturally on Earth. Some of the other polymorphs of ice are likely to turn up on the ice-covered moons of other planets, such as Jupiter.

Note that some materials occur in nature as native elements; that is, they are not combined in compounds with other elements. Minerals that can occur in this form include some metals, such as gold (Au) and silver (Ag), and some nonmetals, such as sulfur (S) and graphite and diamond (C).

SILICATES: THE MOST IMPORTANT ROCK FORMERS

Not only are silicates the most common minerals, but they also have an unusual diversity of atomic structures. To explain this phenomenon, we begin by looking at the silica anion itself, in which four oxygen atoms are tightly bonded to the single silicon atom. As shown in Figure

2.14, if we draw a stick figure, joining the centers of the oxygen atoms with lines, we would get a regular tetrahedron (i.e., a pyramid with a triangular base). The small ionically bonded silicon atom occupies the space at the center of the tetrahedron, and the oxygen atoms occupy the corners.

Two silica tetrahedra can bond by sharing an oxygen atom, in a process called *polymerization*. Through polymerization, even larger complex ions can form, consisting of rings, chains, sheets, or three-dimensional frameworks of tetrahedra. Different common cations, such as calcium (Ca^{2+}), aluminum (Al^{3+}), magnesium (Mg^{2+}), iron (Fe^{2+}), sodium (Na^+), and others, can fit into the spaces or *interstices* between the polymerized silica tetrahedra. The identity of a silicate mineral is determined by three factors: how the silica tetrahedra occur within the mineral, that is, whether they are single or polymerized; which cations are present; and how the cations are distributed throughout the crystal structure. FIGURE 2.15 shows the common silicate mineral families. For example, polymerization in mica produces a sheet, and the pronounced cleavage of mica is parallel to the sheets. In quartz the three-dimensional framework, shown in FIGURE 2.16, is equally strong in all directions, so quartz is a hard, tough mineral that lacks cleavage.

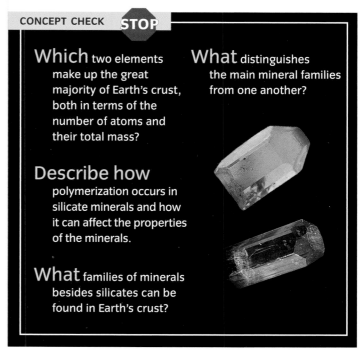

CONCEPT CHECK STOP

Which two elements make up the great majority of Earth's crust, both in terms of the number of atoms and their total mass?

What distinguishes the main mineral families from one another?

Describe how polymerization occurs in silicate minerals and how it can affect the properties of the minerals.

What families of minerals besides silicates can be found in Earth's crust?

Living in a Mineral World

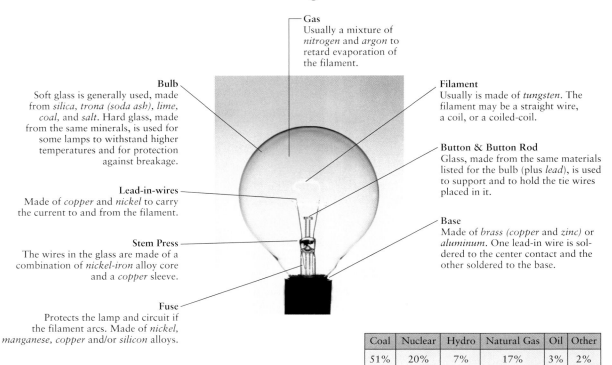

How Many Minerals and Metals Does It Take to
Make a Light Bulb?

Gas
Usually a mixture of *nitrogen* and *argon* to retard evaporation of the filament.

Bulb
Soft glass is generally used, made from *silica*, *trona (soda ash)*, *lime*, *coal*, and *salt*. Hard glass, made from the same minerals, is used for some lamps to withstand higher temperatures and for protection against breakage.

Filament
Usually is made of *tungsten*. The filament may be a straight wire, a coil, or a coiled-coil.

Button & Button Rod
Glass, made from the same materials listed for the bulb (plus *lead*), is used to support and to hold the tie wires placed in it.

Lead-in-wires
Made of *copper* and *nickel* to carry the current to and from the filament.

Base
Made of *brass (copper* and *zinc)* or *aluminum*. One lead-in wire is soldered to the center contact and the other soldered to the base.

Stem Press
The wires in the glass are made of a combination of *nickel-iron* alloy core and a *copper* sleeve.

Fuse
Protects the lamp and circuit if the filament arcs. Made of *nickel*, *manganese*, *copper* and/or *silicon* alloys.

Coal	Nuclear	Hydro	Natural Gas	Oil	Other
51%	20%	7%	17%	3%	2%

Even a simple appliance like a lightbulb contains materials that must be extracted from different minerals mined in different counties around the world. Many of these materials can be recycled, but none of them can be replenished.

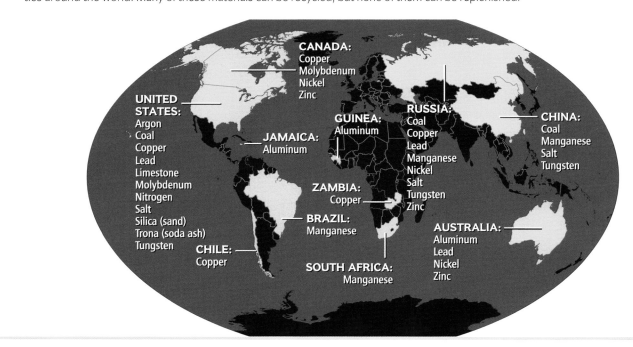

CANADA:
Copper
Molybdenum
Nickel
Zinc

UNITED STATES:
Argon
Coal
Copper
Lead
Limestone
Molybdenum
Nitrogen
Salt
Silica (sand)
Trona (soda ash)
Tungsten

GUINEA:
Aluminum

JAMAICA:
Aluminum

RUSSIA:
Coal
Copper
Lead
Manganese
Nickel
Salt
Tungsten
Zinc

CHINA:
Coal
Manganese
Salt
Tungsten

ZAMBIA:
Copper

BRAZIL:
Manganese

CHILE:
Copper

SOUTH AFRICA:
Manganese

AUSTRALIA:
Aluminum
Lead
Nickel
Zinc

Rocks: A First Look

LEARNING OBJECTIVES

Explain the difference between a rock and a mineral.

Identify the three major families of rocks.

Explain what holds rocks together and why some rocks are more cohesive than others.

Rocks are the words that tell the story of Earth's long history; minerals are the letters that form the words. An important distinction between minerals and rocks is that rocks are *aggregates*; this means that rocks are collections of minerals (and sometimes other types of particles) stuck together or intergrown. Rocks usually consist of several different types of minerals, but sometimes they are made of just one type of mineral. In any case, a rock will contain many grains of the constituent mineral or minerals.

IGNEOUS, SEDIMENTARY, AND METAMORPHIC ROCKS

Rocks are grouped into three large families, according to the processes that form them (see Amazing Places on the following two pages). Within each of these families, called *igneous, sedimentary,* and *metamorphic* rock, there is a range of possible *mineral assemblages*—the types and relative proportions of minerals that constitute the rock. Mineral assemblages help identify rocks (for example, whether it is a granite or a basalt), and they reveal much about the geologic environment in which a particular rock formed.

Igneous rocks

Igneous rocks (named from the Latin *ignis,* meaning fire) form by the cooling and solidification of **magma**. Magma, or molten rock, often contains mineral grains and dissolved gases. Depending on its environment, magma may cool slowly or very rapidly. When it reaches a temperature that is suffi- ciently low, mineral grains begin to crystallize from the magma, just as ice crystals form in water as it cools. The physical properties of the rock will differ, depending on whether the cooling process is slow, giving the crystals a lot of time to grow, or fast. We will have much more to say about these differences in Chapter 6.

Sedimentary rocks

The next major rock family forms when mineral and rock particles are transported by water, wind, or ice and then deposited. This fragmented, transported, and deposited material is called *sediment*. In certain geologic circumstances, under conditions of low pressure and low temperature near Earth's surface, sediment may be transformed into **sedimentary rocks**. "Low temperature" and "low pressure" here are by geologic standards; they may still be higher than the pressure and temperature we live at. Sediment- and soil-forming processes and transportation processes are discussed in detail in Chapter 7, and sedimentary rocks and rock-forming processes are examined in Chapter 8.

Metamorphic rocks

The third major rock family is **metamorphic rock**, whose original sedimentary or igneous form and mineral assemblage have been changed as a result of exposure to high temperature, high pressure, or both. The term *metamorphic* comes from the Greek *meta,* meaning change, and *morphe,*

---

rock A naturally formed, coherent aggregate of minerals and possibly other nonmineral matter.

igneous rocks Rocks that form by cooling and solidification of molten rock.

magma Molten rock underground.

sedimentary rocks Rocks that form under conditions of low pressure and low temperature near the surface.

metamorphic rocks Rocks that have been altered by exposure to high temperature, high pressure, or both.

Igneous rock

This rounded granite boulder on Mt. Desert Island in Acadia National Park, Maine, was transported to its current location by a glacier. The boulder is sitting on a thick platform of granite that was once part of an enormous magma chamber underlying a volcano. The volcano is now gone and its top has eroded away, leaving the solidified remnants of the magma chamber exposed. The weathered surfaces of these granites make them appear buff-colored.

Granite is an igneous rock. It forms from magma that cools and solidifies slowly, deep in the crust. The minerals in this sample are quartz (gray), potassium feldspar (light pink), plagioclase feldspar (white), and biotite (black). The same minerals are present in the sample of gneiss, shown above, but the overall appearance of the two rocks is quite different.

www.wiley.com/college/murck

Metamorphic rock ▶

The beautiful Lauterbrunnen Valley in Switzerland lies within a great mountain range—the Alps—where the rocks have been uplifted and chemically and physically altered by enormous tectonic forces. In this photo the Lauterbrunnen Falls cascade down a steep cliff of metamorphic rock.

◀ **Sedimentary rock**

This remarkable landscape is part of Bryce Canyon National Park in Utah. The horizontal rock layers are mainly sandstones and limestones. The tall spires, called hoodoos, and the deeply incised crevices between them, are the result of erosion over many, many years. Bryce Canyon today is a desert, but the grains of sediment in these rocks were originally deposited in an environment that was alternately hot and dry (sandstone) or covered by a shallow sea (limestone).

◀ This photo shows a sample of gneiss (pronounced "nice"), a metamorphic rock similar to the rocks that form many of the cliffs in the Swiss Alps. This rock may have started as either a sedimentary or igneous rock, but it has been altered by heat and pressure so that a new set of minerals has developed, along with its banded appearance. The minerals present are potassium feldspar (light pink), plagioclase (white), quartz (gray), and biotite (black). The minerals are similar to those in the igneous rock granite, shown on facing page, but the overall appearance is quite different.

◀ Sandstone is a sedimentary rock that consists largely of grains of quartz, with traces of other minerals. The grains became rounded as they were transported by running water to the place where they were deposited. After being deposited in layers, the grains eventually became cemented together and turned into sandstone.

The inset samples are all 7 cm wide and 4 cm high.

Recrystallization FIGURE 2.17

When a wolf walked across this snow-covered ground in Siberia, its paws compressed the snow and caused the ice crystals in the snow to become compacted and to recrystallize. Later, wind eroded away the loosely packed snow. But the compacted ice in the footprints was more resistant to erosion, and as a result the footprints now stand out above the snow surface. A similar process of compaction and recrystallization, over a longer period of time and at higher temperatures and pressures, creates sedimentary rock out of loose sediment.

meaning form—hence, change of form. Metamorphism occurs in a variety of tectonic and geologic environments. It can occur as an end result of the processes of burial and compaction through which sediments become transformed into rocks. It can also occur when rocks are subjected to the extreme pressures and temperatures associated with mountain building along convergent plate boundaries. Temperature-induced metamorphism can occur when rocks are heated—essentially baked—by nearby magmas. How and why metamorphism occurs, the kinds of metamorphic rocks, and their characteristic mineral assemblages are discussed in detail in Chapter 10.

WHAT HOLDS ROCKS TOGETHER?

The minerals in some rocks are held together with great tenacity, whereas in others they are easily broken apart. While the forces that hold minerals together are chemi-cal, the forces that hold rocks together are partly mechanical and partly chemical. Igneous and metamorphic rocks are the most cohesive because both types contain intricately interlocked minerals. During the formation of igneous and metamorphic rocks, the growing minerals crowd against each other, filling all spaces and forming an intricate, three-dimensional jigsaw puzzle that will not pull apart very easily. A similar interlocking of grains holds together steel, ceramics, and bricks.

The forces that hold the grains of sedimentary rocks together are less obvious. Sediment is a loose aggregate of particles, and it must be transformed into sedimentary rock by one or more rock-forming processes, discussed in detail in Chapter 7. These processes tend to cause the individual grains of the sediment to become interlocking. Sedimentary rocks can form by *compaction*, during which the mineral grains in the sediment are squeezed and compressed by the weight of overlying sediment, becoming compacted into an interlocking network of grains. Secondly, water may deposit new minerals such as calcite, quartz, or iron oxide into the open spaces that act as a *cement*. A third way that sedimentary rocks are held together is *recrystallization*. As sediment becomes deeply buried and the temperature rises, mineral grains begin to recrystallize. The growing grains interlock and form strong aggregates. The process is the same as that which occurs when ice crystals in a snow pile recrystallize into a compact mass of ice (**FIGURE 2.17**).

CONCEPT CHECK STOP

What is a "mineral assemblage"?

What are the differences between igneous, sedimentary, and metamorphic rocks?

How can a loose collection of sediment get turned into a cohesive rock?

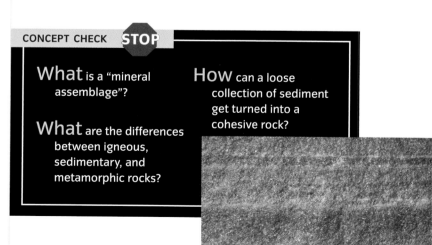

1 Elements and Compounds

1. **Elements** are the most fundamental of all naturally occurring substances because they cannot be separated into chemically distinct materials. The minute particles that make up all matter, including elements, are **atoms**. All atoms of a particular element are the same.

2. Atoms are made of smaller particles called *protons*, *neutrons*, and *electrons*, which have no independent chemical properties. Protons are electrically positive and electrons are electrically negative. The protons and neutrons reside in the atom's *nucleus*. The number of protons identifies the element, and is called the *atomic number*. The sum of protons and neutrons is the *mass number*. Most elements have several different *isotopes*, which differ in the number of neutrons. Ordinarily an atom has equal numbers of protons and electrons, but it may gain or lose electrons, in which case it becomes electrically charged and is called an **ion**.

3. **Compounds** consist of multiple elements, and therefore multiple types of atoms. The smallest unit that has the properties of a given compound is a **molecule**. Molecules in a compound are held together by **bonds**. The most common form of bonding is *ionic bonding*, caused by the electrostatic attraction of two oppositely charged ions. Other forms of bonding are *covalent*, *metallic*, and *Van der Waals bonding*.

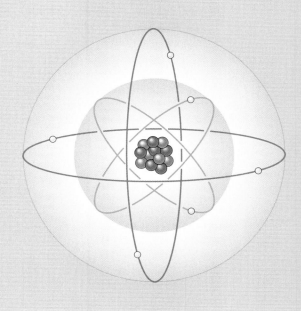

2 What Is a Mineral?

1. **Minerals** are *naturally occurring solids*, each with a unique **crystal structure**. Minerals are formed by *inorganic processes* and must have a *fixed chemical composition*. One exception to the last rule is *atomic substitution*, in which other atoms of like size and ionic charge may be substituted for specific atoms in a mineral.

2. The visible crystal form of a mineral is a direct consequence of its atomic lattice. In some cases, a mineral will not have enough room to grow identifiable crystals, but the underlying geometric lattice remains the same. *Crystal faces* bound a crystal.

3. Several properties can be used to tell minerals apart. One of the most easily visible is a mineral's **luster**, or the quality and intensity of light that reflects from it. Measuring the angles of the faces may also be useful in identifying a crystal as a mineral, for though crystal size and overall shape in a mineral may vary, *Steno's Law* states that the angle between corresponding faces remains constant. Other factors that can aid in mineral identification include the distinct external shape, or **habit**; a mineral's resistance to scratching, known as its **hardness**; how a mineral cleaves when broken, known as **cleavage**; and its **density**, or *specific gravity*. Color may also be useful but is often misleading. However, when a mineral is rubbed on a *streak plate*, it produces a thin layer of powdered mineral, known as its **streak**, which is more reliable than color for identification.

3 Mineral Families

1. The distribution of elements in Earth's crust is far from even. Only 12 elements are present at a level of more than one part in a thousand by mass, and of these oxygen and silicon dominate. As a result, the number of naturally occurring minerals is relatively small, and the number of important **rock-forming minerals** is even smaller—only about 30 or so.

2. **Silicate minerals**, based on the $(SiO_4)^{4-}$ anion, are the most abundant family of rock-forming minerals. They can adopt a variety of crystal structures because of the ability of the *silica anion* to link together. These structures include chains, sheets, and three-dimensional lattices. **Oxide minerals** are the next most abundant family. Other important mineral families include *sulfides* and *sulfates*, *carbonates*, and *phosphates*. Less abundant minerals are called **accessory minerals**, and they usually do not affect the properties of the rock they are found in. Nevertheless, these minerals are frequently of great economic importance as sources of metal ore.

3. Two minerals can have the same chemical formula, but different crystal structures. These different forms are *polymorphs*. Other minerals occur in nature as *native elements* that are uncombined with other elements.

4. Silicates are the most common minerals. Large, complex ions can be formed by *polymerization*, a process of bonding based on the sharing of an oxygen atom. Three factors affect the identity of a silicate mineral: whether the silicate tetrahedra are single or polymerized, which cations are present, and how the cations are distributed.

4 Rocks: A First Look

1. **Rocks** are *aggregates*, or complex *mineral assemblages*, that are sometimes mixed with other materials such as natural glass or organic matter.

2. Rocks come in three major families. **Igneous rocks** are formed by the solidification of **magma**, either slowly beneath Earth's surface or rapidly at the surface. **Sedimentary rocks** are formed at or near the surface by the deposition of many layers of sediment. **Metamorphic rocks** start as either igneous or sedimentary rocks but change their form and sometimes their mineral assemblage as a result of high temperature, high pressure, or both.

3. Igneous and metamorphic rocks are held together by the interlocking of their grains. The loose particles of sedimentary rocks are held together either by *compaction*, during which the mineral grains are forced together by the pressure of overlying sediment, *cementing* of open spaces within a rock, or *recrystallization*, a process that occurs when growing grains interlock because of increasing pressure and heat.

KEY TERMS

CRITICAL AND CREATIVE THINKING QUESTIONS

1. When astronauts brought back rock samples from the Moon, the minerals present were mostly the same as those found on Earth. Can you think of reasons why this might be so? Would you expect minerals on Mars or Venus to be the same, or at least very similar, to those on Earth?

2. When a volcano erupts, spewing forth a column of hot volcanic ash, the ash particles are tiny fragments of solidified magma that slowly fall down to Earth's surface, forming a layer of sediment. Would a rock formed from cemented particles of volcanic ash be igneous or sedimentary? Can you think of other circumstances that might form rocks that are intermediate between two of the major rock families?

3. Which of the following materials are minerals, and why (or why not)? Water; beach sand; diamond; wood; vitamin pill; gold nugget; fishbone; emerald.

4. The minerals calcite and aragonite have the same chemical formula ($CaCO_3$) but different crystal structures. Are they polymorphs? The materials halite (NaCl) and galena (PbS) have the same geometric patterns in their crystal structures but different compositions. Are they polymorphs?

5. Do some research to find out whether any valuable ores are mined in your area. What are the rocks and minerals involved?

What is happening in this picture ?

The geologist is part of a team studying the feasibility of establishing a platinum mine near Stillwater, Montana. He is looking at a core sample through a hand lens, one of the simplest and most useful tools in any geologist's backpack. What features might he be looking for?

1. A(n) _____ is an atom that has gained or lost one or more electrons and has a net electric charge.

    a. molecule
    b. isotope
    c. element
    d. ion
    e. compound

2. On this illustration, locate and label the following parts of the atom:

    proton   nucleus
    electron   first energy level electron shell
    neutron   second energy level electron shell

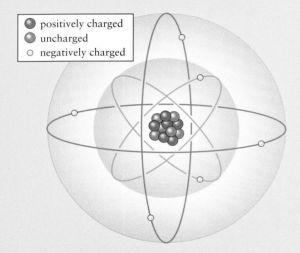

○ positively charged
○ uncharged
○ negatively charged

3. In _____, electrons from different atoms are shared and the force of this sharing forms the strongest of chemical bonds.

    a. ionic bonding
    b. covalent bonding
    c. metallic bonding
    d. Van der Waals bonding

4. In _____, the mobility of electrons in the outermost shell allows materials with these types of bonds to act as good conductors of electricity and heat.

    a. ionic bonding
    b. covalent bonding
    c. metallic bonding
    d. Van der Waals bonding

5. To be considered a mineral, a substance must

    a. have a specific chemical composition.
    b. be formed by inorganic processes.
    c. be a naturally formed solid.
    d. have a characteristic crystal structure.
    e. have all of the characteristics listed above.

6. Volcanic glass is not considered a mineral because it

    a. is amorphous (that is, it lacks a crystal structure).
    b. is not naturally occurring.
    c. does not have enough silicon or oxygen in its chemical composition.
    d. All of the statements are correct.

7. The photograph shows natural samples of the mineral corundum. What best explains the striking difference in color between the red (ruby) and the blue (sapphire) samples?

    a. polymorphism
    b. small variations in composition
    c. differences in crystalline structure
    d. polymerization

8. Which of the following minerals is the hardest?

    a. quartz
    b. calcite
    c. muscovite
    d. feldspar

9. Which of the following physical properties is the least useful in identifying many varieties of common minerals?

    a. cleavage or fracture
    b. hardness
    c. color
    d. density

10. _____ is the most abundant element by atomic weight in Earth's continental crust.

   a. Silicon
   b. Iron
   c. Calcium
   d. Aluminum
   e. Oxygen

11. Earth's crust is mostly composed of a small number of rock-forming minerals because

   a. polymorphs are common.
   b. of the overwhelming abundance of oxygen and silicon.
   c. of a lack of carbon in igneous rocks.
   d. All of the above statements are true.

12. Gold is an example of a(n) _____, quartz is a(n) _____, and pyrite is a(n) _____.

   a. native element…silicate mineral…sulfide mineral
   b. oxide mineral…sulfide mineral…phosphate mineral
   c. phosphate mineral…sulfide mineral…carbonate mineral
   d. oxide mineral…silicate mineral…phosphate mineral

13. On the illustration, label each silicate structure with its proper name. For each structure give an example of a mineral (name and chemical formula) and indicate the prominent cleavage.

Structure 1

Structure 2

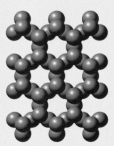

Structure 3

14. The rocks shown in the photographs are examples of

   a. volcanic rock.
   b. plutonic rock.
   c. sedimentary rock.
   d. metamorphic rock.

15. In the rocks depicted in the photographs, what is most likely holding the mineral grains together?

   a. Interlocking crystals formed from cooling magma
   b. Compaction and cementation between the mineral grains
   c. Interlocking crystals formed from the alteration of pre-existing rocks by heat and pressure
   d. Van der Waals bonding

How Old Is Old? The Rock Record and Deep Geologic Time

3

What's in a name? Say the word "Jurassic" and most people probably think of *Jurassic Park*, the blockbuster movie by Steven Spielberg. Next, they would recall the stars of the movie—the terrifying *Tyrannosaurus rex* and the fast and wily velociraptors (for the movie, modeled on close relative *Deinonychus*).

They probably wouldn't think about the Jura Mountains in Switzerland *(right)*. These lushly wooded limestone hills contain millions of fossils of extinct sea creatures called ammonites, which lived in coiled seashells reminiscent of the modern nautilus. In the nineteenth century, geologists gradually realized that these creatures—and the rocks in which they were fossilized—all came from a particular period in the geologic past. They named it the Jurassic Period, after the mountain range where they found some of their best and most distinctive samples of ammonites. It is one of the delightful ironies of science that the fossil chosen to define the period was not a mighty dinosaur, but a humble sea creature.

(By the way, no tyrannosaurs or velociraptors lived in the Jurassic Period, though many other dinosaurs did. The name "Cretaceous Park" would have been more accurate.)

Fossils and rocks don't come with a date stamped on them. So how do geologists tell when an ancient animal lived, or when a particular layer of rocks was deposited? We will describe several methods in this chapter. It has taken geologists two centuries to piece together Earth's chronology, and they are still refining it.

Global Locator

EUROPE

SWITZERLAND

60

Relative Age

The first scientific attempts to determine the numerical extent of geologic time were made almost three centuries ago. Geologists speculated that they might be able to estimate the time needed to erode away a mountain range by measuring the amount of sediment transported by streams. Their attempts were imprecise, but the inescapable conclusion was that Earth must be millions of years old because of the great thickness of sedimentary rocks. One of the founders of geology, a Scot named James Hutton, was so impressed by the evidence that in 1788 he wrote that for Earth there is "no vestige of a beginning, no prospect of an end."

The geologists who followed Hutton agreed with his conclusion that Earth must be very ancient, but they lacked a precise way to determine exactly how long ago a particular event occurred. The only thing they could do was figure out the sequence of past events. They could thus establish the **relative ages** of rock formations or other geologic features, which means that they could determine whether a particular formation or feature was older or younger than another formation or feature. By doing so, they took the essential first steps toward unraveling

relative age
The age of a rock, fossil, or other geologic feature relative to another feature.

stratigraphy
The science of rock layers and the process by which strata are formed.

Earth's geologic history. Relative ages are derived from four basic principles of **stratigraphy**, generally attributed to Hutton and Steno (see Chapter 2).

STRATIGRAPHY

In places where you can find large exposed rock formations, such as the Badlands of South Dakota and the American Southwest, you will often see that the rocks have a banded appearance (see FIGURE 3.1). These bands are called **strata** (an individual band is called a *stratum*), from the Latin word for "layer." The bands are often horizontal, but it is not at all unusual to see them tilted or bent.

All of the rocks in a typical stratified formation are sedimentary, deposited over the eons by water. This observation leads to the first of four key principles of stratigraphy: the *principle of original horizontality*, which states that water-laid sediments are deposited in horizontal strata. You can test this principle yourself. Shake up a bottle of muddy water so that all of the particles are suspended. Let the bottle stand and then examine the result—the mud will be deposited at the bottom of the bottle in a horizontal layer. Therefore, whenever we observe water-laid strata that are bent, twisted, or tilted so that they are no longer horizontal, we can infer that some tectonic force must have disturbed the strata after they were deposited (see Figure 3.1B and C).

A second key to stratigraphy, which, like the first, is based on common sense, is the *principle of stratigraphic superposition*. This principle states that in any undisturbed sequence of strata, each stratum is younger than the stratum below it and older than the stratum above it (see What a Geologist Sees on page 64).

A third principle for determining relative ages, the *principle of cross-cutting relationships*, states that a stratum must always be older than any feature that cuts or disrupts it. If a stratum is cut by a fracture, for example,

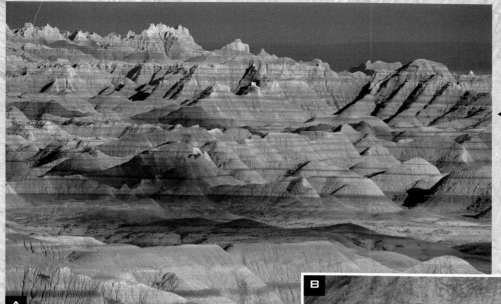

All of these strata were originally deposited as horizontal layers; the strata in (B) and (C) were disrupted by later tectonic activity.

◀ A Horizontal strata in Badlands State Park, South Dakota.

B Tilted strata in Telfer Gold Mine, Great Sandy Desert, Western Australia. The tiny figure of a person walking along a mining road gives an idea of scale. ▼

◀ C Folded strata in Hamersley Gorge, Western Australia.

Global Locator

The Principle of Stratigraphic Superposition

The Grand Canyon is considered one of the natural wonders of the world—not surprising that nearly 5 million people visit it every year. To a geologist, the Grand Canyon is more than just a scenic stop. It is also a beautiful example of the geologic record from millions of years ago.

Kaibab and Toroweap Limestones
Coconino Sandstone
Hermit Shale
Supai Sandstone
Redwall Limestone
Muav Limestone
Bright Angel Shale
Tapeats Sandstone
Grand Canyon Supergroup

A panoramic view of the Grand Canyon from the Toroweap Overlook shows a nearly 2-kilometer thickness of horizontal, sedimentary strata lying on top of older strata that were tilted and tectonically deformed before the horizontal strata were deposited. According to the principle of stratigraphic superposition, each stratum is younger than the layer below it (inset). Think of a pile of newspapers as an analogy. If a paper is added to the pile each day, the paper on the bottom will always be the oldest, with the most recent on top. Geologists have named each stratum in this section of the Grand Canyon after its predominant rock type and the location in the Grand Canyon where the best outcroppings occur (such as the Toroweap Overlook). The tilted strata, called the Grand Canyon Supergroup, are now known to be much older than the horizontal ones, but stratigraphy can only tell us the relative ages.

the stratum itself is older than the fracture that cuts across it, as shown in FIGURE 3.2. When magma fills a fracture, the result is a vein of rock that cuts across the strata. In this case, too, the sedimentary rocks must be older than the cross-cutting vein. Similarly, a "foreign rock" (called a *xenolith* or an *inclusion*) that is encased within another rock unit must pre-date the rock that surrounds it.

Finally, the *principle of lateral continuity* of sedimentary strata is based on the idea that sediments are deposited in continuous layers. A layer of sediment will extend horizontally as far as it was carried by the water that deposited it. Layers of sediment do not terminate abruptly, though they get thinner and ultimately pinch out altogether at their farthest edges. The same is true of sedimentary rocks; they may thin or pinch out laterally, but if a stratum terminates abruptly against another rock unit then it must have been cut by that unit, which is therefore younger.

GAPS IN THE RECORD

Nineteenth-century geologists also tried to estimate the **numerical ages** of rocks—the exact number of

---

numerical age
The age of a rock or geological feature in years before the present.

unconformity
A substantial gap in a stratigraphic sequence that marks the absence of part of the rock record.

---

years that have elapsed since a given stratum was deposited. For example, observations might suggest that it would take 1 year for 10 centimeters of sediment to accumulate. If a rock lies 3 meters (300 cm) below the ground surface under younger sedimentary strata, then its numerical age could be estimated to be about 3000 years (300 cm $\times$ 10 cm/year). Three assumptions must be true for this approach to work. First, the rate of sedimentation must have been constant while the layers were deposited. Second, the thickness of sediment must have been the same as the thickness of the sedimentary rock that eventually formed from it. Finally, all strata must be *conformable*. This means that each layer was deposited on the one below it without any interruptions—in other words, there are no depositional gaps in the stratigraphic record.

There are problems with each of these assumptions. Rates of sedimentation vary greatly. Sediments are often compressed while they are turning into rock, resulting in a much thinner stratum than was originally deposited. The conformity assumption also fails; gaps in the stratigraphic record, or **unconformities**, are common and occur for a variety of reasons

---

Principle of cross-cutting relationships FIGURE 3.2

A These fractures are younger than the strata they cut, according to the principle of cross-cutting relationships. The fractures are in a sequence of sandstone strata in Merseyside, United Kingdom.

B At Three Valley Gap in Alberta, Canada, layers of metamorphic rock are sliced by two darker, cross-cutting igneous rock formations called *dikes*. A geologist sees four stages: deposition of sediment;

formation of sedimentary rocks; metamorphism into a rock called *gneiss*; intrusion of the dikes. By the principle of cross-cutting relations, the dikes must be the youngest feature.

Unconformities: gaps in the rock record FIGURE 3.3

Any boundary that represents a gap in the sedimentary record is called an unconformity. This diagram illustrates the formation of three common types of unconformities: ① nonconformity, ② angular unconformity, and ③ disconformity.

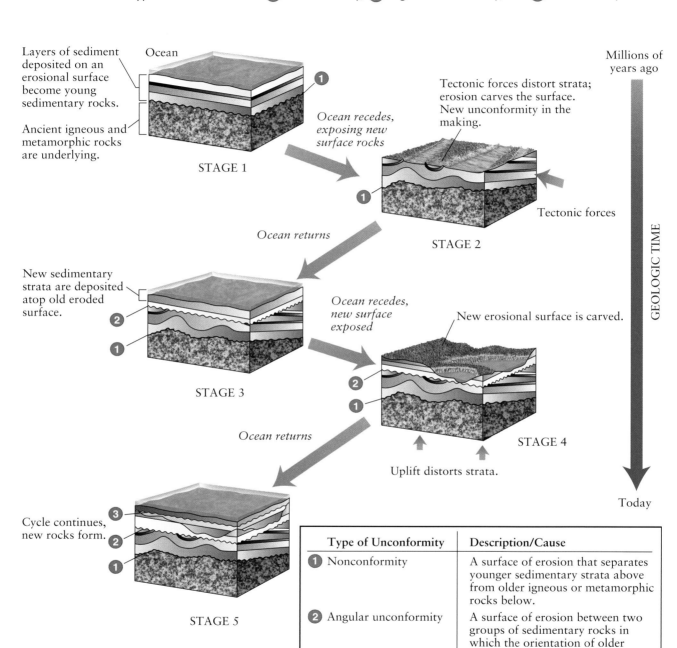

Layers of sediment deposited on an erosional surface become young sedimentary rocks.

Ancient igneous and metamorphic rocks are underlying.

Ocean

STAGE 1

Ocean recedes, exposing new surface rocks

Tectonic forces distort strata; erosion carves the surface. New unconformity in the making.

Tectonic forces

STAGE 2

Ocean returns

Millions of years ago

New sedimentary strata are deposited atop old eroded surface.

STAGE 3

Ocean recedes, new surface exposed

New erosional surface is carved.

STAGE 4

Uplift distorts strata.

Ocean returns

GEOLOGIC TIME

Today

Cycle continues, new rocks form.

STAGE 5

Type of Unconformity	Description/Cause
① Nonconformity	A surface of erosion that separates younger sedimentary strata above from older igneous or metamorphic rocks below.
② Angular unconformity	A surface of erosion between two groups of sedimentary rocks in which the orientation of older strata, below, are at an angle to younger strata, above.
③ Disconformity	A surface of erosion in which the orientation of older strata, below, are parallel to younger strata, above.

www.wiley.com/
college/murck

Most fossils begin as hard plant or animal parts, such as the bones of this 23-cm-long Pachypleurosaurus, a marine reptile that lived about 230 million years ago, and was preserved in rocks now found in Switzerland.

These fossils, found in Sweden, record the slithering, sliding, and digging of some marine creatures, identity unknown, about 530 million years ago.

▲ Other fossils record the impressions of soft tissue on mud that later solidified. The lobe-shaped fins of fish like this *Eusthenopteron*, preserved here in 380-million-year-old rocks found in Quebec, eventually evolved into legs.

(see FIGURE 3.3). In sum, of numerical ages determined by early geologists from the thickness of stratigraphic sequences were not very accurate.

FOSSILS AND CORRELATION

Many strata contain remains of plants and animals that were incorporated into the sediment as it accumulated. **Fossils** usually consist of "hard parts" like shells, bones, or wood whose forms have been preserved in sedimentary rocks (see FIGURE 3.4). In some cases the imprints of soft animal tissues, like skin or feathers, or the leaves and flowers of plants have been preserved. Even the preserved tracks and footprints of animals are considered to be fossils. Many fossils found in

paleontology
The study of fossils and the record of ancient life on Earth; the use of fossils for the determination of relative ages.

geologically young strata look similar to plants and animals living today (see FIGURE 3.5). The farther down in the stratigraphic sequence we go, the more likely we are to find fossils of extinct plants or animals, and the less familiar they seem.

Nicolaus Steno, whom we introduced in Chapter 2 in connection with crystals, was also one of the first scientists to conclude that fossils were the remains of ancient life. He published his ideas in a landmark paper in 1669, in which he also stated the principles of stratigraphic superposition and original horizontality. His conclusions were ridiculed at the time, but by the next century the idea of the plant and animal origin of fossils was widely accepted.

The study of fossils, called **paleontology**, is hugely important for understanding the history of life on

A

B

Ancient and modern
FIGURE 3.5

A fossilized imprint of a *Lebachia* (A), a conifer that is found in rocks about 250 million years old, strongly resembles a modern Norfolk pine (B).

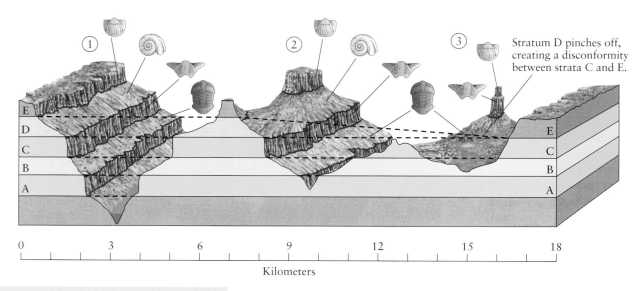

Fossils and correlation FIGURE 3.6

Geologists can use fossils to correlate strata at localities (1, 2, and 3) that are many kilometers apart. Strata B, C, D, and E have fossil assemblages that are different from one another, but consistent among the three sites. Note that stratum D is missing at locality 3, because E directly overlies C. Why?

Either D was never deposited there, or it was deposited and later removed by erosion. Either way, the boundary between C and E is a disconformity.

Earth, and we will have much more to say about it in Chapter 14. Aside from their biologic interest, fossils also have great practical value to geologists. About the same time that James Hutton was working in Scotland, a young surveyor named William Smith was laying out canal routes in southern England. As the canals were excavated, Smith noticed that each group of strata contained a specific assemblage of fossils. In time, he could look at a specimen of rock from any sedimentary layer in southern England and name the stratum and its position in the sequence of strata. This skill enabled him to predict what kind of rock the canal excavators would encounter and how long it would take to dig through it. It also earned him the nickname "Strata."

The stratigraphic ordering of fossil assemblages is known as the principle of *faunal and floral succession.*

■ correlation
A method of equating the ages of strata that come from two or more different places.

Faunal means animals, floral means plants; succession means that new species succeed earlier ones as they evolve. William Smith's practical discovery turned out to be of great scientific importance. Geologists soon demonstrated that the faunal succession in

northern France is the same as that found by Smith in southern England. By the middle of the nineteenth century, it had become clear that faunal succession is essentially the same everywhere. Thus Smith's practical observations led to a means of worldwide **correlation**, which is the most important way of filling the gaps in the geologic record (see FIGURE 3.6).

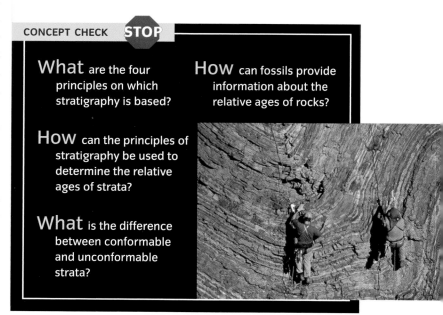

CONCEPT CHECK STOP

What are the four principles on which stratigraphy is based?

How can fossils provide information about the relative ages of rocks?

How can the principles of stratigraphy be used to determine the relative ages of strata?

What is the difference between conformable and unconformable strata?

The Geologic Column

W orldwide stratigraphic correlation (see FIGURE 3.7) was one of the greatest successes of nineteenth-century science. It meant that a gap in the stratigraphic record in one place could be filled using evidence from somewhere else. Through worldwide correlation, nineteenth-century geologists assembled the **Geologic Column**, or *stratigraphic time scale*, a composite diagram showing the succession of all known strata, fitted together in chronological order, on the basis of their fossils and other evidence of relative age (see FIGURE 3.8 on the following page).

EONS AND ERAS

Even the most cursory inspection of the Geologic Column reveals how closely our understanding of strata is intertwined with the history of life. The vast majority of Earth's history is divided into three **eons** in which fossils are extremely rare or nonexistent, and in which geologic information is similarly sparse: the *Hadean* ("beneath the earth"), *Archean* ("ancient"), and *Proterozoic* ("early life") Eons. The Archean is roughly the period when single-celled life developed, and the Proterozoic is when multi-celled, soft-bodied organisms first emerged. We now know that each eon spanned several hundred million years of time, as shown in Figure 3.8; however, the nineteenth-century geologists who first worked out the Geologic Column had no way of determining this information.

At the beginning of the fourth and current eon, called the *Phanerozoic* ("visible life"), the fossil record suddenly becomes much more detailed, thanks to the appearance of the first animals with hard shells and skeletons. Thus, geologists have divided the Phanerozoic Eon into three shorter units called **eras**: the *Paleozoic* ("ancient life"), *Mesozoic* ("middle life"), and *Cenozoic* ("recent life"). These eras were separated by major extinction events, when more than 70 percent of the species on Earth perished (see FIGURE 3.9 on page 71).

Geologic Column The succession of all known strata, fitted together in relative chronological order.

Markers of geologic time FIGURE 3.7

These ammonite fossils are encased in limestone and came from the Jura Mountains in Switzerland. Such fossils are typical of those used to correlate strata from place to place. Rocks anywhere in the world that contain the same species of ammonite can be reliably dated to the Jurassic Period (named after the Jura Mountains). Other species of marine invertebrates can be used to correlate rocks from other periods. While dinosaur and other vertebrate fossils may attract more attention, marine invertebrate fossils are the ones that have been most helpful in allowing scientists to reconstruct the history of our planet because they are more common and widely distributed than dinosaur fossils.

The Geologic Column in words

FIGURE 3.8

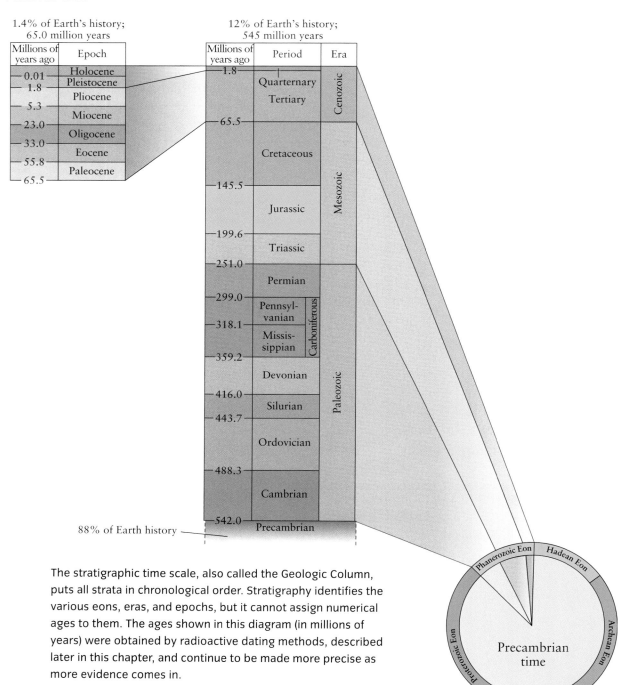

1.4% of Earth's history;
65.0 million years

Millions of years ago	Epoch
0.01	Holocene
1.8	Pleistocene
	Pliocene
5.3	
	Miocene
23.0	
	Oligocene
33.0	
	Eocene
55.8	
	Paleocene
65.5	

12% of Earth's history;
545 million years

Millions of years ago	Period		Era
1.8	Quarternary		Cenozoic
	Tertiary		
65.5			
	Cretaceous		Mesozoic
145.5			
	Jurassic		
199.6			
	Triassic		
251.0			
	Permian		Paleozoic
299.0			
	Pennsyl-vanian	Carboniferous	
318.1			
	Missis-sippian		
359.2			
	Devonian		
416.0			
	Silurian		
443.7			
	Ordovician		
488.3			
	Cambrian		
542.0	Precambrian		

88% of Earth history

Precambrian time

Phanerozoic Eon   Hadean Eon

Proterozoic Eon   Archean Eon

The stratigraphic time scale, also called the Geologic Column, puts all strata in chronological order. Stratigraphy identifies the various eons, eras, and epochs, but it cannot assign numerical ages to them. The ages shown in this diagram (in millions of years) were obtained by radioactive dating methods, described later in this chapter, and continue to be made more precise as more evidence comes in.

The Geologic Column in pictures
FIGURE 3.9

Bats

Primitive woodpecker

Messelobunodon

Pangolin

Tiny horses

Jewel beetle

Termites and ants

Anteater

www.wiley.com/college/murck

Cenozoic Era
In the Cenozoic Era, birds and mammals have flourished. In this scene from 15 million years ago in the Tertiary Period, we can see some possible ancestors of primates.

Bald cypress trees

Edmontosaurus

Anchiceratops

Cockroach

Magnolia

Ground beetle

Plant beetle

Mesozoic Era
The Mesozoic Era saw the rise of dinosaurs, which were the dominant vertebrates (animals with backbones) on land for many millions of years. If you look closely in the corners, you can see two harbingers of the future that first appeared in the Mesozoic Era: the first magnolia-like flowering plants *(lower left)* and the first shrew-like mammals *(lower right)*.

Hyneria

Hynerpeton

Paleozoic Era
During the Paleozoic Era, the evolution of life progressed from marine invertebrates (animals without backbones) to fish, amphibians, and reptiles. In this scene from 350 million years ago during the Carboniferous Period, a lobe-finned fish like the one in Figure 3.4 comes to life *(foreground)* along with some early amphibians whose fins have evolved into legs, but who still have a fishlike tail.

The Cambrian explosion FIGURE 3.10

Only the sketchiest remains are left of any life forms from Earth's earliest eons. But in the Cambrian Period, life suddenly exploded in a profusion of bizarre and now extinct forms. Trilobites (*below*) ranged from 1 millimeter in size to the largest known specimen, 72 centimeters.

Anomalocaris was the most fearsome predator of the Cambrian seas, a swimming creature up to 1 meter long.

Opabinia, which among other oddities had five eyes, was about 7 centimeters long and would have been a tasty morsel for *Anomalocaris*.

PERIODS AND EPOCHS

Geologists have divided the three eras of the Phanerozoic Eon into shorter units called **periods**. Some of them are delineated by massive die-outs, or extinctions, of species. Geologists named the periods in a somewhat haphazard manner; with some named for geographic locations (e.g., Jurassic, from the Jura Mountains in Switzerland where this layer was first studied) and others named for the characteristics of their strata.

The earliest period of the Paleozoic Era, the Cambrian Period, is especially noteworthy. Not only is this when animals with hard shells first appeared in the geologic record, but in the opinion of many paleontologists it was a time of unparalleled diversity, a phenomenon called the "Cambrian Explosion" (see FIGURE 3.10). Rocks that formed before the Cambrian Period sometimes contain microscopic, soft-bodied fossils, but strata cannot be differentiated on the basis of the fossils they contain. Thus geologists often lump this enormous 4-billion-year swath of time and the rocks deposited then into one category, **Precambrian**.

Even periods, as we now know, lasted for tens of millions of years. Thus geologists find it helpful to split them into still smaller divisions called **epochs**. The names of the epochs of the Tertiary and Quaternary Periods are somewhat more familiar than others because humans and their ancestors emerged during these epochs, and the names sometimes appear in the popular press. These recent epochs are not defined by extinction events, but according to the percentage of their fossils that are represented by still-living species. Many plant and animal fossils found in Pliocene strata, for example, have still-living counterparts, but fossils in Eocene strata have few still-living counterparts.

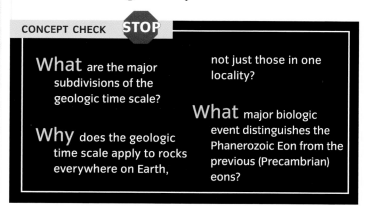

CONCEPT CHECK STOP

What are the major subdivisions of the geologic time scale?

Why does the geologic time scale apply to rocks everywhere on Earth,

not just those in one locality?

What major biologic event distinguishes the Phanerozoic Eon from the previous (Precambrian) eons?

Numerical Age

The scientists who worked out the Geologic Column were tantalized by the challenge of numerical time. They wanted to know Earth's age, how fast mountain ranges rise, how long the Paleozoic Era lasted, and most challenging of all, when life first appeared, and how long humans have inhabited Earth.

EARLY ATTEMPTS

Several methods for solving the problem of numerical time were proposed during the nineteenth century. All of them were unsuccessful, but the reasons for their failure are nonetheless quite illuminating—and they help explain where our currently accepted estimates came from (see FIGURE 3.11 on the following page). As previously mentioned, an early approach involved estimating rates of sedimentation and multiplying by the thickness of stratigraphic sections. Unfortunately, the resulting estimates for Earth's age varied too widely to be useful, from 3 million to 1.5 billion years.

Early scientists recognized that rivers carry dissolved substances to the sea from the erosion of rocks on land. In 1715, Edmund Halley, for whom Halley's Comet is named, suggested that one could measure the rate at which salts are added to the sea by river input and calculate the time needed to transport all the salts now present in the sea. Halley did not carry out his own suggestion, and it was not until 1899 that John Joly made the necessary measurements and calculations. Joly estimated that the ocean, and, therefore, Earth were 90 million years old. Unfortunately, neither Halley nor Joly realized that the ocean, like other parts of the Earth system, is an open system. Salt is removed by reactions between seawater and volcanic rocks on the seafloor, as well as the evaporation of seawater in isolated basins. The addition and removal of salts have balanced each other for hundreds of millions (even billions) of years.

The nineteenth century also saw the publication of Charles Darwin's *On the Origin of Species* in 1859. The book intensified the ongoing debate over numerical ages. Darwin understood that the evolution of new species by natural selection must be a very slow process that needed vast amounts of time. The first edition of his book contained a rough estimate of Earth's age, based on the erosion rate of a mountain range, of at least 300 million years.

However, William Thomson (Lord Kelvin), a leading physicist and contemporary of Darwin, emphatically rejected Darwin's estimate. Kelvin used the laws of thermodynamics to calculate how long Earth has been a solid body. Kelvin made two assumptions: Earth was once completely molten, and no heat was added to it after it formed. Once it had cooled enough to form a solid outer layer, heat would escape only by conduction (the same way that heat moves through the wall of a coffee cup). Kelvin measured the current-day rate of heat loss and extrapolated backward to figure out the age of Earth's solid crust. He arrived at a figure of 20 million years. Even Darwin conceded, "Thomson's views on the recent age of the world have been for some time one of my sorest troubles," and removed his age estimate from later editions of *On the Origin of Species*.

Darwin died before geologists vindicated him. Kelvin's mistake was his assumption that no heat had

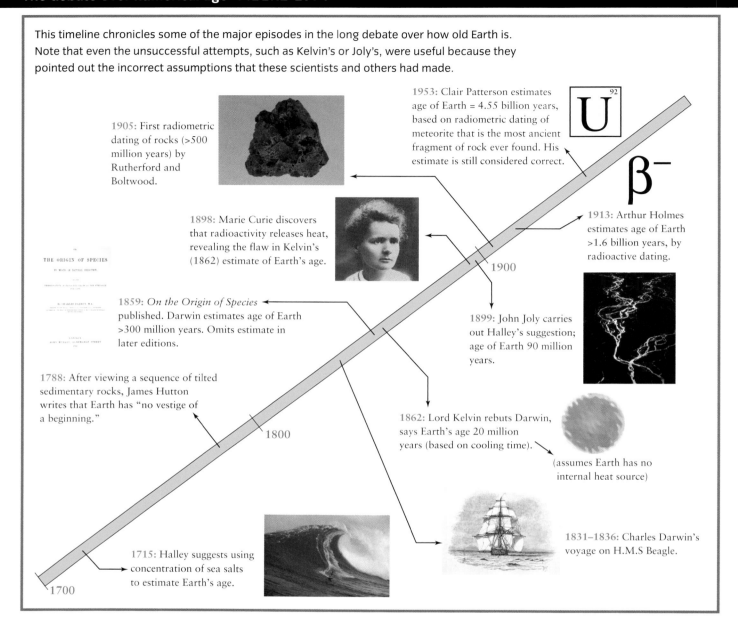

This timeline chronicles some of the major episodes in the long debate over how old Earth is. Note that even the unsuccessful attempts, such as Kelvin's or Joly's, were useful because they pointed out the incorrect assumptions that these scientists and others had made.

1905: First radiometric dating of rocks (>500 million years) by Rutherford and Boltwood.

1953: Clair Patterson estimates age of Earth = 4.55 billion years, based on radiometric dating of meteorite that is the most ancient fragment of rock ever found. His estimate is still considered correct.

1898: Marie Curie discovers that radioactivity releases heat, revealing the flaw in Kelvin's (1862) estimate of Earth's age.

1913: Arthur Holmes estimates age of Earth >1.6 billion years, by radioactive dating.

1859: *On the Origin of Species* published. Darwin estimates age of Earth >300 million years. Omits estimate in later editions.

1899: *John Joly carries out Halley's suggestion; age of Earth 90 million years.*

1788: After viewing a sequence of tilted sedimentary rocks, James Hutton writes that Earth has "no vestige of a beginning."

1862: Lord Kelvin rebuts Darwin, says Earth's age 20 million years (based on cooling time).

(assumes Earth has no internal heat source)

1831–1836: Charles Darwin's voyage on H.M.S Beagle.

1715: Halley suggests using concentration of sea salts to estimate Earth's age.

1700 1800 1900

---

been added to Earth's interior since its formation. He did not know about **radioactivity**, a natural physical process that releases heat. Though Earth is a closed system (as we explained in Chapter 1), it does have inputs and outputs of energy, and Kelvin had missed a key input.

What was needed to solve the problem of numerical geologic time was a way to

> **radioactivity** A process in which an element spontaneously transforms into another isotope of the same element, or into a different element.

measure events by some process that runs continuously, is not reversible, is not influenced by such factors as chemical reactions and high temperatures, and leaves a continuous record without any gaps. The discovery of radioactivity not only proved that Kelvin's assumptions were wrong, but by fortunate chance, also provided the breakthrough needed to measure absolute time.

RADIOACTIVITY AND NUMERICAL AGES

To explain how radioactivity allows us to determine the ages of rocks, we need to go back to the fundamentals of chemistry. Chapter 2 stated that most chemical elements have two or more *isotopes* that have the same number of protons per atom but a different number of neutrons per atom. To put it another way, each isotope of an element has the same atomic number but a different mass number.

Most naturally occurring isotopes have stable nuclei. However, a number of them—such as carbon-14 and potassium-40—are unstable. They spontaneously release particles from their nuclei. In the process, they change their mass number, their atomic number, or both (see FIGURE 3.12). Any isotope that spontaneously undergoes such nuclear change is said to be *radioactive*, and the process of change is referred to as *radioactive decay*.

Because radioactive decay involves the nucleus, it is *not* a chemical process. In fact, it releases far more energy than any chemical reaction, which explains why it is still keeping Earth's interior hot after more than 4 billion years. The decay of radioactive isotopes is not influenced by any chemical process or by any conditions of heat or high pressure within Earth. Thus radioactive decay is a perfect built-in geologic clock, at least for rocks that contain radioactive isotopes. Furthermore, because each radioactive isotope has its own rate of decay, a rock that contains several different radioactive isotopes has numerous built-in clocks that can be checked against each other.

Rates of Decay   In any radioactive decay system, the number of original radioactive *parent atoms* continuously decreases, while the number of nonradioactive *daughter atoms* produced by the radioactive decay continuously increases. For this reason, many of the

Radioactive decay FIGURE 3.12

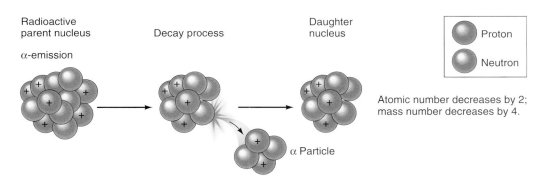

A radioactive nucleus releases an alpha-particle, which consists of two protons and two neutrons.

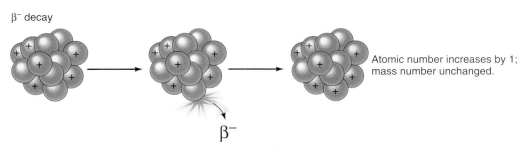

A radioactive nucleus releases a beta-particle, and one of its neutrons turns into a proton.

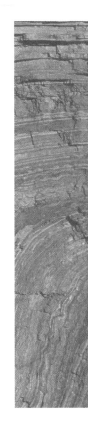

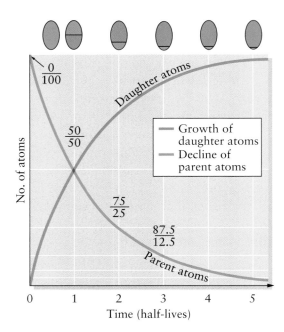

Radioactivity and time FIGURE 3.13

This graph illustrates the basic decay law of radioactivity. Suppose that a given isotope has a half-life of 1 hour. If we started with a sample consisting of 100 percent radioactive parent atoms, after an hour only 50 percent of the parent atoms would remain and an equal number of daughter atoms would have formed. At the end of the second hour, another half of the parent atoms would be gone, so there would be 25 percent parent atoms and 75 percent daughter atoms. After the third hour, another half of the parent atoms would have decayed, leaving 12.5 percent parent atoms and 87.5 percent daughter atoms. Note that the total number of atoms, parent plus daughter, remains constant. This fact is the key to the use of radioactivity to measure geologic time.

radioactive isotopes that were present when Earth was formed have decayed away because they were short-lived. However, radioactive isotopes that decay very slowly are still present.

All decay rates follow the same basic law: The *proportion* of parent atoms that decay during each unit of time is always the same (see FIGURE 3.13). Proportion means a fraction, or percentage, not a whole number. The rate of radioactive decay is determined by the **half-life**.

Radiometric dating: Not one clock but many

Following the discovery of radioactivity by Becquerel in 1896, and further work by Marie and Pierre Curie, the first estimates of the ages of rocks using radioactive decay were made in 1905. The long-hoped-for "rock clock" was finally available. The results were, and continue to be, remarkable. **Radiometric dating** has revolutionized the way we think about Earth and its long history.

A radioactive rock clock usually measures the amount of time that has elapsed since the minerals in the rock crystallized. When a new mineral grain forms—for example, a grain of feldspar in cooling lava—all the atoms in the grain become locked into the crystal structure and isolated from the environment outside the grain. In a sense, the atoms in the mineral grain, including any radioactive atoms, are sealed in an atomic time capsule. The trapped radioactive parent atoms decay to daughter atoms at a rate determined by the half-life.

In the simplest case, if no daughter atoms were present in the mineral at the time of formation, we can use Figure 3.13 to work backward and determine how long ago the time capsule was sealed. For example, if the mineral crystal contained some daughter atoms at the time of formation, the process is more difficult. Geologists have developed several ways to estimate the initial contamination of a sample by daughter atoms. Once that is done, and provided we know the half-life of the radioactive parent, it is a simple matter to calculate how long ago the mineral crystallized. Geologists use different isotopic systems to study rocks, fossils, and biologic materials of different ages and compositions (see FIGURE 3.14).

Radiometric dating has been particularly useful for determining the ages of igneous rocks, because the mineral grains in an igneous rock form at the same time

half-life The time needed for half of the parent atoms of a radioactive substance to decay into daughter atoms.

radiometric dating The use of naturally occurring radioactive isotopes to determine the numerical age of minerals, rocks, or fossils.

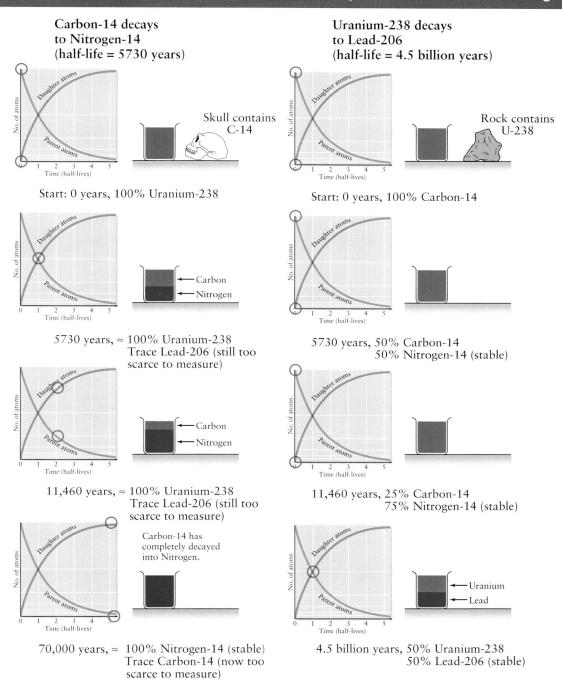

**Carbon-14 decays
to Nitrogen-14
(half-life = 5730 years)**

Skull contains
C-14

Start: 0 years, 100% Uranium-238

5730 years, ≈ 100% Uranium-238
Trace Lead-206 (still too
scarce to measure)

Carbon
Nitrogen

11,460 years, ≈ 100% Uranium-238
Trace Lead-206 (still too
scarce to measure)

Carbon
Nitrogen

Carbon-14 has
completely decayed
into Nitrogen.

70,000 years, ≈ 100% Nitrogen-14 (stable)
Trace Carbon-14 (now too
scarce to measure)

**Uranium-238 decays
to Lead-206
(half-life = 4.5 billion years)**

Rock contains
U-238

Start: 0 years, 100% Carbon-14

5730 years, 50% Carbon-14
50% Nitrogen-14 (stable)

11,460 years, 25% Carbon-14
75% Nitrogen-14 (stable)

Uranium
Lead

4.5 billion years, 50% Uranium-238
50% Lead-206 (stable)

Long-lived isotopes such as uranium-238 are especially useful for dating ancient rocks. Short-lived isotopes such as carbon-14 are useless for dating samples that are more than about 70,000 years old, which include many of the rocks that are of interest to geologists. However, carbon-14 is very useful for dating geologically recent items of biologic origin, such as human remains and artifacts. Geologists have also used it to study the latest ice age, by estimating the ages of wood samples taken from trees killed by the advancing ice sheet.

as the rock that contains them. On the other hand, most sedimentary rocks consist largely of mineral grains that were formed long before the strata that contain them were deposited. Radiometric dating will tell how old the grains are but not when the strata were deposited. This makes it difficult to directly date most sedimentary rocks and to infer the numerical age of ancient life forms fossilized in the sediments.

Numerical time and the Geologic Column

As geologists worked out the Geologic Column, they found many locations where layers of solidified lava and volcanic ash are interspersed with sedimentary strata. Through radiometric dating, they could determine the numerical ages of the lavas and volcanic ash and thereby bracket the ages of the sedimentary strata (see FIGURE 3.15).

Through a combination of geologic relations and radiometric dating, scientists have been able to fill in all the dates in the Geologic Column as shown in Figure 3.8. The scale is continually being refined, so the numbers given in the figure are considered the best currently available but are subject to change. Further work will make the numbers more precise. It is a tribute to the work of geologists during the nineteenth century that the Geologic Column they established by ordering strata according to their relative ages has been fully confirmed by radiometric dating. At the same time, it is humbling to see how one wrong assumption caused Lord Kelvin to be more than 4 billion years off in his estimate of Earth's age.

MAGNETIC POLARITY DATING

Time is so central to the study of Earth that geologists are always seeking new ways to estimate ages. An exciting newer method of dating, developed in the 1960s, involves **paleomagnetism**, the study of Earth's past magnetic field (FIGURE 3.16). Both igneous and sedimentary rocks "lock in" information about the magnetic field at their time of formation.

Earth's magnetic field reverses its polarity at irregular intervals, but on average, about once every half million years. This means that the magnetic pole that

> ■ **paleomagnetism** The study of rock magnetism in order to determine the intensity and direction of Earth's magnetic field in the geologic past.

Radiometric dating and the Geologic Column FIGURE 3.15

This idealized drawing shows how radiometric dating can be used to bracket the ages of sedimentary strata. The conduit for the lava flow labeled A can be dated directly by radiometric means. The conduit cuts through strata 1, 2, 3, and 4, so it must be younger than all of them. Igneous intrusion B can also be dated by radiometric means. It cuts through strata 1, 2, and 3 but not 4, and hence must be older than layer 4. Thus the sedimentary rocks in layer 4 are between 20 million and 34 million years old.

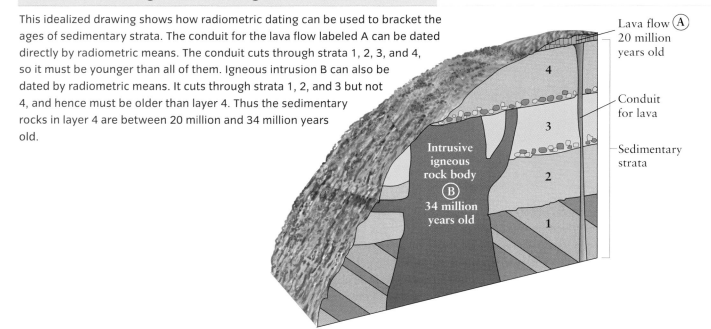

Lava flow (A)
20 million years old

Conduit for lava

Sedimentary strata

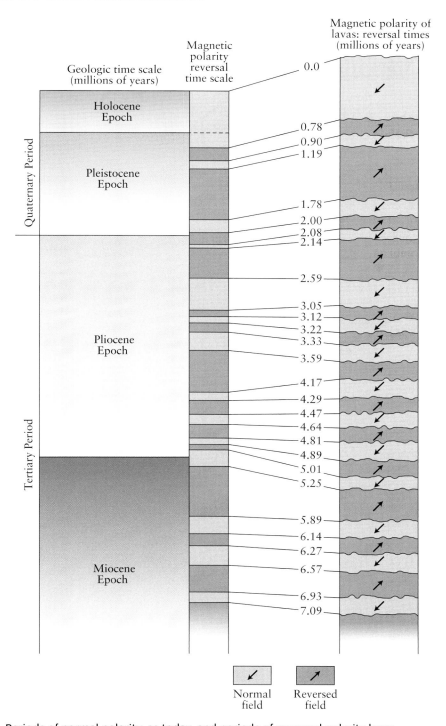

Periods of normal polarity, as today, and periods of reversed polarity have been identified and dated, using radiometric dating of lavas, back to the beginning of the Jurassic Period, about 200 million years ago. This figure shows the most recent 7 million years.

Dating Human Ancestors

The Haddar region of northern Ethiopia has been a fertile site for finding fossils of ancient human ancestors, part of a larger group called *hominids*. Many of the fossils have been found in ancient stream gravels, where they were tumbled and battered by long-ago floodwaters. The fossil-bearing gravels are interlayered with sediments that give good magnetic signals.

The problem is to know where in the magnetic polarity time scale the Haddar sediment falls. Geologists answered this question by using the Potassium–Argon (K–Ar) method to establish the ages of a lava flow that lay below some of the hominid fossils and a layer of volcanic ash that lay above them. The two radiometric dates determined the magnetic reversal ages unambiguously and indicated that early hominids lived in the region until about 3.15 million years ago, during the Pliocene Epoch. Through many studies like this, it is slowly becoming clear that the hominid genus *Homo* (which includes our own species, *Homo sapiens*) evolved a little more than 2 million years ago from an older genus called *Australopithecus*. (Two reconstructed *Australopithecus* skulls are shown at bottom right.)

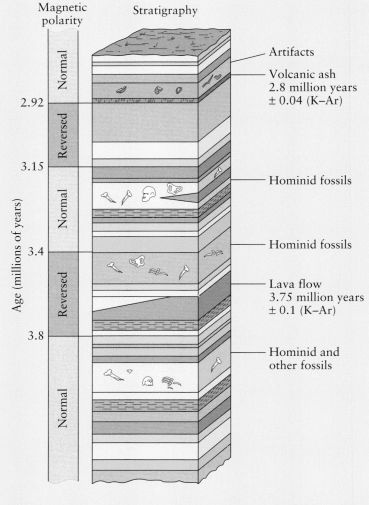

Magnetic polarity | Stratigraphy

Normal

Artifacts

Volcanic ash
2.8 million years
± 0.04 (K–Ar)

2.92

Reversed

3.15

Normal

Hominid fossils

Age (millions of years)

3.4

Hominid fossils

Reversed

Lava flow
3.75 million years
± 0.1 (K–Ar)

3.8

Hominid and
other fossils

Normal

NATIONAL GEOGRAPHIC

had been in the northern hemisphere moves to a position near Earth's south pole, and the magnetic pole that had been in the south moves to the north. Note, however, that Earth's geographic North and South poles of rotation do not change position. (Earth's magnetic field is explained in additional detail in Chapter 4.) Scientists are still working out details of how or why **magnetic reversals** happen. The two important points are that the reversal happens quickly by geologic time standards, and any iron-bearing mineral in a sedimentary or igneous rock retains or "remembers" the magnetic polarity of Earth at the time that the rock was formed—that is, a change in the magnetic field does not affect already formed minerals. Through a combination of radiometric dating and magnetic polarity measurements, it has been possible to establish a time scale of magnetic polarity reversals dating back to the Jurassic Period (see Figure 3.16). Still earlier reversals are a subject of ongoing research.

Correlation on the basis of magnetic reversals differs from other geologic correlation methods. One magnetic reversal looks just like any other in the rock record. When evidence of a magnetic reversal is found in a sequence of rocks, the problem lies in knowing which of the many reversals it actually represents. When a continuous record of reversals can be found, starting with the present, it is simply a matter of counting backward. But if not, the technique must be combined with stratigraphic and radioactive dating techniques (see Case Study on page 80). Chapter 4 discusses how magnetic polarity studies played a crucial role in the development of plate tectonic theory.

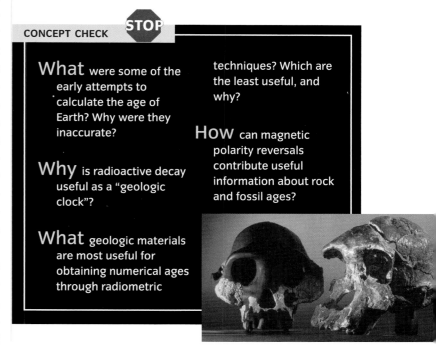

CONCEPT CHECK STOP

What were some of the early attempts to calculate the age of Earth? Why were they inaccurate?

Why is radioactive decay useful as a "geologic clock"?

What geologic materials are most useful for obtaining numerical ages through radiometric techniques? Which are the least useful, and why?

How can magnetic polarity reversals contribute useful information about rock and fossil ages?

The Age of Earth

LEARNING OBJECTIVES

Explain why the oldest rocks are not necessarily the same age as the planet.

Explain why scientists currently believe Earth is about 4.56 billion years old.

Throughout this book, we mention examples of actual rates of geologic processes. This would not be possible without the numerical dates obtained through radiometric dating and other numerical age methods. In fact, more than any other contribution by geologists, the ability to determine numerical dates has changed the way humans think about the world and the immensity of geologic time.

Now that we know how to determine the numerical ages of rocks, can we determine Earth's age? It's not as easy as you might think. The continual recycling of Earth's surface by erosion and plate tectonics means that very few, if any, remnants of Earth's

Earth's oldest rocks FIGURE 3.17

The Acasta gneiss in northern Canada was formed 4.0 billion years ago. It is the most ancient body of rock so far discovered on Earth.

original crust remain. Of the many radiometric dates obtained from Precambrian rocks, the oldest is about 4.0 billion years (see FIGURE 3.17). Although no older rocks than this have been found, an individual mineral grain from a sedimentary rock in Australia has been dated to 4.4 billion years, so it is conceivable that igneous rocks older than 4.0 billion years may someday be located.

However, geologists believe that there is still a gap between the age of these oldest mineral grains and the age of Earth. There is strong evidence that Earth formed at the same time as the Moon, the other planets, and meteorites. Through radiometric dating, it has been possible to determine the ages of meteorites and of "moon dust" brought back by astronauts. The *Apollo* astronauts found rocks and individual grains of Moon

dust that are believed to be pieces of the Moon's original crust. Such rocks should be abundant because the Moon has been geologically much less active than Earth.

Meteorite ages are especially valuable because some meteorites have remained virtually unaltered since the formation of the solar system. Melting and other types of geologic alteration will reset the radiometric clock. However, some meteorites, such as the Allende Meteorite, which fell to Earth in the Mexican state of Chihuahua on February 8, 1969, belong to a rare category called *carbonaceous chondrites*, which are believed to contain unaltered material from the formation of the solar system. It is "carbonaceous" because it contains tiny amounts of carbon (about 3 parts per 1000). Some of the carbon is in chemical compounds called amino acids—organic components

A cosmic interloper FIGURE 3.18

The Allende Meteorite, which fell to Earth in Mexico, is one of the most famous meteorites in history. Note the white spots on the meteorite. Some of these inclusions, which are slightly older than the black carbonaceous material around them, are more than 4.6 billion years old, making them the oldest objects of any kind ever found on Earth.

that are essential for life. The dark, fine-grained part of the meteorite is mostly olivine, with a few flecks of metallic iron and some carbon. The clumps of white material are oxides of calcium and aluminum and are thought to be among the first matter to condense from the gas cloud from which the Solar System formed (FIGURE 3.18). The white clumps are older than Earth itself.

The ages of many of these objects cluster closely around 4.56 billion years. Planetary scientists therefore believe that Earth, and indeed the Sun's entire planetary system, formed at that time.

Today, more than two centuries after Hutton, it is widely agreed that Earth's age is approximately 4.56 billion years. When will it cease to exist? Astronomers tell us that billions of years in the future, the Sun will become a red giant, at which point it will expand and engulf Earth. However, Hutton is still correct in one sense: Earth's history is profound and geologic evidence shows no prospect of an end.

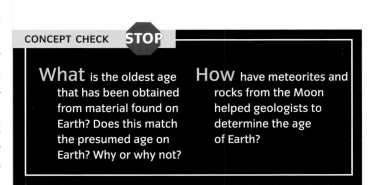

CONCEPT CHECK STOP

What is the oldest age that has been obtained from material found on Earth? Does this match the presumed age on Earth? Why or why not?

How have meteorites and rocks from the Moon helped geologists to determine the age of Earth?

Tapeats Sandstone

angular unconformity

Grand Canyon Supergroup

Standing at the rim of the Grand Canyon, you can see more than 2 billion years of Earth's history, preserved in the rocks. All three kinds of unconformities can be found here. The upper layers in this photograph were all deposited during the Paleozoic Era. Some of the contacts between different-colored parallel strata are disconformities, and they record the rising and ebbing of seas over this part of the North American continent. Below the Tapeats Sandstone (arrow) you can see an angular unconformity, which also represents a major time gap between Precambrian rocks (deposited about 825 million years ago) and Cambrian rocks (deposited less than 545 million years ago). Finally, there is a nonconformity (not visible in the photo) between the lowermost sedimentary layer of the Grand Canyon Supergroup, the Bass Limestone, and the Vishnu Formation, a foundation of metamorphic and igneous rocks that once lay at the base of an ancient mountain range that eroded away long before the Rocky Mountains came into existence.

As a geology student you owe it to yourself not to stop at the rim, as most tourists do, but to descend to the river level and get a close-up look at 2 billion years of Earth's history. You might be lucky and see trilobite tracks, like these, in the Tapeats Sandstone. The tracks were made when trilobites (like the one shown in Figure 3.10) extended their legs sideways, pulled in mud, then, under the safety of their hard shells, picked over the mud for food.

NATIONAL GEOGRAPHIC

1 Relative Age

1. Geologists study the chronologic sequence of geologic events, that is, their **relative age**. Relative age is derived from **stratigraphy**, the study of rock layers and how those layers are formed.

2. There are four basic principles of stratigraphy. **Strata**, or sedimentary rock layers, are horizontal when they are deposited as water-laid sediment *(principle of original horizontality)*. Strata accumulate in sequence, from the oldest on the bottom to the youngest on the top *(principle of stratigraphic superposition)*. A rock stratum is always older than any geologic feature, such as a fracture, that cuts across it *(principle of cross-cutting relationships)*. Finally, rock strata extend outward horizontally; they may thin or pinch out at their farthest edges, but they generally do not terminate abruptly unless cut by a younger rock unit *(principle of lateral continuity)*.

3. **Numerical age**, the exact number of years of a geologic feature, is more difficult to find. One difficulty that arises is that the sequence of strata in any particular location is not necessarily continuous in time. An **unconformity** is a break or gap in the normal stratigraphic sequence. It usually marks a period during which sedimentation ceased and erosion removed some of the previously laid strata. The three common types of unconformities are nonconformities, angular unconformties, and disconformities.

4. **Correlation** of strata is the establishment of the time equivalence of strata in different places. **Fossil** assemblages, usually consisting of hard shells, bones, and wood, have been the primary key to correlation of strata across long distances. The study of fossils and the record of ancient life on Earth is called **paleontology**. The principle of *faunal and floral successions* (animals and plants, respectively) is the stratigraphic ordering of fossil assemblages.

2 The Geologic Column

1. The **Geologic Column**, a *stratigraphic time scale*, is a composite diagram that shows the succession of all known strata, arranged in chronological order of formation, based on fossils and other age criteria.

2. The Geologic Column is divided into several different units of time, called **eons**, **eras**, **periods**, and **epochs**. The majority of Earth's history is divided into three eons, in which fossils are very rare or nonexistent. Those eons, each spanning several hundred million years, are the *Hadean*, *Archean*, and *Proterozoic*. The fourth and most recent eon, the *Phanerozoic*, is the only eon in which fossils are abundant. Very dramatic changes in fossil assemblages occur between the three eras of the Phanerozoic eon—the *Paleozoic*, *Mesozoic*, and *Cenozoic*—which were separated by major extinction events. The earliest period of the Paleozoic Era is especially important, as this was a time of unique diversity. This period is known as the "Cambrian Explosion." Rocks formed before the Cambrian Period cannot be differentiated by the fossil record; these rocks are considered to be *Precambrian*.

12% of Earth's history; 545 million years

Millions of years ago	Period	Era
1.8	Quarternary	Cenozoic
	Tertiary	
65.5	Cretaceous	Mesozoic
145.5	Jurassic	
199.6	Triassic	
251.0	Permian	Paleozoic
299.0	Pennsyl-vanian	Carboniferous
318.1	Missis-sippian	
359.2	Devonian	
416.0	Silurian	
443.7	Ordovician	
488.3	Cambrian	
542.0	Precambrian	

3 Numerical Age

1. Finding the age of Earth continued to be of interest to many scientists. Different theories were proposed by Halley, Joly, Darwin, Lord Kelvin, and several other prominent scientists of the later 1800s and early 1900s. Though many of these theories were shown to be incorrect, each was an important step in finding a method of numerical dating.

2. **Radioactivity** is the process in which an element spontaneously transforms itself into another *isotope* of the same element, or into a different element, through the release of particles and heat energy. The *radioactive decay* of isotopes of chemical elements provides a basis for radiometric dating, which gives values for the numerical ages (age in years) of rock units and thus values for numerical dates of geologic events. Because radioactive decay is not influenced by chemical processes or by heat and high pressure in Earth, it is an extremely accurate gauge of numerical age.

3. **Radiometric dating** is based on the principle that in any sample containing a radioactive isotope, half of its atoms of that isotope will change to daughter atoms in a specific length of time, called the **half-life**. (The *proportion* of parent atom decay during a unit of time is always the same.) Radioactive isotopes with a long half-life, such as uranium, are most useful for dating rocks. Carbon-14, which has a much shorter half-life, is most useful for dating organic materials of relatively recent origin (less than 70,000 years).

4. Though radiometric dating is primarily useful for igneous rocks, a complementary technique called magnetic polarity dating works for sedimentary rocks, too. Magnetic polarity dating involves **paleomagnetism**, the study of reversals in Earth's magnetic field. As yet, **magnetic reversals**, or periods of time in which Earth's magnetic polarity reverses itself, are not fully understood.

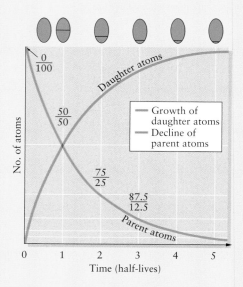

4 The Age of Earth

1. Through measures of numerical age, it has become clear that most of Earth's history took place in Precambrian time. The oldest Earth rocks discovered are about 4.0 billion years old.

2. Earth is not a good place to look for the oldest rocks in the solar system. Earth's surface has been subjected to a lot of geologic activity. This has reset some radiometric clocks and destroyed the earliest fragments of the crust. Samples from the Moon and from meteorites indicate that the solar system formed about 4.56 billion years ago, and by inference this is also the age of Earth.

- relative age p. 62
- stratigraphy p. 62
- numerical age p. 65
- unconformity p. 65

- paleontology p. 67
- correlation p. 68
- Geologic Column p. 69
- radioactivity p. 74

- half-life p. 76
- radiometric dating p. 76
- paleomagnetism p. 78
- magnetic reversal p. 81

CRITICAL AND CREATIVE THINKING QUESTIONS

1. Do the same principles of stratigraphy apply on the Moon as they do on Earth? Bear in mind that the geologic processes on the Moon have been very different from those on Earth. If you had to determine the relative age of features on the Moon, based entirely on satellite photographs, what would you look for and how might you proceed?

2. Check the area in which you live to see if there is an excavation —perhaps one associated with a new building or road repair. Visit the excavation and note the various layers, the paving (if the excavation is in a road), and the soil below the surface. Is any bedrock exposed beneath the soil?

3. How old are the rock formations in the area where you live and attend college or university? How can you find out the answer to this question?

4. Choose one of the geologic periods or epochs listed in Figure 3.8 and find out all you can about it: How are rocks from that period identified? What are its most characteristic fossils? Where are the best samples of rocks from your chosen period found?

5. Do some research to determine the ages of the oldest known fossils. What kind of life forms were they?

What is happening in this picture ?

This skier is hauling a sled past a cliff face on Ellesmere Island, Canada. Why do you think the rock strata in the background tilt at such a steep angle? Why are they wavy instead of straight?

1. A _____ is the age of a one rock unit or geologic feature compared to another.

   a. relative age
   b. numerical age

2. The principle of cross-cutting relationships says that

   a. water-borne sediments are deposited in nearly horizontal layers.
   b. a sediment or sedimentary rock layer is younger than the layers below it and older than the layers that lie above.
   c. a rock unit is older than a feature that disrupts it, such as a fault or igneous intrusion.
   d. a sediment or sedimentary rock layer is older than the layers below it and younger than the layers that lie above.

3. The _____ states that water-borne sediments are deposited in nearly horizontal layers.

   a. law of superposition
   b. principle of faunal succession
   c. principle of original horizontality
   d. principle of cross-cutting relationship

4. In a conformable sequence

   a. each layer must have been deposited on the one below it without any interruptions.
   b. there must not be any depositonal gaps in the stratigraphic record.
   c. Answers a. and b. are correct.
   d. None of the above is true.

5. An unconformity represents

   a. a gap in the stratigraphic record.
   b. a period of erosion or nondeposition.
   c. Both a. and b.
   d. None of the above.

6. On this illustration, label each unconformity as one of following:

   nonconformity   angular unconformity
   disconformity

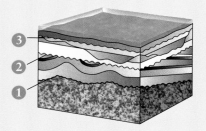

7. Fossils found in strata

   a. are the records of ancient life.
   b. allow the correlation of strata separated by many miles.
   c. have been useful to geologists in creating the Geologic Column.
   d. All of the above statements are correct.

8. The three eras that make up the Phanerozoic Eon are the

   a. Hadean, Archean, and Proterozoic.
   b. Paleozoic, Mesozoic, and Cenozoic.
   c. Triassic, Jurassic, and Cretaceous.
   d. Pliocene, Pleistocene, and Holocene.

9. The most distinctive changes in the fossil record occur across the boundaries between

   a. periods.
   b. eras.
   c. epochs.

10. The dinosaurs were dominant during _____.

   a. the Cenozoic Era
   b. the Mesozoic Era
   c. the Paleozoic Era
   d. the Precambrian time

11. Label the two decay sequences depicted in this illustration as either alpha emission or beta decay. For each decay sequence, also label the following:

> parent nucleus   alpha-particle
> daughter nucleus   beta-particle

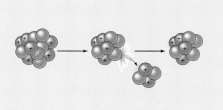

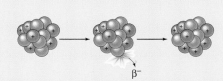

12. Potassium-40 is a naturally occurring radioisotope that decays to Argon-40 and is common in many rocks of the continental crust. The half-life of Potassium-40 is 1.3 billion years. Assuming no contamination, what would be the age of a sample that contained a 3:1 ratio of Potassium-40 to Argon-40?

a. 1.3 billion years.
b. 650 million years.
c. 2.6 billion years.
d. 325 million years.

13. If the sample indicated above showed evidence that it had been heated by contact with a more recent lava flow, what would be the likely error in the determined age?

a. The sample would appear too young.
b. The sample would appear too old.
c. Rocks are a "closed system"; there would be no error.

14. A gravel deposit containing an important hominid tooth fossil is found in a field location in Northern Ethiopia. The gravel deposit has fragments of volcanic rocks dated at 3.75 million years ± 0.1 by Potassium–Argon dating (K–Ar) and is known from stratigraphy to be younger than a 2.8 million year ± 0.06 (K–Ar) volcanic ash deposit. Sediments interlayered with the fossil-bearing gravels have good magnetic signals and have a normal polarity. Given the figure in the second column for the region, what is the most likely date for the hominid fossil?

a. Between 3.8 and 3.4 million years.
b. Between 3.4 and 3.15 million years.
c. Exactly 3.75 million years.
d. Younger than 2.92 million years.

Question 14

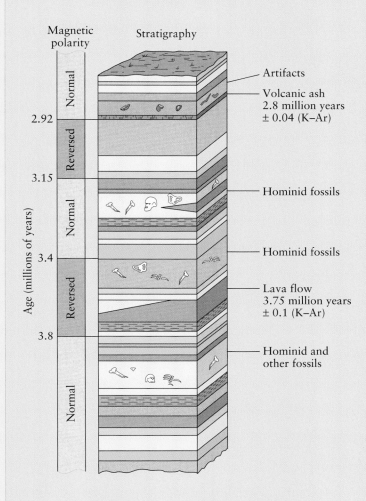

15. Earth is not considered a good place to look for the oldest rocks in the solar system because

a. contamination from atmospheric tests of nuclear weapons have contaminated the crust of Earth.
b. Earth's magnetic field interferes with the radiometric clocks in most igneous rocks.
c. melting has reset radiometric clocks in the rocks of Earth's crust.
d. the earliest crustal rocks have been destroyed by geologic activity.
e. All of the above are true.
f. Answers c. and d. are correct.

Plate Tectonics

The peak of Ama Dablam, in the Himalaya of Nepal, basks in the late afternoon sunshine. For mountain climbers and explorers, the Himalaya are the top of the world. Many of the peaks here are more than 7000 meters tall. The tallest of them all, of course, is Mount Everest, which has become a favorite, and sometimes fatal, challenge for thousands of mountaineers since Edmund Hillary and Tenzing Norgay first scaled it in 1953.

Only two people—a Swiss engineer named Jacques Piccard and a Navy lieutenant, Donald Walsh—have ever been to the lowest point on Earth's surface. The Marianas Trench, near Guam, bottoms out at 11,000 meters below sea level—deep enough that it could swallow up Mount Everest. To get there, Piccard had to build a special submarine designed to resist the intense pressure of the deep ocean.

Amazingly, both the Himalaya and the Marianas Trench formed as a result of the same geologic process—the collision of two lithospheric plates. In one case, the plates compressed each other and thrust up the world's highest mountain range. In the other case, one plate dove beneath the other, dragging the seafloor down with it.

The imperceptibly slow motion of plates, known as plate tectonics, is constantly reshaping our world. Over millions of years, it has restructured and relocated continents, built mountain ranges, and opened up new oceans where none were before. Plate tectonics unites all of the world's geologic features into one system.

Global Locator

90

A Revolution in Geology

LEARNING OBJECTIVES

Describe the supercontinent called Pangaea.

Identify the early arguments for and against Wegener's theory of continental drift.

Explain how paleomagnetism provided the definitive evidence for continental drift.

Define seafloor spreading.

S cientific revolutions challenge us to look at the world in a new way. They turn accepted ideas upside down. However, they don't happen overnight. For example, for millennia, people thought that Earth was the center of the universe, with the Sun and all the planets revolving around it. In 1543, Nicolaus Copernicus argued it was the other way around—that the planets revolve around the Sun. It took decades of debate, new inventions like the telescope, and finally the persuasive writing of such scientists as Galileo and Newton to convince astronomers that Copernicus was right.

Geology had a revolution in the 1960s, when geologists discovered new evidence for a 50-year-old idea called *continental drift.* Just like Copernicus's theory, the hypothesis that continents move provoked controversy at first. But in recent decades, it has become a true scientific revolution, called the *theory of plate tectonics,* which has changed our view of the planet. It has encouraged us to think of Earth as a system in constant flux, a dynamic world whose appearance has changed considerably over the eons. Early geologists gathered information about Earth and its processes painstakingly, one piece at a time. Plate tectonics shows us how the pieces fit together.

WEGENER'S THEORY OF CONTINENTAL DRIFT

In 1912, a German meteorologist named Alfred Wegener began lecturing and writing scientific papers about **continental drift**. Wegener's idea was that the continents had once been joined together in a single "supercontinent," which he called Pangaea (pronounced Pan-JEE-ah), meaning "all lands" (see FIGURE 4.1). He suggested that Pangaea had split into fragments like pieces of ice floating on a pond and that the continental fragments

had slowly drifted to their present locations.

Wegener marshalled a good deal of evidence in favor of the continental drift hypothesis. Nevertheless, his proposal created a storm of protest in the international scientific community, one that continued for decades. Some of the criticisms were well founded. Contemporary geologists could not envision any reasonable mechanism by which continents could be moved. The terminology suggests continents somehow "adrift" in a sea of water. But remember that the continents are anchored in solid rock, and a drifting continent would have to plow through or across a seafloor also made of solid rock. Even Wegener could not explain how this could happen.

Let's put this problem aside for a moment, and look at the evidence as the geologists of the time did. No single piece of evidence is conclusive on its own. It took all of these arguments (and more) to convince geologists several decades later that Wegener was right.

> **continental drift** The slow, lateral movement of continents across Earth's surface.

THE PUZZLE-PIECE ARGUMENT

It's easy to see from a map that the Atlantic coastlines of Africa and South America seem to match, almost like puzzle pieces. But is this an accident, or does it truly support the hypothesis that the continents were once joined together?

To decide whether the continents really do match, we must first note that there is more to a continent than meets the eye. The Atlantic coasts of South America and Africa, like those of many continents, do not terminate abruptly at the shoreline but slope gently

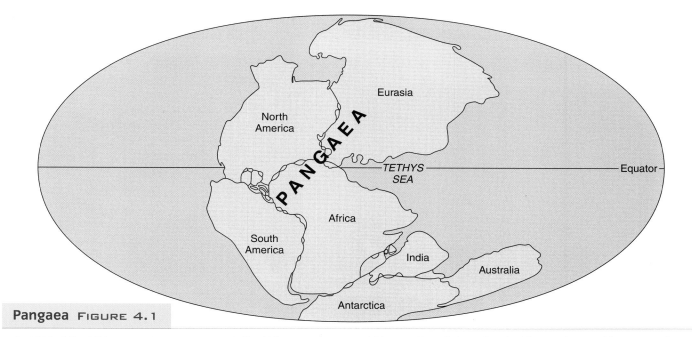

Pangaea FIGURE 4.1

In 1915, Alfred Wegener drew a map much like this one, showing the distribution of the continents about 300 million years ago in the Carboniferous Period. Wegener proposed that the continents at that time were joined in one "supercontinent," which he named Pangaea. The apparent close fit of the coastlines of the modern continents, particularly Africa and South America, had been observed by Leonardo da Vinci and Francis Bacon as early as the 1500s. Modern reconstructions of Pangaea differ from Wegener's version in some respects (for example, Africa is located farther south on this map than on Wegener's original), but his basic idea was correct.

seaward. This gently sloping land, part above sea level and part below, is called the **continental shelf**. At a water depth of about 100 meters, there is an abrupt change of slope called the *shelf break*, below which lies the steeper **continental slope** (see FIGURE 4.2A on the next page). The edge of a continent is defined as being halfway down the continental slope, where the continental crust meets the oceanic crust.

As FIGURE 4.2B shows, when we fit together South America and Africa along the true edges of the continents, we get an even better match than we might expect. In the "best-fit" position, the average gap or overlap between the two continents is only 90 kilometers. In addition, the most significant overlaps consist of sedimentary or volcanic rock that were added *after* the continents are thought to have split apart.

MATCHING GEOLOGY

The close fit between Africa and South America suggests they were once joined. But a jigsaw puzzle requires both the shapes and the designs to match. If the continents were once joined together, we should find similar geologic features on both sides of the join. However, matching the geology of rocks on opposite sides of an ocean is more difficult than you might imagine. Erosion and rock formation since the breakup of Pangaea may have destroyed or covered up some of the evidence, so we will be putting together a puzzle with some pieces missing and others defaced.

A starting point is to see if the ages of similar rock match up across the ocean. In Wegener's time, the technique of radiometric dating was just being developed, so it was not easy to determine the exact age of a rock. But now we know that there is, indeed, some similarity in the ages of rocks and correlation between rock sequences on both sides of the ocean. As shown in FIGURE 4.3A on page 95, the match is particularly good between rocks about 550 million years old in northeast Brazil and West Africa.

We can also check for the continuity of mountain chains. FIGURE 4.3B on page 95 shows a

A Revolution in Geology   93

What is the "true" edge of a continent? FIGURE 4.2A

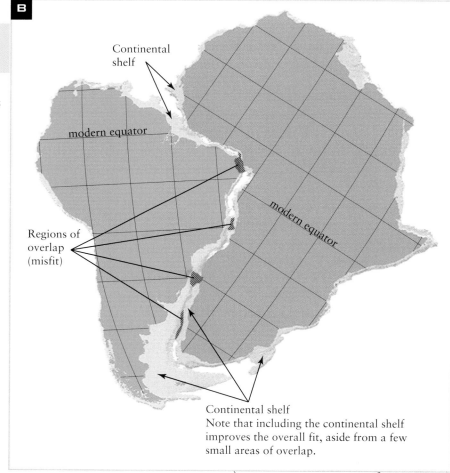

A

Defined edge of continent
Continental shelf
Continental slope
Continental rise
North American plate
Midocean ridge
African plate
Sea level
Kilometers
0
5
10
Sediment
Abyssal plain
Central rift valley
Oceanic crust
Continental crust

The edge of a continent usually contains several components. The shallow and gently sloped *continental shelf* lies just off-shore. The steeper *continental slope* leads down to the bottom of the ocean. At the base of the continental slope lies the *continental rise*, which is usually composed of a layer of sediment. The main part of the ocean floor is the flat *abyssal plain*. In the center of the ocean, between two continents, we often find a submarine mountain range called a *midocean ridge*. This structure turns out to be key to the mystery of what causes continental drift.

How well do the continents fit?
FIGURE 4.2B

This map reconstructs the fit of South America and Africa along the *true* edges of the continents—not along the shoreline, but the submerged edge of the continental crust. The darkly shaded areas show overlap. Note that the inclusion of the shallow off-coastal areas makes the fit better, not worse. Wegener did not have access to this information.

Note also that Africa and South America have rotated in different ways, a point we will return to later in the chapter.

B

Continental shelf
modern equator
modern equator
Regions of overlap (misfit)
Continental shelf
Note that including the continental shelf improves the overall fit, aside from a few small areas of overlap.

Older than 550 million years
About 550 million years old
Younger than 550 million years

Close match in rock ages

Africa

Modern Brazil

South America

How well do the ages match? FIGURE 4.3A–B

◀ **A** When the continents are rotated back together, the ages of rock units generally match, particularly in the regions of northeast Brazil and West Africa.

B In this reconstruction of the northern part of Pangaea, mountain belts of similar ages match up.
▼

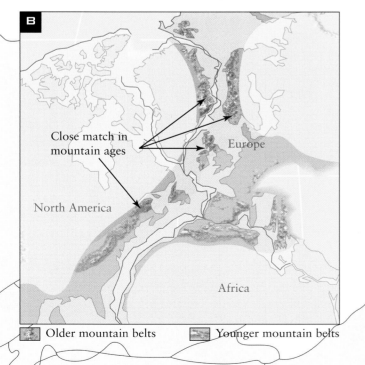

Close match in mountain ages

Europe

North America

Africa

Older mountain belts   Younger mountain belts

What evidence did glaciers leave? FIGURE 4.3C–D

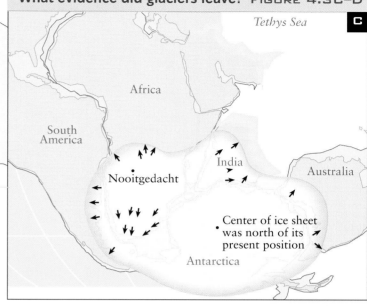

Tethys Sea

Africa

South America

India

Australia

Nooitgedacht

Center of ice sheet was north of its present position

Antarctica

▲ **C** In this reconstruction of the southern continents of Pangaea, we have shown the rough extent of glacial deposits. Small arrows indicate the direction in which the ice was moving during the glaciation, as deduced from evidence like grooves and scratches in the bedrock.

D As glaciers advance, the sediment and rocks contained in them grind and abrade the bedrock, producing a polished and grooved surface. This polished surface, at Nooitgedacht in South Africa (see map above), was created by the Pangaean ice sheet. The grooves (and the pen, upper right) indicate the direction of ice movement. ▶

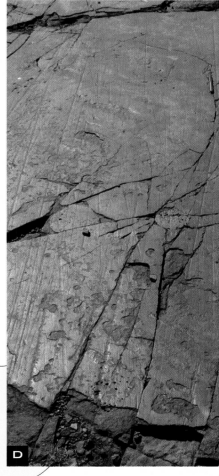

A Revolution in Geology   95

reconstruction of the northern part of the supercontinent Pangaea. Notice, again, how mountain chains of similar ages seem to line up when the continents are moved back into this position. The oldest portions of the Appalachian Mountains, extending from the northeastern United States through eastern Canada, match up with the Caledonides of Ireland, Britain, Greenland, and Scandinavia. A younger part of the Appalachians lines up with a belt of similar age in Africa and Europe.

The deposits left by ancient continental ice sheets also match up across continental joins. In South America and Africa there are thick glacial deposits of the same age (Permian-Carboniferous), which match almost exactly when the continents are moved back together (see FIGURE 4.3C on the previous page).

As glacial ice moves, it cuts grooves and scratches in underlying rocks and produces folds and wrinkles in soft sediments (see FIGURE 4.3D on the previous page). These features provide evidence not only of the extent of glaciation but also of the direction the ice was moving during the glaciation. When Africa and South America are moved back together, the direction of ice movement on both continents is consistent, radiating outward from the center of the former ice sheet. It's hard to imagine how such similar glacial features could have been created if the continents had not been joined together. The glacial deposits suggest also that Africa and South America were once closer to the South Pole than they are today, and had cooler climates.

MATCHING FOSSILS

If Africa and South America were really joined at one time, with the same climate and matching geologic features, then we should expect that in ice-free intervals they would have been inhabited by the same plants and animals. To check this hypothesis, Wegener looked at fossils. He found that some communities of plants and animals apparently evolved together until the time that Pangaea split apart, and after that time they evolved separately.

For example, Wegener found fossils of an ancient tree, *Glossopteris*, in matching areas of southern Africa, South America, Australia, India, and Antarctica (see FIGURE 4.4A and B). Could the seeds of this plant have been carried by wind or water from one

How well do the fossil records match?
FIGURE 4.4

▲ A The tongue-shaped fossil leaves shown here came from a tree of Carboniferous age called *Glossopteris* (named after the Greek word for "tongue"). Similar fossils have turned up in Africa, India, Australia, and South America, providing strong evidence that these regions were once contiguous.

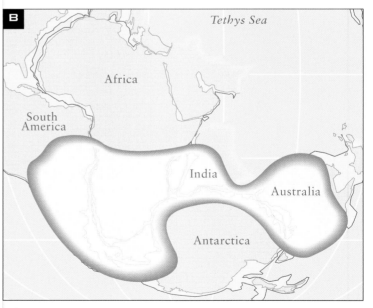

▲ B This map outlines the locations where *Glossopteris* fossils have been found. The territories match well when the continents are moved back to their probable locations in the Carboniferous Era.

continent to another? Probably not. The seeds of *Glossopteris* were large and heavy, and probably would not travel far on the wind or water currents. Not only that, *Glossopteris* flourished in a cold climate; it would not have thrived in the warm present-day regions where its fossil remains are found. This, too, is consistent with the idea that these continents were once joined together in a more southerly location.

Certain animal fossils, too, match up well. The fossil remains of *Mesosaurus*, a small reptile from the Permian Period, are found both in southern Brazil and in South Africa. The types of rocks in which the fossils are found are very similar. *Mesosaurus* did swim, but was too small (about half a meter long) to swim all the way across the ocean. Fossil remains of certain types of earthworms also occur in areas that are now widely separated. Since an earthworm could not have hopped across a wide ocean, the landmasses in which they lived must once have been connected.

APPARENT POLAR WANDERING PATHS

Wegener and his few supporters gathered more and more evidence in support of continental drift, but many scientists remained unconvinced. Wegener died in 1930 without seeing a resolution to the debate.

In the 1950s, *paleomagnetism* emerged as a new tool for studying Earth's history, and it was destined to break the intellectual logjam. Recall from Chapter 3 that when magma cools and solidifies into rock, grains of iron-bearing minerals become magnetized and take on the prevailing polarity (north-south orientation) of Earth's magnetic field at that time. We explained in Chapter 3 how this fact allows us to detect reversals in Earth's magnetic field. It also allows us to reconstruct the apparent location of Earth's past magnetic poles.

In the 1950s, geophysicists studying paleomagnetic pole positions found evidence suggesting that Earth's magnetic poles had wandered all over the globe over the last several hundred million years. They plotted the pathways of the poles on maps like FIGURE 4.5, and called the phenomenon *apparent polar wandering*. Geophysicists found this puzzling because they knew that the magnetic poles could not actually wander very far from Earth's axis of rotation. Even more puzzling,

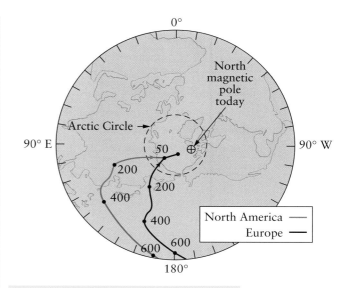

Wandering poles? FIGURE 4.5

This map traces out the *apparent* path of the magnetic North Pole over the last 600 million years. The numbers indicate millions of years before the present. Rocks in North America point to the apparent magnetic poles shown on the red curve, and rocks in Europe point to the apparent poles shown on the blue curve. The *actual* magnetic pole probably never wandered far from the geographic North Pole (center). Its apparent motion is created by the motion of the continents, which carried magnetically oriented rocks along with them.

geologists found that the path of apparent polar wandering measured from North American rocks differed from that of European rocks. Earth certainly cannot have two different magnetic North Poles at the same time! The only solution is that the magnetic pole is fixed and the continents themselves moved, carrying the rocks with them. In that case, the apparent motion of the poles would be an illusion, like the apparent motion of trees when you drive by them. In fact, the apparent polar wandering path of a continent, determined from rocks of various ages, provides a historical record of the position of that continent relative to the magnetic poles.

Notice also that the apparent wandering paths for Europe and North America actually look quite similar from 600 million years ago to 200 million years ago. In fact, if you rotated Europe and America so that they were side by side, the two paths would overlap each other exactly. This indicates, once again, that Europe and America were moving as a single continent during that time.

The missing clue: seafloor spreading

By the early 1960s, many clues had been amassed in support of continental drift. The theory made testable predictions, such as matching geology and fossils, and those predictions had been confirmed. The theory was also consistent with evidence that was not known during Wegener's lifetime—the polar wandering paths. Even so, many scientists were still skeptical. So far, all the evidence was circumstantial. The ripping apart and relocation of an entire continent should have left signs that would still be visible today.

Then, three decades after Wegener's death, geologists found the evidence they needed—at the bottom of the sea. The missing clue turned up when oceanographers applied the then-new technique of magnetic polarity dating to rocks at the bottom of the Atlantic Ocean. They were astounded to find that parts of the seafloor consist of magnetized rocks with alternating bands of normal and reversed polarities. The bands are hundreds of kilometers long. More importantly, they are symmetrical on either side of a centerline that coincides with the ridge running down the middle of the Atlantic Ocean. In other words, if you could fold the seafloor in half along the midocean ridge, the bands on either side would match. Later evidence would show that not only the paleomagnetic polarities but also the ages of the rocks were symmetrically banded, and mirrored on either side of the ridge (**FIGURE 4.6**), with the youngest rocks closest to the ridge itself.

The symmetrical pattern of magnetic bands provided powerful support for a hypothesis, first proposed in 1960, called **seafloor spreading**. According to this hypothesis, the midocean ridge is a place where the seafloor has split apart and the rocks are moving away from one another (see **FIGURE 4.7**). When magma from below wells up into the crack, it solidifies into new volcanic rock on the seafloor. Over time, the expanding seafloor operates as a conveyer belt, carrying the newly magnetized bands of rock away from the ridge in either direction.

The emergence of hot volcanic rocks and resultant heating of ocean water along midocean ridges can result in extreme environments of high temperatures, very low light levels, and unusual chemistry. It was discovered fairly recently that some exotic life forms

> **seafloor spreading** The processes through which the seafloor splits and moves apart along a midocean ridge, and new oceanic crust forms along the ridge.

Banded rocks on the seafloor FIGURE 4.6

This map shows the ages of magnetically banded rocks on either side of the Mid-Atlantic Ridge. The numbers indicate the ages of the rocks in millions of years. The youngest bands lie along the midocean ridge, and the oldest are far away from the ridge, indicating that the seafloor has been spreading over time.

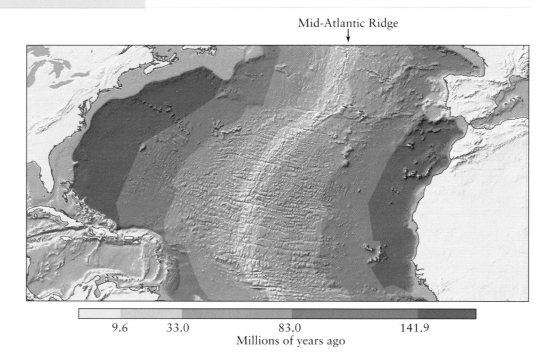

Mid-Atlantic Ridge

9.6   33.0   83.0   141.9
Millions of years ago

Seafloor spreading FIGURE 4.7

Lava extruding along a midocean ridge forms new oceanic crust. As the lava cools, it becomes magnetized with the polarity of Earth's magnetic field at the time. As the plates on either side of the midocean ridge move apart from one another, successive bands of oceanic crust have alternating normal and reversed polarities. The resulting magnetic bands are symmetrical on either side of the midocean ridge.

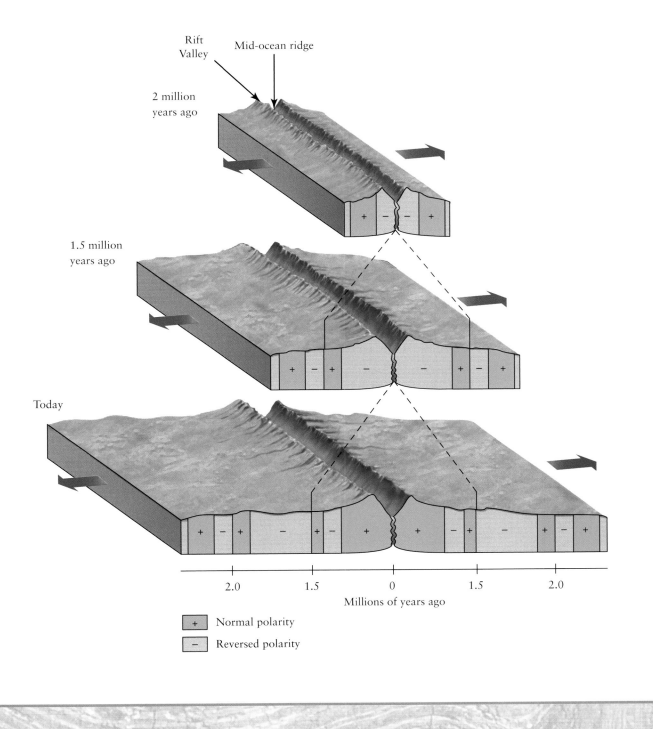

Rift Valley

Mid-ocean ridge

2 million years ago

1.5 million years ago

Today

2.0   1.5   0   1.5   2.0

Millions of years ago

+ Normal polarity

− Reversed polarity

The extreme environment of the midocean ridge
FIGURE 4.8

The unique geologic conditions at a midocean ridge create one of the most bizarre ecosystems on Earth. West of Vancouver Island, Canada, the tiny Juan de Fuca Plate is separating from the Pacific Plate. Two kilometers beneath the ocean surface, the rift releases hot gases such as hydrogen sulfide and methane through the "black smokers" seen in the background. These chimneys are made of minerals that precipitate from the 400°C water as it contacts cooler ocean water. Pogonophoran tubeworms, up to 2.5 meters in length (seen in the foreground), live symbiotically with bacteria that metabolize the gases belching out of the black smokers. The worms must thrive in this environment; they have a life span of 250 years.

survive and even thrive in these harsh environments (see FIGURE 4.8).

The decisive piece of evidence for seafloor spreading is that the ages of seafloor rocks increase with distance from the ridge. The youngest rocks are found along the center of the ridge, where new molten material wells up (as shown in Figure 4.6). This final piece of evidence convinced the great majority of geologists that seafloor spreading is real. The continents are not trundling along on top of a static ocean floor, which would be physically impossible. Instead, they are conveyed in opposite directions by adynamic ocean floor that is constantly spreading and replenishing itself. Ironically, it was the geophysicists—the group that had most vigorously opposed Wegener's ideas—who found the final piece of evidence that proved Wegener correct. The many investigations seeking to test the continental drift hypothesis provide a good example of the scientific method at work.

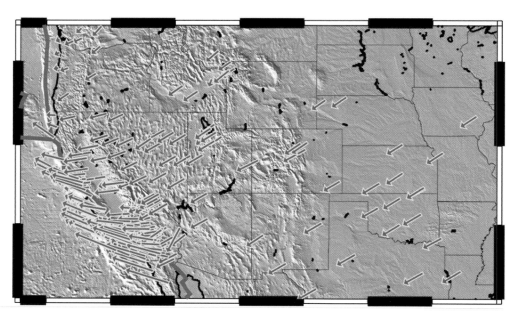

Measuring plate motion using GPS FIGURE 4.9

The arrows show surface motions from a continuous series of GPS measurements. Note that points on the same plate are generally moving in the same direction, confirming that the plates are rigid. In this figure, GPS stations on the North American plate are moving slowly southwest, while those on the Pacific plate are moving more rapidly northwest. The speeds are too slow for humans to be aware of (1 to 10 cm per year) but are easily detectable by GPS sensors.

Recently, an even more direct piece of evidence for continental plate motion became available. In the 1980s, the U.S. Department of Defense declassified the data from a network of satellites that form the Global Positioning System (GPS). GPS receivers help motorists, hikers, boaters, and pilots find their exact location quickly and easily, and they now can be used to track continental drift. Geologists have attached GPS receivers permanently to the ground, so that they do not move unless the ground moves. This allows them to plot maps of Earth's surface velocities (see FIGURE 4.9). The GPS data confirm that the continents have been moving—at least for the last 15 years! Of course, the other evidence presented in this chapter shows that they have been moving for at least 600 million years.

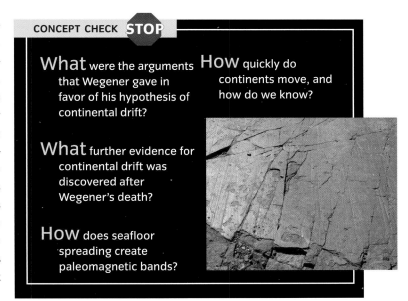

CONCEPT CHECK STOP

What were the arguments that Wegener gave in favor of his hypothesis of continental drift?

What further evidence for continental drift was discovered after Wegener's death?

How does seafloor spreading create paleomagnetic bands?

How quickly do continents move, and how do we know?

The Plate Tectonic Model

Once the realities of continental drift and seafloor spreading were established, researchers began seeking a mechanism that caused the motions. A lot of previously puzzling phenomena in geology suddenly began to make sense. For instance, geologists could now explain the location of mountain ranges and deep ocean trenches. They could explain why, in some places (such as the Tibetan Plateau), Earth's crust seems to be squeezed together, while in other places (such as the East African Rift Valleys) it seems to be pulling apart. They could also understand the distribution of earthquakes and volcanic activity around the planet, which is far from uniform (see FIGURE 4.10 on pages 102–103). Other chapters in this book discuss each of these developments. The unifying theory that emerged from all the research is called **plate tectonics**. Note that we call it a *theory* rather than a *hypothesis*. A hypothesis is a tentative explanation. Before it can become a theory, it must be supported by extensive experimentation and observation. A theory is an explanatory model that is supported—in this case, quite strongly—by a lot of scientific evidence.

plate tectonics
The movement and interactions of large fragments of Earth's lithosphere, called plates.

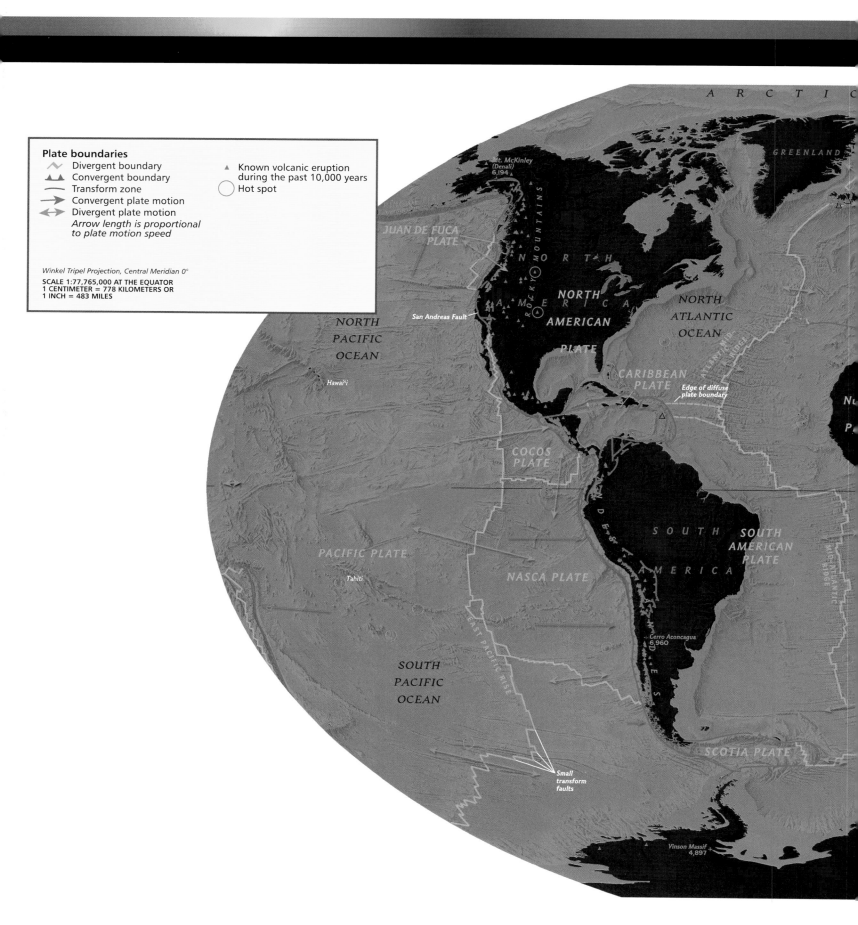

Plate boundaries

~~ Divergent boundary
▲▲ Convergent boundary
— Transform zone
→ Convergent plate motion
↔ Divergent plate motion
*Arrow length is proportional
to plate motion speed*

▲ Known volcanic eruption
   during the past 10,000 years
◯ Hot spot

Winkel Tripel Projection, Central Meridian 0°

SCALE 1:77,765,000 AT THE EQUATOR
1 CENTIMETER = 778 KILOMETERS OR
1 INCH = 483 MILES

ARCTIC

GREENLAND

Mt. McKinley
(Denali)
6,194

PRINCE

JUAN DE FUCA
PLATE

ROCKY MOUNTAINS

N O R T H

NORTH

NORTH
ATLANTIC
OCEAN

A M E R I C A

NORTH
AMERICAN
PLATE

ATLANTIC MID-

San Andreas Fault

NORTH
PACIFIC
OCEAN

Hawai'i

CARIBBEAN
PLATE

Edge of diffuse
plate boundary

Nu

P

COCOS
PLATE

A N D

S O U T H

SOUTH
AMERICAN
PLATE

PACIFIC PLATE

Tahiti

NASCA PLATE

E S

A M E R I C A

MID-ATLANTIC RIDGE

EAST PACIFIC RISE

Cerro Aconcagua
6,960

SOUTH
PACIFIC
OCEAN

Small
transform
faults

SCOTIA PLATE

Vinson Massif
4,897

This map illustrate Earth's major and minor plates and their relationships to earthquakes, volcanic activity, mountain ranges, oceanic trenches, and other geologic features.

PLATE TECTONICS IN A NUTSHELL

Earth's *lithosphere*, or rocky outer layer, is very thin relative to Earth as a whole. The solid rock that makes up the lithosphere is strong, but it lies on top of a vast mantle of hotter, weaker material that is constantly in motion (albeit very slow motion). The layer directly below the lithosphere, called the *asthenosphere*, is especially weak because it is close to the temperature at which melting begins. (The lithosphere and asthenosphere were introduced in Chapter 1 and will be discussed in much greater detail in Chapter 5.) The relationship between these two layers is a condition called *isostasy*, which means that the lithosphere is essentially "floating" onthe asthenosphere, like a sheet of glass floating on water.

If you place a very thin, cool, hard shell on top of hot, ductile material that is moving around, what would you expect to happen? It's almost a certainty that the shell will crack, and that is exactly what has happened to Earth's lithosphere. It has broken into a set of enormous rocky fragments we call **plates**. Today there are six large plates, extending for thousands of kilometers, and a number of smaller ones (see Figure 4.10).

Driven in part by the underlying motion of the asthenosphere, these plates collide, split apart, and slide past one another. This generates a lot of geologic activity, such as earthquakes and mountain building. Much of this occurs along the boundaries at which plates interact with one another, typically marked by huge fractures, or **faults**.

Types of plate margins

Interactions between lithospheric plates occur mainly along their edges. There are three fundamentally different ways in which they can interact. They can move away from each other (*diverge*); they can move toward each other (*converge*); or they can slide past each other along a long fracture. We will take a look at examples of each of these types of margins.

Divergent margins (see FIGURE 4.11A), also called *rifting* or *spreading centers*, occur where two plates

---

■ **fault** A fracture in Earth's crust along which movement has occurred.

■ **divergent margin** A boundary along which two plates move apart from one another.

■ **convergent margin** A boundary along which two plates come together.

■ **subduction zone** A boundary along which one lithospheric plate descends into the mantle beneath another plate.

■ **transform fault** A fracture in the lithosphere where two plates slide past each other.

---

are moving apart. They can occur either in continental or oceanic crust. In East Africa, for example, the African plate is being stretched and torn apart, creating long *rift valleys*. Eventually a new ocean may form in the widening rift; a modern example of this is the Red Sea. Where oceanic crust is splitting apart, the result is a *midocean ridge*.

Convergent margins occur where two plates move toward each other. This leads to different types of margins, depending on whether the boundary is between two oceanic plates, two continental plates, or one of each.

When one continent meets another continent along a convergent margin, they crumple upwards and downwards as the lithosphere thickens, in a *collision zone*. The Himalaya (see FIGURE 4.11B), Earth's highest mountain range, was thrust up in just this way by an ongoing violent collision over the last 50 million years between the Australian-Indian plate and the Eurasian plate.

Another kind of convergent margin occurs when one or both plates are oceanic. In this case, one plate will typically slide beneath the other plate, plunging into the asthenosphere, where water released from the wet rocks of the seafloor promotes formation of magma. **Subduction zones** are marked by very deep *oceanic trenches*— the deepest points in the ocean—and, on the surface, by lines of volcanoes formed as a result of melting in the mantle, generated by water released from the subducting plate (see FIGURE 4.11C).

Transform faults occur where two plates slide past each other, grinding and abrading their edges as they do so. The San Andreas Fault in California, shown in FIGURE 4.11D, is a well-known example: it separates the northwesterly moving Pacific plate from the adjacent North American plate, which is moving toward the southwest.

Transform faults are much more common in the ocean floor, where hundreds of them run perpendicular to the midocean ridges (see Figure 4.10). This causes the plates to have very complicated, jagged boundaries in which spreading centers alternate with transform faults.

Types of plate margins FIGURE 4.11

www.wiley.com/
college/murck

A Divergent margin

A The only place in the world where a mid-ocean ridge rises to the surface is in Iceland. The rift valley seen here lies directly atop the Mid-Atlantic Ridge, a divergent margin where the North American and Eurasian plates are gradually moving apart. The valley is getting wider at a rate of about 2 cm per year, and part of it has filled in with water, creating Lake Thingvallavarn (at right). The rift itself extends in width off the right side of the photo.

Rift valley

B Convergent margin

Collision zone

D The constant grinding of the Pacific plate and the North American plate have made the area near the San Andreas Fault, a transform fault margin, highly unstable. There have been several notable earthquakes related to this fault, including the 1906 earthquake that destroyed much of San Francisco. (For more on this earthquake and others, see Chapter 5.)

▼

▲ **B** In this photograph, the snow-covered Himalaya is on the right, the lower-lying Indian peninsula is off to the left, and the convergent margin lies roughly on the border between them.

▼ **C** The string of volcanoes marching down the spine of the island of Java—and continuing in the islands to the east—is typical of an ocean–ocean convergent margin. Some of the more famous volcanoes are labeled.

C Convergent margin

Subduction zone

Transform fault margin

Subduction zone

Krakatau

Galunggung   Merapi

Semeru

Agung   Tambora

The Plate Tectonic Model   105

There is no solid evidence to suggest that Earth has either grown or shrunk significantly since its formation. This means that the total area of Earth's crust stays constant. For every square kilometer of crust that is created at a midocean ridge, therefore, somewhere else a square kilometer is consumed by subduction at a convergent margin. Overall, there is a balance between creation and destruction in the tectonic cycle. Convergent and divergent plate margins are summarized in FIGURE 4.12.

Plate margins: A summary FIGURE 4.12

Several different types of plate margins are illustrated schematically in this diagram. In reality, these different types of margins would not be so close to one another. At middle left and middle right are divergent margins, which may occur in continental or oceanic crust. At far left, center, and far right are three types of convergent margins. Continent–continent convergent margins thicken the lithosphere and cause high mountain chains to form.

At ocean–continent and ocean–ocean margins, one plate subducts underneath the other. The subducting oceanic plate, which typically carries ocean floor sediments, has a relatively low melting temperature because of its high water content (Chapter 6). Therefore it will begin to melt at a relatively shallow depth. The melt rises to the surface, creating a line of volcanoes called a *volcanic arc* along the edge of the overriding plate.

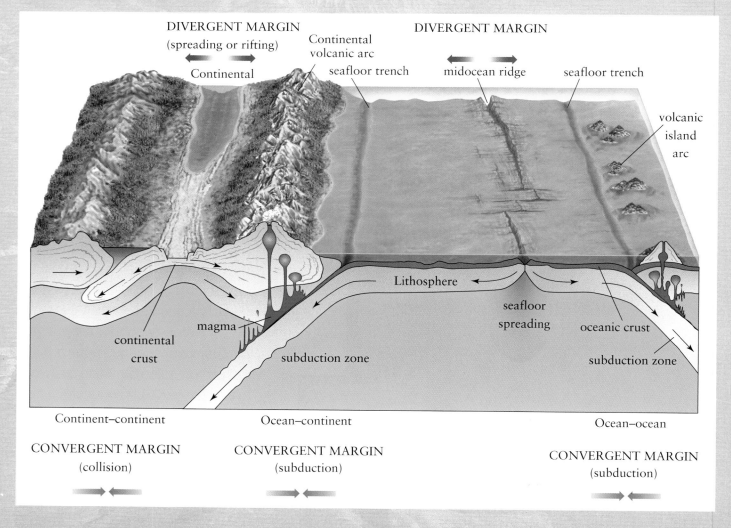

Earthquakes and plate margins Earthquakes and volcanic eruptions, which occur primarily along plate margins, are the most obvious manifestation of active plate interactions (see Figure 4.10).

Earthquakes occur along faults, where huge blocks of rock are grinding past each other (this topic is discussed in greater detail in Chapter 5). Tectonic motions produce directional pressure, which causes rocks on either side of a large fracture to move past each other. The movement is rarely smooth; usually the blocks stick because of friction, which slows their movement. Eventually, the friction is overcome and the blocks slip abruptly, releasing pent-up energy with a huge "snap"—an **earthquake**.

The actual location beneath the surface where the earthquake begins is called the **focus**. This should not be confused with the better-known **epicenter**, which is the point on Earth's surface that lies directly over the focus. The depths of foci provide useful information about the characteristics of the plate margin (see FIGURE 4.13).

Earthquake depths FIGURE 4.13

Earthquakes occur at depths from a few kilometers to several hundred kilometers below the ground surface. The depth of the quake tells geologists a lot about the characteristics of the plate margin.

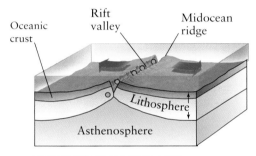

DIVERGENT BOUNDARY

At divergent margins, earthquakes tend to be fairly weak and shallow. Earthquakes can only occur in rock that is cold and brittle enough to break; at a midocean ridge, this means they cannot be very deep.

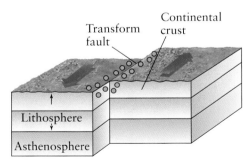

TRANSFORM FAULT BOUNDARY

Transform fault margins have shallow earthquakes, but they can be very powerful.

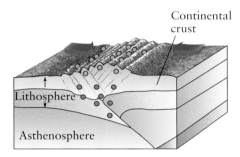

CONTINENTAL COLLISION BOUNDARY

In collision zones the earthquakes can be deep and also very powerful.

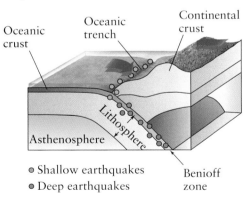

- Shallow earthquakes
- Deep earthquakes

SUBDUCTION ZONE BOUNDARY

The deepest and most powerful quakes occur in subduction zones. Here, an oceanic plate moves downward relative to a continental plate. The earthquake foci are shallow near the oceanic trench, but become deeper along the descending edge of the subducting plate. These zones of shallow- and deep-focus earthquakes, called *Benioff zones*, first alerted scientists to the phenomenon of subduction.

THE SEARCH FOR A MECHANISM

Although virtually all geologists accept the basic theory of plate tectonics, some questions remain. What, exactly, drives plate motion? How does the mantle interact with the crust? What initiates subduction? Scientists have a basic understanding of these processes, but the details have not been completely worked out. Thermal motion in the mantle is at least partially responsible for the motion of plates. This thermal motion, in turn, results from the release of heat from Earth's interior. Let's take a closer look at some of the complexities of Earth's heat-releasing processes.

Earth's internal heat Earth gives off heat for two main reasons. First, it is slowly cooling off from its initial formation processes, including formation of a molten iron core. Second, heat is constantly being generated by the decay of radioactive elements in the interior, primarily uranium, potassium, and thorium. If Earth did not release heat into outer space, the entire interior would eventually melt.

Some of Earth's heat is released through **conduction**. This is a gentle and slow process similar to what you feel when you hold a cup of hot coffee in your hands. The heat moves through the wall of the cup by conduction, a gradual transfer of energy from atom to atom.

When you boil water in a pot on a stove, you will see the water churning around in big circles called *convection cells* (see **FIGURE 4.14A**). A mass of hot water at the bottom is slightly less dense than the cooler water at the top, and hence it will rise. When it reaches the surface, it sheds its heat, moves sideways as it cools, and then sinks back down to the bottom, where it is reheated. This mechanism of heat transfer, is more efficient than conduction. The convection cells act like couriers, carrying

---

Mantle convection FIGURE 4.14

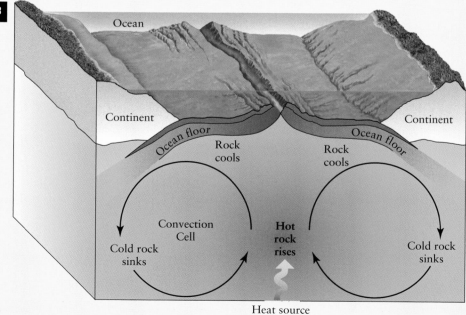

A

Water cools

Cold water sinks | Hot water rises | Cold water sinks

Convection Cell

Convection inside Earth drives the lithospheric plates on the surface.

◀ **A** An everyday example of convection can be seen when you boil a pot of water. The water closest to the burners is hotter than the rest of the water. As it heats up it becomes less dense and rises to the top. At the surface it cools down and moves sideways to make room for the hot water rising beneath it. As the water at the surface cools, it becomes denser and sinks.

B The same process happens in ▶ Earth's mantle on a much grander scale and over a much longer time. Hot rock rises slowly and plastically from deep inside Earth, then cools, flows sideways, and sinks. The relation between convection cells and lithospheric plates is far more complex than what we see, for example, in a pot of boiling water.

B

Ocean

Continent | Ocean floor | Rock cools | Rock cools | Ocean floor | Continent

Convection Cell

Cold rock sinks | Hot rock rises | Cold rock sinks

Heat source within the Earth

The tectonic cycle FIGURE 4.15

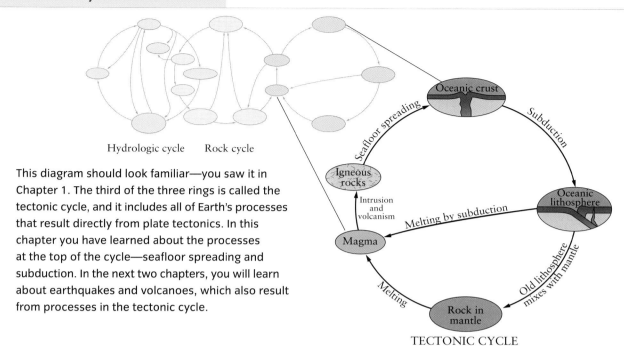

Hydrologic cycle   Rock cycle

This diagram should look familiar—you saw it in Chapter 1. The third of the three rings is called the tectonic cycle, and it includes all of Earth's processes that result directly from plate tectonics. In this chapter you have learned about the processes at the top of the cycle—seafloor spreading and subduction. In the next two chapters, you will learn about earthquakes and volcanoes, which also result from processes in the tectonic cycle.

Oceanic crust

Seafloor spreading

Subduction

Igneous rocks

Intrusion and volcanism

Oceanic lithosphere

Melting by subduction

Magma

Old lithosphere mixes with mantle

Melting

Rock in mantle

TECTONIC CYCLE

heat directly from the burners to the top of the pan instead of passing the heat along from atom to atom.

Convection as a driving force Even though Earth's mantle is composed mostly of solid rock, it too releases heat through **convection**. Solid rock, if it is hot enough, can behave as a flowing viscous fluid, just as the solid ice in a glacier flows. Rock deep in the mantle heats up and expands, becoming buoyant. Very, very slowly it moves upward, in huge convection cells (see FIGURE 4.14B). Just like the water in the boiling pot, the hot rock moves laterally near the surface as it sheds its heat.

> **convection** A form of heat transfer in which hot material circulates from hotter to colder regions, loses its heat, and then repeats the cycle.

This lateral movement of rock in the asthenosphere is believed to be one of the causes of the motion of lithospheric plates.

Convection in the mantle is complex and incompletely understood. Many challenging questions remain unanswered. For example, does the whole mantle convect as a unit, or does the top part convect separately from the bottom, creating rolls upon rolls? In subduction zones, are lithospheric plates pushed down or pulled down, or do they sink into the asthenosphere under their own weight? (Plates with subduction edges move faster than plates without subduction edges, which suggests that the sinking plate edge is pulling the rest of the plate.) How do midocean ridges start, and what role do they play in plate motion? There appears to be a force that pushes plates apart, away from the midocean ridge. What are the shapes and distribution of convection cells? We know, for example, that hot rocks sometimes do not travel in neatly packaged cells, and that they may rise in long, thin blobs called *plumes* (see Amazing Places on pages 110–111). Do the plumes originate in the middle of the mantle, or perhaps even deeper, where the core meets the mantle? Geologists continue to actively research questions such as these.

The tectonic cycle Recall that in Chapter 1 we introduced the concept of interacting cycles in the Earth system (see Figure 1.8). In this chapter we have introduced you to the **tectonic cycle**, on the right-hand side of the diagram in FIGURE 4.15. New material is

tectonic cycle

Movements and interactions in the lithosphere, by which rocks are cycled from the mantle to the crust and back; includes earthquakes, volcanism, and plate motion, driven by convection in the mantle.

constantly added to oceanic crust by volcanism along divergent plate boundaries, while subduction consumes an equal amount of oceanic crustal material at convergent margins. Because of these processes, the seafloor renews itself on the order of every 200 million years.

Continental crust lasts much longer than oceanic crust. Because of its lower density (greater buoyancy), it cannot easily be subducted. The recycling of continental crust involves both plate tectonics and erosion. We will therefore return to this subject in Chapter 7, when we discuss the middle "ring" in Figure 4.15, which includes all processes having to do with erosion.

Plate tectonics affects all life on Earth, sometimes in subtle ways. It influences climate through the distribution of continents and ocean basins, for example. These changes are slow but profound; they can lead to ice ages or periods of unusual warmth, both of which strongly affect the evolution of species. Other effects are more obvious, such as major earthquakes or volcanic eruptions.

In the beginning of this chapter, we described plate tectonics as a "unifying" theory of how the Earth system works. This is because the plate-tectonic model brings together many diverse observations of Earth's geologic features and unifies them into a single, reasonably straightforward "story." As a model, plate tectonics is truly representative of the Earth system science approach, because it illustrates how internal Earth processes are integrated with all other parts of the Earth system.

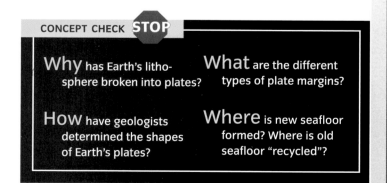

CONCEPT CHECK STOP

Why has Earth's lithosphere broken into plates?

What are the different types of plate margins?

How have geologists determined the shapes of Earth's plates?

Where is new seafloor formed? Where is old seafloor "recycled"?

Amazing Places: The Hawaiian Islands

A growing island chain

One of the most striking geologic features of the Hawaiian Islands, first noted by geologist and explorer James Dwight Dana on an expedition to the islands in 1840–1841, is the increasing ages of the islands from southeast to northwest. Lovely Hanalei Bay (FIGURE A), on the north shore of Kauai, the northernmost of the major islands, shows highly eroded volcanic formations. By contrast, on the south end of the big island of Hawaii (FIGURE B), an ongoing eruption of the Kilauea volcano has added new land to the island almost ceaselessly from 1983 to the present day. Just 15 kilometers south of Hawaii, the next Hawaiian island is already being born. Loihi seamount (FIGURE C), an underwater volcano that appears as a raised feature in this topographical map of the seafloor, already has a peak that is only 1 kilometer below the ocean surface (shown in red), and it is still growing. Loihi erupted as recently as 1996.

The youngest Hawaiian Island FIGURE B

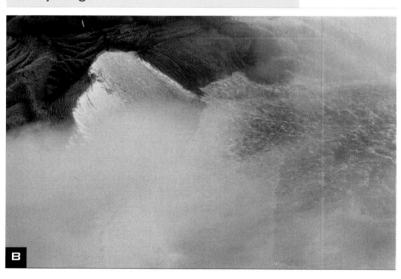

B

Even though Hawaii is thousands of kilometers from the nearest plate boundary, the processes that have formed the islands, as well as their differing ages, have a great deal to do with plate tectonics. The big island of Hawaii is currently sitting above a long, thin plume of hot material, called a *hot spot*, rising from deep in the mantle. The plume itself is stationary, but the lithospheric plate above it is moving to the northwest. The plume provides magma for the currently active volcanoes of Mauna Loa and Kilauea, and the undersea volcano of Loihi. The older islands, such as Kauai, once lay above the hot spot but moved off it as the Pacific Plate moved northwest. With no new volcanic additions to their land mass, they will eventually erode into the sea.

The oldest Hawaiian island FIGURE A

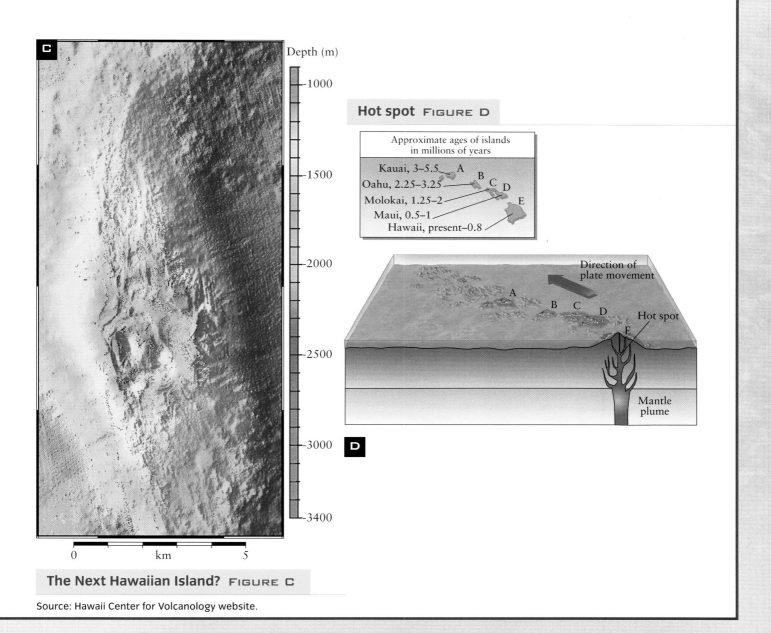

Depth (m)

-1000

-1500

-2000

-2500

-3000

-3400

0   km   5

The Next Hawaiian Island? FIGURE C

Source: Hawaii Center for Volcanology website.

Hot spot FIGURE D

Approximate ages of islands
in millions of years

Kauai, 3–5.5 ——— A
Oahu, 2.25–3.25 ——— B C D
Molokai, 1.25–2 ———   E
Maui, 0.5–1 ———
Hawaii, present–0.8 ———

Direction of
plate movement

A

B C D

E   Hot spot

Mantle
plume

D

CHAPTER SUMMARY

1 A Revolution in Geology

1. A revolution in geology began almost 100 years ago with the hypothesis that the continents have not always been in their present positions but moved into their present-day positions after the breakup of a "supercontinent" known as *Pangaea*. This became known as the **continental drift** hypothesis. At first it was quite controversial because scientists could not envision a satisfactory physical mechanism that could move the continents through solid rock.

2. The evidence for continental drift included the fit between the coastlines of continents. The evidence is stronger when the fit is made along the "true" edge of the continent—the **continental shelf**. In addition, geologic features such as mountain ranges, ancient glacial deposits, and rock types match very closely on both sides of the Atlantic Ocean, which now separates these continents. Matching fossils can be found on both sides of the ocean. Many are from plants and animals that could not have crossed a wide body of water and that would have thrived at different latitudes than the present location of the fossils. In the 1950s, *apparent polar wandering* was added to the list of evidence in favor of the theory of continental drift.

3. The most significant piece of evidence in support of continental drift was the discovery of bands of magnetized rock on the seafloor with alternating normal and reversed polarities, aligned symmetrically on either side of the *midocean ridges*. The only plausible explanation for this discovery was **seafloor spreading**. It was quickly realized that continents and adjacent seafloor move together. This evidence was decisive because it finally led to a mechanism that could account for the continents' movement. At midocean ridges, the ocean floor is constantly replenished, pushing older rocks apart. At deep sea trenches old ocean floor sinks back into the mantle.

4. Global positioning satellites (GPS) can now measure the speed and direction of continental drift directly and precisely, which shows that the continents are still moving today at the rate of several centimeters per year.

2 The Plate Tectonic Model

1. According to **plate tectonics**, the lithosphere has fragmented into several large **plates**. These plates essentially float on an underlying layer of hot, ductile rock called the *asthenosphere*. The asthenosphere is in slow, but constant, motion and this forces the more rigid *lithospheric* plates to move around, collide, split apart, or slide past each other. At present there are six major lithospheric plates and many smaller ones.

2. Plate tectonics predicts three different kinds of interactions along plate margins. At **divergent margins,** two plates move apart. If the boundary occurs in oceanic crust, a divergent margin coincides with a *midocean ridge*. When the divergent boundary occurs in continental crust, it produces long and relatively straight *rift valleys*. Eventually, if the divergence continues for a long enough time, the rift valley will become wide and deep enough to form a sea or an ocean.

3. At **convergent margins,** two plates move toward each other. When an oceanic plate collides with another oceanic plate or with a continental plate, one oceanic plate slides beneath the other plate. This creates a **subduction zone,** which is marked by a deep *oceanic trench* and a great deal of volcanic activity. A second kind of convergent margin occurs where two continental plates collide over an ancient subduction zone forming a *collision zone*. Continental crust is generally too buoyant to subduct, so instead the lithosphere crumples and thickens, building up giant mountain ranges.

4. At **transform fault margins,** which are fractures in the lithosphere, two plates slide past each other. Transform faults are common on the ocean floor, where they run perpendicular to midocean ridges.

5. Many **earthquakes** and active volcanoes are located along plate margins. Studies of the locations and depth of earthquakes enable scientists to determine the shapes of lithospheric plates and the type of margin between the plates. Earthquakes occur as a result of motion of rocks along a **fault**. The **focus**, where the motion starts, is underground; the **epicenter** is the place on the surface that lies directly above the focus.

6. The release of heat from Earth's interior creates huge **convection** cells, which are at least partly responsible for driving plate motion. Convection brings hot rock up from deep in the mantle and recycles cold rock back into the mantle. Many unanswered questions remain about *convection cells*; for example, hot rock also rises in *plumes* that do not seem to be related to the convection cells.

7. Plate tectonics unifies the processes of continental drift, seafloor spreading, mountain building, faulting, earthquakes, and volcanism into the **tectonic cycle**. The tectonic cycle creates and recycles oceanic crust on a time scale of roughly 200 million years. Continental crust lasts much longer, and its recycling involves processes such as weathering and erosion.

- **continental drift** p. 92
- **seafloor spreading** p. 98
- **plate tectonics** p. 101
- **fault** p. 104

- **divergent margin** p. 104
- **convergent margin** p. 104
- **subduction zone** p. 104
- **transform fault margin** p. 104

- **convection** p. 109
- **tectonic cycle** p. 110

CRITICAL AND CREATIVE THINKING QUESTIONS

1. Why was the discovery of paleomagnetic bands on the Atlantic Ocean floor such an important turning point in the acceptance of the theory of plate tectonics? Suppose you could rewrite history so that the satellite measurements of continental drift came first. Would the discovery of seafloor spreading still have been important?

2. What are some of the important questions about plate tectonics that remain unanswered today?

3. We have called plate tectonics a scientific revolution. What other scientific revolutions do you know about? How was plate tectonics similar to, or different from, other revolutions, such as the Copernican revolution in astronomy?

4. Why do geologists call plate tectonics a "unifying theory"?

5. What is the tectonic environment of the place where you live or attend school? Do you live near a plate margin (what type?) or in the middle of a plate (which one?)?

What is happening in this picture ?

A 1971 eruption on Hawaii filled an old crater called Mauna Ulu and created a lava lake. As the lava cooled, geologists were treated to a spectacular "mini" version of plate tectonics, with plates made of recently hardened lava. In these two pictures, can you spot a convergent margin, a divergent margin, and a transform fault that joins two divergent margins?

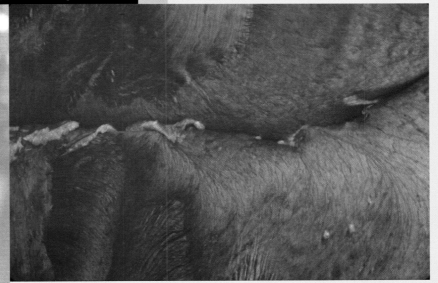

1. The work of geologists over the years has supported Wegener's contention that the current continental masses were assembled into a single super continent, which Wegener called

    _____ .

    a. Pangaea
    b. Transantarctica
    c. Gondwana
    d. Tethys
    e. Laurasia

2. Which of the following lines of evidence, supporting continental drift, was not used by Wegener when he first proposed his hypothesis?

    a. The apparent fit of the continental margins of Africa and South America
    b. Ancient glacial deposits of the southern hemisphere
    c. The apparent polar wandering of the north magnetic pole
    d. Close match of ancient geology between West Africa and Brazil
    e. Close match of ancient fossils on continents separated by ocean basins

3. Analysis of apparent polar wandering paths led geophysicists to conclude

    a. that Earth's magnetic poles have wandered all over the globe in the last several hundred million years.
    b. that the continents had moved because it is known that the magnetic poles themselves are essentially fixed.
    c. that the apparent wandering path of a continent provides a historical record of the position of that continent over time.
    d. Both b and c are correct.

4. _____ is the process through which oceanic crust splits and moves apart along a midocean ridge and new oceanic crust forms.

    a. Continental drift
    b. Paleomagnetism
    c. Seafloor spreading
    d. Continental rifting

5. This illustration is a map showing the age of the seafloor, across the northern extent of the Atlantic Ocean. The Mid-Atlantic Ridge can be seen stretching roughly north-south (in the yellow band) down the middle of the map. Yellow through red colors show rocks of similar age; number them from 1 (oldest) through 5 (youngest).

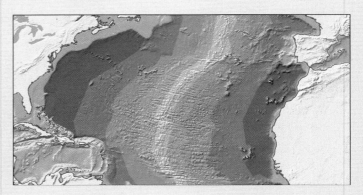

6. _____ technology has allowed scientists to measure the movement of continental crust.

    a. Global positioning systems (GPS)
    b. Seismic recording
    c. Magnetometer
    d. Gravity meter

7. This map shows radiometric ages for the Hawaiian island chain in the middle of the Pacific plate. These islands formed over a hot spot in Earth's mantle. Draw an arrow on the map showing the direction of movement of the Pacific plate over this hot spot as indicated by the ages of the islands in the Hawaiian chain.

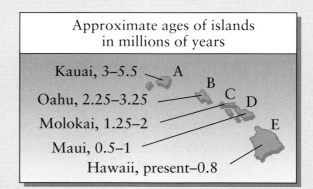

Approximate ages of islands in millions of years

Kauai, 3–5.5   A
   B
Oahu, 2.25–3.25   C   D
Molokai, 1.25–2   E
Maui, 0.5–1
Hawaii, present–0.8

8. At a _____ two lithospheric plates slide past one another horizontally.

  a. divergent boundary
  b. transform fault boundary
  c. subduction zone boundary
  d. continental collision boundary

9. At a _____ oceanic crust is consumed back into the asthenosphere.

  a. divergent boundary
  b. transform fault boundary
  c. subduction zone boundary
  d. continental collision boundary

10. At a _____ new oceanic crust forms along mid-ocean ridges.

  a. divergent boundary
  b. transform fault boundary
  c. subduction zone boundary
  d. continental collision boundary

11. A _____ is a convergent margin along which subduction is no longer active and high mountain ranges are formed.

  a. divergent boundary
  b. transform fault boundary
  c. subduction zone boundary
  d. continental collision boundary

12. _____ is the horizontal movement and mutual interaction of large fragments of Earth's lithosphere.

  a. Continental drift
  b. Polar wandering
  c. Paleomagnetism
  d. Chemosynthesis
  e. Plate tectonics

13. Heat from the solid mantle is released through a process of _____ .

  a. polar wandering.
  b. paleomagnetism.
  c. convection.
  d. magnetic reversal.

14. This illustration shows depicting different types of plate boundaries. For each block diagram label the appropriate plate boundary from the following list:

Divergent Boundary   Continental Collision Boundary
Transform Fault Boundary Subduction Zone Boundary

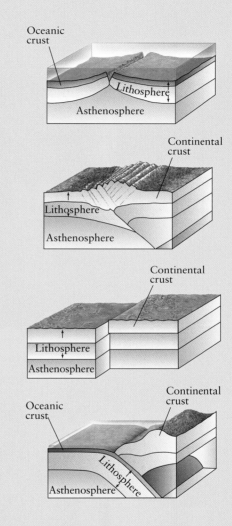

15. On the block diagrams indicate the locations of earthquakes for each type of plate boundary. Use a red dot to show shallow focus earthquakes and a blue dot to show the location of deep focus earthquakes.

Earthquakes and Earth's Interior

On December 26, 2004, the most powerful earthquake in 40 years—and the third-most powerful of the last century—shook the Indian Ocean floor, about 160 kilometers off the island of Sumatra. Earthquakes on the ocean floor are common, but usually they are noticed only by geologists. However, this earthquake—the Sumatra-Andaman earthquake of 2004—will be long remembered, because it caused the largest and deadliest tsunami in history.

The quake began when part of the Indian Plate, which is subducting under the Eurasian Plate, suddenly slipped downward approximately 15 meters. The motion pushed the seafloor up as much as 5 meters on the Eurasian side. Unlike most earthquakes, which are over in seconds, the slippage continued for 10 harrowing minutes as the fault broke, section by section, for 1200 kilometers to the north, toward the Andaman Islands. On the surface of the ocean, killer waves generated by the sudden movement of the ocean floor swept toward Indonesia, Thailand, Sri Lanka, and India. When they reached shore, the waves grew to 20 or 30 meters, and swept far inland, obliterating everything in their path. The photograph here shows a ground-level view of the destruction in Sumatra, and the satellite views (inset) show houses, roads, and bridges in Banda Aceh, on the island of Sumatra, before and after the tsunami.

Although earthquakes and tsunamis are common in the Indian Ocean, there was no warning system in place when this disaster occurred. The resulting devastation caused hundreds of billions of dollars in damages and at least 275,000 deaths.

All earthquakes, including the destructive Sumatra-Andaman earthquake, give geologists a window into Earth's internal structure. We cannot prevent earthquakes, but scientists are getting better at understanding and preparing for their after-effects, including devastating tsunamis. The knowledge we gain from such events can allow us to prepare more effectively for future disasters.

Global Locator

ASIA

Sumatra

BEFORE

AFTER

Earthquakes and Earthquake Hazards

LEARNING OBJECTIVES

Explain the connection between earthquakes and plate tectonics.

Describe the theory of elastic rebound.

Identify several earthquake-related hazards.

Compare short-term prediction and long-term forecasting of earthquakes.

The Sumatra-Andaman earthquake occurred in a subduction zone, where the oceanic lithosphere of the Indian Plate is being subducted beneath the Eurasian Plate (see **FIGURE 5.1**). It was the fifth giant subduction zone earthquake since 1900. The others were in Kamchatka, Russia, in 1952; the Aleutians in 1957; southern Chile in 1960; and Prince William Sound, Alaska, in 1964.

This association of subduction zones and large earthquakes suggests that the motion of the plates is somehow responsible for earthquakes. But plate motion is very gradual, typically on the order of a few centimeters per year. Why, then, should earthquakes be so sudden, and the big ones so catastrophic? The science of **seismology** works to answer this and related questions.

seismology The scientific study of earthquakes and seismic waves.

EARTHQUAKES AND PLATE MOTION

Most earthquakes are caused by the sudden movement of stressed blocks of Earth's crust along a fault. If the rocks could slide past one another smoothly, like the parts of a well-oiled engine, big quakes like the Sumatra-Andaman quake would not happen. In the real world, smooth sliding is rare; friction between the huge blocks of rock causes them to seize up, bringing the motion along that part of the fault to a temporary stop. While the fault remains locked by friction, energy continues to build up as a result of the plate motion, causing rocks adjacent to the jammed section to bend and buckle. Finally, the stress becomes great enough to overcome the friction along the fault. All at once the blocks slip, and the pent-up energy in the rocks is released as the violent

Megathrust earthquakes FIGURE 5.1

Major earthquakes at the boundary between a subducting plate and an overriding plate are sometimes called "megathrust" quakes. The world's five most powerful earthquakes since 1900 (shown on this map) have been of this variety.

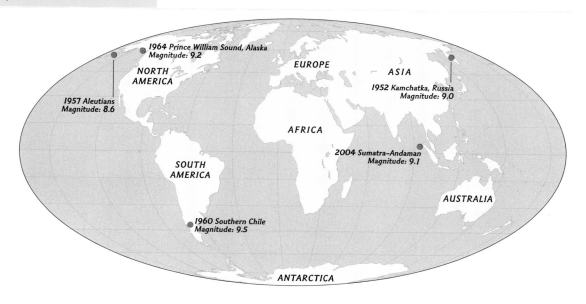

1964 Prince William Sound, Alaska
Magnitude: **9.2**

NORTH AMERICA

EUROPE

ASIA

1952 Kamchatka, Russia
Magnitude: **9.0**

1957 Aleutians
Magnitude: **8.6**

AFRICA

2004 Sumatra–Andaman
Magnitude: **9.1**

SOUTH AMERICA

AUSTRALIA

1960 Southern Chile
Magnitude: **9.5**

ANTARCTICA

Evidence of lateral and vertical fault motion FIGURE 5.2

A When these orange trees were planted on land that lies over the San Andreas Fault in southern California, the rows were straight. In 1938 earthquake motion along the fault displaced the trees significantly. Arrows show the direction of movement of the plates.

B The second most powerful earthquake on record, the Alaska "Good Friday" earthquake, struck the Anchorage area on March 27, 1964. The vertical motion along the fault amounted to several meters in some places. In this photograph, officials examine damage in the Turnagain Heights area.

tremors of an earthquake. This cycle of slow buildup of energy followed by abrupt movement along a fault repeats itself many times, so earthquakes tend to happen again and again on the same fault.

Although movement along a large fault may eventually total many kilometers, this distance is the sum of numerous smaller slips happening over many millennia. In some places these small slips (collectively called *creep*) are frequent, though imperceptible to humans. Nevertheless, they can create visible distortion of surface features, as shown in FIGURE 5.2A.

Earthquakes can also cause vertical dislocations of the ground surface (see FIGURE 5.2B). The largest abrupt vertical displacement on record occurred in 1899 at Yakutat Bay, Alaska, when a long stretch of the Alaskan shore was suddenly lifted 15 meters above sea level during a major earthquake.

The elastic rebound theory
The initial vertical or horizontal motion, dramatic as it may appear, is often not what does the most damage during an earthquake. It is the sustained shaking of the ground that destroys buildings, bridges, and cities, sometimes many kilometers away from the location of the quake.

In 1910, Harry Fielding Reid, a member of a commission appointed to investigate the infamous 1906 San Francisco earthquake that destroyed much of the city, proposed the most widely accepted explanation for the shaking. Reid's **elastic rebound theory** suggests that rocks, like all solids, are *elastic* (within limits). This means that they will stretch or bend when subjected to stress, and snap back when the stress is removed, which happens when the two blocks on either side of a fault manage to overcome friction and slip past one another. But the strained rocks on either side of the fault don't just snap back and stop. Like a guitar string after it is plucked, they continue vibrating. These vibrations are called **seismic waves**. Like sound waves from a guitar string, they can travel a long distance from their place of origin. (We will investigate the scientific concepts of *stress* and *strain* in greater depth in Chapter 9.)

elastic rebound theory The theory that continuing stress along a fault results in a buildup of elastic energy in the rocks, which is abruptly released when an earthquake occurs.

seismic wave An elastic shock wave that travels outward in all directions from an earthquake's source.

The first evidence to support Reid's elastic rebound theory came from studies of the San Andreas Fault, a large, complex fault in California that generated the 1906 quake and the 1989 "World Series" quake in Oakland. Beginning in 1874, scientists from the U.S. Coast and Geodetic Survey had been measuring the precise positions of many points both adjacent to and distant from the fault. As time passed, movement of the points revealed that at some places the two sides of the fault were smoothly slipping in opposite directions. Near San Francisco, however, the fault appeared to be locked by friction and did not reveal any slip. Then, on April 18, 1906, the two sides of this locked section of fault shifted abruptly (see FIGURE 5.3). The elastically stored energy in the rocks was released as the crust snapped to its new position, creating a violent earthquake. Reid's measurements after the quake revealed that the bending, or strain, stored in the crust had disappeared.

Horizontal motion along fault blocks FIGURE 5.3

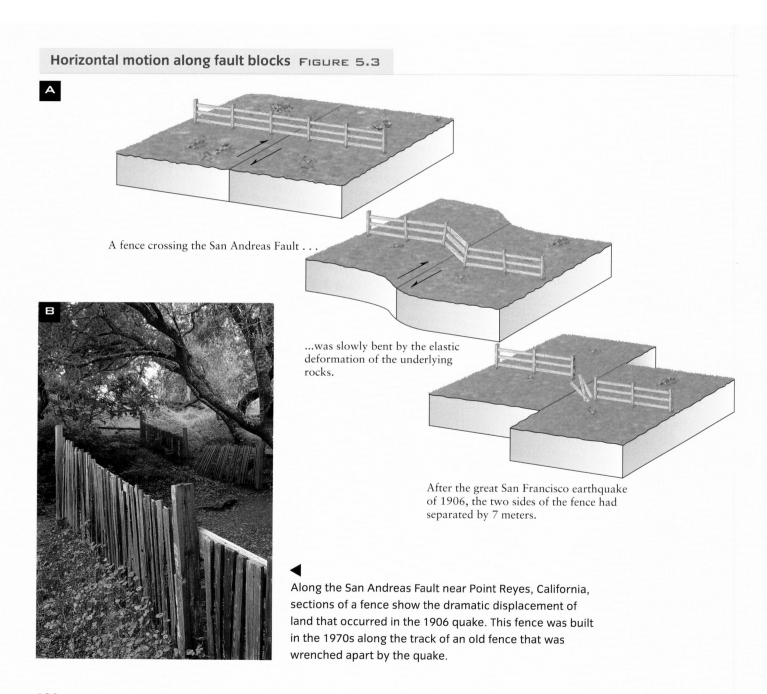

A

A fence crossing the San Andreas Fault . . .

...was slowly bent by the elastic deformation of the underlying rocks.

After the great San Francisco earthquake of 1906, the two sides of the fence had separated by 7 meters.

B

Along the San Andreas Fault near Point Reyes, California, sections of a fence show the dramatic displacement of land that occurred in the 1906 quake. This fence was built in the 1970s along the track of an old fence that was wrenched apart by the quake.

How a tsunami was unleashed FIGURE 5.4

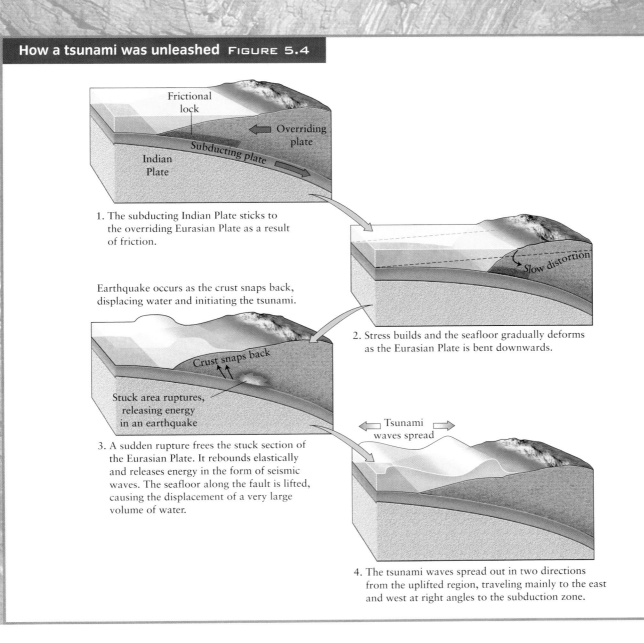

1. The subducting Indian Plate sticks to the overriding Eurasian Plate as a result of friction.

Earthquake occurs as the crust snaps back, displacing water and initiating the tsunami.

3. A sudden rupture frees the stuck section of the Eurasian Plate. It rebounds elastically and releases energy in the form of seismic waves. The seafloor along the fault is lifted, causing the displacement of a very large volume of water.

2. Stress builds and the seafloor gradually deforms as the Eurasian Plate is bent downwards.

4. The tsunami waves spread out in two directions from the uplifted region, traveling mainly to the east and west at right angles to the subduction zone.

The elastic rebound theory also explains why the Sumatra-Andaman earthquake of 2004 lifted the seafloor and caused a giant tsunami (see FIGURE 5.4).

EARTHQUAKE HAZARDS AND PREDICTION

Each year more than a million earthquakes occur around the world. Fortunately, only a few are large enough, or close enough to major population centers, to cause much damage or loss of life. A great deal of research focuses on earthquake prediction and hazard assessment. Geologists are working hard to improve their forecasting ability to the point where effective and accurate early warnings can be issued. Let's look briefly at the hazards associated with earthquakes and at efforts to predict them.

Earthquake hazards   Earthquakes can cause total devastation in a matter of seconds. The most disastrous quake in history occurred in Shaanxi Province, China, in 1556, killing an estimated 830,000 people. The earthquake caused the caves in which most of the population lived to collapse. In all, nineteen earthquakes in history have caused 50,000 or more deaths apiece.

The Sumatra-Andaman Tsunami of 2004

Tsunamis are uncommon compared with other secondary hazards associated with earthquakes, but in the final week of 2004 the world witnessed their devastating potential as never before.

FIGURE A shows the progress of the tsunami from 0 to 3 hours after the Sumatra-Andaman earthquake. In the open sea, tsunamis are barely noticeable; the tsunami wave velocity may be hundreds of kilometers per hour and the wavelength many kilometers, but the peak wave height is typically half a meter or less. Upon reaching the shore, the wave begins to "hit bottom," slows down, and may build up to colossal heights; wave heights of 20 to 30 meters were reported in Sumatra.

The frequency of tsunamis is greatest around the Pacific Ocean, where an extensive circum-Pacific tsunami early warning system has been effective in reducing fatalities. As a result of the tsunami generated by the Sumatra-Andaman earthquake, research is being directed to the installation of early warning systems for the Indian Ocean. Research is also proceeding on ways to reduce the damage caused by monster waves.

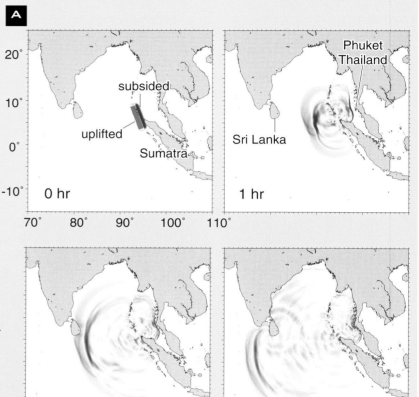

In FIGURE B, a tsunami crashing into a populated coast is simulated in miniature in a research laboratory in Oregon; the experiment is based on a vulnerable area in Puget Sound.

Ground motion, with the resulting collapse of buildings, bridges, and other structures, is usually the most significant *primary hazard* to cause damage during an earthquake. In the most intense quakes, the surface of the ground can be observed moving in waves. Sometimes large cracks and fissures open in the ground. Where a fault breaks the ground surface, buildings can be split, roads disrupted, and anything that lies on or across the fault broken apart. To make matters worse, movement on one part of a fault can cause stress along another part of the fault, which in turn slips, generating another earthquake, called an *aftershock*. Aftershocks triggered by large earthquakes tend to be on the same fault system as the original quake, though they may be quite far from the original location, causing the damage to be spread more widely as time passes. The 1992 Landers earthquake, near Los Angeles, triggered major aftershocks at fourteen locations, some of them hundreds of kilometers away.

Ground motion is not the only source of damage in an earthquake. Sometimes the aftereffects, or *secondary hazards* related to an earthquake, can cause even more damage than the original quake. Examples of secondary hazards that can be initiated by earthquakes include landslides, fires, ground liquefaction, and tsunamis. (See the Case Study and FIGURE 5.5 on the next page.)

Earthquake prediction

Charles Richter, inventor of the Richter scale for quantifying the severity of earthquakes, once said, "Only fools, charlatans, and liars predict earthquakes." Today, unfortunately, this is still more or less correct: No one can predict the exact magnitude and time of occurrence of an earthquake. However, scientists' understanding about seismic mechanisms and the tectonic settings in which earthquakes occur has improved greatly since Richter's time, and advances in modern seismology may yet prove him wrong.

There are two aspects to the problem of earthquake prediction. *Short-term prediction* would identify the precise time, magnitude, and location of an earthquake in advance of the actual event, providing an opportunity for authorities to issue an early warning. *Long-term forecasting* involves the prediction of a large earthquake years or even decades in advance of its occurrence.

Short-term prediction and early warning Unfortunately, the short-term prediction of earthquakes has not been very successful to date. Attempts at short-term prediction are based on observations of anomalous *precursor phenomena*—that is, unusual activity preceding and leading up to the occurrence of an earthquake. For example, the magnetic or electrical properties of the rock could change, the level of well water could drop, or the amount of radon gas in the groundwater could rise in advance of an earthquake, any of which may indicate unusual activity in the underlying rock. Strange animal behavior, glowing auras, and unusual radio waves have also been reported as precursors near the sites of large earthquakes; there are plausible scientific explanations for these. Small cracks and fractures can develop in severely strained rock and cause swarms of tiny earthquakes— *foreshocks*—that may presage a big quake.

The most famous successful earthquake prediction, made by Chinese scientists in 1975, was based on slow tilting of the land surface, fluctuations in the magnetic field, and numerous foreshocks that preceded a large quake that struck the town of Haicheng. Half the city was destroyed, but because authorities had evacuated more than a million people beforehand, only a few hundred were killed.

However, less than two years after the prediction of the Haicheng earthquake, the devastating 1976 T'ang Shan earthquake struck with no apparent precursory activity. With an official death toll of 240,000 (and unofficial reports suggesting many more deaths), it rivaled the Shaanxi earthquake of 1556 as the most disastrous in history. In 1976, and still today, short-term prediction and early warning of earthquakes remain elusive goals for seismologists.

Long-term forecasting Long-term earthquake forecasting is based mainly on our understanding of the tectonic cycle and the geologic settings in which earthquakes occur. In places where earthquakes are known to occur repeatedly, such as along plate boundaries, seismologists have detected patterns in the recurrence intervals of large quakes. Because historical records seldom go back as far as seismologists would like, they also use the information provided by **paleoseismology**.

■ **paleoseismology** The study of prehistoric earthquakes.

Earthquake-related secondary hazards FIGURE 5.5

A ◀ Landslide
Huascaran, Peru

B ◀ Open fissure
Golcuk, Turkey

▼ Fire
San Francisco, California

C

▼ Tsunami, Kalutara Beach, Sri Lanka ▼

D Before

E During

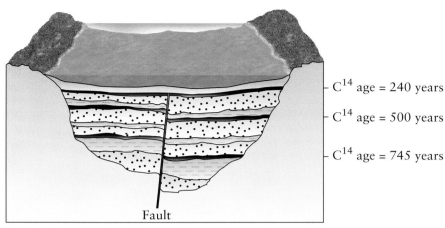

- C^{14} age = 240 years

- C^{14} age = 500 years

- C^{14} age = 745 years

Fault

[:::] Sand layer

[—] Clay layer

[■] Carbon-containing layer

Evidence of ancient quakes

FIGURE 5.6

Layers of sediment were offset by ancient earthquakes. The carbon-rich layers of sediment have yielded carbon-14 dates, which help scientists pinpoint the ages of ancient quakes.

Prehistoric quakes leave evidence in the stratigraphic record, such as vertical displacement of sedimentary layers, indications of liquefaction, or horizontal offset of geologic features (see **FIGURE 5.6**). If the pattern of recurrence suggests regular intervals of, say, a century between major quakes, it may be possible to predict within a decade or two when a large quake is due to happen next in that location.

By studying ancient earthquakes, scientists have identified a number of *seismic gaps* around the Pacific Rim. A seismic gap is a place along a fault where a large earthquake has not occurred for a long time, even though tectonic movement is still active and stress is building. Some geophysicists consider seismic gaps to be the places most likely to experience large earthquakes.

Long-term forecasting has met with reasonable success. Seismologists know where most (but not all) hazardous areas are. They can calculate the probability that a large earthquake will occur in a particular area within a given period. They have a theory of earthquake generation that successfully unites their predictions and observations in the context of plate tectonic theory. Forecasting helps people who live in seismically active areas to plan and prepare, well in advance of a major event. If short-term prediction could advance as much as forecasting has done, many lives could be saved.

DESIGNING FOR EARTHQUAKE SAFETY

It is important to realize that the most powerful quakes are not necessarily the deadliest, and vice versa. The death toll depends to a great extent on the population of the affected region, and how well prepared they are for a major quake. Every earthquake hazard, from fires to tsunamis, can be reduced in severity (though not eliminated) by proper planning and preparedness (see **FIGURE 5.7** on the next page). For example, skyscrapers and bridges can be built with reinforced concrete and large counterweights to help them resist shaking. Some of the tragic loss of life in the tsunami of 2004 was especially preventable because one to two hours elapsed between the time of the earthquake off Sumatra and the arrival of the waves in Sri Lanka and India. Early warning systems would probably not have helped Sumatra or Thailand because they were so close to the tsunami's source, but an early warning system could have given people in more distant locations the opportunity to move to higher ground.

CONCEPT CHECK STOP

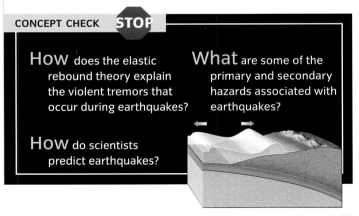

How does the elastic rebound theory explain the violent tremors that occur during earthquakes?

What are some of the primary and secondary hazards associated with earthquakes?

How do scientists predict earthquakes?

A Engineers and scientists are collaborating to find ways to make the built environment safer during earthquakes.

A

UTILITY LINES RUPTURE
Major quakes can cut unprotected electric power lines, water mains, and gas lines, adding fire and flooding to a city's miseries.

HOUSES FALL
Older wood-frame houses are shaken off their foundations, sending chimneys, roofing, and unsecured items toppling onto residents.

MID-RISES COLLAPSE
The most vulnerable structures are often unreinforced masonry mid-rises, whose rigid walls tend to crack and crumble rather than sway.

HIGHWAYS BUCKLE
Roads and bridges give way under relentless shaking and swaying. Even where decks don't collapse, damage can slow rescue efforts.

6 BRACED STEEL SKELETONS

1 BOLTED FOUNDATION

3 ISOLATED FOUNDATION

2 FLEXIBLE TUNNEL JOINTS

Building for Protection

Earthquake-resistant methods and materials have consistently improved since serious study of earthquakes began in the early 1900s. In many of the world's at-risk urban centers, high-rises and homes now regularly weather seismic events and stand firm as the ground rolls around them. In cities that enforce strict construction standards, new structures of all kinds—bridges, tunnels, stadiums—are designed from the start to withstand at least some shaking. But experts still worry about those built decades ago, before quake resistance became common. For such structures, engineers have developed a host of innovations that help hold things together when quakes strike.

1 HOUSES
Several retrofits can help homeowners outlast an earthquake. Wood-frame houses should be bolted to their foundations; in masonry homes, walls should be reinforced with fiber mesh and tied to each other and the roof. Water heaters and other appliances can be strapped or bolted down to prevent tipping, and cabinet doors can be latched shut to keep their contents from spilling out.

2 SUBWAY LINES
Tunnels are most vulnerable where they run from one ground material into another, say from rock to soil.

Flexible joints let tunnels bend with quake motion and resist breaking.

3 MID-RISES
Engineers can reinforce the masonry walls of mid-rises with sprayed concrete. They can also isolate the structure from its foundation, hoist it onto steel and rubber pads, and insert dampers to absorb shock and steady the building.

4 UTILITY LINES
Plunging underground like concrete veins, common utility ducts (CUDs) carry lifelines like water pipes and electricity cables. CUDs move with the ground during a quake, reducing damage, and allow for quicker repairs.

5 HIGHWAYS AND BRIDGES
Driving micropiles—long, pipe-like anchors—through bridge foundations boosts stability. Bridge columns are strengthened by encasing them in steel jackets or fiber mesh. Roads can be preserved by anchoring loose earth to rock on shoulders, reducing slides.

6 HIGH-RISES
High-rises are buttressed with braces and shock absorbers bolted to inner steel skeletons. This allows movement but prevents catastrophic swaying.

NATIONAL GEOGRAPHIC

B Japanese schoolchildren practice what to do in case of an earthquake.

C Apartment buildings in Niigata, Japan, fell over after an earthquake caused liquefaction of the ground.

D A major quake in Kobe, Japan, caused highways and bridges to buckle and collapse.

HIGH-RISES RATTLE
Designed to sway, skyscrapers generally don't collapse but do sustain damage. Windows explode, girder welds crack, fires erupt.

BUILDINGS SINK
When soil is saturated, quakes can turn solid ground into a molasses-like mix that causes buildings to lean or even topple.

5 REINFORCED COLUMNS

4 PROTECTED UTILITY LINES

The Science of Seismology

Seismologists can quickly locate an earthquake anywhere on Earth and tell how strong it is. Seismologists are also very good at telling the difference between earthquakes and other seismic disturbances, such as explosions and landslides.

SEISMOGRAPHS

The earliest known **seismographs** (also called *seismometers*) were invented in China in the second century (see FIGURE 5.8A). The first seismographs in Europe were invented much later, in the nineteenth century. Modern seismographs provide a printed or digital record of seismic waves, called a **seismogram** (see FIGURE 5.8B).

The most advanced seismographs measure the ground's motion optically

> **seismograph** An instrument that detects and measures vibrations of Earth's surface.

> **seismogram** The record made by a seismograph.

Ancient and modern seismographs FIGURE 5.8

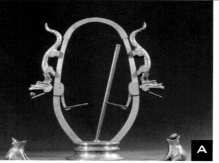

A Ancient Chinese seismograph. ▲

B Seismographs use the principle of *inertia*—the resistance of a heavy mass to motion. In this schematic diagram, seismic waves cause the support post and the roll of paper to vibrate back and forth. However, the large mass attached to the pendulum, and the pen attached to it, barely move at all. It looks to an observer as if the pen is moving, but in reality it is the paper that moves underneath the pen to create the seismogram.
▼

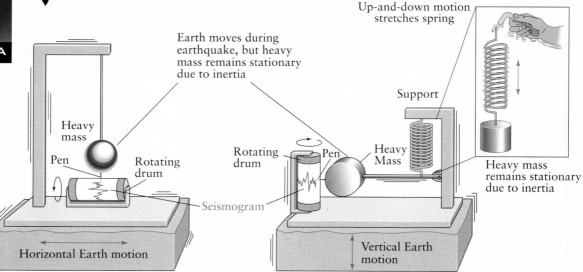

Earth moves during earthquake, but heavy mass remains stationary due to inertia

Heavy mass

Pen

Rotating drum

Seismogram

Horizontal Earth motion

Up-and-down motion stretches spring

Support

Rotating drum

Pen

Heavy Mass

Heavy mass remains stationary due to inertia

Vertical Earth motion

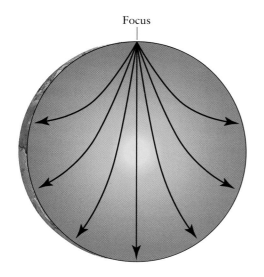

Focus

The energy released during an earthquake travels through Earth from its source (the focus). If Earth were of uniform density throughout, the waves would travel in straight lines. However, rock density increases with depth as a result of increasing pressure. Seismic waves travel faster through denser rocks; hence they travel more quickly at greater depths. This increase in velocity with depth causes seismic wave paths to be curved, rather than straight. (This diagram is not completely accurate, because the increase in rock density and seismic velocity with depth is not smooth; you will see a more detailed diagram later in the chapter.)

and amplify the signal electronically. Vibrations as tiny as one hundred-millionth (10^{-8}) of a centimeter can be detected. Indeed, many instruments are so sensitive that they can sense vibrations caused by a moving automobile many blocks away.

SEISMIC WAVES

body wave A seismic wave that travels through Earth's interior.

surface wave A seismic wave that travels along Earth's surface.

focus The location where rupture commences and an earthquake's energy is first released.

The energy released by an earthquake is transmitted to other parts of Earth in the form of seismic waves. The waves pass through in the form of elastic deformations of the rocks; they leave no record behind them once they have passed, so they must be detected while they pass. The waves, which include both **body waves** and **surface waves**, travel outward in all directions from the earthquake's **focus** (see **Figure 5.9**).

Body waves   Body waves can be further subdivided into two types. **Compressional waves** (see **Figure 5.10A** on the next page) can pass through solids, liquids, and gases. They have the highest velocity of all

seismic waves—typically 6 km/s in the uppermost portion of the crust—and thus they are the first waves to arrive and be detected by a seismograph after an earthquake. For this reason they are called **P waves** (or *primary waves*).

Shear waves, the other type of body wave, travel through materials by generating an undulating motion in the material (see **Figure 5.10B**). Solids tend to resist a shear force and bounce back to their original shape afterward, whereas liquids and gases do not. Without this elastic rebound, there can be no wave. Therefore, shear waves cannot be transmitted through liquids or gases. This has important consequences for the interpretation of seismic waves, as you will soon see. Shear waves do not travel as fast as compressional waves do: "only" 3.5 km/s. Because they arrive at a seismograph after the P waves from the same earthquake, they are called **S waves** (or *secondary waves*).

compressional wave A seismic body wave consisting of alternating pulses of compression and expansion in the direction of wave travel; P wave or primary wave.

shear wave A seismic body wave in which rock is subjected to side-to-side or up-and-down forces, perpendicular to the wave's direction of travel; S wave or secondary wave.

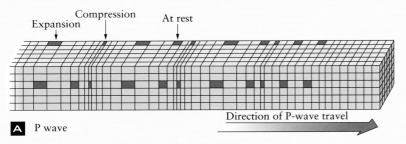

A P wave

A A compressional wave alternately squeezes and stretches the rock as it passes through. The grid is intended to help you visualize how the rock responds. All the divisions in the grid start out square, but the wave alternately squeezes them down to narrow rectangles, then stretches them to long rectangles.

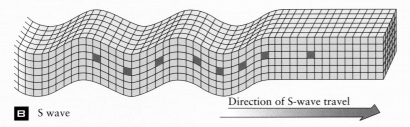

B S wave

B A shear wave causes the rock to vibrate up and down, like a rope whose end is being shaken. In this case the squares do not expand or contract but do get distorted, changing shape alternately from a square to a parallelogram and back to a square again.

Seismic body waves FIGURE 5.10

C P and S waves travel outward from the focus, generating waves that travel along the surface.

Surface waves

Surface waves travel along or near Earth's surface, like waves along the surface of the ocean. They travel more slowly than P and S waves, and they pass around Earth, rather than through it. Thus, surface waves are the last to be detected by a seismograph. FIGURE 5.11A shows a typical seismogram, in which the P waves' arrival is first, followed by the arrival of the S waves, and finally by that of the surface waves. Surface waves are responsible for much ground shaking and structural damage during major earthquakes.

| **epicenter** The point on Earth's surface directly above an earthquake's focus. |

LOCATING EARTHQUAKES

The **epicenter** or surface location of an earthquake can be determined through simple calculations, provided that at least three seismographs have recorded the quake. The first step is to find out how far each seismograph is from the source of the earthquake. The greater the distance traveled by the seismic waves, the more the S waves will lag behind the P waves. Thus, the lag time between the P and S waves on a seismogram (see FIGURE 5.11B) provides geologists the necessary distance information.

After determining the distance from each seismograph to the source of the earthquake, the seismologist draws a circle on a map, with the seismic station at the center of the circle. The radius of the circle is the distance from the seismograph to the focus. It is a circle because the seismologist knows only the distance, not the direction. When this information is calculated and plotted for three or more seismographs, the unique point on the map where the three circles intersect is the location of the epicenter (see Figure 5.11B). This process is called *triangulation*.

A Seismogram of a typical earthquake.

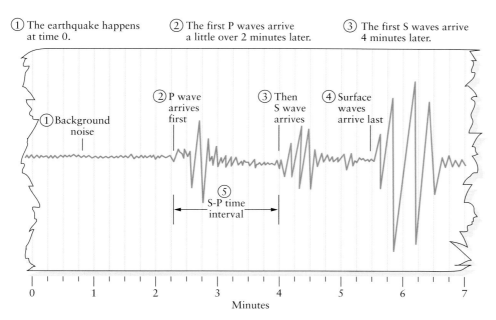

① The earthquake happens at time 0.

② The first P waves arrive a little over 2 minutes later.

③ The first S waves arrive 4 minutes later.

① Background noise

② P wave arrives first

③ Then S wave arrives

④ Surface waves arrive last

⑤ S-P time interval

Minutes

④ The surface waves, which travel the long way around Earth's surface, arrive last.

⑤ The S-P interval, here slightly less than 2 minutes, tells the seismologist how far away the earthquake was.

B The method of *triangulation*. If three seismic stations—shown here in Stockholm, Honolulu, and Manila—record an earthquake, each one can independently determine its own distance from the focus of the quake, thus generating a circle on which the epicenter of the earthquake must lie. The three circles have a unique intersection point, which is the location of the epicenter: in this case Kobe, Japan (the site of a major earthquake in 1995).

The black arrows show the calculated distance from the quake to each seismic station, based on measured S-P intervals.

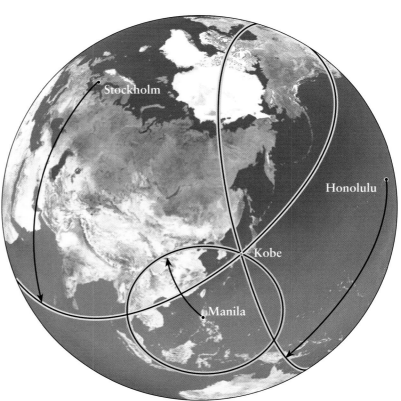

Stockholm

Honolulu

Kobe

Manila

The Science of Seismology 131

MEASURING EARTHQUAKES

Geologists use several different scales to quantify the strength or *magnitude* of an earthquake, by which we mean the amount of energy released during the quake. The most familiar of these is the **Richter magnitude scale**. It has been superseded for most purposes by the *moment magnitude*, which uses the same scale but is computed in a somewhat different way.

The Richter magnitude scale

Charles Richter developed his famous magnitude scale in 1935. Though it was not the first earthquake intensity scale, it was an important advance because it used data from seismographs rather than subjective estimates of damage.

> **Richter magnitude scale** A scale of earthquake intensity based on the recorded heights, or amplitudes, of the seismic waves recorded on a seismograph.

Also, it compensated for the distance between the seismograph and the focus. This means that each seismic station will (in principle) calculate the same magnitude for a given earthquake, no matter how far from the epicenter it may be located. In contrast, a scale based on estimates of the damages sustained will yield the highest magnitude closest to the epicenter, where the damage is greatest. The *Modified Mercalli Intensity Scale* is an example of an earthquake scale based on damage estimates, ranging from I (felt only by instruments, even near the epicenter) to XII (complete devastation near the epicenter).

The Richter scale is *logarithmic*, which means that each unit increase on the scale corresponds to a tenfold increase in the amplitude of the wave signal. Thus, a magnitude 6 earthquake has a wave amplitude ten times larger than that of a magnitude 5 quake. A magnitude 7 earthquake has a wave amplitude 100 times larger (10×10) than that of a magnitude 5 quake.

However, even this comparison understates the difference, because the amount of damage done by an earthquake is more closely related to the amount of energy released in the quake. Each step in the Richter scale corresponds roughly to a 32-fold increase in energy (see What a Geologist Sees). The actual amount of damage done by a quake will also depend, of course, on local conditions—how densely populated the area is, how the buildings are constructed, how deep the focus is, and how severe the secondary effects are.

Moment magnitude

Seismologists today determine magnitudes using both the Richter and **moment magnitude**, which are calculated using different starting assumptions. Richter scale calculations are based on the assumption that an earthquake focus is a point. Therefore, the Richter scale is best suited for earthquakes in which energy is released from a relatively small area of a locked fault. In contrast, the calculation of seismic moment takes account of the fact that energy may be released over a large area. A classic example was the Sumatra-Andaman earthquake of 2004, when a 1200-kilometer length of fault moved. Though the method of calculation is different, the scales are the same because they measure the same thing—the amount of energy released. In either system, magnitude 9 is catastrophic, whereas magnitude 3 is imperceptible to humans.

> **moment magnitude** A measure of earthquake strength that is based on the rupture size, rock properties, and amount of displacement on the fault surface.

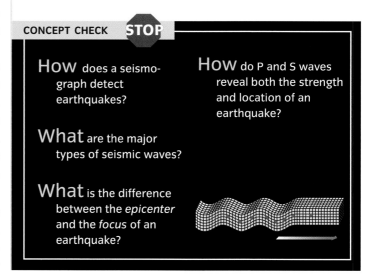

CONCEPT CHECK STOP

How does a seismograph detect earthquakes?

What are the major types of seismic waves?

What is the difference between the *epicenter* and the *focus* of an earthquake?

How do P and S waves reveal both the strength and location of an earthquake?

Richter Magnitude: A Logarithmic Scale

The energy released in an earthquake increases exponentially with its magnitude. A magnitude 6 earthquake releases as much energy as the atomic bomb dropped on Hiroshima, the largest ever used in combat. A magnitude 7 quake would be equivalent in energy to about 32 Hiroshima bombs, and a magnitude 8 quake would be equivalent to 32 x 32, or about 1000, of them. A magnitude 9 quake, such as the Sumatra-Andaman quake, is equivalent to 32 x 32 x 32, or about 32,000 bombs.

Parkfield, CA, 2004

◀

Richter magnitude 6

Damage on surface: small objects broken, sleepers awake (Mercalli intensity ≈VII)
Energy released: about the same as one atomic bomb

▶

Richter magnitude 7

Damage on surface: some walls fall, general panic (Mercalli intensity ≈IX)
Energy released: about the same as 32 atomic bombs

Kobe, Japan, 1995

◀

Richter magnitude 8

Damage on surface: wide destruction, thousands dead (Mercalli intensity ≈XI)
Energy released: about the same as 1000 atomic bombs

San Francisco, CA, 1906

Studying Earth's Interior

Although earthquakes are significant to society because of the damage they can cause, they also have benefits from a scientific perspective. They provide us with some of our most detailed information about Earth's interior—including parts that we can never hope to observe directly.

When scientists cannot study something by direct sampling, a second method comes to the forefront: indirect study or *remote sensing*. Some familiar objects—including the human eye—are actually remote-sensing devices. A camera, for instance, is a remote sensing instrument that collects information about how an object reflects light. Medical techniques such as X-rays allow doctors to study the inside of the body remotely without opening it up surgically.

The seismic waves from an earthquake are much like X-rays, in the sense that they enter Earth near the surface, travel all the way through it and emerge on the other side. They travel along different paths depending on the different kinds of materials they encounter. We will first discuss what earthquakes reveal about Earth's structure, and then describe other sources of information about Earth's interior.

Process Diagram

Seismic waves in Earth's interior FIGURE 5.12

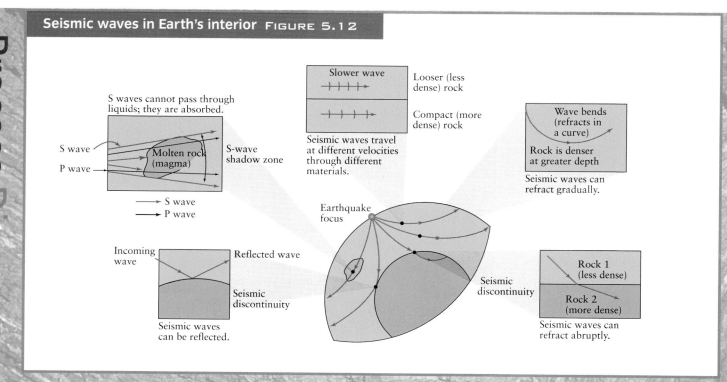

S waves cannot pass through liquids; they are absorbed.

S wave
P wave
Molten rock (magma)
S-wave shadow zone

→ S wave
→ P wave

Slower wave
Looser (less dense) rock

Compact (more dense) rock

Seismic waves travel at different velocities through different materials.

Wave bends (refracts in a curve)
Rock is denser at greater depth

Seismic waves can refract gradually.

Incoming wave
Reflected wave

Seismic discontinuity

Seismic waves can be reflected.

Earthquake focus

Seismic discontinuity

Rock 1 (less dense)
Rock 2 (more dense)

Seismic waves can refract abruptly.

HOW GEOLOGISTS LOOK INTO EARTH'S INTERIOR—SEISMIC METHODS

Before 1906, scientists' understanding of seismic waves was limited. But that year, British geologist Richard Dixon Oldham first identified the difference between P waves and S waves and then suggested an explanation for the complicated patterns recorded after an earthquake: Underneath thousands of kilometers of solid rock, he postulated, Earth has a liquid core. Oldham's theory, now universally accepted by geologists, is illustrated in FIGURE 5.12.

When a major earthquake strikes, the tremors are measured at seismic stations around the world, and the arrival times of P waves and S waves can be used to analyze the kinds of rock that the seismic waves passed through. Even a crude diagram of Earth's structure shows that the pattern of arrival times is quite complex (as shown in Figure 5.12).

How do seismic waves tell us what is inside Earth? Because Earth's interior is not homogeneous, there are boundaries between different materials. Seismic waves behave differently, depending on the properties of the materials they pass through—most importantly, whether the material is a liquid or a solid. Three distinct things can happen to seismic waves when they meet such a boundary, called a **seismic discontinuity** (see Figure 5.12):

■ **seismic discontinuity** A boundary inside Earth where the velocities of seismic waves change abruptly.

1. They can be refracted, or bent, as they pass from one material into another. This is the same thing that happens to light waves when they pass from air to water.

2. They can be reflected, which means that all or part of the wave energy bounces back, like light from a mirror.

3. They can be absorbed, which means that all or part of the wave energy is blocked.

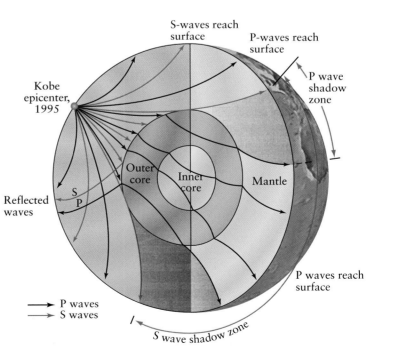

The combination of the various behaviors (left) creates a complex pattern of arrival times of seismic waves at distant locations following an earthquake. Note especially the *shadow zones*, where P waves and S waves are not observed.

Refraction, **reflection**, and absorption all play a role in Oldham's model (see Figure 5.12). P waves are bent dramatically when they pass from the mantle to the outer core. This bending creates a ring-shaped *P-wave shadow zone* on the opposite side of Earth from the earthquake. S waves, on the other hand, are blocked completely by the outer core, because shear waves cannot pass through liquid. This creates an even larger *S-wave shadow zone*, as well as providing firm evidence that Earth has a liquid core.

If this picture looks complicated to you, imagine how confusing it was to Oldham and his contemporaries, who were figuring it out for the first time. We have presented this story *deductively*—that is, proceeding from a model to a conclusion. But many major discoveries in science are *inductive*—they begin with observed effects, and scientists try to infer a hypothesis that explains them. This is much more difficult to do, and justifiably confers some fame on the scientists who succeed.

refraction The bending of a wave as it passes from one material into another material, through which it travels at a different speed.

reflection The bouncing back of a wave from an interface between two different materials.

Seismic tomography   As more sophisticated seismic equipment was developed and scientists began to make more detailed observations, they discovered more boundaries and layers within Earth. Today seismologists use seismic waves to probe Earth's interior in much the same way that doctors use X-rays and CAT scans to probe the interior of a human body. In CAT (*computer-aided tomography*) scanning, a series of X-rays along successive planes can be used to create a three-dimensional picture of the inside of the body. Similarly, *seismic tomography* allows seismologists to superimpose many two-dimensional seismic "snapshots" to create a three-dimensional image of the inside of Earth. These

Beneath the surface FIGURE 5.13

Geologists don't always have to wait for earthquakes to be able to study Earth's interior. For shallow, near-surface studies, they can use explosive charges (or large "thumper trucks") as the source for seismic waves, and then map out the underground rock layers with seismic tomography. Here, two oil prospecting geologists in Texas are looking at seismic images made from data collected in Alaska.

techniques have also helped us to understand more about how plate tectonics works. They also allow scientists to map the locations of seismic discontinuities, the distribution of hot and cold masses, and the distribution of dense and less dense materials inside Earth (See FIGURE 5.13).

HOW GEOLOGISTS LOOK INTO EARTH'S INTERIOR—OTHER METHODS

Earthquakes have provided a great deal of information about Earth's interior, but geologists have many other tools and techniques that allow them to study the deepest parts of our planet. Some of these tools, like the use of seismic information, involve indirect or remote observation of materials and processes deep within the planet. Others are more direct, and give geologists access to actual samples from deep in the crust, and even from the mantle.

Direct observation: drilling and xenoliths

Perhaps the most obvious tool for the retrieval and study of samples from Earth's interior is drilling. To date, Earth's deepest mine (in South Africa) is 3.6 kilometers deep, and the deepest hole ever drilled (in the Kola Peninsula of Russia) reached a depth of just over 12 kilometers. Recall from Chapter 1 that Earth's crust varies from an average thickness of 8 kilometers for oceanic crust to an average of 45 kilometers for continental crust. Therefore, a 12-kilometer hole sounds just about right for sampling the top part of the mantle—or does it?

The problem with drilling is that areas where the crust is thin tend to have high heat flow. In other words, if you try to drill a hole through thin oceanic crust all the way to the mantle, you will quickly encounter temperatures that could destroy your drilling equipment. Another problem with oceanic crust is that it's deep under water, which makes drilling more difficult. Thus the only place where the rocks are both accessible and cool enough for drilling to great depth is precisely where the crust is very thick—on the continents. The hole in the Kola Peninsula went through more than 12 kilometers of thick continental crust and never even came close to reaching the mantle. Thus, although drilling has yielded much interesting and useful information about the composition and properties of the crust, it hasn't even penetrated to the outermost seismic discontinuity in the mantle.

If we can't obtain samples from deep within Earth by reaching in to retrieve them, perhaps we can wait for them to come to us. This does happen, in two different ways. Molten rock, or *magma*, is formed in the upper portions of Earth's mantle, in areas where the temperature is high enough. By studying magma that originates at depth and erupts to the surface, scientists learn about the temperature, pressure, and composition of the mantle in the region where the magma formed. Furthermore, as magma rises toward the surface, it often breaks off and carries with it fragments of the unmelted surrounding rock. We call these fragments **xenoliths**, from the Greek words *xenos* (foreigner) and *lithos* (stone). A xenolith that reaches Earth's surface is a sample of the deep crust or mantle, accessible for direct scientific study. (See What a Geologist Sees on the next page.)

Indirect observation: methods from physics, astronomy, and chemistry

The availability of indirect or remote techniques for the study of Earth's interior has increased considerably since the dawn of the Space Age. Many of these techniques are also used to study other planets in the solar system. We have already discussed the single most important tool from geophysics for the study of Earth's interior: the study of seismic waves generated by earthquakes. Another important geophysical method relies on the study of Earth's magnetic field.

Magnetism is a force created either by permanent magnets (ferromagnets) or by moving electrical charges. We can try thinking of Earth as having a huge dipole bar magnet with north and south poles at its center, offset slightly from the geographic north and south poles. The problem with this analogy is that solids, including bar magnets, lose their magnetism at temperatures above a critical transition temperature, called the *Curie point*, which is specific to each material. The Curie point for iron is about 770°C, but we know that the temperature deep inside Earth is *much* higher than this—at least 5000°C.

Diamonds: Messengers from the Deep

A Diamonds form only at the extremely high pressures found at depths of 100 to 300 kilometers. The different colors in this diamond—nick named the "Picasso" diamond—show various zones of growth. The diamond is approximately 1 millimeter across, and the colors are revealed by a special type of photography that highlights small variations in composition.

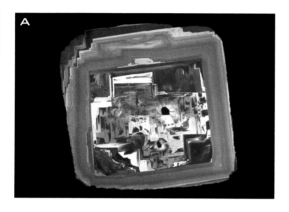

B The beautiful uncut diamond below is the 253.7-carat Oppenheimer Diamond from the Smithsonian Institution; it was originally discovered in South Africa in 1964.

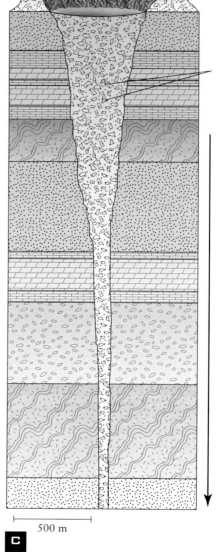

Magma vent is circular when viewed from above.

Xenoliths of mantle rock

Pipe extends 150–200 km down into mantle

500 m

C To reach the surface from such great depths, diamonds must be carried by an eruption of unusual ferocity. These eruptions leave behind a long, cone-shaped tube of solidified magma, called a *kimberlite pipe*. Although most people treasure diamonds for their beauty and luster, geologists treasure them also as "messengers"—samples from an otherwise inaccessible region of Earth's interior.

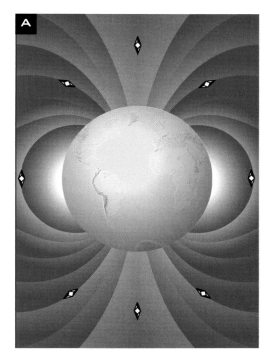

Earth's magnetic field FIGURE 5.14

A Earth is surrounded by a magnetic field, which causes a compass needle to point north. More precisely, the needle is aligned along the field lines that lead to the north and south magnetic poles, which are almost—but not exactly—aligned with Earth's north and south geographic poles.

B This is a photograph of the aurora borealis, or northern lights, as seen from Fairbanks, Alaska. This phenomenon is caused by charged particles from the Sun entering Earth's atmosphere at high latitudes along magnetic field lines.

Since there can't be a giant bar magnet inside the planet, moving electrical charges must be responsible for generating Earth's magnetic field (see FIGURE 5.14). Physicists have shown that the movement of an electrically conducting liquid inside a planet could generate a self-sustaining magnetic field, much like a rotating coil of wire in an electric motor. This is consistent with the observation from seismology that at least the outer part of Earth's core is liquid. However, molten rock is not a good enough electrical conductor to generate a magnetic field in this manner; for this and other reasons, geologists believe that the liquid outer core is made of molten iron and nickel. This is consistent with evidence from meteorites, discussed later.

We can also gain a certain amount of information about any planet's interior—including Earth's—from astronomical observations. The first step is to determine the planet's *mass*. This can be deduced from the planet's gravitational influence on other planets and satellites. Second, we need to know the *diameter* of the planet. Knowing the dimensions of the planet, and its shape (in the case of Earth, a very slightly flattened sphere), it is a simple matter to figure out its volume and average density (mass divided by volume).

What do these kinds of measurements reveal about Earth's interior? For one thing, we can determine whether material is distributed evenly throughout the planet. The rocks at Earth's surface are very light (low-density) compared to the planet as a whole. Surface rocks have an average density of about 2.8 g/cm^3, whereas Earth's overall density is 5.5 g/cm^3. (For comparison, water has a density of 1 g/cm^3 at 4°C.) For the planet as a whole to have such a high density, with such low-density rocks at the surface, there must be a concentration of denser material somewhere inside the planet. The density of this material must be greater than 10 g/cm^3, which is consistent with the densities of iron meteorites.

A third way to study Earth's interior is to analyze the building blocks that formed it. Planetary scientists have discovered that most (though not all) meteorites were formed at about the same time and in the same part of the solar system as Earth. Some of these meteorites are *primitive*—that is, they have remained unaffected by melting and other geologic processes since the beginning of the solar system. These meteorites give scientists an idea of the overall composition of the solar system and its constituent bodies. Other meteorites—the *irons*, *stony-irons*, and some kinds of *stony meteorites*—show signs of melting and differentiation, and may be more representative of Earth's core and mantle. It is highly significant that a core with the composition of a typical iron meteorite (mostly iron and nickel) would bring Earth's overall density up to the observed value of 5.5 g/cm³.

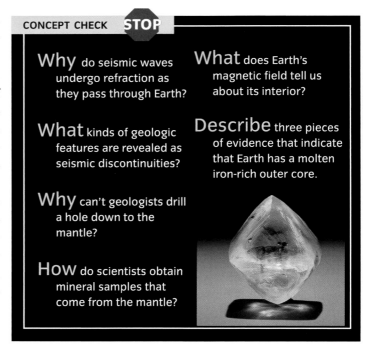

CONCEPT CHECK STOP

Why do seismic waves undergo refraction as they pass through Earth?

What kinds of geologic features are revealed as seismic discontinuities?

Why can't geologists drill a hole down to the mantle?

How do scientists obtain mineral samples that come from the mantle?

What does Earth's magnetic field tell us about its interior?

Describe three pieces of evidence that indicate that Earth has a molten iron-rich outer core.

A Multi-Layered Planet

LEARNING OBJECTIVES

Define crust, mantle, and core.

Describe the composition of the crust, mantle, and core.

Describe the layering within the mantle and core.

By piecing together information from all of these sources, geologists have arrived at a very detailed understanding of Earth's interior. Let's take a brief tour, starting at the top and working down to the innermost layers. As we proceed, keep in mind that some boundaries inside Earth mark the transition between layers with differing composition, whereas others separate layers with the same composition but differing physical properties.

THE CRUST

Earth's outermost compositional layer is the **crust**. The thickness of the crust varies greatly, from an average of 8 kilometers for oceanic crust to an average of 45 kilometer for continental crust (see **FIGURE 5.15**). Even at its thickest spots, the crust is extremely thin compared with Earth as a whole. It's like a thin, brittle eggshell, or about the same relative thickness as the glass of a light bulb.

The composition of the crust also varies from place to place. About 95 percent of the crust is igneous rock or metamorphic rock derived from igneous rock. In general, the rocks of the crust are lighter (less dense) than the material of Earth's interior, because the crust is composed of material that "floated" to the top during planetary differentiation.

crust The outermost compositional layer of the solid Earth; part of the lithosphere.

The very outermost layer of the crust, both on land and on the ocean floor, is quite different from the crust as a whole. This surface layer—the ground surface we see around us every day—consists of about 75 percent sediment and sedimentary rock, formed by the constant action of erosion, weathering, and deposition.

The boundary that separates the crust from the mantle was the next discontinuity to be discovered after the liquid core. It was named the *Mohorovičić discontinuity* after the seismologist who discovered it in 1909, but it is usually called the *Moho* for short. Mantle rocks, being denser and compositionally different from crustal rocks, transmit P waves much more quickly. The Moho is thus an example of a boundary between two layers of rock that have different compositions and densities but similar physical characteristics (rigidity).

Inside view of Earth FIGURE 5.15

This diagram shows Earth's internal structure, and summarizes what we have learned from seismic studies and other direct and indirect observations. The upper part of the cutaway shows compositional layers and the lower part shows layers with differing rock properties. Note that the boundaries between zones that differ in strength, such as the rigid lithosphere and the more plastic asthenosphere, do not always coincide with compositional boundaries.

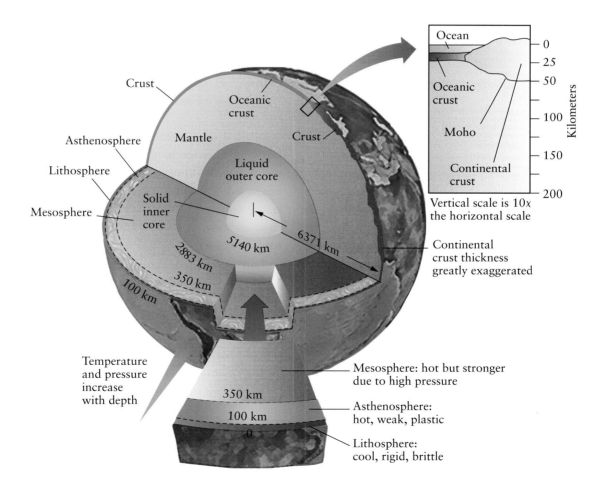

THE MANTLE

The **mantle** extends from the Moho to the core (see Figure 5.15). About 80 percent of Earth's volume is contained in the mantle. Geologists believe, on the strength of evidence from xenoliths, meteorites, and seismic analyses, that the mantle consists mainly of iron- and magnesium-silicate minerals. The upper part of the mantle has a composition similar to that of *peridotite*, an igneous rock not typically found in the crust, which consists mainly of the minerals olivine and pyroxene.

Seismic studies have revealed boundaries within the mantle, but not all of them are compositional boundaries. Extending from about 100 to 350 kilometers below the surface is a layer called the **asthenosphere**, from the Greek words meaning "weak sphere." In this zone, some of the rocks are very near the temperatures at which rock melting begins, so they have the consistency of butter or warm tar. The composition of the asthenosphere appears to be the same as that of the mantle just above and below it. Thus, the asthenosphere is a layer whose distinctiveness is based on its physical properties—reduced rigidity—rather than its composition.

The outermost 100 kilometers of Earth, which includes the crust and the part of the uppermost mantle just above the asthenosphere, is called the **lithosphere** (see Figure 5.15). The rocks of the lithosphere are cooler, more rigid, and much stronger than the rocks of the asthenosphere. In plate tectonic theory, it is the entire lithosphere—not just the crust—that forms the plates. These plates can move around because they rest on the underlying weaker rocks of the asthenosphere, which slowly deform and flow like an extremely thick, viscous liquid.

The rest of the mantle, from the bottom of the asthenosphere (at about 350 kilometers depth) down

> **mantle** The middle compositional layer of Earth, between the core and the crust.

> **asthenosphere** A layer of weak, ductile rock in the mantle that is close to melting but not actually molten.

> **lithosphere** Earth's rocky, outermost layer, comprising the crust and the uppermost part of the mantle.

Seismic discontinuities in the mantle FIGURE 5.16

Earth's mantle is not uniform, but has several seismic boundaries within it. We know the boundaries exist because P waves and S waves slow down or speed up abruptly and are refracted or reflected at these boundaries. However, the exact nature of the boundaries is still not completely understood.

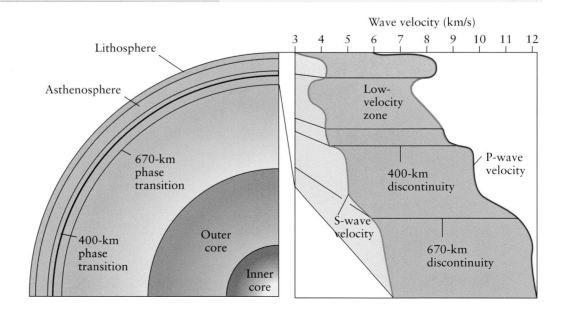

to the core-mantle boundary, is the **mesosphere**. Although temperatures in the mesosphere are very high, the rocks are a bit stronger than in the asthenosphere because they are so highly compressed. Additional seismic discontinuities exist within the mesosphere, with transitions at about 400 kilometers and at about 670 kilometers below the surface (see FIGURE 5.16). These discontinuities are not well understood; they do not seem to be compositional boundaries, but rather to result from changes in physical properties. For example, when the mineral olivine is squeezed at a pressure equal to that found at a depth of 400 kilometers, the atoms rearrange themselves into a more compact structure or *polymorph* of olivine (see Chapter 2). Perhaps this change to a more compact form causes the 400-kilometer seismic discontinuity.

It is important to remember that the mantle is mostly solid rock, except for small pockets of melt in the asthenosphere. We know that the mantle must be solid, because both P waves and S waves can travel through it. Nevertheless, pressures and temperatures deep within Earth are so high that even solid rock can flow, in very, very slow convection currents, as described in Chapter 4. The mechanism of flow is similar to that of ice flowing in a glacier, and the rate of flow is roughly the same as the rate of growth of human fingernails. Seismologists have recently detected hot regions in the mantle that may coincide with the rising limbs of convection cells that help drive plate motion.

THE CORE

At a depth of 2883 kilometers, there is a huge decrease in the velocity of P waves, and the velocity of S waves drops to zero. This is the core-mantle boundary (see Figure 5.15), which represents a change in both the composition and physical properties of the rocks. The **core**, the innermost of Earth's compositional layers, is the densest part of Earth. It consists of

core Earth's innermost compositional layer, where the magnetic field is generated and much geothermal energy resides.

material that "sank" to the center during the process of planetary differentiation. As discussed earlier, geologists believe that Earth's core is composed primarily of iron-nickel metal.

As mentioned earlier, the S-wave shadow zone (and other evidence) tells us that the **outer core**, from 2883 kilometers to 5140 kilometers depth, must be liquid. For three decades geologists thought the core was a homogeneous liquid, but in 1936 the Danish seismologist Inge Lehmann showed that the core, too, has layers. She detected faint seismic waves within the P-wave shadow zone that had reflected off the **inner core**. The pressure in the inner core is so great that iron must be solid there, in spite of the very high temperature. We know this from high-pressure experiments on iron. The main difference between the inner and outer core is thus the physical state, rather than composition. As heat escapes from the core and works its way to the surface, the core is gradually crystallizing. Thus the solid inner core must be growing larger, although very slowly.

It should be clear by now that what happens deep in the interior of Earth profoundly affects the surface. The release of heat from the interior is an important driving force for plate tectonics, which, in turn, is the major uplifting force in shaping Earth's varied landscapes and topographies.

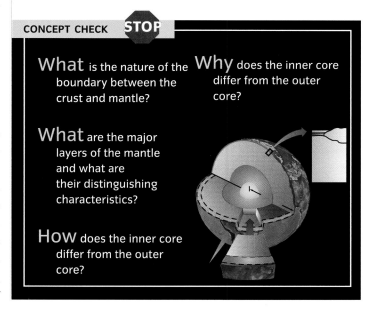

CONCEPT CHECK **STOP**

What is the nature of the boundary between the crust and mantle?

Why does the inner core differ from the outer core?

What are the major layers of the mantle and what are their distinguishing characteristics?

How does the inner core differ from the outer core?

Global Locator

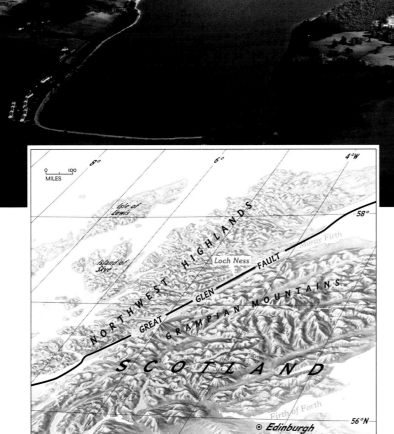

▲
Aerial view of Loch Ness, looking southwest from its northeasterly end. The loch is 37 kilometers in length and occupies the deeply eroded trench of the Great Glen Fault.

NATIONAL GEOGRAPHIC

Though it is most famous for its fabled inhabitant, the "Loch Ness Monster," this Scottish lake has a fascinating geologic history. It lies on the Great Glen Fault, which 300 million years ago was part of a plate margin and was as active as the San Andreas Fault is today. Over time the rocks on the north side of the fault moved at least 100 kilometers to the left (see inset). Movement on the fault crushed rocks along its margins. These rocks were particularly susceptible to erosion during the most recent ice age. Thus, they have eroded into a long and remarkably straight valley that runs along the fault. Today this valley is occupied by several lakes, including Loch Ness.

1 Earthquakes and Earthquake Hazards

1. The subject of **seismology** relates earthquakes to the processes of plate tectonics. Although the motion of tectonic plates is very gradual, friction causes the rocks in the crust to jam together for long periods, then to break suddenly and lurch forward, causing an earthquake to occur. Earthquakes can cause large vertical or horizontal displacements of the ground, but much of the damage they cause comes as a result of the violent shaking that accompanies the displacement.

2. The shaking motion experienced during an earthquake can be explained by the **elastic rebound theory**, which says that the energy stored in bent and deformed rocks is released as **seismic waves**. After the earthquake, the rocks return to their previous state.

3. In many cases the destructiveness of earthquakes is magnified by *secondary hazards*, such as fires, landslides, soil liquefaction, and tsunamis. Proper building design and earthquake preparedness can greatly reduce the loss of life from earthquakes and secondary hazards.

4. *Short-term forecasting* of earthquakes is still very unreliable. Scientists have concentrated their efforts on finding *precursor phenomena*, such as *foreshocks*, changes in water level, release of gases, and even unusual animal behavior—but with limited success. However, *long-term forecasting* can provide a good idea of which regions are at risk. One of the main tools of long-term forecasting is **paleoseismology**, which reveals when past earthquakes occurred in a given region, as well as the periodicity and magnitudes of past earthquakes.

2 The Science of Seismology

1. **Seismographs** produce recordings of seismic waves that are called **seismograms**. In a basic seismograph, a pen is attached to a heavy suspended mass. Seismic waves cause the paper to shake while the pen stays still and traces a wavy line on the vibrating paper.

2. Earthquakes produce three main types of seismic waves: **compressional** or **P waves** (*primary waves*), **shear** or **S waves** (*secondary waves*), and a variety of **surface waves**. Compressional and shear waves are called **body waves** because they travel through Earth's interior.

3. Compressional waves travel faster than shear waves, and hence arrive at seismographs first. The difference in arrival times between the P and S waves allows seismologists to compute the distance, but not the direction, to the **focus** of the earthquake. To determine the precise location of the **epicenter**, seismologists need measurements from three separate seismic stations. They can then determine the location by *triangulation*.

4. The **Richter** and **moment magnitudes** are measures of earthquake intensity that can be determined regardless of the distance to the earthquake or the amount of damage done. Both are *logarithmic* scales, in which each unit of magnitude corresponds roughly to a 10-fold increase in the amplitudes of seismic waves, but a 32-fold increase in the amount of energy released by the earthquake. The Richter magnitude assumes all of the energy is released from a point source—an assumption that is seldom valid for large quakes. In general, therefore, geologists prefer to use the moment magnitude to describe the sizes of earthquakes.

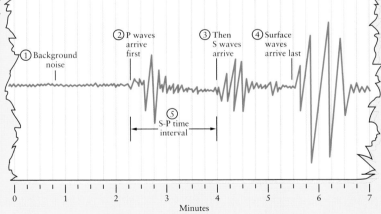

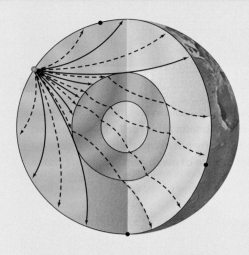

3 Studying Earth's Interior

1. After an earthquake, seismic waves travel downward into Earth's interior as well as upward and along the surface. Seismic waves travel at different velocities through different materials, and they change velocity and direction when they pass from one material to another material with different physical and/or compositional properties. This understanding of seismic waves has allowed seismologists to identify many **seismic discontinuities** in Earth's interior.

2. Seismic discontinuities can result from either a change in composition or a change in physical properties of the material. They may **refract**, **reflect**, or even block seismic waves. P waves are strongly refracted, or bent, when they pass from the **mantle** to the **core**. S waves, on the other hand, are completely blocked by the core. This discovery provides evidence that Earth has a liquid outer core.

3. Geologists use a variety of remote or indirect techniques to understand Earth's interior, in addition to the study of naturally occurring earthquakes. They can set off explosive charges at the surface and use *seismic tomography*, analogous to medical tomography, to detect seismic discontinuities underground.

4. Other sources of information about Earth's interior include drilling; the *magnetic field*; the *mass* and *diameter* of Earth; and *meteorite* studies. Evidence points to the likelihood that Earth has a dense core that consists mainly of iron and nickel.

5. So far, geologists have been unable to drill deep enough to sample the mantle directly. However, some mineral samples from the mantle come to the surface as *xenoliths*, carried along by magma that rises to the surface.

4 A Multi-Layered Planet

1. Earth's three main compositional layers are the **crust**, the **mantle**, and the **core**. Each of these layers creates a seismic discontinuity, and this is how geologists can determine their thickness.

2. The crust consists of solid rock that is mostly igneous, with a thin veneer of sediment and sedimentary rock at the surface. It varies in average thickness from 8 kilometers (for oceanic crust) to 45 kilometers (for continental crust).

3. The mantle contains several discontinuities that are caused by differences in physical property rather than composition. The most important discontinuities are the boundaries between the **lithosphere**, the **asthenosphere**, and the **mesosphere**. The lithosphere is about 100 kilometers thick; the asthenosphere begins about 100 kilometers beneath the surface and ends at a depth of about 350 kilometers; and the mesosphere extends from a depth of 350 to 2883 kilometers.

4. The core has two layers, a liquid *outer core* and an *inner core* that is solid—despite its high temperature—because of the extremely high pressure. The boundary between them is a physical boundary, not a compositional boundary. The outer core begins at a depth of 2883 kilometers and extends to a depth of 5140 kilometers. The inner core, which was first discovered by its reflection of P waves, is almost certainly growing very gradually as Earth cools down from its formation.

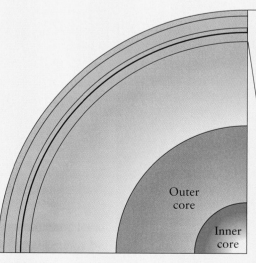

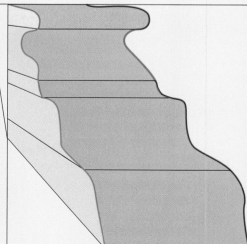

KEY TERMS

- seismology p. 118
- elastic rebound theory p. 119
- seismic wave p. 119
- paleoseismology p. 123
- seismograph p. 128
- seismogram p. 128
- body wave p. 129
- surface wave p. 129

- focus p. 129
- compressional wave p. 129
- shear wave p. 129
- epicenter p. 130
- Richter magnitude scale p. 132
- moment magnitude p. 132
- seismic discontinuity p. 135
- refraction p. 136

- reflection p. 136
- crust p. 140
- mantle p. 142
- asthenosphere p. 142
- lithosphere p. 142
- core p. 143

CRITICAL AND CREATIVE THINKING QUESTIONS

1. Use the elastic rebound theory to describe what happens to rocks at the focus just before, during, and after an earthquake.

2. Why is short-term prediction of earthquakes so much less successful than long-term prediction? Why do you think seismologists are extremely cautious about making predictions? Do you think it will ever be possible to predict earthquakes accurately? Research your answer.

3. If you were asked to determine the exact shape and size of Earth, how would you go about it? What would you do differently if you were not allowed to use Space Age technology such as satellite photographs and orbital data?

4. Some of the boundaries inside Earth represent transitions between layers with differing compositions, whereas others represent transitions between layers with different physical states. Find out more about these different layers, and draw a detailed diagram to show the layering.

5. Which of the techniques used to study Earth's interior could also be used to study other planets? Which ones cannot, and why? Scientists know more about the surface of the Sun than about the interior of our own planet; why do you think this is so?

What is happening in this picture ?

This photograph shows a stream in Carrizo Plains, California. The stream makes an abrupt turn to the right and then a 90-degree turn to the left. What reason can you suggest for this stream's strange behavior?

1. At which type of plate boundary have the largest recorded earthquakes occurred?

   a. divergent boundaries
   b. transform fault boundaries
   c. subduction zone boundaries
   d. continental collision boundaries

2. According to the elastic rebound model, earthquakes are caused by

   a. the slow release of gases from the athenosphere.
   b. the sudden release of energy stored in rocks through continuing stress.
   c. the sudden movement of otherwise stable tectonic plates.
   d. the rapid release of gases from the asthenosphere.

3. _____ and the resulting collapse of buildings, bridges and other structures are usually the most significant primary hazards to cause damage during an earthquake.

   a. Fire
   b. Tsunami
   c. Ground liquefaction
   d. Ground shaking

4. _____ can provide a good idea of which regions are at risk for severe earthquakes.

   a. Short-term forecasting
   b. Long-term forecasting
   c. Unusual animal behavior studies
   d. Studies of groundwater levels

5. Body waves

   a. move through Earth's interior.
   b. cannot penetrate Earth's liquid outer core.
   c. move along Earth's surface, causing great destruction.
   d. Both b and c are correct.

6. Illustrations A and B depict two different types of seismic waves. Which of the following statements can be made about these two seismic waves?

   a. The wave depicted in A is a P wave and has a greater velocity through Earth's crust than other types of seismic waves.
   b. The wave depicted in A is an S wave and has a greater velocity through Earth's crust than other types of seismic waves.
   c. The wave depicted in B is a P wave and has a greater velocity through Earth's crust than other types of seismic waves.
   d. The wave depicted in B is an S wave and has a greater velocity through Earth's crust than other types of seismic waves.

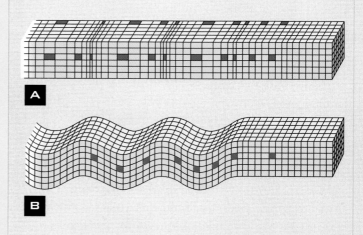

7. This illustration shows a seismogram of a hypothetical earthquake. On the seismogram, label the following:

   S – P interval   First arrival of P wave
   First arrival of S wave   Background noise
   First arrival of surface waves

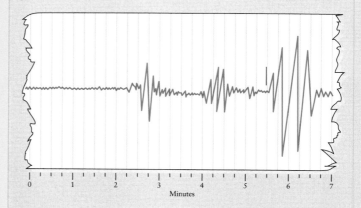

8. Using seismograms from three different seismic recording stations A, B, and C, you determine the epicenter of an earthquake. Stations A and B both had an S – P interval of 3 seconds while C had an S – P interval of 11 seconds. Which of the following statements most accurately depicts the location of the epicenter?

   a. The epicenter is closest to Station A and equally far from B and C.

   b. The epicenter is closest to Station B and equally far from A and C.

   c. The epicenter is closest to Station C and equally far from A and B.

   d. The epicenter is equally close to A and B and farthest from Station C.

9. For the earthquake mentioned above, which seismic recording station would have recorded the P wave first?

   a. Station A

   b. Station B

   c. Station C

   d. Both stations A and B would have recorded the P wave before station C.

10. A magnitude 8 earthquake releases approximately _____ times more energy than a magnitude 7 event.

   a. 2   c. 20   e. 31.5

   b. 10   d. 21.5

11. Moment magnitude differs from Richter magnitude

   a. because the Richter magnitude assumes earthquakes are generated at a point source, whereas moment magnitude takes into account that earthquakes can be generated over a large area of rupture.

   b. because the moment magnitude assumes earthquakes are generated at a point source, whereas Richter magnitude takes into account that earthquakes can be generated over a large area of rupture.

   c. in that moment magnitude uses Roman numerals to designate strength of an earthquake.

   d. in that Richter magnitude uses Roman numerals to designate strength of an earthquake.

12. When seismic waves reach a discontinuity inside Earth's interior,

   a. they can be refracted, or bent, as they pass from the first material into the second.

   b. they can be reflected, which means that all or part of the wave energy bounces back.

   c. they can be absorbed, which means that all or part of the wave energy is blocked by the second material.

   d. All of the above statements are correct.

13. On this illustration, label Earth's internal structure using the following terms:

mantle	asthenosphere	inner core
lithosphere	outer core	continental crust
oceanic crust	moho	mesosphere

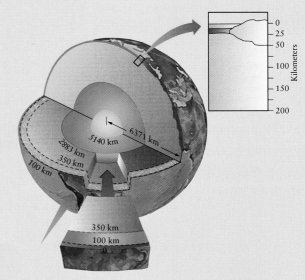

14. Earth's mantle is composed of _____ which surrounds a(n) _____ core.

   a. rock that contains iron- and magnesium-silicate minerals; iron-nickel metallicy

   b. iron-nickel alloy; rocky iron- and magnesium-silicate

   c. rock that contains iron- and magnesium-silicate minerals; molten

   d. molten rock; solid iron-nickel metallic

15. The asthenosphere is a layer whose distinctiveness from the rest of the mantle is based on its

   a. differences in composition.

   b. reduced rigidity.

   c. increased rigidity.

   d. relatively low temperature.

Volcanoes and Igneous Rocks

The eruption of Mount Pinatubo, on the island of Luzon in the Philippines on June 15, 1991, was the second-largest volcanic eruption of the twentieth century and the largest in a densely populated area. The top of the mountain exploded (main photo), blasting a gaping hole 2.5 kilometers in diameter and propelling volcanic ash and sulfurous gases more than 30 kilometers into the atmosphere. The cloud lingered in the stratosphere and lowered worldwide temperatures for the next year by half a degree.

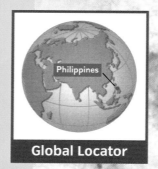

Global Locator

By unhappy coincidence, a typhoon was bearing down on the island at the time of the eruption. The rain-soaked ash caused many roofs to collapse. It formed loose, unstable mud that continued to flow and slide downhill for months, burying towns, wiping out bridges and ultimately causing more damage than the eruption itself. The two inset photos show the town of Bamban, 30 kilometers away from Mt. Pinatubo, one month (top) and three months after the eruption (bottom).

Despite the damage it did, the Mt. Pinatubo eruption killed relatively few people. Thanks to early warnings from geologists, most of the area around the volcano had been evacuated. Although 847 people died from the effects of the eruption (including the mudslides afterward), scientists estimate that 5,000 to 20,000 lives were saved by the timely evacuation.

This chapter will explore the processes that lead to volcanic eruptions, the kinds of rocks they form, the ways in which geologists can tell when an eruption might be coming, and the reasons some volcanoes erupt violently and others do not.

Volcanoes and Volcanic Hazards

LEARNING OBJECTIVES

Identify several different categories of volcanic eruptions.

Explain why stratovolcanoes tend to erupt explosively, whereas shield volcanoes tend to erupt nonexplosively.

Describe how volcanic features such as calderas, geysers, and fumaroles arise.

Identify the hazards of volcanoes and the ways in which they can have a beneficial effect.

Describe how scientists monitor volcanic activity.

F or many people, the thought of a **volcano** conjures up visions of fountains of **lava** spurting up into the air and pouring out over the landscape (FIGURE 6.1A). Although it's true that most volcanoes produce at least some liquid lava, many other types of materials can emerge from volcanoes as well, such as fragments of rock, glassy volcanic ash, and gases. Stored-up gases can cause a volcano to explode, covering the surrounding area with a catastrophic shower or flow of ash and broken rock (FIGURE 6.1B and C). Or gases can seep out silently and poison a whole town overnight, as you will discover in the Case Study. The different kinds of eruptions and the volcanoes they build have much to do with the physical properties of the **magma** that lies at their source. We will begin our

volcano A vent through which lava, solid rock debris, volcanic ash, and gases erupt from Earth's crust to its surface.

lava Molten rock that reaches Earth's surface.

magma Molten rock, which may include fragments of rock, volcanic glass and ash, or gas.

NATIONAL GEOGRAPHIC

Different eruption styles, different hazards FIGURE 6.1

A

◄ **A** Nonexplosive Eruptions: Hawaii's volcanic eruptions, such as this eruption of Kilauea that began in 1983, are very photogenic and pose minimal danger to humans, although they can cause extensive property damage if lava flows into populated areas.

B

◄ **B** Explosive Eruptions: The May 18, 1980, eruption of Mount St. Helens in Washington was much more violent than the relatively harmless eruptions of Kilauea in Hawaii. The entire top of Mount St. Helens was destroyed in a cataclysmic explosion, as seen in these before- (**B**) and after- (**C**) pictures.

▼

C

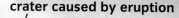
crater caused by eruption

In **C** you can see the large crater where the top of the mountain used to be. At least 57 people were killed in the eruption. ▶

discussion by taking a look at some of the different kinds of volcanoes.

ERUPTIONS, LANDFORMS, AND MATERIALS

Eruption types

Although all volcanoes are dangerous at some level, as Figure 6.1 illustrates, there is a range of eruption styles. We can distinguish among several types of eruptions on the basis of their explosiveness and the materials they produce. These differences are also reflected in the kind of terrain they build. FIGURE 6.2 on the following page shows several types of volcanoes, arranged in order from least to most explosive. The eruptive style of volcanoes can change from year to year, month to month, or even from one hour to the next; there is no such thing as a completely safe volcano or a completely predictable one.

Hawaiian eruptions consist of very runny lava that flows easily from a volcanic vent. These flows gradually build up to form broad, flat volcanoes with very gently sloping sides, called **shield volcanoes** (Figure 6.2A). They can grow to enormous size, and resemble a warrior's shield lying flat. Mauna Kea and Mauna Loa, on the big island of Hawaii, rise more than 10 kilometers from their bases (6 kilometers below sea level) to their peaks, 4 kilometers above sea level. That makes them the tallest mountains on Earth, measured from base to peak.

Sometimes lava rises to the surface through long fissures rather than central craters. These fissures can produce vast, flat lava plains called *flood basalts* or *basalt plateaus* (Figure 6.2B). Shield volcanoes, like Mauna Loa, often display some fissure activity as well.

Strombolian eruptions (Figure 6.2C) are more explosive than Hawaiian eruptions. The volcano may eject showers of lava or rock hundreds of feet in the air. This type of eruption creates cones of loose volcanic rock called *spatter cones* or *cinder cones*.

Vulcanian eruptions are much more explosive (Figure 6.2D). They propel billowing clouds of ash to a height of 10 kilometers or so. They also produce **pyroclastic flows** of hot cinders and ash that sweep down the mountainside like an avalanche. These travel much faster than flowing lava, and they are the most dangerous consequence of a volcanic eruption.

The most violent and famous eruptions in history are *Plinian*, named after Pliny the Elder, a Roman scholar who died during the eruption of Mt. Vesuvius in 79 A.D. (Figure 6.2E). They produce ash columns that reach into the stratosphere (20 kilometers or more), and create pyroclastic flows. Both Vulcanian and Plinian eruptions tend to build steep-sided volcanoes, called **stratovolcanoes** (FIGURE 6.3 on page 155).

What causes the diversity of eruption types? The answer lies mostly in the kind of magma that provides the source for the volcano. Two factors are important: the **viscosity** of the magma and the amount of gas dissolved in it. If gas is present in the magma, it must escape somehow. If the magma is fairly runny, the dissolved gas will escape relatively easily. The lava may bubble and fountain dramatically, especially at the beginning of an eruption, but the volcano will not explode. However, if the magma is relatively thick or viscous, it is harder for gas bubbles to form and escape. When the gas finally does escape, it usually vents explosively (Figure 6.3).

pyroclastic flow
Hot volcanic fragments (tephra) that are buoyed by heat and volcanic gases, and flow very rapidly.

stratovolcano A volcano composed of solidified lava flows interlayered with pyroclastic material. Such volcanoes usually have steep sides that curve upward.

viscosity The degree to which a substance resists flow; a less viscous liquid is runny, whereas a more viscous liquid is thick.

shield volcano
A broad, flat volcano with gently sloping sides, built of successive lava flows.

Shield volcano profile (gentle slopes) due
to fluid, low-silica lava, and few pyroclastics

Mauna Loa

Mauna
Kea

A Mauna Kea and Mauna Loa are shield volcanoes formed by
Hawaiian eruptions.

Fluid, low-silica magma
Extensive, nearly flat lava flow
No apparent volcano (lava
erupted from fissures)

B These flood basalts are on the Snake River Plains, Idaho.

Incandescent lava and volcanic
bombs, variable magma type

C In 2002, Mount Etna in Sicily experienced a Strombolian
eruption.

Pyroclastic flows, gas-
and silica-rich magma

D The 1993 Vulcanian eruption of Mount Mayon in the Philippines included pyroclastic flows.

Stratovolcano

Ash column, pyroclastic flows (not
seen here), gas- and silica-rich magma

E The Plinian eruption of Mount St. Helens in May 1980 produced
an ash column and released destructive pyroclastic flows and debris avalanches down the steeply sloping sides of the volcano.

How a stratovolcano is formed FIGURE 6.3

The classic volcano profile is that of a stratovolcano, which builds up over time from pyroclastic flows released in explosive eruptions and from less frequent lava flows. This creates steep-sided volcanoes like Mount Fuji (inset), a national symbol of Japan. It is a sleeping beauty now, but the steep upward-curving profile is a sign of a violent past.

www.wiley.com/college/murck

1. Magma is very viscous, preventing the escape of gas bubbles. The trapped gas and upward movement of magma create increasing pressure inside the volcano. Thick deposits of pyroclastic material, like these, are often a sign of a past violent explosion—and a sign of possible future eruptions.

2. Gas continues to build and magma rises until the pressure causes an explosion. Small bits of lava and rock (called tephra) are ejected in all directions.

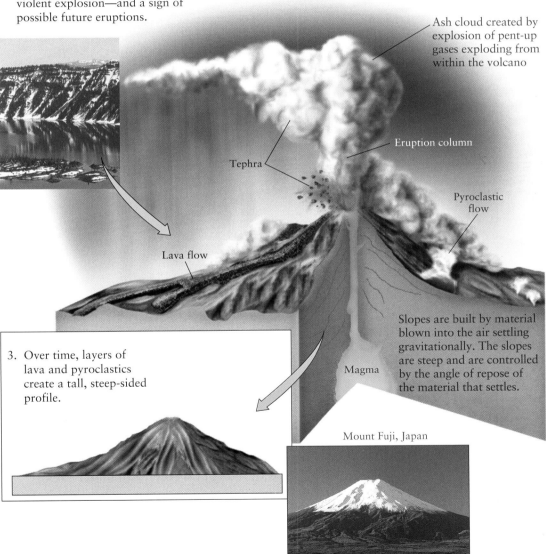

Ash cloud created by explosion of pent-up gases exploding from within the volcano

Eruption column

Tephra

Pyroclastic flow

Lava flow

3. Over time, layers of lava and pyroclastics create a tall, steep-sided profile.

Magma

Slopes are built by material blown into the air settling gravitationally. The slopes are steep and are controlled by the angle of repose of the material that settles.

Mount Fuji, Japan

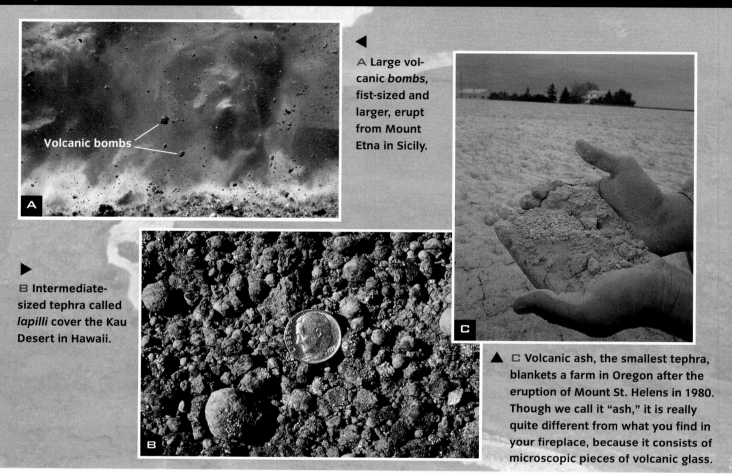

◀ A Large volcanic *bombs*, fist-sized and larger, erupt from Mount Etna in Sicily.

Volcanic bombs

A

▶ B Intermediate-sized tephra called *lapilli* cover the Kau Desert in Hawaii.

B

C

▲ C Volcanic ash, the smallest tephra, blankets a farm in Oregon after the eruption of Mount St. Helens in 1980. Though we call it "ash," it is really quite different from what you find in your fireplace, because it consists of microscopic pieces of volcanic glass.

A fragment of rock ejected during a volcanic eruption is called a *pyroclast* (from the Greek words meaning "fire broken"). Collectively, all the ejecta from a volcano are known as *tephra*, and they can range from car-sized rocks to ultrafine *volcanic ash* whose individual particles can only be seen under a microscope (**FIGURE 6.4**).

Loose pyroclasts are often welded together during the eruption, or cemented together afterwards, forming pyroclastic rocks. These rocks are called *agglomerates* when the tephra particles are large and *tuff* when the particles are small. Spatter cones and tephra cones are made of exactly this kind of rock. On the other hand, stratovolcanoes have a somewhat more complicated structure, with alternating layers of pyroclastic material and solidified lava flows. Their height stems from the layers of hardened lava, which act as a cement that holds the pyroclasts together.

Other Volcanic Features

Near the summit of most volcanoes is a *crater*, the funnel-shaped depression from which gas, tephra, and lava are ejected. Some volcanoes have a much larger depression known as a *caldera*, a roughly circular, steep-walled basin that may be several kilometers in diameter. Calderas form when the chamber of magma underlying a volcano partially empties due to eruption, and the unsupported roof of the chamber collapses under its own weight. Crater Lake in Oregon (**FIGURE 6.5**) occupies a caldera 8 kilometers in diameter that formed after an immense eruption about 6,600 years ago. Tephra deposits from that eruption can still be seen in Crater Lake National Park and over a vast area of the northwestern United States and southwestern Canada.

Volcanoes do not necessarily become inactive after a major eruption. If magma begins to enter the

Crater Lake FIGURE 6.5 ▶

Beautiful Crater Lake, Oregon, the deepest lake in the United States, is all that remains of a once-lofty stratovolcano that geologists have named Mount Mazama. Wizard Island, a small tephra cone in the middle of the lake, formed by resurgent activity after the collapse that created the caldera.

chamber again, it may lift the floor of the caldera or crater and form a *resurgent dome*. The caldera of Mount St. Helens contains a dome that has been growing since the eruption of 1980.

When volcanism finally ceases, the magma chamber still contains hot, (though not necessarily molten) rock, for hundreds of thousands of years. When groundwater comes into contact with this hot rock, it heats up and may create a *thermal spring*. Many such springs have turned into famous health spas. Some thermal springs have a natural system of plumbing that allows intermittent eruptions of water and steam. These are called *geysers*, a name that comes from the Icelandic word *geysir*, meaning "to gush" (FIGURE 6.6). Finally, some volcanic vents emit only gas—usually water vapor that's sometimes mixed with foul-smelling sulfur compounds. These features are known as *fumaroles*.

HAZARDS AND PREDICTION

Like other natural hazards, such as earthquakes, a volcanic eruption has *primary effects*, which are directly caused by the eruption itself; *secondary effects*, which are indirectly triggered by the eruption; and *tertiary effects*, which are long-lasting or even permanent changes brought about by the eruption. Even though many effects of volcanism are harmful, some, such as volcanic ash supplying nutrients to the soil, are beneficial.

The Great Geysir FIGURE 6.6 ▼

The geyser from which all other geysers take their name is located in Iceland, which is a literal hotbed of geothermal activity.

Lava flow FIGURE 6.7

This house in Kalapana, Hawaii, is about to succumb to the slow but unstoppable advance of a lava flow in June 1989. The grass of the lawn burns on contact with the molten rock.

Primary effects
Most volcanoes produce at least some lava flows. Because people are usually able to outrun them, lava flows typically cause more property damage than injuries. In Hawaii, where Kilauea has erupted almost continuously for more than two decades, homes, cars, roads and forests have been buried by lava, but not a single life has been lost (**FIGURE 6.7**). It is sometimes possible to control a lava flow, at least partially, with retaining walls or a water spray, but otherwise nothing can be done to stop an eruption from occurring.

The greatest threats to human life during volcanic eruptions do not come from lava, but from pyroclastic flows and volcanic gases. Unlike slowly moving lava, pyroclastic flows move extremely fast and can easily outrace a running (or even a driving) human. The most destructive pyroclastic flow in the twentieth century (in terms of lives lost) occurred on the island of Martinique in 1902, when an avalanche of searing ash descended Mont Pelée at a speed of more than 160 km/hr and killed 29,000 people. (There were two survivors.) In 79 A.D., the Italian towns of Pompeii and Herculaneum were buried under hot pyroclastic material, entombing the bodies and buildings in a natural time capsule (**FIGURE 6.8**). However, most of these people were dead already, due to another hazard of

volcanic eruptions: poisonous gases. More recently, at least 1700 people and 3000 cattle lost their lives when poisonous gas erupted from a volcano at Lake Nyos in Cameroon (see Case Study). In 1783 the Laki eruption in Iceland released so much acidic gas that nearly a third of the people and half the domestic animals in the country perished.

Secondary effects
Secondary effects, related to volcanic activity but not a direct result of it, include fires (which are often caused by lava flows) and flooding (which may happen if a river channel is blocked or a crater lake bursts). The famous eruption at Krakatau, Indonesia, in 1883 claimed most of its victims due to a tsunami, the same kind of ocean wave that can also be caused by earthquakes. Volcanoes can also produce *volcanic tremors*, a type of seismic activity that helps scientists predict eruptions but rarely poses a threat itself.

Victim of Mt. Vesuvius FIGURE 6.8

This is not an actual body but a plaster cast of a citizen of Pompeii, Italy, who was killed during the eruption of Mount Vesuvius in 79 A.D. Death was caused by poisonous gases; then the body was encased by pyroclastic material. Over the centuries the body decayed, but a mold of its shape remained in the tephra. The cast was made when the natural mold was discovered during modern archaeologic expeditions.

Lakes of Death in Cameroon

Two small volcanic lakes in a remote part of Cameroon, a country in central Africa (A), made international news in the mid-1980s when they emitted lethal and invisible clouds of carbon dioxide from deep beneath their surface. The first gas discharge, which occurred at Lake Monoun (not shown) in 1984, asphyxiated 37 people. The second, which occurred at Lake Nyos in 1986, released a highly concentrated cloud of carbon dioxide that killed more than 1700 people. Both occurred at night during the rainy season, both involved volcanic crater lakes, and both are likely to happen again if technological intervention is not successful.

After the incidents, scientists discovered that the lakes had huge reservoirs of carbon dioxide dissolved and stored beneath unstable, density-layered lake waters. Some minor event—a landslide, or perhaps nothing more than winds at the surface—disturbed the layers, and the stratified column of water turned over, allowing approximately 100 million cubic meters of carbon dioxide to bubble to the surface in just two

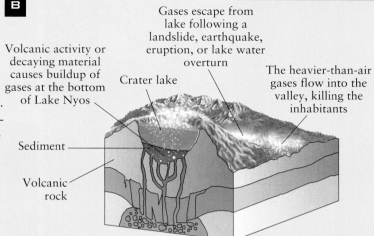

B

Volcanic activity or decaying material causes buildup of gases at the bottom of Lake Nyos

Gases escape from lake following a landslide, earthquake, eruption, or lake water overturn

Crater lake

The heavier-than-air gases flow into the valley, killing the inhabitants

Sediment

Volcanic rock

hours (B). Because carbon dioxide is heavier than air, it flowed down the mountainside in a ground-hugging layer that displaced the oxygen that both cattle (C) and people needed to breathe.

The supply of gas at the lake bottom is constantly being replenished. Scientists estimated that Lake Monoun could experience another violent degassing event within 10 years, and Lake Nyos is at risk within 20 years. It may be possible to siphon off the excess gas by installing a subsurface network of pipes. A prototype system was tested successfully at Lake Monoun in 1992. However, Lake Nyos poses a more complicated problem because of its greater depth and its fragile stratification. Engineers must be very careful not to initiate the very situation that they are trying to prevent.

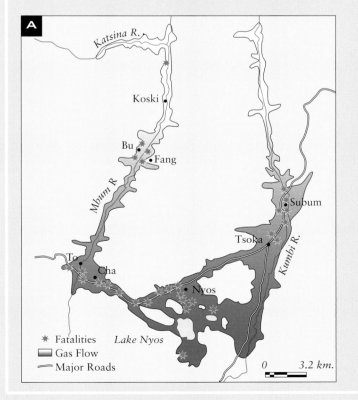

A

Katsina R.

Koski

Bu
Fang

Mbum R.

Subum

Tsoka

Kumbi R.

To
Cha

Nyos

* Fatalities *Lake Nyos*
▭ Gas Flow
〰 Major Roads

0   3.2 km.

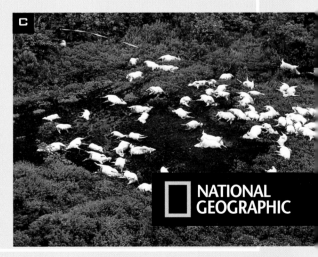

C

NATIONAL GEOGRAPHIC

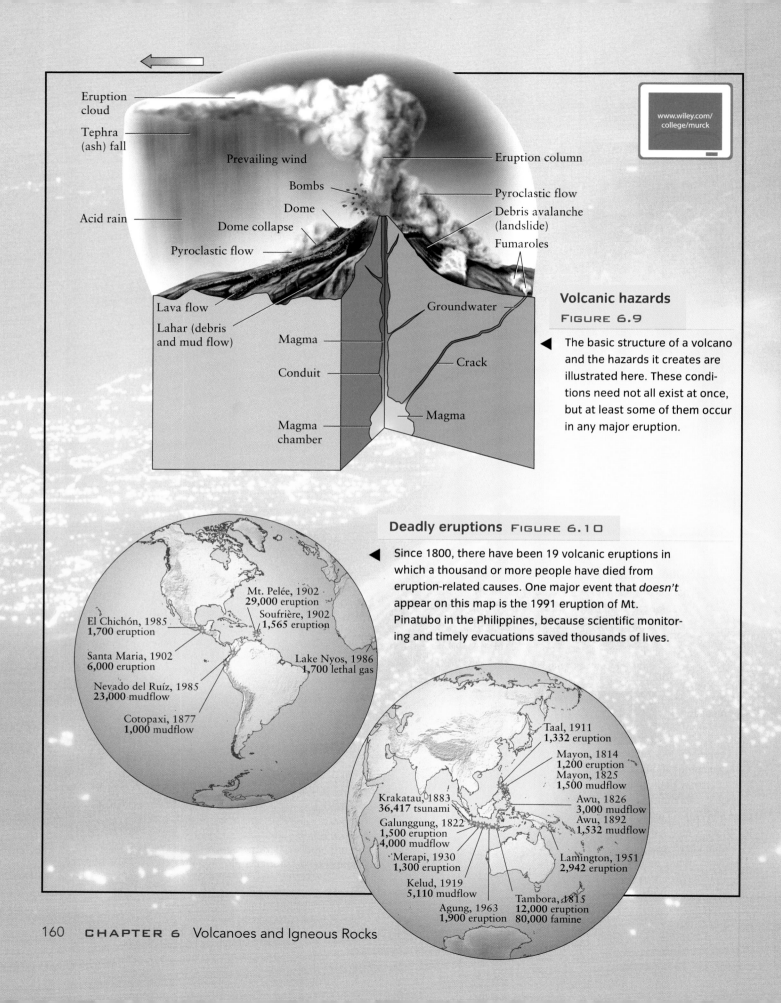

Eruption cloud

Tephra (ash) fall

Prevailing wind

Acid rain

Bombs

Dome

Dome collapse

Pyroclastic flow

Lava flow

Lahar (debris and mud flow)

Magma

Conduit

Magma chamber

Eruption column

Pyroclastic flow

Debris avalanche (landslide)

Fumaroles

Groundwater

Crack

Magma

www.wiley.com/college/murck

Volcanic hazards
FIGURE 6.9

◄ The basic structure of a volcano and the hazards it creates are illustrated here. These conditions need not all exist at once, but at least some of them occur in any major eruption.

Deadly eruptions FIGURE 6.10

◄ Since 1800, there have been 19 volcanic eruptions in which a thousand or more people have died from eruption-related causes. One major event that *doesn't* appear on this map is the 1991 eruption of Mt. Pinatubo in the Philippines, because scientific monitoring and timely evacuations saved thousands of lives.

Mt. Pelée, 1902
29,000 eruption

Soufrière, 1902
1,565 eruption

El Chichón, 1985
1,700 eruption

Santa Maria, 1902
6,000 eruption

Nevado del Ruíz, 1985
23,000 mudflow

Lake Nyos, 1986
1,700 lethal gas

Cotopaxi, 1877
1,000 mudflow

Taal, 1911
1,332 eruption

Mayon, 1814
1,200 eruption
Mayon, 1825
1,500 mudflow

Awu, 1826
3,000 mudflow
Awu, 1892
1,532 mudflow

Krakatau, 1883
36,417 tsunami

Galunggung, 1822
1,500 eruption
4,000 mudflow

Merapi, 1930
1,300 eruption

Kelud, 1919
5,110 mudflow

Lamington, 1951
2,942 eruption

Agung, 1963
1,900 eruption

Tambora, 1815
12,000 eruption
80,000 famine

fissure

crater

Volcanoes and climate FIGURE 6.11

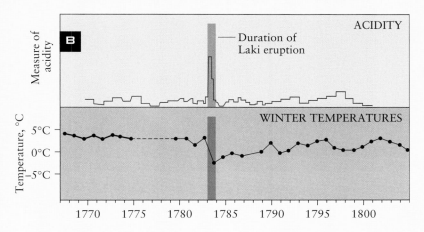

A The fissure eruption of Laki, a volcano in Iceland, lasted from 1783 to 1784 and was the largest flow of lava in recorded history.

B In the winter after Laki's eruption, the average temperature in the northern hemisphere was about 1°C below normal. In the eastern United States, the decrease was closer to 2.5°C. At the same time, ice cores from Greenland record a dramatic spike in acidity, due to acid precipitation.

Mudslides have often been a major cause of volcano-related deaths. When volcanic ash mixes with snow at the volcano's summit (or vice versa, when rain falls on recently deposited volcanic ash), it can start a deadly mudflow called a *lahar*. In the chapter introduction you saw a lahar from the eruption of Mt. Pinatubo. Lahars can occur months after the eruption. A related phenomenon is a volcanic *debris avalanche,* in which many different types of material such as mud, pyroclastic material, and downed trees are mixed together. A devastating debris avalanche caused much of the damage from the 1980 eruption of Mount St. Helens.

FIGURE 6.9 summarizes the various primary and secondary hazards from an eruption, and FIGURE 6.10 shows where the most deadly eruptions of the past two centuries occurred. Note that in many cases the secondary effects are responsible for the greatest loss of life.

Tertiary and beneficial effects

Volcanic activity can change a landscape. Eruptions can block river channels and divert the flow of water. They can dramatically alter a mountain's appearance, as in the 1980 eruption of Mount St. Helen's (Figure 6.1). They can form new land, such as the black sand beaches of Hawaii, which are made of dark pyroclastic fragments, or the volcanic island of Surtsey, which emerged from the ocean near Iceland in 1963 and is composed of both lava flows and pyroclastic cones.

Volcanoes can also affect the climate on a regional and global scale (see FIGURE 6.11). Major eruptions can cause toxic and acidic rain, spectacular sunsets, or extended periods of darkness. Sulfur dioxide, a common gaseous emission of volcanoes, forms small droplets or *aerosols.* If they get into the stratosphere, these aerosols spread around the world, absorb sunlight, and cool Earth's surface. An example from recent times is the 1815 eruption of Tambora in Indonesia, which caused three days of near darkness as far away as Australia. The following year was so cool in Europe and North America that it was called "the year without a summer." Farther back in time the eruption of flood basalts, such as the Deccan Traps in India and the Siberian Traps in Russia, may have caused or contributed to several of the mass extinctions that divide geologic periods.

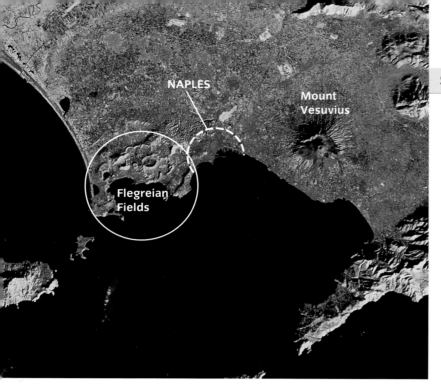

Satellite monitoring FIGURE 6.12

This false-color satellite image shows the area around Mount Vesuvius (center right) and the Bay of Naples, Italy, a densely populated region. Recent lava flows show up bright red in this image, which records infrared radiation (i.e., heat). Older lavas and volcanic ash show up as shades of yellow and orange. The dark blue and purple region at the head of the bay is the city of Naples. West of Naples lies a cluster of smaller volcanoes called the Flegreian Fields. By comparing successive satellite images, geologists can detect changes in ground temperature.

Not all the effects of volcanoes are negative, and it is no accident that people live near active volcanoes. Periodic volcanic eruptions renew the mineral content of soils and replenish their fertility. Volcanism also provides geothermal energy and some types of mineral deposits.

Predicting eruptions

It isn't possible to stop volcanic eruptions, but it is sometimes possible to predict them. The first step in prediction is to identify a volcano as active, dormant, or extinct. An *active* volcano has erupted within recorded history; a *dormant* one has not erupted in recent history. A volcano is called *extinct* when it shows no signs of activity and is deeply eroded. Mount Pinatubo in the Philippines had been dormant for about 500 years prior to its awakening in 1991.

Another important step in prediction is identifying the volcano's past eruptive style. For example, Mount Pinatubo is surrounded by thick deposits of pyroclastic material, a sign that the volcano erupted violently in the past. Subduction zone volcanoes such as Mount St. Helens are more likely to erupt explosively than shield volcanoes and fissure eruptions. The type of rock that has solidified from past eruptions, either silica-rich or silica-poor, also indicates a volcano's style of eruption.

When a volcano shows warning signs of increasing activity, scientists monitor it more closely. Some of these warning signs include changes in the shape or elevation of the ground, such as bulging, swelling, or the formation of a dome. The presence of these features suggests that the underground reservoir of magma is growing. The release of gases can also be a warning sign, as can changes in the temperature of crater lakes, well water, or hot springs. A sudden increase in local seismic activity is also a warning sign. Geologists monitor volcanic activity by using tiltmeters to detect bulging, satellite images, and devices that identify gas emissions or changes in the temperature of the ground (FIGURE 6.12).

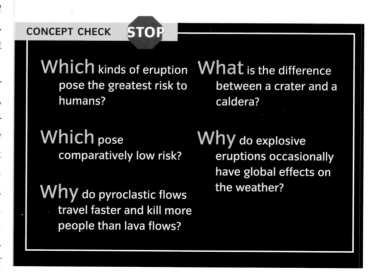

CONCEPT CHECK STOP

Which kinds of eruption pose the greatest risk to humans?

What is the difference between a crater and a caldera?

Which pose comparatively low risk?

Why do explosive eruptions occasionally have global effects on the weather?

Why do pyroclastic flows travel faster and kill more people than lava flows?

Why and How Rocks Melt

U nderneath every active volcano lies a reservoir of magma, called a *magma chamber*. Understanding volcanism includes understanding how rocks melt to become magma. Fortunately, rocks can be melted artificially as well as naturally (**FIGURE 6.13**). We can thus learn about the behavior of molten rock from laboratory experiments.

At Earth's surface, rock begins to liquefy when heated to a temperature between about 800°C and 1000°C. However, rock (unlike ice, for example) typically consists of many different minerals, each with its own characteristic melting temperature. Thus we cannot talk about a single melting point for a rock. Complete melting is commonly attained by about 1200°C. Two other factors also strongly affect the melting temperature: pressure and the presence of water in the rock.

HEAT AND PRESSURE INSIDE EARTH

If you descend into a mine, it becomes apparent that the farther down you go, the hotter it gets. The rate at which temperature increases with depth, called the geothermal gradient, is quite different underneath continental surfaces than it is under the seafloor. Continental crust is thick, and the temperature underneath it increases more gently, at an average rate of about 6.7°C per kilometer, reaching 1000°C at a depth

Molten rock: artificial vs. natural FIGURE 6.13

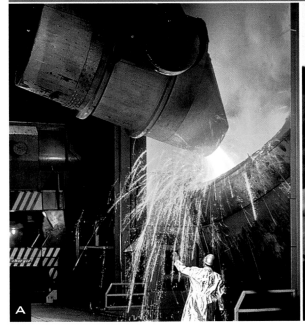

A In a steel mill, workers heat metal ores to the melting point in order to separate the metal from the surrounding rock.

B A geologist in a protective suit measures the temperature of lava erupting from Mauna Loa, Hawaii. Bright orange, yellow, and white lava is hotter, whereas dull red, brown and black colors indicate cooler lava.

Geothermal gradient FIGURE 6.14

A Temperature increases with depth. The dashed lines are *isotherms*, lines of equal temperature. Notice how the lines "sag" underneath the continental crust, because the rate of increase of temperature is slower there.

B This represents the same information as **A**, in a graph. Earth's surface is at the top, so depth (and pressure) increase as you move down. The dashed curve shows the geothermal gradient under oceans, and the solid curve shows the gradient under continental crust. Note that the two curves merge (and the isotherms become level) below 200 kilometers.

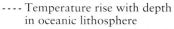
- - - - Temperature rise with depth in oceanic lithosphere
—— Temperature rise with depth in continental lithosphere

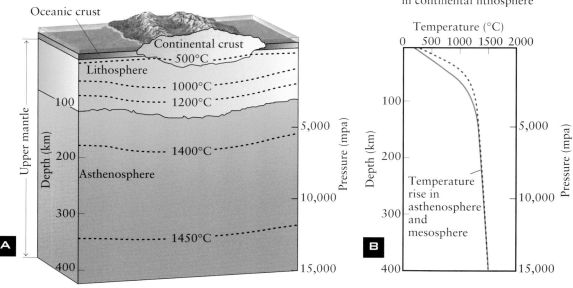

of 150 kilometers. Underneath the ocean floor, the rate of increase is about twice as rapid. The temperature increases by 13°C per kilometer, reaching 1000°C at a comparatively shallow depth of 80 kilometers (FIGURE 6.14). Below the asthenosphere-lithosphere boundary, the geothermal gradient becomes more gradual (0.5°C/km) and the temperature difference between suboceanic and subcontinental rock disappears.

As you can see in Figure 6.14, the temperature in the upper mantle is higher than the temperature at which most rocks melt at Earth's surface. Yet the upper mantle is mostly solid. How is this possible?

The answer is that the pressure also rises very dramatically with increasing depth, and increasing pressure causes rock to resist melting (FIGURE 6.15A). For example, albite, a common rock-forming mineral (a feldspar), melts at 1104°C at the surface. At a depth of 100 kilometers, the pressure is 35,000 times greater than it is at sea level. At that pressure, the melting temperature of albite rises to 1440°C, which still slightly exceeds the normal temperature at that depth. Thus albite remains solid when it is beneath the surface.

The presence of water (or water vapor) in a rock dramatically reduces its melting temperature (FIGURE 6.15B). By analogy, as anyone who lives in a cold climate knows, salt can melt the ice on an icy road because a mixture of salt and ice has a lower melting temperature than pure ice. Similarly, a mineral-and-water mixture has a lower melting temperature than the dry mineral alone.

The effect of water on the melting of a rock becomes particularly important in subduction zones, where water is carried down into the mantle by oceanic crust, as described in Chapter 4.

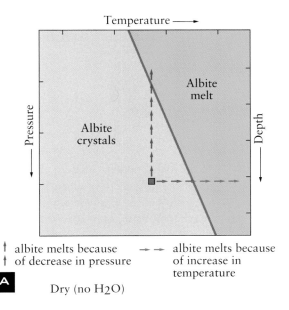

albite melts because
of decrease in pressure

albite melts because
of increase in
temperature

A Dry (no H2O)

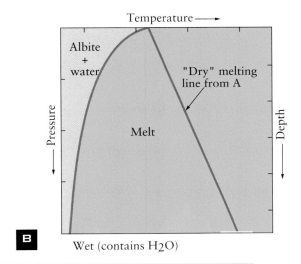

B Wet (contains H2O)

Effects of temperature and pressure on melting FIGURE 6.15

A The melting temperature of a dry mineral (albite, in this case) increases at high pressures. A mineral at depth (shown by the small square) can melt in two different ways: either by an increase in temperature (red arrows) or by a decrease in pressure (blue arrows). The latter effect is called *decompression melting*, and it is an important reason why many magmas stay molten all the way to the surface. B The melting temperature of a mineral in the presence of water typically decreases as pressure increases. This is exactly the opposite of what happens to dry minerals. Magmas containing dissolved water typically solidify before they reach the surface.

Fractional melting Because rocks are composed of many different minerals, they melt over a range of temperatures. This means that the boundary between solid and melt is not crisp, but blurry, as in FIGURE 6.16 on the following page. When the temperature rises enough for part of the materials in a rock to melt and part to remain solid, it becomes a **fractional melt**. Only if the temperature continues to increase or the pressure decrease, will the rock melt completely. **Fractionation**, an important process that can lead to the development of a diversity of rock types, is caused by fractional melting.

> ■ **fractional melt** A mixture of molten and solid rock.
>
> ■ **fractionation** Separation of melted materials from the remaining solid material during the course of melting.

Magma and lava As mentioned earlier, molten rock above or below ground is called *magma*. When magma reaches the surface, it is called *lava*. A lot of magma never reaches the surface, but instead remains underground, trapped in a magma chamber, until it crystallizes and hardens to igneous rock. We cannot study magma underground in its natural setting, but we can study lava and we can experiment with synthetic magma. From our direct observations of lava, we know that magmas differ in *composition, temperature,* and *viscosity.*

Composition Most magma is dominated by silicon, aluminum, iron, calcium, magnesium, sodium, potassium, hydrogen, and oxygen—Earth's most abundant elements. Oxygen combines with the others to form oxides, such as SiO_2, Al_2O_3, CaO, and H_2O. Silica, or SiO_2, usually accounts for 45 percent to 75 percent of the magma, by weight. In addition, a small amount of dissolved gas (between 0.2% and 3% of the magma, by weight) is usually present, primarily water vapor (H_2O) and carbon dioxide (CO_2). Despite their low abundance, these gases strongly influence the properties of magma. The proportion of silica (SiO_2) also has a strong effect on the magma's appearance and properties.

Process Diagram

Effect of temperature and pressure on rocks FIGURE 6.16

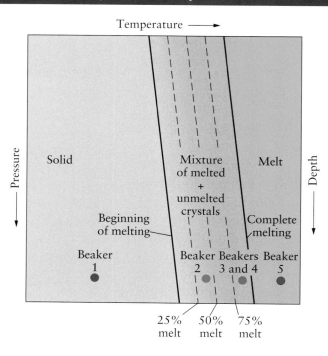

◀ Because almost all rocks contain a mixture of materials, they do not melt all at once, at a single temperature; instead, there is a range of temperatures and pressures in which they contain a mixture of melted and unmelted crystals. The dots on this diagram show the stages in the melting process, which are illustrated further below.

▼ The process of fractional melting can lead to another effect called fractionation (illustrated for convenience in a laboratory beaker rather than buried in Earth's mantle).

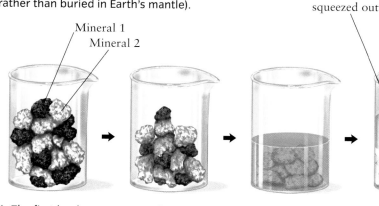

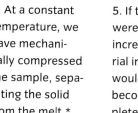

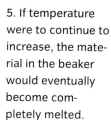

Mineral 1
Mineral 2

Melt, rich in Mineral 1 is squeezed out

Remaining solid is rich in Mineral 2 (high-temperature component)

1. The first beaker shows a mixture of two minerals. At a low temperature, both are solid.

2. As the temperature increases, mineral 1 (the dark mineral) begins to melt.

3. Mineral 1 has totally melted and has dissolved some of mineral 2 in the process; remainder of mineral 2 remains solid.

4. At a constant temperature, we have mechanically compressed the sample, separating the solid from the melt.*

5. If temperature were to continue to increase, the material in the beaker would eventually become completely melted.

*In the lithosphere, this mechanical separation can occur as a result of the tectonic motion of plates. If the melt now cooled again, we would have two separate deposits, one of mineral 1 plus a little of mineral 2, crystallizing from the melt, and one of pure mineral 2, the solid remnant in Beaker 4.

A Lavas at high temperature and low silica content tend to flow more freely. This stream of low-viscosity (runny) lava was erupted in Hawaii in 1983. The temperature of the lava was about 1100°C.

B Two strikingly different lava flows are visible in this scene from Kilauea volcano in Hawaii. The smooth, ropy rocks on which the geologist is standing formed from a low-viscosity lava such as that shown in **A**. The rough, chunky rocks that the geologist is sampling came from a more viscous, slow-moving flow that erupted 40 years later. Geologists use Hawaiian words, calling the first kind of flow *pahoehoe* and the second *aa*.

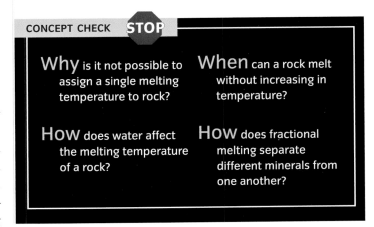

Temperature   We know from direct measurements at erupting volcanoes that lavas vary in temperature from about 750°C to 1200°C. From laboratory experiments with synthetic magma, geologists know that magma temperatures in the mantle must rise as high as 1400°C. They also know that magmas with high H_2O contents tend to melt at lower temperatures.

Viscosity   All magma is liquid and has the ability to flow, but magmas differ to a marked extent in how readily they flow. This is certainly true for lavas. As shown in FIGURE 6.17, some lavas are very fluid, almost like a stream of water. But other lavas creep along slowly and inexorably, like molasses. In particular, lavas with high silica content tend to flow slowly because of the tendency of silica molecules to polymerize, or form long chains (see Chapter 2). Thick, slow-moving lavas have high viscosity, and these lavas have the greatest tendency to erupt explosively.

CONCEPT CHECK **STOP**

Why is it not possible to assign a single melting temperature to rock?

When can a rock melt without increasing in temperature?

How does water affect the melting temperature of a rock?

How does fractional melting separate different minerals from one another?

Cooling and Crystallization

LEARNING OBJECTIVES

Distinguish between volcanic and plutonic rocks.

Identify different textures of igneous rocks and **explain** the physical processes that produce them.

Identify the three main types of volcanic rock and their plutonic equivalents.

Explain how different igneous rocks may form from the same magma, through fractional crystallization.

Whereas melting influences the properties of magma, cooling and **crystallization** influence the properties of igneous rock. For example, the *rate of cooling* determines how large the individual mineral grains in the rock will grow. Grain size affects the appearance or *texture* of the rock. The *composition* of the magma determines the final mineral assemblage in the solidified rock. Let's examine each of these factors more closely.

RATE OF COOLING

Even with a cursory look at samples of igneous rocks, you can see that some contain easily visible mineral grains, and others do not. Rate of cooling is the main factor that distinguishes the two groups. **Volcanic rocks** form at Earth's surface, where lava contacts much cooler air or water. They solidify so quickly that large mineral grains have no time to form. **Plutonic rocks** form when magma crystallizes deep underground. This process is much slower, and therefore gives the mineral grains time to grow larger. Rocks with different grain sizes have easily distinguishable textures.

Rapid cooling: volcanic rocks and their textures
Sometimes lava cools so rapidly that mineral grains do not have a chance to form at all. The resulting volcanic rock is not crystalline but *glassy* (**FIGURE 6.18A**). More often, mineral grains do form in solidifying lavas, but they are extremely small, and can only be seen under magnification. (See What a Geologist Sees, p. 172.)

Rocks with this very fine-grained texture are said to be *aphanitic* (**FIGURE 6.18B**).

Some volcanic rocks, with a texture known as *porphyritic* (**FIGURE 6.18C**), consist of large mineral grains embedded in an aphanitic matrix. This happens when the magma starts to crystallize at depth and grow large grains before it erupts, and the suspended large grains erupt along with the liquid lava. Once erupted, the remaining liquid lava cools rapidly to form an aphanitic matrix in which are embedded the large mineral grains that formed earlier.

Dissolved gases, too, can affect the texture of volcanic rock. An erupting lava may froth and bubble; if the froth is blasted into the air and cools quickly it forms *pumice*, which is a glassy mass full of bubbles (called *vesicles*). As lava cools, the viscosity increases and it becomes increasingly difficult for gas bubbles to escape. When the lava finally solidifies into rock, the last bubbles to form may become trapped. In basaltic lava, this process can create a volcanic rock with lots of bubble holes that looks like Swiss cheese. Such a rock is called a *vesicular basalt*.

Slow cooling: plutonic rocks and their textures
Unlike the minerals in volcanic rocks, those in plutonic rocks usually have time to form mineral grains that can be readily seen by the unaided eye. This coarse-grained texture is said to be *phaneritic* (**FIGURE 6.19**). Exceptionally large mineral grains (sometimes up to several meters!) typically form in the last stage of crystallization of a plutonic rock body, when gases build up in the remaining magma. The vapor facilitates the growth of large crystals, because chemicals can migrate quickly to the growing crystal faces.

crystallization
The process whereby mineral grains form and grow in a cooling magma (or lava).

volcanic rock
An igneous rock formed from lava.

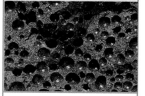

plutonic rock
An igneous rock formed underground from magma.

Volcanic rock textures FIGURE 6.18

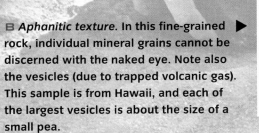

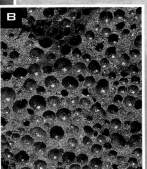

A *Glassy texture.* This photograph shows obsidian obelisks from a Mayan grave in Guatemala. Though these were sculpted by humans, a shiny, curvy appearance is typical for volcanic glass.

B *Aphanitic texture.* In this fine-grained rock, individual mineral grains cannot be discerned with the naked eye. Note also the vesicles (due to trapped volcanic gas). This sample is from Hawaii, and each of the largest vesicles is about the size of a small pea.

C *Porphyritic texture.* These volcanic rocks from Nevada contain large mineral grains, or *phenocrysts*, suspended in an aphanitic material called the *groundmass*. The largest grains visible in the photo are approximately 6 mm in length.

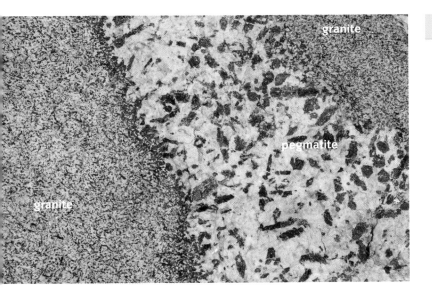

Plutonic rock and textures FIGURE 6.19

Two distinct textures of plutonic rock can be seen in this granite specimen from California's Sierra Nevada. The two outside layers have *phaneritic* texture, with small but visible grains of plagioclase, potassium feldspar, quartz (white), and biotite (black). Sandwiched between them is a vein of pegmatite, which contains the same minerals but in much larger grains. *Pegmatite* is a rock with mineral grains larger than 2 cm.

CHEMICAL COMPOSITION

Geologists subdivide the most common igneous rocks into three broad categories, based on their silica contents. Rocks that contain large amounts of silica (about 70% SiO_2 by weight) are usually light-colored. Geologists call them *felsic* (a word formed from "feldspar" and "silica"), because feldspar is the most common mineral found in them. At the other end of the scale are rocks called *mafic* (a word formed from "magnesium" and "ferric," or iron-rich), which contain large amounts of dark-colored minerals rich in magnesium and iron. They are usually lower in silica content (about 50% SiO_2 by weight). Finally, igneous rocks with about 60% SiO_2 by weight are said to be *intermediate*.

Volcanic and plutonic equivalents TABLE 6.1

Grain Size →

Silica Content ↑

Silica Content of Magma	Resulting Volcanic Rocks		Resulting Plutonic Rocks	
High (= 70%–75%)	*Rhyolite* lies at the felsic, high-silica end of the scale, and consists largely of quartz and feldspars. It is usually pale, ranging from nearly white to shades of gray, yellow, red, or lavender.		*Granite,* the plutonic equivalent of rhyolite, is common because felsic magmas usually crystallize before they reach the surface. It is found most often in continental crust, especially in the cores of mountain ranges.	
Intermediate (= 60%)	*Andesite* is an intermediate-silica rock, with lots of feldspar mixed with darker mafic minerals such as amphibole or pyroxene. It is usually light- to dark gray, purple, or green.		*Diorite* is the plutonic equivalent of andesite, an intermediate-silica rock.	
Low (= 45%–50%)	*Basalt,* a mafic rock, is dominant in oceanic crust, and the most common igneous rock on Earth. Large, low-viscosity lava flows from shield volcanoes and fissures are usually basaltic. Dark-colored pyroxene and olivine give it a dark gray, dark green, or black color.		*Gabbro* is the plutonic equivalent of basalt, a low-silica rock.	

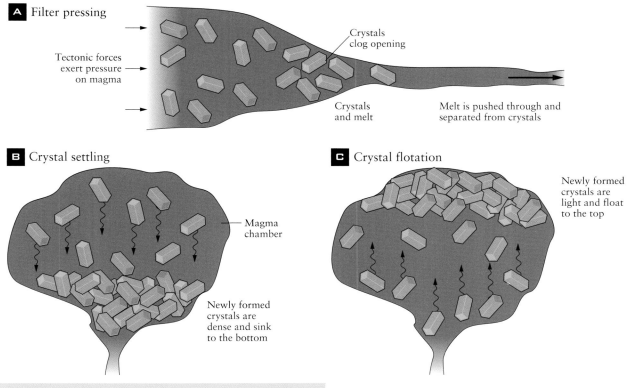

A Filter pressing

Tectonic forces exert pressure on magma

Crystals clog opening

Crystals and melt

Melt is pushed through and separated from crystals

B Crystal settling

Magma chamber

Newly formed crystals are dense and sink to the bottom

C Crystal flotation

Newly formed crystals are light and float to the top

Separating crystals from melts FIGURE 6.20

A In *filter pressing*, magma is squeezed through a small opening by tectonic forces. Only the liquid gets through, and the newly formed crystals are left behind.

B In *crystal settling*, the first minerals to crystallize are denser than the melt and may sink to the bottom.

C In *crystal flotation*, the first crystals are lighter than the liquid, and may rise to the top.

Geologists organize rock types in two ways: either by grain size or by silica content. The results are shown in TABLE 6.1. The rocks on the left are volcanic (aphanitic and fine-grained), and the ones on the right are plutonic (phaneritic and coarse-grained). The ones on the top contain the most silica and are lightest in color, and the ones on the bottom contain the least silica and are darkest in color.

The composition of an igneous rock is closely related to the composition of its parent magma. Basaltic magma flows freely, tends to contain less trapped gas, and erupts less explosively. Rhyolitic magma, because of its high silica content, flows less freely, traps more gas, and thus tends to erupt explosively. In this way, the

fractional crystallization
Separation of crystals from liquids during crystallization.

rock content of a volcano gives an important clue to its history; that is, a volcano's rocks show how it has erupted in the past and how it is likely to erupt in the future.

FRACTIONAL CRYSTALLIZATION

Although the six igneous rocks discussed in Table 6.1 are the most common, there are literally hundreds of other kinds of igneous rocks on Earth. The reason for this diversity is that a single magma source can *differentiate* into several kinds of igneous rock through **fractional crystallization**, a sort of reversal of fractional melting (FIGURE 6.20).

Crystallization occurs over the same range of temperatures as melting, and the last minerals to melt are the first to crystallize. As shown in Figure 6.20, the newly formed crystals can become separated from the remaining magma in several different ways. The result is a rock and a magma with different compositions, both of them different from the original magma. When you combine the different magma compositions, rates of cooling, and modes of fractional crystallization, you can see why there are so many different types of igneous rocks.

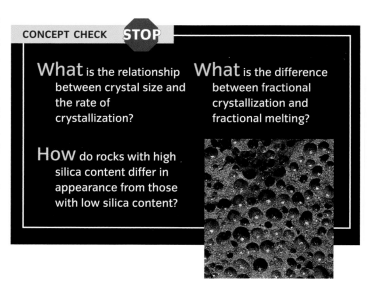

What a Geologist Sees

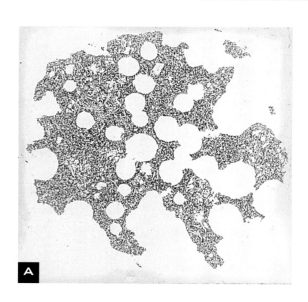

Putting Rocks Under a Microscope

Geologists can identify the minerals in an aphanitic rock by studying it under a microscope.

A The specimen of aphanitic volcanic rock from Figure 6.18B (page 169) has been cut and polished into a wafer that is thin enough to allow light to pass through. Geologists call these rock slices *thin sections*. This magnified thin section is about 0.03 mm thick, as thin as a piece of tissue paper, and about 3.5 mm across.

B Using polarized light, geologists can see individual mineral grains and identify them by their color and other optical properties. Using a microscope to study thin sections is a common technique in geology to analyze rocks of all types for many purposes.

Plutons and Plutonism

Although they may be less familiar to you than volcanic rocks, plutonic rocks give rise to some very dramatic geologic formations, known as **plutons** (FIGURE 6.21). Plutons are always *intrusive* bodies, different from the rock that surrounds them. They always originate as magma underground, but they are exposed as plutons at the surface by erosion.

BATHOLITHS AND STOCKS

Plutons are named according to their shapes and sizes. The largest type of pluton is a **batholith** (from the Greek words meaning "deep rock"). Some batholiths exceed 1000 kilometers in length and 250 kilometers in width (FIGURE 6.22 on the following page).

Where they are visible at the surface due to erosion, the walls of batholiths tend to be nearly vertical. This early observation led geologists to believe that batholiths extend downward to the base of Earth's crust. However, geophysical measurements suggest that this perception is incorrect. Most batholiths seem to be only 20 to 30 kilometers thick.

A smaller version of a batholith, only 10 kilometers or so in its maximum dimension, is called a *stock*. In some cases, as shown in Figure 6.21, a stock may be associated with a batholith that lies underneath it.

pluton Any body of intrusive igneous rock, regardless of size or shape.

batholith A large, irregularly shaped pluton that cuts across the layering of the rock into which it intrudes.

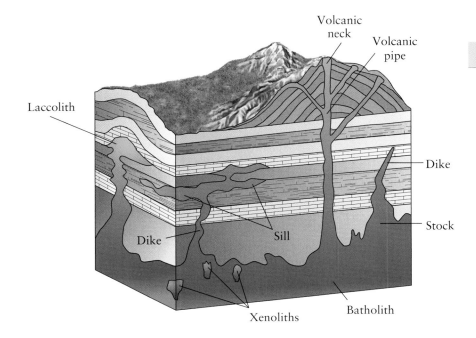

Volcanic neck

Volcanic pipe

Laccolith

Dike

Dike

Sill

Stock

Xenoliths

Batholith

Plutons FIGURE 6.21

This diagram shows the origin of various forms taken by plutons. Note the vertical volcanic necks; sills parallel to the layering in the surrounding rocks; and dikes, which cut across the surrounding rock layers. In every case, the magma intrudes into previously existing rock.

Most stocks and batholiths are granitic or between granite and diorite in composition. The magma that forms batholiths probably results from extensive fractional melting of the lower continental crust. Despite their huge size, the magma bodies that form batholiths migrate upward, squeezing into preexisting fractures and pushing overlying rocks out of their way.

DIKES AND SILLS

Smaller plutons tend to take advantage of fractures in the rock. Two of the most obvious indicators of past igneous activity are **dikes** and **sills** (FIGURE 6.23). A dike forms when magma squeezes into a cross-cutting fracture and then solidifies. If the magma intrudes between two layers and is parallel to them, it forms a sill. Sometimes this intrusion will cause the overlying rock to bulge upward, forming a mushroom-shaped pluton called a *laccolith*. As shown in Figure 6.21, all of these intrusive forms may occur as part of a network of plutonic bodies.

Dikes and sills can be very large. For instance, there is a large and well-known sill-like mass, made of gabbro, in the Palisades, the cliffs that line the Hudson River opposite New York City. The Palisades Intrusive Sheet is about 300 meters thick. It formed from multiple charges of magma intruded between layers of sedimentary rock about 200 million years ago. The sheet is visible today because tectonic forces raised that portion of the crust upward, and then the covering sedimentary rocks were largely removed by erosion.

As Figure 6.21 shows, plutonic rocks can also be connected with volcanoes. Beneath every volcano lies a complex network of channels and chambers through which magma reaches the surface. When a volcano becomes extinct, the magma in the channels solidifies into various kinds of plutons. A *volcanic pipe* is the remnant of a channel that originally fed magma to the volcanic vent; when exposed by erosion, it is called a *volcanic neck* (Figure 6.23C).

Batholiths FIGURE 6.22

Because batholiths are so immense, we cannot show one in a single photo. However, we can illustrate them on a map. The Coast Range batholith of southern Alaska, British Columbia, and Washington dwarfs the largest batholiths of Idaho and California. Many of the individual stocks and batholiths shown on this map are just the exposed tops of much larger intrusive bodies that lie underground.

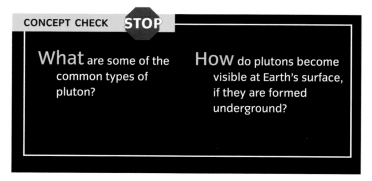

CONCEPT CHECK **STOP**

What are some of the common types of pluton?

How do plutons become visible at Earth's surface, if they are formed underground?

A A dike of gabbro cuts across horizontally layered sedimentary rocks in Grand Canyon National Park, Arizona.

dike

B The sill is the middle piece of this rock "sandwich," a layer of dark brown gabbro intruded between layers of sedimentary rock above and below, in Big Bend National Park, Texas.

sill

C This volcanic neck called Devil's Tower, in Wyoming, is all that remains of an ancient, eroded volcano. You might remember this location for the role it played in the movie *Close Encounters of the Third Kind.*

A Volcanic ash continued to erupt from the crater for nine hours.

pumice plain

terminus of debris avalanche

C Here you see hummocks left by the largest debris avalanche in recorded history.
Immediately after the eruption, this entire landscape was a barren gray. The terminus of the debris avalanche, where the flowing material stopped and hardened, dwarfs a passing truck in this photo.

B Weeks after the eruption, not a living thing could be found in the vicinity of the volcano. Here, a forester inspects a forest that was flattened instantaneously by the blast wave, as if the trees were just blades of grass.

Mount St. Helens is a young and active volcano that has erupted many times in its 37,000-year history. Even so, geologists and nearby residents alike were stunned by the violence of its eruption in 1980. The mountain blew off its top and most of its northern flank. The initial blast rattled windows all over the state of Washington, flattened trees as far as 27 kilometers away, and killed 57 people.

If you visit the Mount St. Helens National Volcanic Monument, you will see many signs of revival amidst vistas of incredible desolation.

D Today the volcano is quiescent, but you may be able to spot steam rising from the resurgent dome in the center of the crater. This is a reminder that Mount St. Helens may one day (if it remains active) grow back to its former height and violently explode or erupt again.

CHAPTER SUMMARY

1 Volcanoes and Volcanic Hazards

1. **Volcanoes** eject a wide variety of materials, including **lava** (molten rock, or **magma**, that has reached Earth's surface), gases, *volcanic ash,* larger pebbles, and rocks. The solid fragmental ejecta are collectively called *tephra.* Large tephra particles are called volcanic *bombs,* intermediate-sized *lapilli,* and the smallest, volcanic ash. When tephra is consolidated into a rock it is called an *agglomerate* if the particles are large, *tuff* if they are small.

2. The diversity of volcanic eruption types is mostly due to two factors, the **viscosity** of the magma, and the amount of gas present in it. The viscosity of the magma, in turn, depends on its chemical composition, temperature, and gas content. **Shield volcanoes** and fissure eruptions tend to be quiet and nonexplosive. They gradually build up the volcano through a series of lava flows. **Stratovolcanoes** tend to erupt explosively. They are built up from a series of layers of lava and pyroclastic material.

3. Funnel-shaped *craters* and much larger depression known as *calderas* are common features at the summits of volcanoes. New magma may push up the floor of a crater or caldera and form a *resurgent dome.* If groundwater in the vicinity of an active magma chamber becomes heated, it may form *thermal springs* or *geysers.*

4. Direct, or *primary*, volcanic hazards of volcanoes are those directly caused by the volcanic eruption. They include **pyroclastic flows**, lava flows, and poisonous gases. *Secondary* hazards are those triggered by the eruption, such as *volcanic tremors* or *lahars.* Some extremely large eruptions produce *tertiary effects*, long-lasting and even permanent changes brought about by the eruption. One example is a worldwide drop in temperature because of *aerosols* in the upper atmosphere. Tertiary effects can also be beneficial, such as the creation of new land.

5. For purposes of prediction, volcanoes are categorized as *active*, *dormant*, or *extinct*. Volcanologists can often tell an eruption is coming because of changes in seismic activity, gas emissions, ground and water temperature, and slope of a volcano's sides.

2 How and Why Rocks Melt

1. Both temperature and pressure increase with depth. The increase of temperature with depth is called the **geothermal gradient**; the rate of increase with depth is less through and under the continental crust than it is through and under the oceanic crust.

2. Minerals can melt in two ways: by an increase in temperature or a decrease in pressure (known as *decompression melting*). The presence of water in a rock typically lowers its melting point.

3. The minerals in a rock melt at different temperatures, so rocks do not have a single melting point; rather, they melt over a range of temperatures. A **fractional melt** is a body of rock in which some materials have melted and others have not. Fractional melting often separates minerals through a process called **fractionation**. When a rock begins to melt, therefore, only a small volume of melt—a **partial melt**— forms at first. In some circumstances, the melt may become segregated from the remaining solid rock (for example, by *filter pressing*), which may then continue to melt on its own. The separated material will have a different mineral composition from the original rock.

4. Magmas can be distinguished from one another by **composition**, **temperature**, and **viscosity**. Those with high silica contents usually melt at lower temperatures and have higher viscosities.

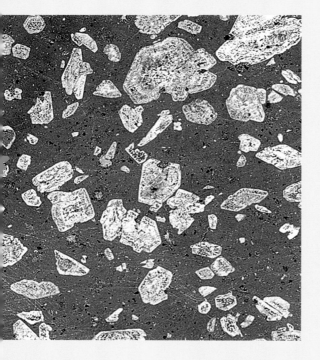

3 Cooling and Crystallization

1. **Crystallization,** the process whereby mineral grains form and grow in a cooling magma or lava, influences the final properties of the igneous rock, such as its *texture* and grain size. The crystallization process and the rock's final mineral assemblage, in turn, depend on such factors as the magma *composition* and the *rate of cooling.* A rock that cools from the molten state more rapidly will have smaller grains, because the grains do not have as much time to crystallize.

2. Igneous rocks can be classified into **volcanic** and **plutonic** rocks. Volcanic rocks crystallize from lava; plutonic rocks crystallize underground from magma. Because volcanic rocks cool rapidly, their crystals are small, microscopic, or even absent. These rocks are known as *aphanitic.* Plutonic rocks, on the other hand, are *phaneritic,* and have easily visible crystal grains.

3. *Felsic* rocks are typically light in color and have high silica contents, and *mafic* rocks are typically darker, with lower silica contents and a higher content of iron. The most common volcanic rock is *basalt,* a dark, low-silica *(mafic)* rock that is the main constituent of oceanic crust. *Granite,* a light-colored, high-silica *(felsic)* plutonic rock, is an important dominant constituent of continental crust.

4. **Fractional crystallization** occurs when mineral grains become separated from the melt from which they are crystallizing. This separation can happen if the crystals sink to the bottom *(crystal settling)* or float to the top of the magma *(crystal flotation).* If the magma flows through an opening that is too constricted to allow the crystals to pass through, separation by *filter pressing* may occur. The combination of different magma compositions, rates of cooling, and fractional crystallization leads to the great diversity of igneous rock types on Earth.

4 Plutons and Plutonism

1. Plutonic rock bodies, or **plutons,** form underground when magma intrudes into pre-existing rock layers, either cutting through them or flowing between them.

2. Plutons, which occur in a variety of sizes and shapes, can be exposed at the surface when surrounding rock is stripped away by weathering and erosion. The largest type of pluton is a **batholith,** an irregularly shaped igneous body more than 50 kilometers in diameter, which cuts across the rock it intrudes on. A *stock* is a smaller version of batholith. Two smaller types of plutons, **dikes** and **sills,** are indicators of past magmatic activity.

KEY TERMS

- volcano p. 152
- lava p. 152
- magma p. 152
- shield volcano p. 153
- pyroclastic flows p. 153

- stratovolcano p. 153
- viscosity p. 153
- fractional melt p. 165
- fractionation p. 165
- crystallization p. 168

- volcanic rock p. 168
- plutonic rock p. 168
- fractional crystallization p. 171
- pluton p. 173
- batholith p. 173

CRITICAL AND CREATIVE THINKING QUESTIONS

1. For many years scientists debated the reasons for the existence of Earth's geothermal gradient. What explanations can you think of for the hotter temperatures toward the center of Earth? Why is the geothermal gradient steeper under the oceans?

2. Volcanic rock is sometimes called "extrusive" and plutonic rock is sometimes called "intrusive." Why do you think that geologists describe them this way? (You may wish to look up these words in a dictionary.)

3. What factors might prevent magma from reaching Earth's surface?

4. The slopes of active volcanoes tend to be populated. What reasons can you think of for living near a volcano? Do you think the advantages outweigh the disadvantages? Why or why not?

5. Several flood basalt eruptions apparently occurred at roughly the same time as mass extinctions that divide geologic eras or periods. But geologists are not sure yet whether volcanic activity actually causes mass extinctions. What are some arguments for and against this theory?

What is happening in this picture ?

- Pumice, a volcanic rock, is so light that it actually floats in water. How can a rock be lighter than water?

- What does this rock tell us about the magma from which it came?

1. _____ are explosive eruptions characterized by pyroclastic flows and ash plumes that extend into the stratosphere.

   a. Vulcanian
   b. Hawaiian
   c. Plinian
   d. Strombolian

2. _____ consist of low viscosity lava that flows easily from a volcanic vent.

   a. Vulcanian
   b. Hawaiian
   c. Plinian
   d. Strombolian

3. These two photographs show Mauna Loa and Mauna Kea (A) and Mount Fuji volcanoes (B). Of these volcanoes which has the greatest potential for an explosive eruption?

   a. Mauna Loa and Mauna Kea (A)
   b. Mount Fuji (B)
   c. Mauna Kea (A)
   d. They all have equal potential for an explosive volcanic eruption.

4. Which of the volcanoes depicted in the photos is being fed by magma with the highest viscosity?

   a. Mauna Loa (A)
   b. Mount Fuji (B)
   c. Mauna Kea (A)
   d. The magma composition is probably identical for all of these volcanoes.

5. _____ form when the chamber of magma underlying a volcano empties due to eruption and the unsupported roof of the chamber collapses under its own weight.

   a. Fumaroles
   b. Geysers
   c. Calderas
   d. Craters

6. The greatest threats to human life during volcanic eruptions do not come from lava but from _____ and _____.

   a. pyroclastic flows, volcanic gases
   b. pyroclastic flows, ash fall
   c. ash fall, mudflows
   d. ash fall, volcanic gases

7. One way geologists monitor volcanic activity is by studying changes in the shape of volcanic features using _____.

   a. temperature gauges
   b. seismographs
   c. geologic studies of past eruptions
   d. tiltmeters

8. Melting of a rock can occur because of _____ and can be facilitated by the presence of water.

   a. increasing temperature or increasing pressure
   b. increasing temperature or decreasing pressure
   c. decreasing temperature or decreasing pressure
   d. decreasing temperature or increasing pressure

9. These two photographs are close-up views of two igneous rocks. Label each appropriately with the following terms:

   porphyritic texture   volcanic rock
   phaneritic texture   plutonic rock

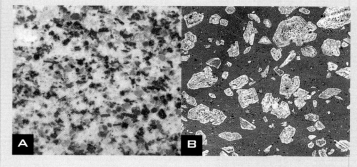

10. Which of the samples depicted in the photographs cooled more slowly? _____

   a. The sample labeled A.
   b. The sample labeled B.
   c. Both rocks cooled quickly.
   d. Both rocks cooled slowly.

11. Which of the samples depicted in the photographs on the facing page records two distinct phases of cooling—slow followed by rapid?

   a. The sample labeled **A**.
   b. The sample labeled **B**.
   c. Neither. Both rocks cooled at the same rate.

12. These six photographs show volcanic rocks paired with their plutonic equivalents. Identify each and label the photograph with the proper rock name from the following list:

basalt	andesite
granite	gabbro
rhyolite	diorite

1. _____   2. _____

3. _____   4. _____

5. _____   6. _____

13. In the process of _____, crystals that have already formed in a magma become separated from the remaining melt.

   a. magmatic redistribution
   b. distributed crystallization
   c. fractional crystallization
   d. fractional melting

14. Label this block diagram, depicting various plutonic bodies, using the following terms:

dike	volcanic neck
sill	stock
batholith	xenoliths

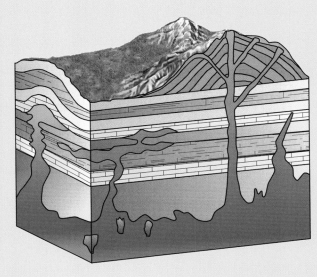

15. How will the plutonic features depicted in the block diagram become exposed at Earth's surface?

   a. through continued volcanic eruptions
   b. through uplift and erosion
   c. through melting of the overlying rock
   d. only after an explosive volcanic eruption

Weathering and Erosion

7

In May 2003, New Hampshire's beloved landmark, the Old Man of the Mountain, fell victim to the forces of nature. The Old Man was a granitic formation, sculpted in part by glacial activity during the last ice age, that bore a conspicuous resemblance to a human face jutting out from a cliff—a sort of Mount Rushmore sculpted by nature. It had been commemorated just three years earlier on the New Hampshire state quarter.

The Old Man's demise can be attributed to the inexorable force of weathering processes. Mere rock cannot resist the combined assault of water, ice, chemical changes, and gravity. Deep inside the Old Man's granite features, water repeatedly froze and thawed, extending numerous minute cracks inside the rock, while at the same time natural erosion chiseled away at the outer edges. After hundreds of years of this environmental wear and tear, the five granite ledges that created the man's profile broke apart and tumbled to the bottom of the hill.

New Englanders were dismayed at the loss of the seemingly timeless monument. Plans are under way to create a museum with high-tech imagery of the formation before its fall.

In reality, rock formations are never timeless or immutable. They are constantly broken down by weathering and erosion; without this process, we would not have soil or beaches. At the same time, new rock is constantly forming in Earth's crust. This constant recycling of material is called "the rock cycle," the main subject of our next four chapters.

Weathering—The First Step in the Rock Cycle

LEARNING OBJECTIVES

Define the rock cycle.

Describe the two major kinds of rock weathering.

Explain the three types of chemical reactions involved in chemical weathering.

Describe the conditions of climate, topography, and rock composition that are most conducive to mechanical and chemical weathering.

arth's surface is a meeting place. It is where the activities of Earth's internally driven processes—plate motion, seismicity, rock deformation, and volcanism—confront the quicker-paced activity of Earth's surface layers, the atmosphere, hydrosphere, and biosphere. The surface layers are the most dynamic parts of the Earth system. The external forces of wind, water, and ice constantly modify the surface, cutting away material here, depositing material there, and sculpting the landscapes that surround us. But the internal forces of the tectonic cycle have an important role, too, as they recycle sedimentary rocks deposited at the surface and thrust them up again. Sometimes, in the process, the rocks change form; that is, they undergo metamorphism, or melt and form magma.

All of this activity is called the **rock cycle**, one of the three great cycles that drive the Earth system, as shown in **FIGURE 7.1**. (This figure also appears in Chapters 1, 4, and 11.) The Earth system has three interconnected parts: the hydrologic cycle, the rock cycle, and the tectonic cycle. The rock cycle is in the middle,

The rock cycle FIGURE 7.1

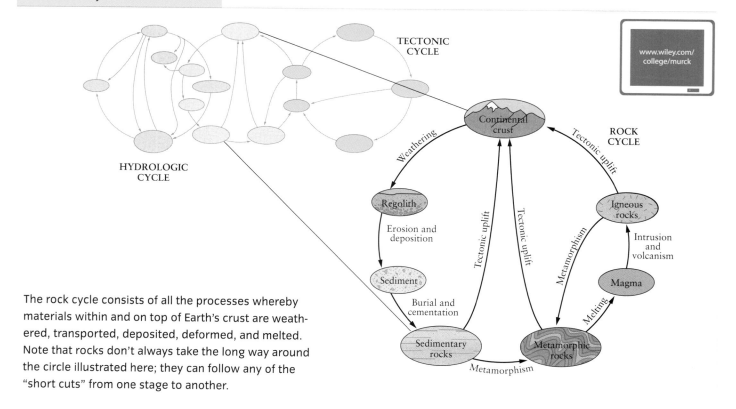

The rock cycle consists of all the processes whereby materials within and on top of Earth's crust are weathered, transported, deposited, deformed, and melted. Note that rocks don't always take the long way around the circle illustrated here; they can follow any of the "short cuts" from one stage to another.

because it is affected by the other two. The tectonic cycle is the subject of Chapters 4 through 6, and the rock cycle is the subject of Chapters 7 through 10.

The rock cycle has no beginning or ending; instead, it is an endless process, powered by Earth's internal heat energy and by incoming energy from the Sun. Nevertheless, we have to jump in somewhere, so we will begin our discussion with **weathering**, the process that breaks bedrock into smaller rock and mineral fragments, and eventually into soils.

HOW ROCKS DISINTEGRATE

Weathering takes place throughout the zone in which materials of the lithosphere, hydrosphere, atmosphere, and biosphere can mix. This zone extends downward below Earth's surface as far as air, water, and microscopic organisms can readily penetrate, and it ranges from 1 meter to hundreds of meters in depth. Rock in the weathering zone usually contains numerous fractures and *pores* (small spaces between mineral grains) through which water, air, and organisms can enter. Given enough time, they produce major changes in the rock (see **FIGURE 7.2**).

From rock to soil FIGURE 7.2

In this photograph taken in South Africa, soil has formed through disintegration of a layer of sedimentary rock. Water and air penetrate through the fractures in the rock and react with the minerals there. The rock near the surface is more heavily weathered than the lower layers, because it has been exposed to more water and air. At the very surface, the rock has completely turned to soil. Note that weathering does not have to be accompanied by erosion. Here, the products of weathering have stayed in exactly the same place where they formed.

> **weathering** The chemical and physical breakdown of rock exposed to air, moisture, and living organisms.

> **regolith** A loose layer of broken rock and mineral fragments that covers most of Earth's surface.

> **soil** The uppermost layer of regolith, which can support rooted plants.

The product of weathering is called **regolith**, from the Greek words meaning "blanket" and "stone." Fragments in the regolith range in size from microscopic to many meters across, but all of them have formed by chemical and physical breakdown of bedrock. As the particles get smaller and smaller, plants become capable of growing roots into the regolith and extracting mineral nutrients from it. At this point, the regolith becomes **soil**.

The breakdown processes involved in weathering fall into two categories. In **mechanical weathering**, the rock physically breaks down into pieces, but there is no change in its mineral content. **Chemical weathering** involves dissolving of minerals or chemical reactions that replace the original minerals with new minerals that are stable at Earth's surface. Although mechanical weathering is distinct from chemical weathering, the two processes almost always occur together, and their effects are sometimes difficult to separate.

MECHANICAL WEATHERING

Rocks in the upper half of the crust are brittle, and like any other brittle material they break when they are twisted, squeezed, or stretched by

> **mechanical weathering** The breakdown of rock into solid fragments by physical processes that do not change the rock's chemical composition.

> **chemical weathering** The decomposition of rocks and minerals by chemical and biochemical reactions.

tectonic forces. Though we cannot always determine the timing and origin of tectonic forces, we can see the results in the form of **joints**. What a Geologist Sees shows a common way by which joints form.

Joints are the main passageways through which rainwater, air, and small organisms enter the rock and lead to mechanical and chemical weathering. They differ from faults (Chapter 5) because there has not been any noticeable slippage along the fracture.

Joints do not always have to be straight. In a process that is called *sheet jointing* or *exfoliation* (see FIGURE 7.3A), large, curved slabs of rock peel off from the surface of a uniformly textured igneous rock. As with other types of joints, sheet jointing may be due to pressure release or a combination of forces that contribute to mechanical weathering.

> **joint** A fracture in a rock, along which no appreciable movement has occurred.

Mechanical weathering takes place in four main ways—through freezing of water, formation of salt crystals, penetration by plant roots, and abrasion. By far the most widespread type of mechanical weathering involves the freezing of water.

Water is an unusual substance. Most liquids contract when they freeze, and the volume of the resulting solid is smaller than the volume of the liquid. However, when water freezes it expands, increasing in volume by about 9 percent. If you put a full, capped bottle of water in the freezer, the bottle will burst when the water freezes, because it cannot contain the larger volume of ice. Wherever temperatures fluctuate around the freezing point for part of the year, water in the ground will alternately freeze and thaw. If the water gets inside a joint in the rock, the freeze–thaw cycles act like a lever prying the rock apart, and eventually the rock shatters. This process is known as **frost wedging** (see FIGURE 7.3B and C).

What a Geologist Sees

Joint Formation

A These heavily jointed rocks are found in the Joshua Tree National Monument in California. Erosion has widened some of the joints to the point of detaching stones from the main rock body.

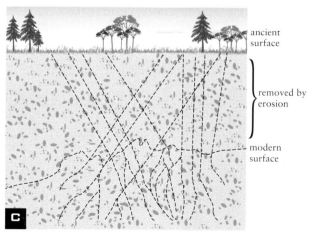

ancient surface

removed by erosion

modern surface

C As the rock rose to the surface and the overlying rock was eroded, the pressure decreased, causing the rock to expand and crack. Later, erosion rounded off and widened the joints even further.

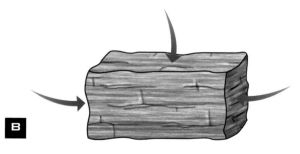

B The rock originally formed underground, where it was subject to great pressure from the overlying and surrounding rocks.

www.wiley.com/
college/murck

A

Smooth, rounded surface where sheets have exfoliated (peeled off) in the past.

▲ A **Sheet Jointing**
Sheet jointing results in curved domes, like the famous Half Dome in Yosemite National Park, California.

B

Frost wedging has widened this joint in the rock.

▲ B **Frost Wedging**
This granite boulder in the San Andres Mountains of New Mexico has been split apart by repeated freezing and thawing of water that penetrated along the joints.

C

▲ C A geology student takes notes on a steep scree slope in Victoria Land, Antarctica. All of the loose rubble in this photograph was separated from the bedrock by frost wedging.

D

▲ D **Root Wedging**
This Ponderosa pine began growing into a crack in an outcrop of granite bedrock. Eventually, a large flake of rock broke away, exposing the tree's root system.

The formation of salt crystals can also cause visible mechanical weathering. Water moving slowly through rock fractures will dissolve soluble material, which may later precipitate out of solution to form salt crystals of various kinds. The force exerted by growing crystals, either within rock cavities or along grain boundaries, can cause rocks to fall apart. Mechanical weathering by crystal growth occurs mostly in desert regions, where compounds such as calcium carbonate and calcium sulfate are precipitated from groundwater as a result of evaporation.

Another type of mechanical weathering is caused by penetration by plant roots. Trees are very resourceful and can grow where there seems to be hardly any soil to sustain them. A tree may become rooted in a crack in the bedrock, eventually widening the crack and wedging apart the bedrock, as shown in **FIGURE 7.3D**. Large trees swaying in the wind can also cause fractures to widen. When trees are blown over, they can cause additional fracturing of rock. Although it is difficult to measure, the total amount of rock breakage caused by plants must be very large.

The final method of mechanical weathering, **abrasion**, is the wearing away of bedrock by loose particles transported by water, wind, or ice. Abrasion is a form of mechanical weathering because it reduces the size of rocks by a physical (rather than chemical) process.

CHEMICAL WEATHERING

Chemical weathering is primarily caused by water that is slightly acidic. As raindrops form and fall through the air, they dissolve atmospheric carbon dioxide.

$$H_2O + CO_2 \rightarrow H_2CO_3$$

Rainwater thus is a weak solution of carbonic acid (H_2CO_3)—a weak version of Perrier water! When weakly

dissolution The separation of a material into ions in solution by a solvent, such as water or acid.

acidified rainwater sinks into the regolith and becomes soil water or groundwater, it may dissolve additional carbon dioxide from decaying organic matter, becoming more strongly acidified. Another way that rainwater can become acidified is by interacting with *anthropogenic* (human-generated) sulfur and nitrogen compounds released into the atmosphere. This produces a phenomenon called *acid rain*. Human-caused acid rain is stronger than natural acid rain, and causes accelerated weathering.

Through **dissolution**, some minerals can be dissolved and completely removed without leaving a residue. Some common rock-forming minerals, such as calcite (calcium carbonate) and dolomite (calcium magnesium carbonate), dissolve in slightly acidified water (see **FIGURE 7.4**).

Water can also alter the mineral content of rock without dissolving all of it. One reaction of special importance in chemical weathering is *ion exchange*. Ions, which are atoms with a positive or negative electric charge, exist both in solution and in minerals. They form when atoms give up or accept electrons (see Chapter 2). The difference is that ions in minerals are tightly bonded and fixed in a crystal lattice, whereas ions in solutions can move about randomly and cause chemical reactions. In ion exchange, hydrogen ions (H^+) from acidic water enter and alter a mineral by displacing larger positively charged ions such as potassium (K^+), sodium (Na^+), and magnesium (Mg^{2+}) (see **FIGURE 7.5**).

Where do the potassium and other ions go after they are replaced by hydrogen ions? Some remain in the

Dissolution of calcite FIGURE 7.4

This marble tombstone, which has stood in a New England cemetery since the early nineteenth century, has gradually been dissolved by rainwater. Over the years, the once sharply chiseled inscriptions have become barely legible. Marble, which contains the soluble mineral calcite, is one of the rock types that are vulnerable to dissolution.

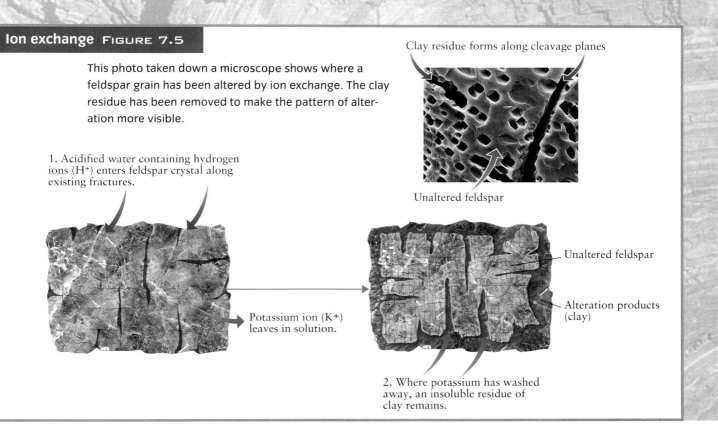

Ion exchange FIGURE 7.5

This photo taken down a microscope shows where a feldspar grain has been altered by ion exchange. The clay residue has been removed to make the pattern of alteration more visible.

Clay residue forms along cleavage planes

Unaltered feldspar

1. Acidified water containing hydrogen ions (H+) enters feldspar crystal along existing fractures.

Potassium ion (K+) leaves in solution.

Unaltered feldspar

Alteration products (clay)

2. Where potassium has washed away, an insoluble residue of clay remains.

groundwater, accounting for the taste of "mineral water" that some people find pleasant and others do not, and some flow out to sea and form part of the ocean's reserve of dissolved salts. On the other hand, if the water evaporates, the dissolved materials can precipitate out again as solid evaporites, such as halite and gypsum.

Another very important process of chemical weathering is *oxidation*, a reaction between minerals and oxygen dissolved in water. Iron and manganese, in particular, are present in many rock-forming minerals. When such minerals undergo chemical weathering, the iron and manganese are released and are immediately oxidized (see FIGURE 7.6). Oxidized iron commonly forms an insoluble yellowish hydrous material called limonite, and manganese forms an insoluble black mineral called pyrolusite.

FACTORS AFFECTING WEATHERING

As FIGURE 7.7 on pages 190–191 shows, many factors influence the susceptibility of a rock to chemical and mechanical weathering. The most important factors are tectonic setting, composition of the rock, rock structure (the abundance of openings such as joints),

Oxidation FIGURE 7.6

The yellow rocks are limonite, an oxidized form of iron that contains water in its chemical structure. When the water is removed, limonite becomes hematite (red rocks), a common ingredient in red tropical soils. The green rocks are a less common iron ore called siderite, an iron carbonate. These rocks were mined in Egypt.

◄ **Tectonic setting:** Young, rising mountain ranges, such as the Himalaya, weather very rapidly. This view of the Indus River in Pakistan shows many boulders that have separated from the bedrock because of mechanical weathering.

Rock structure: The pace of weathering is also strongly affected by the closeness of joints. Sugarloaf Mountain in Rio de Janeiro, Brazil, is a large, unjointed mass of granite, and it stands out against an otherwise deeply eroded landscape. ▼

Unjointed rock weathers slowly.

Steep slope weathers quickly.

Topography: Weathering proceeds more quickly on a steep slope than a gentle one. This rockslide on the island of Madeira in the Atlantic Ocean will expose new bedrock to weathering. ▶

Bacterial activity can promote chemical weathering.

Biologic activity: Animals—even microorganisms—contribute significantly to the breakdown of rocks. Here, a student lifts a mat of algae. The algae feed on a bacterium, *thiobacillus ferrooxidans,* that thrives in acidic runoff from abandoned sulfur- and iron-bearing mines.
◄

Vegetation: Plants contribute to both mechanical and chemical weathering. They tend to hold a deeper regolith in place, which promotes weathering because it retains water. However, the removal of plants through slash-and-burn agriculture, clear-cut logging, or natural landslides can also accelerate weathering, as seen here in a photo of Madagascar.
▼

Resistant quartzite knob weathers slowly.

Deforestation accelerates weathering and erosion.

▲ **Composition:** Different minerals weather at different rates. Calcite weathers quickly by dissolution, and feldspar weathers at an intermediate rate by ion exchange. On the other hand, quartz is very resistant to weathering, because it dissolves very slowly and is not affected by ion exchange. The knob atop Pilot Mountain, in North Carolina, is made of quartzite that has weathered much more slowly than the surrounding sedimentary rock.

Weathering—The First Step in the Rock Cycle   191

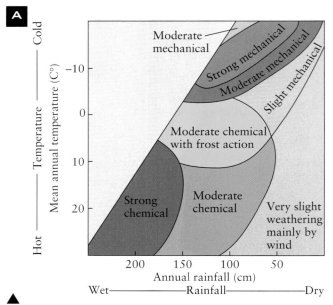

A Different climates cause rocks to weather at different rates and by different types of weathering.

B This map of North and South America illustrates some locations where the climate corresponds to the different zones of weathering shown in A.

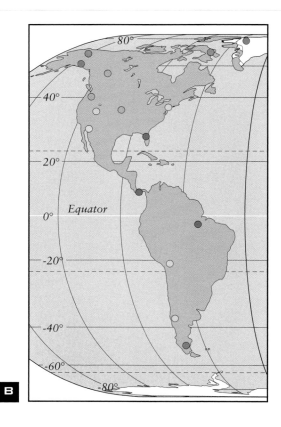

topography, amount of vegetation and biologic activity, and climate (especially temperature and rainfall).

From the perspective of a human lifespan, the chemical weathering of rocks happens very slowly. For example, granite and other hard bedrock surfaces in New England, Canada, and Scandinavia still display polish and fine grooves made by the scraping of glaciers of the last ice age, which ended more than 10,000 years ago. In such regions, which have plentiful rainfall but cool climates, it takes hundreds of thousands of years for a regolith like that shown in Figure 7.2 to develop. In warmer regions, chemical weathering occurs more quickly at the surface and extends to depths of many tens of meters. Through the use of various dating techniques, it has been estimated that deep tropical weathering of 500 meters or more requires many millions and possibly tens of millions of years.

Climate affects the rate of weathering in two different ways. As noted earlier, chemical weathering is more intense and extends to greater depths in a warm, wet, tropical climate than in a cold, dry, Arctic climate.

In cold, dry climates like Greenland or Antarctica, chemical weathering proceeds very slowly. On the other hand, mechanical weathering is fairly rapid in these harsh environments. The only environment where both kinds of weathering proceed very slowly is a hot, dry climate (see FIGURE 7.8).

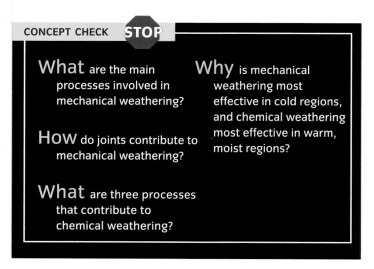

CONCEPT CHECK STOP

What are the main processes involved in mechanical weathering?

How do joints contribute to mechanical weathering?

What are three processes that contribute to chemical weathering?

Why is mechanical weathering most effective in cold regions, and chemical weathering most effective in warm, moist regions?

Products of Weathering

LEARNING OBJECTIVES

Identify three end products of weathering.

Describe the soil horizons found in most soils.

Explain why soils from different climates have different profiles.

Describe the link between human activity and soil erosion.

Have you ever wondered where a boulder in a mountain stream or the sand on a beach or the mud in a swamp actually comes from? The answer is the same in each case: from the mechanical and chemical weathering of rocks exposed at Earth's surface. Chunks of rock freshly broken loose from bedrock are usually similar to the original material. However, with prolonged exposure to chemical weathering, minerals that are stable at higher temperatures and pressures begin to decompose, while new minerals, stable at the conditions of Earth's surface, are formed. For example, the feldspar crystals in granite break down to form **clay** minerals (see **FIGURE 7.9**). Although tiny (typically less than 0.004 mm) clay minerals are one of the most stable types of minerals on Earth's surface. They are a major component of mud (both on land and in the sea).

On the other hand, some minerals, such as quartz, are more resistant to weathering, and hence do not break down into grains as small as

clay A family of hydrous aluminosilicate minerals. The term is also used for tiny mineral particles of any kind that have physical properties like the clay minerals.

Kaolin—from rocks to fine china FIGURE 7.9

The white clay found near Kao Ling, China, has for centuries been used to produce some of the finest porcelain in the world. The same kind of clay has now been found in many other places, but it is still called kaolin in honor of the place that made it famous.

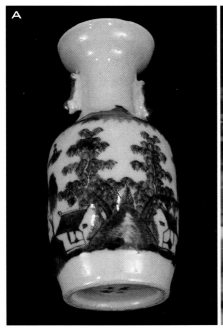

A A Chinese porcelain vase from the fifteenth century

B A worker in Jingdezhan, China, fashions cups out of kaolin clay.

This microscopic view of Earth soil (A) and lunar regolith (B) shows some significant differences. Most importantly, Earth soil contains organic material and hydrous minerals such as clay, while lunar regolith contains none of either.

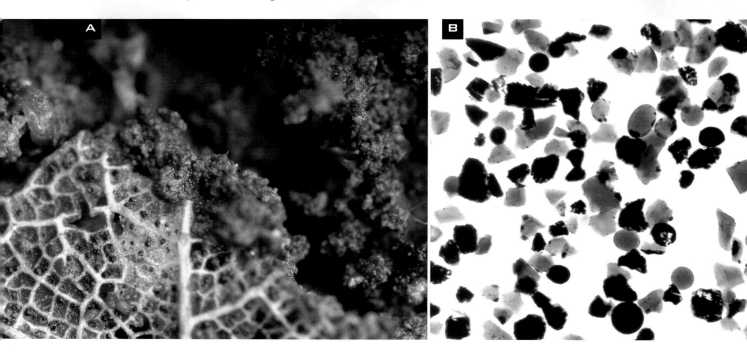

those found in clay. Instead, quartz usually ends up as **sand**, which has grain sizes as coarse as 1–2 mm— roughly 100 to 1000 times larger than the grains in clay. Sediment with grain sizes between these of sand and clay is called **silt**. The grain sizes of particles weathered from bedrock decrease with the distance traveled by the sediment from its source. Some of the finest sand is found at the beach, because it has had an especially long journey from the mountains.

SOIL

Perhaps the most complex product of weathering is also the most familiar: **soil**. It usually contains a mixture of minerals with different grain sizes, all the way from fine clay to coarse sand, along with some material of biologic origin. If you look at soil under a magnifying glass, you will find fragments of **humus**, and possibly some tiny insects and worms. With a very strong microscope you might even observe bacteria living on the humus.

Soil is a complex medium in which all parts interact and play important roles. Humus retains some of the chemical nutrients released by decaying organisms and by the chemical weathering of minerals. Humus is critical to soil fertility, which is the ability of a soil to provide nutrients such as phosphorus, nitrogen, and potassium needed by growing plants. All of the processes that involve living organisms and other soil constituents produce a continuous cycling of plant nutrients between the regolith and the biosphere. Earth is the only planet in the solar system that has true soil. The other rocky bodies in the solar system have blankets of loose rocky material (*regolith*) that have sometimes been pulverized to a very fine texture, but their regoliths lack humus (see FIGURE 7.10).

sand A sediment made of relatively coarse mineral grains.

humus Partially decayed organic matter in soil.

SOIL PROFILES

Soil evolves gradually, from the top down. As erosion removes the top layer, weathering of the underlying material continually creates new soil. When fully developed, soil consists of **soil horizons**, each of which has distinct physical, chemical, and biologic characteristics.

Soil profiles (see FIGURE 7.11) vary considerably, being influenced by such factors as climate, topography, and rock type. However, certain kinds of horizons are common to many profiles.

The uppermost horizon in many soil profiles, the *O horizon*, is an accumulation of organic matter. Below it lies the *A horizon*, which is typically dark in color because of the humus present. An *E horizon*, which is sometimes present below A, is typically grayish in color because it contains little humus and the mineral grains do not have dark coatings of iron and manganese hydroxides. Both the A and E horizons have had the soluble minerals leached out of them. E horizons are most common in the acidic soils of evergreen forests.

The *B horizon* underlies the A horizon (or E, if one is present). B horizons are brownish or reddish in color because of the presence of iron hydroxides that have been transported downward from the horizons above. The B horizon is a zone of accumulation, where materials that were leached from the A horizon are redeposited. Clays are usually abundant in the B horizon. The *C horizon* (commonly known as the *subsoil*) is deepest, consisting of parent rock material in various stages of weathering. Oxidation of iron in the parent rock gives the C horizon a yellowish or rusty color.

Beneath the C horizon lies the unweathered bedrock. Geologists studying the

> **soil horizon**
> One of a succession of zones or layers within a soil profile, each with distinct physical, chemical, and biologic characteristics.

> **soil profile** The sequence of soil horizons from the surface down to the underlying bedrock.

Soil profile FIGURE 7.11

This is a typical sequence of soil horizons that would commonly develop in moist, temperate climates. The A horizon, which lies within reach of plant roots, is commonly called the *topsoil*.

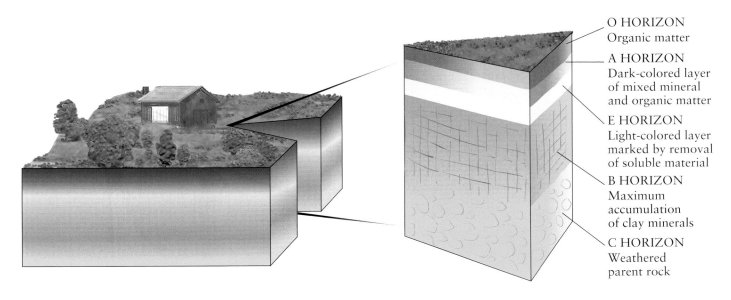

O HORIZON
Organic matter

A HORIZON
Dark-colored layer of mixed mineral and organic matter

E HORIZON
Light-colored layer marked by removal of soluble material

B HORIZON
Maximum accumulation of clay minerals

C HORIZON
Weathered parent rock

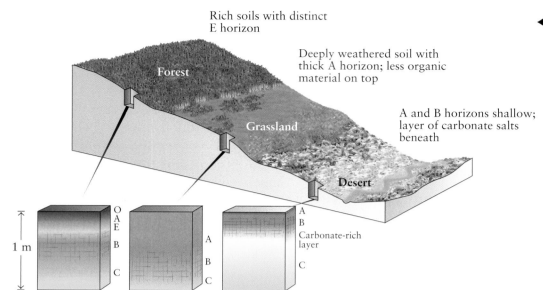

Rich soils with distinct
E horizon

Forest

Deeply weathered soil with
thick A horizon; less organic
material on top

Grassland

A and B horizons shallow;
layer of carbonate salts
beneath

Desert

1 m

O
A
E
B
C

A
B
C

A
B
Carbonate-rich
layer
C

◀ Soil horizons are strongly influenced by the climatic zone in
which they form. For example,
a layer of carbonate minerals
forms in many desert soils
because the hot, dry conditions cause groundwater to
evaporate and precipitate the
dissolved minerals. The precipitates form a hard layer, sometimes called *hardpan* or
caliche. If the hard layer is
near the surface, plant roots
may not be able to grow to
their normal depth.

The photographs show two similar-looking soils from different
locations. The "Steedman profile," left, from Kansas, has a dark
A horizon whose bottom is indicated by the white arrow. The
lighter B horizon extends to the bottom of the picture, and the
C horizon (not seen here) is gray and starts just below this
photo. The "Windsor profile," right, from Connecticut, also has
a dark A horizon and a brown-colored B horizon. The yellowish
brown C horizon is the sandy parent material which was
deposited by glacial meltwaters.

Climate and soils FIGURE 7.12

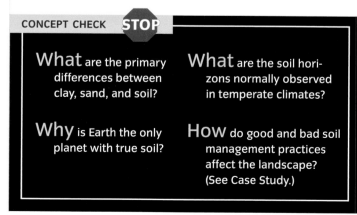

rocks of a region search for samples of "fresh," unweathered bedrock in order to form a true picture of a
rock's original properties. Such samples can be found
in natural outcrops along stream banks, steep sides of
mountains, cliff faces where the soils are thin, or at artificial exposures such as highway road cuts, quarries, or
mines.

As we have emphasized repeatedly in this
chapter, all of Earth's systems interact at the surface.
You might not think the climate above the ground
would have anything to do with the soil profile underground, but in fact they are closely related. As
FIGURE 7.12 shows, the soil profile in an arid region is quite different from a region that gets plenty
of rainfall.

Because it is part of the never-ending rock cycle, soil is not a static thing. Soil can be formed, and it

can be depleted by natural processes. But unlike some
other parts of the rock cycle, it is very strongly affected by human activities, and this makes proper soil
management a vital issue (see Case Study).

CONCEPT CHECK STOP

What are the primary
differences between
clay, sand, and soil?

What are the soil horizons normally observed
in temperate climates?

Why is Earth the only
planet with true soil?

How do good and bad soil
management practices
affect the landscape?
(See Case Study.)

Bad and Good Soil Management

Providence Canyon in Georgia is a gorgeous example of deeply weathered soil, but it is also a dreadful example of poor soil management. In photograph A, you can readily spot the dark brown A horizon, the bright red B horizon that is full of clay, and the paler E horizon. Some people have called Providence Canyon the "Little Grand Canyon," but in reality the two are very different. While the layered appearance of the Grand Canyon is due to the strata of bedrock, the appearance of Providence Canyon is caused by soil alone.

Another difference is age. Providence Canyon, believe it or not, is less than 200 years old. There was no canyon here when settlers from Europe arrived in the early 1800s. But the farmers plowed straight up and down the hills, and the furrows rapidly developed into gullies. By 1850, the gullies were 1 to 2 meters deep. The farmers had to abandon their fields, but by then erosion in the gullies was running amok. The canyon is now more than 50 meters deep. Unfortunately there are many such locations in North America.

To fight erosion, modern farmers use contour plowing (B). Instead of going in straight lines, the furrows follow the contour of the land, slowing runoff and inhibiting the formation of gullies. Crop rotation can also make a significant difference in preventing erosion.

Even so, the erosion of farmland soil is a massive worldwide problem. In the United States, the amount of agricultural soil eroded each year exceeds the amount of replenished soil by about a billion tons. For every kilogram of food we eat, the land loses 6 kilograms of soil. Although there is a small "sustainable farming" movement, we are very far from consuming only as much as we can put back.

NATIONAL GEOGRAPHIC

Erosion and Mass Wasting

A s distinct from *weathering*, **erosion** is a term that describes the transport of regolith from one place to another (weathering happens in place). Both processes can, of course, happen at the same time; for instance, a rock can be abraded (a weathering process) and the particles that break off the bedrock will be transported elsewhere by the wind (an erosion process).

Erosion requires a natural fluid to pick up and transport the eroded material. That requirement differentiates erosion from mass wasting, when regolith moves downslope under the pull of gravity, with no transporting medium required. (We will discuss mass wasting later in this section.)

The fluids that cause the most erosion on Earth are water, wind, and ice. (Even though we usually think of ice as a solid, it does flow when it forms a glacier or ice sheet.) As discussed in Chapter 6, different fluids display differing resistance to flow, or *viscosity*. Ice, which flows so slowly that its motion cannot be seen by the human eye, behaves as an extremely viscous fluid. Water, which flows freely, is much less viscous, and air is the least viscous. A fluid's viscosity, in turn, partly determines whether its flow is *laminar* or *turbulent*. In laminar flow, all fluid particles travel in parallel layers. Turbulent flow is erratic and complex, full of swirls and eddies. Turbulent flow is more effective at picking particles up off the ground. Air flow is almost always turbulent; water flow is usually turbulent except when the velocity of flow is very low.

EROSION BY WATER

Erosion by water begins even before a distinct stream has formed on a slope. This happens in two ways: by impact, which occurs when raindrops hit the ground and dislodge small particles of soil; and by overland flow, which occurs during heavy rains. Overland flow involves water moving as sheets over the ground, not in channels. When the water starts flowing in a channel, particles are moved in several ways: The largest particles, which form the **bed load** (boulders, cobbles, and pebbles), roll or slide along the stream bed due to the force of the flowing water. Smaller (sand-sized) particles move along the stream bed by **saltation** (see FIGURE 7.13).

The particles in the **suspended load**—silt and clay—are small. Although they do not actually float, they do not sink to the bottom as long as the water is flowing. If you took a sample of water from a muddy stream and let it stand, the silt and clay would settle to the bottom as mud. But in the turbulent environment of a stream, the upward-moving currents keep them from sinking. Thus mud deposits form only where velocity decreases and turbulence ceases, as in a lake, the sea, or a reservoir.

Streams also carry a *dissolved load* of soluble materials released by chemical weathering. The dissolved load may also contain organic matter, which accounts for the infamous "black water" found in rivers that drain swamps.

erosion The wearing away of bedrock and transport of loosened particles by a fluid, such as water.

bed load Sediment that is moved along the bottom of a stream.

saltation A mechanism of sediment transport in which particles move forward in a series of short jumps along arc-shaped paths.

suspended load Sediment that is carried in suspension by a flowing stream of water or wind.

NATIONAL GEOGRAPHIC

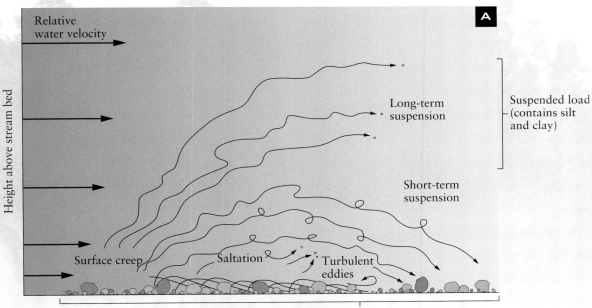

A A stream's bed load consists of the particles that are too heavy to stay suspended in the water. Pebbles creep or roll along the bottom, while sand-sized particles move in small jumps called saltation. Very fine silt and clay particles form the suspended load and give the water a muddy appearance.

B The turbulent flow of this river in Gabon, Africa, enables it to carry a large suspended load.

EROSION BY WIND

Because the density of air is about 800 times less than that of water, air cannot move the large particles that water flowing at the same velocity can move. In exceptional cases such as hurricanes and tornadoes, winds can reach speeds of 300 km/h and sweep up coarse rock particles several centimeters in diameter. In most regions, however, wind speeds rarely exceed 50 km/h, a velocity that is described as a strong wind. As a result, most (at least 75%) of the sediment transported by wind occurs through saltation of sand grains. Only the finest particles, the dust, remain aloft long enough to be moved by suspension. Even so, space-based imaging has recently given geologists a greater appreciation for the massive amounts of material that can be moved by wind (see FIGURE 7.14 on the next page).

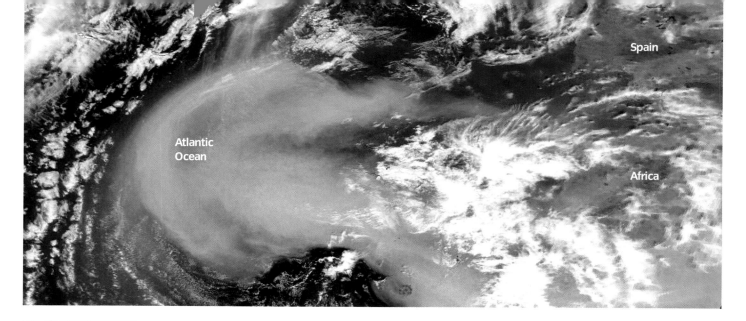

Atlantic Ocean

Spain

Africa

Storms from Africa FIGURE 7.14

In this satellite photo, winds blowing from east to west lift a colossal cloud of dust from the Sahara Desert into the air. The dust can travel all the way across the Atlantic Ocean to America, and it has been implicated in the bleaching of coral reefs off the coast of Florida.

EROSION BY ICE

Ice is a solid. However, it does flow under the influence of gravity, albeit very slowly, in parts of the world where there is enough year-round ice to form a glacier. Compared with water and air, ice is extremely viscous. Glacial ice therefore moves only by laminar flow.

> **glacier** A semi-permanent or perenially frozen body of ice, consisting largely of recrystallized snow, that moves under the pull of gravity.

> **mass wasting** The downslope movement of regolith and/or bedrock masses due to the pull of gravity.

Glaciers play a three-part role in erosion and transport: They act as a plow, a file, and a sled. As a plow, a glacier scrapes up weathered rock and soil and plucks out blocks of bedrock (see FIGURE 7.15A). As a file, the load of sediment rasps away and polishes the rock (see FIGURE 7.15B). As a sled, a glacier carries away the load of sediment acquired by plowing and filing, as well as additional debris that falls onto it from adjacent slopes (see FIGURE 7.15C).

GRAVITY AND MASS WASTING

Landscapes may seem fixed and unchanging, but if you made a time-lapse movie of almost any hillside for a few years you would see that the slope changes constantly as a result of **mass wasting**. There is no such thing as a static hillside—a lesson learned all too often by people who live at the top, the bottom, or on a steep slope. Exactly how movement happens and how fast it happens is controlled by the composition and texture of the regolith and bedrock, the amount of air and water in the regolith, and the steepness of the slope. For convenience, we divide mass wasting into two categories: **slope failures** and **flows**. We illustrat several

> **slope failure** The falling, slumping, or sliding of relatively coherent masses of rock.

> **flow** Any mass-wasting process that involves a flowing motion of regolith containing water and/or air within its pores.

A This rocky debris was deposited at the edge of Matunuska Glacier, Alaska. Note the extreme range in size of the rock fragments, from boulders much larger than the person in the photo down to tiny pebbles.

B This polished and grooved rock surface was produced by the Findelen Glacier in the Swiss Alps. The glacier has retreated in modern times, exposing the weathered rock beneath. A famous mountain, the Matterhorn, is in the background.

C Two ice streams, bearing debris that has fallen from adjacent mountain slopes, merge to form the Kaskawulsh Glacier in the Yukon, Canada. The smooth, parallel streams are a hallmark of laminar flow.

types of slope failures and sediment flows in FIGURES 7.16 and 7.17.

Slope failures occur as one of three basic types. A **fall** (see Figure 7.16A) is a sudden vertical, or nearly vertical, drop of rock fragments or debris. Rockfalls and debris falls are sudden and usually very dangerous. **Slides** (see Figure 7.16B) involve rapid displacement of a mass of rock or sediment in a straight line down a steep or slippery slope. A **slump** (Figure 7.16C) involves *rotational* movement of rock and regolith—that is, downward and outward movement along a curved surface. Slumps often result from poor engineering practices, such as after slopes have been oversteepened for construction of buildings or roads.

Flowing regolith can be either wet or dry. *Slurry flows* (Figure 7.17A) occur when the regolith is saturated with water; they can occur either rapidly or slowly. Rapid slurry flows can move at speeds up to 160 km/h and are very dangerous. Slow slurry flow, a process known as *solifluction*, is common in areas with high rainfall, where soil is thin over bedrock, and in areas where the ground is frozen at depth. Flowing regolith that is not water-saturated is called a *granular flow* (Figure 7.17B). Like slurry flows, granular flows can be either slow or fast. The most common kind of granular flow (and the most common kind of mass wasting) is called **creep**. More rapid granular flows include *debris avalanches*, which are rare, spectacular, and

Slope failures FIGURE 7.16

A Kaibito Canyon in Arizona has been the site of repeated rockfalls, as you can see both from the debris at the base of the cliff and from the scars on the cliff face where rocks have detached themselves in the past. ▼

▲ **B** In this rockslide in the Andes mountains in Argentina, the rocks moved in a roughly straight line from the point of detachment to the valley floor.

C A slump is a slower kind of failure in which the debris ▶ moves rotationally (as shown by the curved arrow). This slumping failure occurred in central California.

extremely dangerous. Debris avalanches often start as a rockfall or slide but gain speed when the material pulverizes and begins to flow downslope like a fluid. Debris avalanches can also be triggered by earthquakes or volcanic eruptions.

TECTONICS AND MASS WASTING

The locations of the world's major historic and prehistoric landslides tend to cluster along belts that lie close to the boundaries between converging lithospheric plates. They do so for two main reasons.

First, the world's highest mountain chains lie at or near plate boundaries, and the rocks of many mountain ranges consist of well-jointed strata that have been strongly fractured and deformed as they were uplifted. Both the joint planes and the bedding surfaces are potential zones of failure. Stratovolcanoes, which are found in the same regions, also tend to have steep slopes conducive to landslides.

Second, most large earthquakes occur along the boundaries between plates where plate margins slide past or over one another. Earthquakes often trigger landslides in areas where the regolith is unstable.

Flows FIGURE 7.17

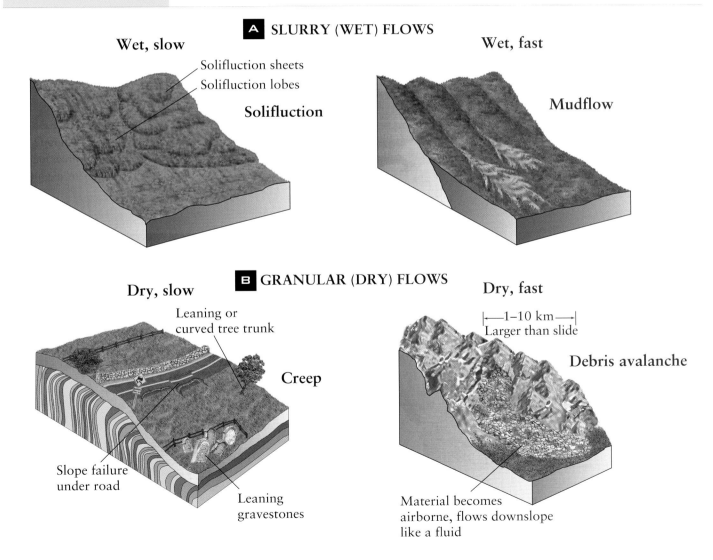

A SLURRY (WET) FLOWS

Wet, slow
Solifluction sheets
Solifluction lobes
Solifluction

Wet, fast
Mudflow

B GRANULAR (DRY) FLOWS

Dry, slow
Leaning or curved tree trunk
Creep
Slope failure under road
Leaning gravestones

Dry, fast
|—1–10 km—|
Larger than slide
Debris avalanche
Material becomes airborne, flows downslope like a fluid

A

B

Earthquakes, particularly those close to plate edges, often trigger landslides. The subduction-related quake of March 28, 1964, the Great Alaska Earthquake, caused many slides.

A In downtown Anchorage, the Turnagain Heights district broke apart and slid toward the sea.

B Farther inland, part of the Chugach Mountains collapsed and caused a flow that covered part of the Sherman Glacier.

Several historic landslides were directly related to major earthquakes (see FIGURE 7.18).

It may seem as if landslides and mass wasting should ultimately level all the world's mountains and leaving the continents as flat, featureless plains. That will never happen, because uplift is always taking place at the same time. For example, at Nanga Parbat, a Himalayan mountain that lies at the boundary between the Indian and Eurasian plates, uplift rates are as high as 5 mm/yr. At this rate, the mountain should increase in altitude by 5000 meters every million years. However, high mountains also mean high erosion rates, and erosion is tearing down the mountain almost as rapidly as it is rising. Much of this destruction is the result of mass wasting.

CONCEPT CHECK STOP

Which three fluids are responsible for most of the erosion on Earth?

What are the differences between a stream's bed load, its suspended load, and its dissolved load?

What is the primary agent that causes mass wasting?

Which kind of mass wasting is most common, and which kinds are most dangerous?

Why is mass wasting most rapid in regions of active tectonic uplift?

Amazing Places: Monadnock—and Monadnocks

Mount Monadnock (FIGURE A), a 1,156 meter peak in New Hampshire, is one of the world's most frequently climbed mountain—no doubt because it is easy to climb yet rewards the climber with a beautiful view of all six New England states. Its name, which came from an Algonquin Indian phrase meaning "mountain standing alone," has actually become a generic term for a mountain that rises out of a surrounding plain. (A synonym used more often by geologists is *inselberg*.) Many of the world's most scenic and best-loved peaks are of this form.

Monadnocks (or inselbergs) are isolated by erosion, either because they are unjointed or because they were made of more resistant material than the surrounding land mass. They are often domed because of exfoliation, like Stone Mountain in Georgia (FIGURE B). They can be made of any of the three rock types. Grand Monadnock is made of schist (a metamorphic rock) and Stone Mountain is made of granite. Another famous example, Uluru (also known as Ayers Rock) in Australia (FIGURE C), is made of sandstone, a sedimentary rock.

As the map shows, there are lots of excellent examples of inselbergs in the United States.

1 Weathering—the First Step in the Rock Cycle

1. The **rock cycle** is the continuous cycle of processes by which rock is formed, modified, transported, decomposed, and re-formed. Most of Earth's surface is covered by a blanket of weathered rock, which we call **regolith**. Regolith fragments can range in size from many meters to microscopic. When small regolith particles are altered by biologic processes, the result is **soil**, a material from which rooted plants can extract nutrients.

2. When rocks are exposed at Earth's surface, they are constantly subjected to **weathering**, the process by which air, water, and microbes break down bedrock into smaller rock and mineral fragments. Weathering extends as far down as air, water, and living organisms can readily penetrate Earth's crust. The two categories of weathering are **mechanical**, the physical breakage of rock, and **chemical**, which involves the removal of some minerals in solution and the transformation of others into new minerals that are stable at Earth's surface. The agents of weathering, such as water from rain or melted snow, enter bedrock along fractures and via *pores*.

3. **Joints** are fractures along which no appreciable movement has occurred; they are the main passageways through which agents of weathering enter the rock and lead to weathering. Mechanical weathering takes place in four main ways: **frost wedging**, or the freezing of water; by the growth of salt crystals in confined spaces; by the prying action of roots; or by **abrasion**, the direct wearing away of bedrock by loose particles transported by moving water, wind, or ice.

4. Chemical weathering involves the removal of some minerals in solution and the transformation of others into new minerals that are stable at Earth's surface. This type of weathering is caused primarily by water that is slightly acidic. *Acid rain* can be naturally occuring or human-generated. Human-generated acid rain, which is created when rain-

water interacts with *anthropogenic* sulfur and nitrogen compounds, is stronger than natural acid rain and causes accelerated weathering.

5. There are three principal processes of chemical weathering—**dissolution**, when minerals dissolve in acidified water; *ion exchange*, in which hydrogen ions from acidic water enter and alter a mineral by displacing larger, positively charged ions; and *oxidation*, a reaction between minerals and oxygen dissolved in water.

6. From a human perspective, chemical and mechanical weathering occur very slowly, over many thousands of years. The effectiveness of weathering depends on the type and structure of rock, the tectonic setting, the steepness of the slope, the local climate, and the amount of biological activity. Chemical weathering is most active in moist, warm climates, whereas mechanical weathering is more active in cold, dry climates.

2 Products of Weathering

1. If given enough time, weathering eventually breaks rock down into very fine particles that can be classified, primarily by size, as **clay**, silt, or **sand**. Clay refers to a family of hydrous aluminosilicate minerals, but it is more commonly used to describe any tiny mineral particles that have similar physical properties to clay minerals. Silt particles are larger than clay, and sand is larger than silt.

2. Soil is regolith, supplemented by organic matter and small organisms, that can support rooted plants. Earth's soil is different from the "soil" of other planets because it contains **humus**, partially decayed oganic matter. Humus retains some chemical nutrients released by decaying organisms and the chemical weathering of minerals. It is critical for soil fertility, because it provides nutrients that plants need to grow, such as phosphorus, nitrogen, and potassium.

O HORIZON
Organic matter

A HORIZON
Dark-colored layer
of mixed mineral
and organic matter

E HORIZON
Light-colored layer
marked by removal
of soluble material

B HORIZON
Maximum
accumulation
of clay minerals

C HORIZON
Weathered
parent rock

3. Soils weather from top to bottom, and develop distinctive **soil horizons** whose properties are a function of the duration, intensity, and nature of the weathering process. In a typical **soil profile**, the *O horizon* is the topmost layer of accumulated organic matter. The *A horizon* is rich in humus; the A and *E horizons* are layers from which soluble material, especially iron and aluminum, has been lost through leaching. Farther down, clay minerals accumulate in the *B horizon*. Deeper still, the *C horizon* consists of slightly weathered parent rock. Soil profiles can vary greatly from location to location, as they are strongly affected by climate.

3 Erosion and Mass Wasting

1. **Erosion** involves the removal and transport of regolith through the combined actions of ice, water, wind, and gravity. This is different from weathering, which happens in place, though both processes can occur at the same time. Most processes of erosion involve a fluid that picks up and transports material.

2. Both air and water move particles by the process of **saltation**, a mechanism of sediment transport in which particles move forward in a series of short hops along arc-shaped paths. Only the smaller sand-sized particles move along the bottom of a stream by saltation, whereas the larger particles move by rolling or sliding. These larger particles are known as the **bed load**; lighter particles carried in suspension form the **suspended load**; and dissolved ions released by chemical weathering form the *dissolved load*. The dissolved load may also contain organic matter.

(continued)

3. Air and water carry suspended particles most effectively through *turbulent* flow. Turbulent flow is dynamic, nonlinear, and generally more effective at picking particles up off the ground. A **glacier** is a permanent body of ice consisting largely of recrystallized snow. Glaciers have very high viscosities and particles in a glacier are moved by *laminar* flow. All fluid particles travel in parallel layers in laminar flow. Glaciers play a significant role in erosion and transport by acting in three ways: as plows, as files, and as sleds.

4. **Mass wasting** is the *en masse* downslope movement of rock or regolith under the pull of gravity. In contrast to other types of erosion, the materials moved in mass wasting do not need to be transported by a fluid. **Slope failures** involve downslope movement of relatively coherent masses of rock or regolith. Slope failures can be one of three types: a **fall**, or a sudden vertical drop of rock fragment or debris; a **slide**, which involves rapid displacement of a mass of rock or sediment in a straight line down a steep slope; or a **slump**, a slower type of slope failure that involves *rotational* movement of rock.

5. **Flows** are mixtures of regolith and water or air. Flowing regolith can be wet or dry. Wet flows include *slurry flows* that can occur rapidly or slowly—a slow slurry flow is also known as *solifluction*. *Granular flows* can be rapid, as in debris avalanches, or slow. The most common type of granular flow, which is also the least noticeable because of its slow motion, is called **creep**.

6. Both weathering and erosion are controlled by climate and topography. Mass-wasting processes—especially landslides—tend to be particularly frequent along plate boundaries, where earthquakes commonly act as triggering mechanisms.

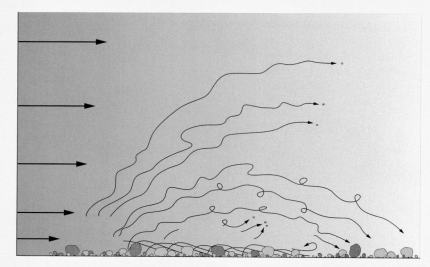

- **weathering** p. 185
- **regolith** p. 185
- **soil** p. 185
- **mechanical weathering** p. 185
- **chemical weathering** p. 185
- **joint** p. 186

- **dissolution** p. 188
- **clay** p. 193
- **sand** p. 194
- **humus** p. 194
- **soil horizon** p. 195
- **soil profile** p. 195

- **erosion** p. 198
- **bed load** p. 198
- **saltation** p. 198
- **suspended load** p. 198
- **glacier** p. 200
- **mass wasting** p. 200

- **slope failure** p. 200
- **flow** p. 200
- **creep** p. 203

CRITICAL AND CREATIVE THINKING QUESTIONS

1. Look around for evidence of mechanical and chemical weathering. How might you determine their relative importance in your area?

2. Many features have recently been discovered on Mars that are suggestive of erosion by water. What is the evidence that Mars once had a hydrosphere? How long ago? Where did the water go?

3. The Moon lacks an atmosphere, a hydrosphere, and a biosphere, but when the first astronauts landed, they discovered that the Moon has a deep regolith. How might the regolith have been formed, and how does it differ from Earth's regolith?

4. What kinds of mass-wasting processes occur where you live? Can you identify any evidence that would suggest how rapidly or how slowly mass wasting is moving regolith downslope? Look especially for signs of creep, which occurs almost everywhere. Some clues are bent tree trunks, curved fences, lobes of soil on grassy slopes, and tilted gravestones.

5. Keep an eye out for the structures in your town used to stabilize slopes or protect property from mass wasting. Are the slopes in your area engineered, or have they been left more or less in their natural state? Where you find retaining walls, do they appear to have stabilized the slope as intended?

What is happening in this picture ?

Mudflows are a particularly rapid and dangerous form of mass wasting. In January 2005, 400 thousand tons of mud cascaded down on the California town of La Conchita, killing 10 people. From a geologic point of view, a town should never have been built in this location. Why? What do you think might have caused the mudslide?

1. On this illustration, locate and label the following processes in the rock cycle:

weathering
melting
tectonic uplift
metamorphism

burial and cementation
erosion and deposition
intrusion and volcanism

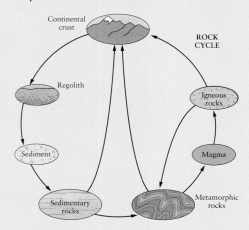

2. In _____ rock breaks down into solid fragments by physical processes that do not change the rock's chemical composition.

a. chemical weathering
b. mechanical weathering
c. mass wasting
d. erosion

3. This illustration shows the chemical weathering of a common feldspar mineral. The process shown depicts what kind of chemical reaction?

a. strong chemical
b. moderate chemical
c. dissolution
d. ion exchange

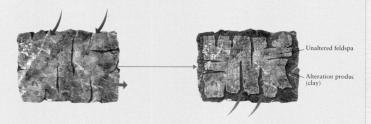

4. Death Valley in California is one of the hottest and driest spots in North America. Summer air temperatures commonly reach 50°C and rainfall averages less than 5 centimeters a year. In this desert environment, what type of weathering would you expect to find?

a. strong chemical weathering
b. strong mechanical weathering
c. moderate chemical weathering
d. moderate mechanical weathering
e. very slight weathering, mostly by wind

5. Sediments found on Earth's surface

a. are one product of the mechanical weathering process.
b. are one product of the chemical weathering process.
c. are the result of a combination of mechanical and chemical weathering processes.
d. None of the above statements is correct.

6. A dark-colored layer of mixed mineral and organic matter defines the

a. O soil horizon.
b. A soil horizon.
c. E soil horizon.
d. B soil horizon.
e. C soil horizon.

7. This illustration shows three climatic zones and accompanying soil profiles. Draw a line on the illustration linking soil profiles to the correct climatic zone.

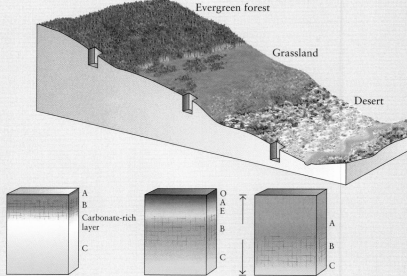

8. _____ is one method modern farmers can use to fight soil erosion.

   a. Uphill plowing
   b. Plowing along natural drainage systems
   c. Contour plowing
   d. Downhill plowing

9. _____ involves the removal and transport of regolith through the combined actions of ice, water, wind, and gravity. This is different from _____, which happens only in place, though both processes can occur at the same time. The process of _____ further alters Earth's surface by the downslope displacement of regolith due to the pull of gravity.

   a. Erosion ... mass wasting ... weathering
   b. Weathering ... erosion ... mass wasting
   c. Mass wasting ... weathering ... erosion
   d. Erosion ... weathering ... mass wasting
   e. Mass wasting ... erosion ... weathering

10. _____ is dynamic, nonlinear, and generally more effective at picking up particles off the ground than is _____, in which particles travel in parallel layers.

   a. Viscous flow ... laminar flow
   b. Viscous flow ... turbulent flow
   c. Laminar flow ... turbulent flow
   d. Turbulent flow ... laminar flow

11. Air and water tend to carry suspended particles most effectively through

   a. a combination of viscous and laminar flow.
   b. turbulent flow.
   c. laminar flow.
   d. a combination of turbulent and laminar flow.

12. _____ transport regolith through laminar flow.

   a. Streams
   b. Winds
   c. Debris flows
   d. Glaciers

13. _____ are a form of slope failure involving rapid displacement of a mass of rock or sediment in a straight path down a steep or slippery slope.

   a. Rock falls
   b. Slumps
   c. Slides
   d. Slurries

14. _____ involve rotational movement of rock or regolith.

   a. Rock falls
   b. Slumps
   c. Slides
   d. Slurries

15. This illustration shows block diagrams of mass wasting of hill slopes through sediment flow. Identify each block diagram based on processes related to rate and degree of wetness from the following list:

   wet, slow   dry, slow
   wet, fast   dry, fast

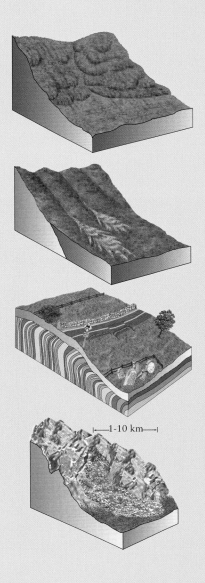

1-10 km

From Sediment to Sedimentary Rock

8

The Omo River basin of Ethiopia is presently a hot, dry, inhospitable place. Yet it is here that geologists and archaeologists have discovered the oldest known fossil remains of our species, *Homo sapiens*. The two skulls (bottom inset), estimated to be 195,000 years old, were discovered in the sedimentary rock formation seen in the background of this photograph.

How do we determine the age of the skulls? And how could the cradle of humanity be a region that has hardly any inhabitants today? Often the answers to such questions come not from the fossils, but from the stream sediments in which they are buried (top inset). These sediments are interlayered with volcanic ash, which can be dated by the methods described in Chapter 3. The sediments also tell us about the environment the early humans lived in. From the color, size, and rounding of the sediment particles, now preserved in solid rock, geologists can conclude that the Omo region was once much wetter than it is today. There were lakes and flowing streams, and the region supported lots of vegetation and small animals, whose fossilized remains have also been encased in the sedimentary rocks. In short, it was an ideal location for our species to survive and proliferate.

Sediments and sedimentary rocks are the best archives we have of Earth's past environments. It is humbling to realize that humans have now been around long enough—at least in certain parts of the world—to become a part of that archive.

Omo 1   Omo 2

Sediments and Sedimentation

Nearly every geologic process leaves its mark in the sedimentary record. Tectonic forces form mountains and basins, which in turn control where the source material for sediment comes from as well as the sites of **deposition**. Climate controls the way that rocks weather and how sediments are transported. To understand the complicated record these events leave, we first have to understand the different kinds of sediments. Geologists separate sediments into three broad categories: *clastic, chemical,* and *biogenic.*

deposition The laying down of sediment.

CLASTIC SEDIMENT

Clastic sediment derives its name from *clasts,* individual grains of mineral or fragments of rock. Clasts range in size from large boulders down to clay particles finer than flour. In fact, the size of the clasts is the primary basis for classifying clastic sediments and the sedimentary rocks they eventually form (see **FIGURE 8.1**). Other characteristics used to describe clastic sediment include the shape, angularity, and size range of grains (see **FIGURE 8.2**).

clastic sediment Sediment formed from fragmented rock and mineral debris produced by weathering and erosion.

From clasts to rocks FIGURE 8.1

	A Gravel	**B** Sand	**C** Silty mud	**D** Clayey mud
SEDIMENT	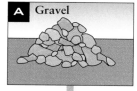	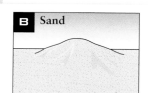	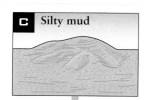	
...WITH COMPRESSION AND TIME, CAN BECOME...	A sediment with pea-sized or larger particles is called *gravel.* When gravel is cemented, the rock so formed is a **conglomerate**.	*Sand* consists of somewhat smaller particles, about the size of a pinhead. When compacted and cemented, sand becomes **sandstone**.	Sediment with even finer particles, the size of a grain of table salt, is called *silt.* The corresponding rock type is **siltstone**.	The finest sedimentary particles, the size of flour or smaller, are called *clay.** The corresponding rock type is **shale** or **mudstone**.
ROCK				
	This shows pebble-rich strata of conglomerates interbedded with sandstone; the area photographed is about 1 m wide.	This sandstone is about 7.5 cm across. The colors of depositional layers vary because of different iron contents.	This siltstone sample is about 8 cm across. Note the impressions of fossil marine shells.	This shale sample is about 10 cm across. The colors are caused by differing contents of organic matter.

*Note that "clay" in this context refers only to particle size, not composition.

SORTING

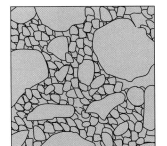

Very poorly sorted   Moderately sorted   Very well sorted

A In some sediments, all the particles are nearly the same size. Such sediments are called *well-sorted*, and usually have been transported by water and wind. Other sediments, such as those transported by ice or by mass wasting, are a jumble of particles of different sizes; these are poorly sorted or even unsorted.

ROUNDNESS

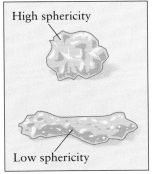
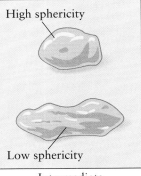
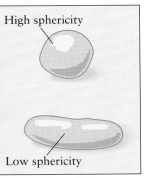

Angular   Intermediate   Rounded

B Individual particles may take a variety of shapes, from *rounded* to *angular*. Note the distinction between roundness and *sphericity*; even an angular particle can have high sphericity, which simply means that it is not much longer than it is wide.

C *Till*, like this deposit in Alaska, is an ice-transported sediment that is usually poorly sorted, of low sphericity, and angular in shape.

D Quartz sand, such as this (magnified) sample from Wisconsin, tends to be well-sorted, with high sphericity and roundness as a result of prolonged weathering and erosion.

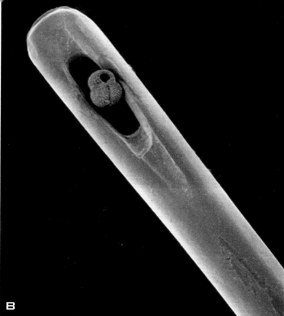

Chemical and biogenic sediments FIGURE 8.3

A Utah's Bonneville Salt Flats is one of the most desolate landscapes on Earth. These chemical sediments, which contain magnesium and potassium chloride in addition to ordinary salt, have been formed over the last 14,000 years by the evaporation of the prehistoric Lake Bonneville. The Great Salt Lake, a remnant of Lake Bonneville, covers less than one-tenth of the former lake's area. B Inside the eye of this needle is the shell of a foraminifer, a one-celled plankton that is abundant in the ocean. The shells of foraminifera cover roughly one-third to one-half of the ocean floor. Over time these biogenic sediment deposits are buried and converted into limestone or chalk.

Volcaniclastic sediments are another kind of clastic sediment. What makes them different is that all of the clasts are volcanic in origin. As discussed in Chapter 6, explosive volcanic eruptions blast out large quantities of fragments during an eruption. An old saying explains the uniqueness of volcaniclastic sediments: "They are igneous on the way up but sedimentary on the way down." Because the fragments were hot when formed, they are also called *pyroclasts* (from the Greek *pyro*, meaning "fire"). Pyroclasts are also classified by size, but the terminology is different (clasts of various sizes are called *bombs*, *lapilli*, and *ash* from largest to smallest).

CHEMICAL AND BIOGENIC SEDIMENTS

All surface water and groundwater contains dissolved chemicals that eventually find their way to a lake or sea. No natural water on or in Earth is completely free from dissolved matter. When dissolved matter precipitates from sea or lake waters, **chemical sediment** is the result. This precipitation can happen in two ways. First, plants and animals living in the water can alter its chemical balance. For example, increasing the amount of carbon dioxide dissolved in the water will cause calcium carbonate to precipitate. Many limestones are formed this way.

Second, if an inland sea is subjected to an increasingly warm and dry climate, or if the inflow of fresh water is restricted for some reason, evaporation may exceed the input of fresh water. The sea may then become so shallow and saline that salts that were dissolved in the water will begin to precipitate as solids. Modern examples of this process are found in the Aral Sea (Uzbekistan), Mono Lake, California, and Utah's Great Salt Lake (see FIGURE 8.3A).

Biogenic sediment is composed of the remains

> **chemical sediment**
> Sediment formed by the precipitation of minerals dissolved in lake, river, or sea water.

> **biogenic sediment**
> Sediment that is primarily composed of plant and animal remains, or precipitates as a result of biologic processes.

of plants and animals. This includes the hard parts of large animals, such as shells, bones, and teeth, as well as fragments of plant matter, such as wood, roots, and leaves (see FIGURE 8.3B). Two of the most common rock types that come from biogenic sediments are limestone, formed of the calcium carbonate skeletons of marine invertebrates, and coal, formed from partially decomposed terrestrial plant material. (Note that limestone can be either biogenic or chemical in origin.)

Sedimentary Rocks

L ithification is another step in the rock cycle. Sediment originates from the erosion of pre-existing rocks that have been broken down through weathering (Chapter 7). When the sediment is deposited and buried, it lithifies to become new rock. In this section, we discuss how lithification occurs and how the appearance of the new rock depends both on the materials and on the process by which it is created.

lithification The group of processes by which loose sediment is transformed into sedimentary rock.

ROCK BEDS

When you look at an outcrop of sedimentary rock, such as the one shown in FIGURE 8.4, one of the first things you will notice is the **bedding**. As we discuss in Chapter 3, the banded appearance comes from the fact that sedimentary particles are laid down in distinct strata. Over time, the mineral composition of the sediments in a particular location may change, or they may be transported or deposited in different ways. This will cause the adjacent strata to look different. The boundary between adjacent strata is called a **bedding surface**. It is the presence of bedding and bedding surfaces that indicates that a rock was once sediment.

bedding The layered arrangement of strata in a body of sediment or sedimentary rock.

bedding surface The top or bottom surface of a rock stratum or bed.

NATIONAL GEOGRAPHIC

Layers of rock: bedding FIGURE 8.4

The Bungle Bungle Range in northwestern Australia derives its unique coloration from layers of sandstone that have different permeabilities. Algae grow in the more permeable strata, tinting the rock black.

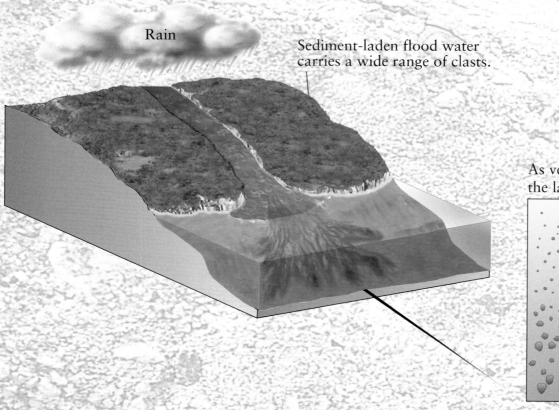

Rain

Sediment-laden flood water carries a wide range of clasts.

As velocity of flow drops, the largest clasts settle first.

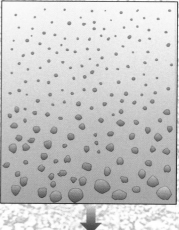

The appearance of a bed can say a lot about its origin. The center stratum in this rock from California is a *graded bed*, which is produced very quickly—by a single flood. Floods or rainfall cause a swollen river to flow faster and pick up a mixture of fine and coarse sediment. When this rapid, sediment-laden flow reaches a lake or the ocean, its velocity drops, and the larger clasts in the sediment settle to the bottom first. After the sediments harden into a rock, they will form a graded bed with the largest clasts on the bottom, like the one shown here.

fine grains

coarse grains

Sometimes the appearance of a bed can tell geologists a great deal about how it was deposited. For example, in a *graded bed*, the coarse clasts are concentrated at the bottom, grading up to the finest clasts at the top (see FIGURE 8.5). Graded beds often form where a stream or river enters a lake or an ocean.

Turbulent flow in streams, wind, or ocean waves produces a type of bedding called *cross bedding*. In FIGURE 8.6, the thick strata of sandstone contain many thin beds that are inclined with respect to the stratum in which they occur. The direction in which the cross bedding is inclined can tell geologists the direction in which the water or air currents were moving when they deposited the sediments.

LITHIFICATION PROCESSES

In order for newly deposited, loose sediment to be lithified and turn into rock, the individual particles must somehow be bound together into a cohesive unit. After a layer of sediment is buried, either by the accumulation of more sediment or by tectonic processes, it is placed under higher pressure, leading to **compaction**. Compaction is generally the first step in lithification, which can happen in several ways. For example, lithification can occur through **cementation**. Cementation can happen in various ways, such as the evaporation of groundwater under desert conditions. As the water evaporates, chemicals such as silica, calcium carbonate, and iron hydroxide precipitate and cement the grains of sediment together. Another method of lithification is **recrystallization**, which is especially common in limestone formed by coral reefs. Calcium carbonate crystals in the fossilized reef change from their original form (aragonite) to the more stable calcite. In the process of recrystallizing, crystals that were separate can grow together.

compaction
Reduction of pore space in a sediment as a result of the weight of overlying sediments.

cementation
The process in which substances dissolved in pore water precipitate out and form a matrix in which grains of sediment are joined together.

recrystallization
The formation of new crystalline mineral grains from old ones.

Cross bedding in ancient dunes FIGURE 8.6

This sandstone formation from Zion National Park in Utah formed when sand dunes were gradually cemented together. The changing slope of the dunes reflects shifts in the direction of the prevailing winds over time.

From FIGURE 8.7, we can see that pressures caused by sediment accumulation or tectonic forces initiate the lithification process. Although the pressures involved in lithification may be high by everyday human standards, they are low pressures by geologic standards, and low in comparison with the pressures that induce metamorphism (Chapter 9). All of the low-temperature, low-pressure changes that happen to sediment after deposition are collectively called **diagenesis**. They include lithification, as well as processes involving chemical reactions or microbial activity, that are not part of lithification.

Lithification FIGURE 8.7

1. COMPACTION

The weight of accumulating sediment forces the grains together, thereby reducing the *pore space* and forcing water out of the sediment.

Water — Mineral grains

Weight of overlaying sediment

Pore space

Reduced pore space

AND
2. CEMENTATION

Pore water expelled from deeply buried sediments migrates upward toward Earth's surface.

As it rises and cools, ions dissolved in the water precipitate, forming minerals that cement the grains together.

Cement

AND/OR
3. RECRYSTALLIZATION

Pressure causes less stable minerals to rearrange crystals into more stable forms. Aragonite is present in the skeletal structures of living corals and other marine invertebrates.

Over time, aragonite recrystallizes and becomes calcite, which has a different crystalline structure.

Shells made of aragonite

Shells made of calcite

Two of these rocks are sedimentary, and one is igneous. Can you tell which is which?

www.wiley.com/
college/murck

Answers:

A The fossils in this specimen are a give-away. No organism can survive the high temperatures at which igneous rocks are formed, so the presence of ancient shells means the rock (now a limestone) is **sedimentary.**

B Looking at this sample, we see interlocking crystals without any cement. This rock is granite, an **igneous** rock.

C In the magnified view of this sample, we see mineral grains rounded by abrasion. In addition, the grains do not abut one another; they have gaps between them that are filled by cement. This rock is **sedimentary** (in fact, a sandstone held together by a calcite cement).

TYPES OF SEDIMENTARY ROCKS

If a sedimentary rock is made up of particles derived from the weathering and erosion of igneous rock, it may contain many of the same minerals as the source rock. How, then, can we tell that it is sedimentary rather than igneous? In addition to such clues as bedding, the texture of the rock provides evidence (see **Figure 8.8**).

CLASTIC SEDIMENTARY ROCKS

When a clastic sediment lithifies, it produces a rock whose properties reflect the type of sediment it came from. The four basic classes are *conglomerate, sandstone, siltstone*, and *mudstone* or *shale*. These are the rock equivalents of gravel, sand, silt, and mud (see Figure 8.1).

> **conglomerate** A clastic sedimentary rock with large fragments in a finer-grained matrix.

> **sandstone** A medium-grained clastic sedimentary rock in which the clasts are typically, but not necessarily, dominated by quartz grains.

> **mudstone** A very fine-grained sedimentary rock of the same composition as shale but without fissility.

> **shale** A very fine-grained fissile or laminated sedimentary rock, consisting primarily of clay-sized particles.

To be classified as a **conglomerate**, a sedimentary rock must have clasts larger than 2 millimeters. Typically, the large clasts are surrounded by much finer-grained material, called a *matrix*. If the clasts are angular, rather than rounded, the rock is called a *breccia*. The presence of angular clasts means that the sediment has only been transported a short distance and was not subjected to a long abrasion process.

The medium-sized grains in **sandstone** range from 0.05 to 2 millimeters in size. They are usually dominated by quartz, because quartz is a tough mineral that resists weathering. If the sediment has not been transported very far from its source, it may still contain a lot of feldspar or rock fragments. Geologists call such a sandstone *immature*.

Mudstone or *siltstone* consists primarily of silt- and clay-sized particles; these include tiny pieces of rock and mineral grains as well as clay minerals. Mudstone breaks into blocky fragments. **Shale**, which contains a higher proportion of clay-sized particles, is a special subvariety of mudstone that is *fissile*, which means that it splits into sheet-like fragments.

CHEMICAL SEDIMENTARY ROCKS

Chemical sedimentary rocks result from the lithification of chemical sediments. As we have seen, such sediments are formed by the chemical precipitation of minerals from water. Most chemical sedimentary rocks contain only one important mineral, such as calcite, dolomite, gypsum, or halite. These monomineralic compositions, together with the modes of precipitation, form the basis for classification of these rocks.

Chemical sedimentary rocks often form as **evaporites**. Calcite, gypsum, and halite typically form from the evaporation of seawater, while the evaporation of lake water may yield more exotic minerals (such as sodium carbonate and borax). Many evaporite minerals are mined because they have industrial uses, such as gypsum, used to make plasterboard. In addition, most of the salt we eat comes from evaporites.

An unusual but economically important kind of chemical sedimentary rock is a **banded iron formation**. Such rocks are the source of most of the iron mined today. Not only are they valuable for their ore, they tell the story of a critical period in Earth's history. Almost all banded iron formations are about the same age—1.8 to 2.5 billion years old. This strongly suggests that unique conditions existed on Earth during that period (see What a Geologist Sees).

> **evaporite** A rock formed by the evaporation of lake water or seawater, followed by lithification of the resulting salt deposit.

> **banded iron formation** A type of chemical sedimentary rock rich in iron minerals and silica.

BIOGENIC SEDIMENTARY ROCKS

Limestone is the most abundant biogenic sedimentary rock. It is formed from lithified shells and other skeletal material from marine organisms. Some of these organisms build their shells or skeletons from calcite, but most construct them from a mineral called aragonite, which, like calcite is calcium carbonate. During diagenesis, the aragonite is transformed

> **limestone** A sedimentary rock that consists primarily of the mineral calcite.

Black layers are rich in reduced iron (Fe^{2+}).

Red layers are rich in oxidized iron (Fe^{3+}).

White layers are rich in silica.

Close-up of layers in Banded Iron Formation

A Change in the Atmosphere

Banded iron formations, such as this 2.5-billion-year-old stratum in the Hamersley Range in Australia, formed when iron that was dissolved in seawater precipitated as a chemical sediment. Today, seawater contains only slight traces of iron, because the oxygen in the atmosphere reacts with it to form insoluble iron compounds. If the ocean was once rich in dissolved iron, there must have been very little oxygen in the atmosphere at that time.

How did Earth make the transformation from an oxygen-poor atmosphere 2.5-billion years ago to the oxygen-rich atmosphere of today? In some 2.5-billion-year-old rocks there are microscopic fossils of cyanobacteria. These bacteria are thought to be the first organisms on Earth to extract energy from sunlight by photosynthesis, a chemical process that releases oxygen. Scientists hypothesize that algae might have oxygenated Earth's atmosphere very rapidly—and possibly more than once, as they proliferated and then poisoned themselves by producing too much oxygen. Each time the oxygen concentration changed, an iron mineral would precipitate out of the seawater, creating a new band of iron-rich sediments. Eventually, about 2.0 to 1.8 billion years ago, the oxygen level of the atmosphere reached a point where the ocean could no longer retain much iron, and banded iron formations could no longer form.

peat A biogenic sediment formed from the accumulation and compaction of plant remains.

coal A combustible rock formed from the lithification of plant-rich sediment.

into the more stable mineral, calcite, which becomes the main ingredient of limestone.

Calcite is sometimes replaced by the mineral dolomite (a carbonate mineral containing both magnesium and calcium); the resulting rock is called *dolostone*. Another kind of biogenic rock, which consists of extremely tiny particles of quartz, is called *chert*. The quartz in chert does not come from sand, but from the shells of microscopic sea animals.

An important class of biogenic sediment consists of the accumulated remains of terrestrial plants. Over time, and with pressure, this mass gradually becomes **peat**. Eventually, given enough time and pressure, peat may lithify and become **coal**. Lithification (also called *coalification*) involves further compaction, release of water, and slow chemical changes that weld

the plant fragments together, thereby making the coal relatively lower in water and richer in carbon than the original peat.

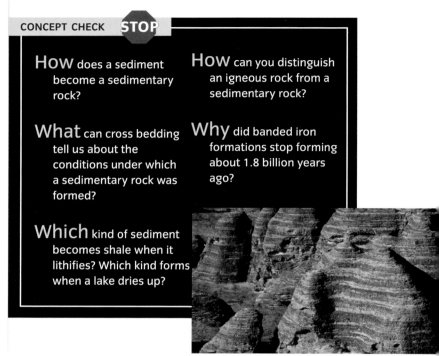

CONCEPT CHECK STOP

How does a sediment become a sedimentary rock?

What can cross bedding tell us about the conditions under which a sedimentary rock was formed?

Which kind of sediment becomes shale when it lithifies? Which kind forms when a lake dries up?

How can you distinguish an igneous rock from a sedimentary rock?

Why did banded iron formations stop forming about 1.8 billion years ago?

Depositional Environments

LEARNING OBJECTIVES

Explain how features like ripple marks, cracks, and fossils can tell geologists about the environment in which a rock originated.

Identify several common sedimentary rock-forming environments.

Define sedimentary facies and how geologists use them to reconstruct Earth's history.

Just as history books record the changing patterns of civilization, sedimentary rocks record the environmental history of our planet. Layers of sediment, like pages in a book, record how environmental conditions have changed throughout Earth's history. Geologists are able to "read" this story by interpreting the evidence in the rocks.

We have already seen that the size, shape, and arrangement of particles in sediment *(texture)*, as

well as the geometry of sedimentary strata, provide evidence about the geologic environment in which sediment accumulates. These and other clues enable us to demonstrate the existence of past oceans, coasts, lakes, streams, deserts, glaciers, and wetlands. As you will learn in this section, ancient sedimentary deposits bear a striking resemblance to sedimentary environments on Earth today. This is an excellent example of the principle of uniformitarianism, discussed in Chapter 1.

INTERPRETING ENVIRONMENTAL CLUES

Patterns formed by currents of water or air moving across sediment can be preserved and later exposed on bedding surfaces. For example, bodies of sand that are being moved by wind, streams, or coastal waves are often rippled; these may be preserved in sandstone as *ripple marks* (FIGURE 8.9A and B). Similarly, *mud cracks* (FIGURE 8.9C and D), fossil tracks (see FIGURE 8.10 on the following page) and even raindrop impacts can be recorded on bedding surfaces, attesting to moist surface conditions at the time they were formed.

Ancient and modern features compared FIGURE 8.9

▲ A Ripples are forming in shallow water near the shore of Ocracoke Island, North Carolina.

▲ B Almost identical ripples are exposed on a bedding surface of a sandstone at Artist's Point, Colorado National Park, Colorado.

▲ C Mud cracks formed on this modern river bed as the river dried up.

▲ D Similarly shaped mud cracks are preserved on the surface of a shale exposed at Ausable Chasm, New York. We can infer that this rock formation was deposited in an intermittently wet environment, such as a seasonal lake bed or a tidal flat. Note that in both (B) and (D) the present-day environment bears no relation to the environment in which the sedimentary rocks were deposited.

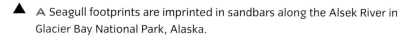

▲ A Seagull footprints are imprinted in sandbars along the Alsek River in Glacier Bay National Park, Alaska.

B Fossilized footprints are preserved in sandstone in the Painted Desert, ▶ near Cameron, Arizona. The animal that left these prints may have been hunting for prey stranded by the falling tide, just like the present-day seagulls in Alaska.

Footprints in the sand FIGURE 8.10

Fossils also provide significant clues about former environments. Some animals and plants inhabit warm, moist climates, whereas others can live only in cold, dry climates. Using the climatic ranges of modern plants and animals as guides, we can infer the general character of the climate in which similar ancestral forms lived (see FIGURE 8.11). Even microscopic fossils are important: The shells of foraminifera can tell us about former temperatures and salinity conditions in the oceans. Fossils are also the basis for determining the relative ages of strata. In fact, fossils have played an essential role in efforts to reconstruct the past 635 million years of Earth's history.

Plants and climate FIGURE 8.11

This fossilized seed fern, found in Greenland, dates from the Triassic Period. Fossils of tropical, moisture-loving plants, such as cycads and palms, can be found in Triassic-aged sedimentary rocks all over the world. The particular fern species in this photograph became extinct after the Triassic, so the fossil can be used to date the sediment in which it was found. And despite the fact that Greenland's modern climate is cooler, the presence of this fossil shows this was not always the case.

The color of fresh, unweathered sedimentary rock can also provide clues to environmental conditions. The color of a rock is determined by the colors of the minerals, rock fragments, and organic matter of which it is composed. Iron sulfides and organic detritus buried with sediment are responsible for most of the dark colors in sedimentary rocks. The presence of these materials implies that the sediment was deposited in an oxygen-poor (reducing) environment. Reddish and brownish colors result mainly from the presence of iron oxides, occurring either as coatings on mineral grains or as very fine particles. These minerals point to oxygen-rich (oxidizing) conditions in the environment.

DEPOSITIONAL ENVIRONMENTS ON THE LAND

Sediments on land can be transported by water, ice, and wind, or moved down-slope under the influence of gravity. They are deposited wherever the carrying capacity of the transporting agent decreases enough for the sediment to settle out. Thus, common environments for the deposition of sediment on land include stream channels and floodplains, lakeshores and bottoms, the margins of glaciers, and areas where the wind is intermittently strong, such as beaches and deserts (see FIGURE 8.12).

Visualizing

Depositional environments FIGURE 8.12

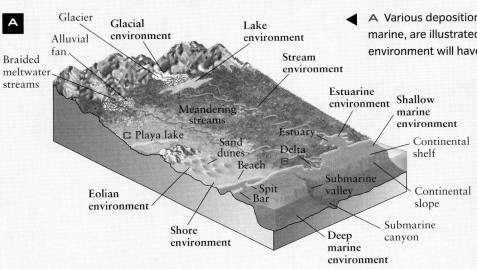

A Various depositional environments, both terrestrial and marine, are illustrated in this schematic diagram. Each kind of environment will have its own unique collection of sediment.

B A river on Russia's Kamchatka Peninsula fans out into a classic delta as it enters the ocean. Deltas are formed where running water enters a large body of standing water, such as a lake or the ocean. ▶

C Storm clouds gather over a dry lake bed, or *playa*, in Nevada. Playas may intermittently be covered with water, but normally they have the mud-cracked, salt-encrusted appearance seen here. The salt deposits formed as a result of a year-round lake called Lake Lahontan that covered this region during the most recent ice age.

Streams   Streams are the main transporters of sediment on land. Sediments deposited by streams differ from place to place depending on the type of stream, strength of flow, and nature of the sediment load (see Figure 8.12). Generally speaking, sediments become better sorted and more rounded with distance from the source. Thus, a typical large, smoothly flowing stream may deposit well-sorted beds of coarse and fine particles as it slowly migrates back and forth across its valley. During spring floods, silt and clay are deposited on the floodplain. In contrast, a large mountain stream flowing down a steep valley will transport a wide range of particle sizes, including large, angular grains and a high proportion of rock fragments. These rocks may contain chemically less-stable minerals, such as amphiboles and feldspars, that have not yet had enough time to undergo chemical weathering. When the stream reaches the front of the mountain and is no longer constrained by valley walls, it may spread out into an *alluvial fan*, in which the sediments range from coarse, poorly sorted gravel upstream to well-sorted, cross-bedded sand downstream.

> **delta** A sedimentary deposit, commonly triangle-shaped, that forms where a stream enters a standing body of water.

Lakes   Sediment that is deposited in a lake (see Figure 8.12) accumulates on the lakeshore and on the lake floor. Lakeshore deposits are generally well-sorted sands. The sediment load of a stream entering a lake will be dropped as the stream's velocity and transporting ability suddenly decrease, forming a **delta**. Inclined, generally well-sorted layers on the front of a delta pass downward and outward into thinner, finer, evenly laminated layers on the lake floor (see Figure 8.12). In arid regions, seasonal lakes or *playas* are common: They fill up with water for part of the year, and then leave evaporite deposits when they dry out.

Lake deposits may appear similar in many ways to marine sediments. Among the distinguishing characteristics are the generally smaller extent of lake deposits (lakes are typically smaller than seas), and the presence of freshwater fossils instead of marine fossils. As you might expect, lake deposits are less common in the geologic record than marine sediments.

Glaciers   Sediment transported by a glacier is either deposited along the glacier's base or released at the margin of the glacier as melting occurs. The sediment may then be subjected to further reworking by running water. Debris that has been deposited directly from ice commonly forms a random mixture of particles that range in size from clay to boulders and consist of all the types of rocks over which the ice has passed. This type of sediment is called *till*. Glacial deposits are discussed in greater detail in Chapter 11.

Wind   Processes related to the wind are referred to as **eolian** (pronounced *ee-OHL-ee-un*), after Aeolus, the Greek god of wind. Sediment carried by the wind tends to be finer than that moved by other erosional agents. Grains of sand are easily transported in places where strong winds are blowing and vegetation is too sparse to stabilize the land surface—for example, seacoasts and deserts (see Figure 8.12A). In such places, the sand may pile up to form dunes composed of well-sorted sand grains, with bedding inclined in the downwind direction (the direction toward which the air is flowing). Using these characteristics, geologists can easily identify ancient dune sand in the rock record, as in Figure 8.6.

> **eolian sediment** Sediment that is carried and deposited by the wind.

Powdery dust that has been picked up and moved by the wind may travel great distances and be deposited thousands of kilometers away. For example, oceanographers have discovered windblown sediment from the Sahara Desert of North Africa at the center of the Atlantic Ocean. A type of yellow-brown windblown sediment is called *loess* (the German word meaning "loose" and pronounced *luhss*). Consisting predominantly of silt, loess is windblown dust transported from desert surfaces, glacial sediments, and glacial stream deposits at times of ice-sheet retreat (see **FIGURE 8.13**).

Soil from the last ice age FIGURE 8.13

The Loess Hills in western Iowa stand out prominently against the flat Missouri River plain in the distance. The most recent ice age produced copious amounts of glacial sediment, which was transported by wind all over midwestern North America.

Here, the loess piled up in large dunes. Though the Loess Hills are too steep for farming, soils developed on loess deposits like these are highly productive for agriculture, and they underlie many of the world's breadbasket regions.

DEPOSITIONAL ENVIRONMENTS IN AND NEAR THE OCEAN

Deltas and estuaries

Deltas were mentioned earlier as a type of deposit that may form where a stream flows into a lake or the sea. Large marine deltas are complex deposits consisting of coarse stream-channel sediments, fine sediments deposited between channels, and still finer sediments deposited on the seafloor.

Where the load of sediment carried by a stream to the sea is smaller, much of it may be trapped in an **estuary** (see Figure 8.12A and B, and FIGURE 8.14A on the following page). In estuaries, coarse sediment tends to settle close to land, while fine sediment is carried seaward. Tiny individual particles of clay carried in suspension settle very slowly to the seafloor. As a transitional environment between land and sea, estuarine sediments often contain a large amount of organic matter, including fossils of organisms from both land and sea.

> **estuary** A semi-enclosed body of coastal water, in which fresh water mixes with seawater.

Beaches

Quartz, the most durable of the common minerals in continental rocks, is a typical component of beach sand. However, not all ocean beaches are sandy. Any beach consists of the coarsest rock particles contributed by the erosion of adjacent sea cliffs, together with materials carried to it by rivers or by currents moving along the shore. Beach sediments tend to be better

Depositional Environments   **229**

sandbar

seagulls

brackish water from mixing of ocean and river water

◀ A Where Stream Meets Sea
Where the Tijuana River (on the border between the United States and Mexico) flows into the Pacific, it deposits sediment—as well as sewage and trash from both sides of the border—into an estuary. Here, seagulls congregate on sandbars formed by sediment.

B Beach ▶
All beaches contain material that has been transported from somewhere else. The unusual color of Hawaii's Green Sand Beach comes from the high concentration of the mineral olivine in its sand (magnified in the inset). The beach is surrounded on three sides by the eroded walls of a cinder cone called Puu Mahana (seen on the left in this photo). The basalt in these walls contains olivine, which washes down onto the beach. The glassy-looking fragments of olivine are not very resistant to weathering, so we would only expect to find a green sand beach very close to its volcanic source.

▲ C Carbonate Shelf
Carbonate sediment accumulates in the warm, shallow offshore waters of these islands in the Bahamas. The sediment shows up as white sandbars and consists of fine skeletal debris of tiny sea creatures.

▲ D Deep-Sea Turbidites
These deep-sea turbidite beds have been tilted, uplifted, and exposed in a wave-eroded bench along the coast of the Olympic Peninsula in Washington.

sorted than stream sediments of comparable coarseness, and they often display cross-stratification. Particles of beach sediment, dragged back and forth by the surf and turned over and over, become rounded by abrasion. Though beach sands are often buff-colored, they don't have to be; they may also be white, black, or even green, reflecting the presence of different minerals (see **FIGURE 8.14B**).

Shelves
Most of the world's sedimentary rocks originate as strata on continental shelves. As fresh water flows out to sea through an estuary or a river mouth, it continues seaward across the continental shelf. It will deposit most of its sand-sized sediment, whether river-derived or formed by erosion of the shore, within about 5 kilometers of the land. However, some sand can be found 100 kilometers or more offshore. Otherwise, continental shelf sediment tends to consist of silty or sandy mud that contains marine fossils.

On the continental shelf of eastern North America, up to a 14-kilometer thickness of fine sediment has accumulated over the last 150 million years. Shelves, in effect, catch weathered continental crust in such a way that it is continually recycled by the processes of plate tectonics and the rock cycle. Due to the abundance of marine life in shelf waters, continental shelf sediments generally contain a high percentage of organic matter.

Carbonate platforms and reefs
Where the climate and surface temperature are warm enough to nurture abundant carbonate-secreting organisms, biogenic sediments composed of calcium carbonate may accumulate on continental shelves or broad, flat carbonate platforms that rise from the seafloor (see **FIGURE 8.14C**). A *reef* is a wave-resistant structure built from the skeletons of marine invertebrates. Reefs are generally restricted to warm, sunlit waters of normal marine salinity.

Marine evaporite basins
In coastal areas with a sufficiently warm and dry climate, ocean water may evaporate fast enough to leave behind the salts that were dissolved in it, as a marine evaporite deposit. These can be distinguished from lake-derived evaporites because they have a different mineral composition.

Marine evaporite deposits are surprisingly common (reflecting the many times that shallow seas have flooded large areas of continents in the past), and underlie as much as 30 percent of the land area of North America.

Turbidites
Thick sediments are found at the foot of the continental slope at depths as great as 5 kilometers beneath the surface of the ocean. The origin of these sediments was difficult to explain until marine geologists discovered that **turbidity currents**—essentially, underwater landslides that originate on the continental shelf—deposit them. These currents, sometimes started by earthquake shaking, rush swiftly down the continental slope at velocities of up to 90 km/h. As a turbidity current reaches the ocean floor, it slows down and deposits a graded layer of sediment called a *turbidite*. At any site on the continental rise, a major turbidity current is a rare event that happens perhaps once every few thousand years. Nevertheless, over millions of years, turbidites can slowly accumulate and form thick deposits consisting of many layers (see **FIGURE 8.14D**).

> ■ **turbidity current**
> A turbulent, gravity-driven flow consisting of a mixture of sediment and water, which conveys sediment from the continental shelf to the deep sea.

Seafloor
As mentioned earlier, a large part of the deep seafloor is covered with biogenic sediments. These deposits come in two principal varieties. *Calcareous ooze*, made of calcium carbonate, comes from the remains of tiny sea creatures. While they are alive these creatures float, but after they die their shells drop to the bottom of the ocean. Calcareous ooze forms at low to middle latitudes, where the water is warm. Other parts of the deep ocean floor are mantled with *siliceous ooze* from silica-secreting organisms. (Siliceous ooze is chemically similar to quartz but differs in terms of mineral structure). This material is most common in the equatorial Pacific and Indian oceans and in a belt encircling the Antarctic continent. These are areas where the biologic productivity of surface waters is high, owing partly to the upwelling of deep ocean water that is rich in nutrients.

Each depositional environment leaves its own kind of sedimentary record, which may change over time. This section shows a variety of depositional environments in which distinctive facies are deposited. On the land surface, they lie side by side. In a vertical section, they lie one above another. Notice the seaward dip of the boundaries between adjacent facies. This indicates that, over time, the boundaries have migrated in the landward direction, due to a rise in sea level.

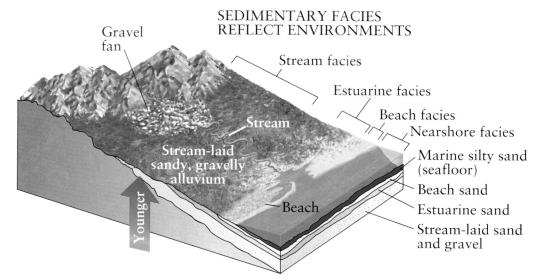

SEDIMENTARY FACIES
REFLECT ENVIRONMENTS

Gravel fan

Stream facies

Estuarine facies

Beach facies

Nearshore facies

Stream

Stream-laid sandy, gravelly alluvium

Marine silty sand (seafloor)

Beach sand

Estuarine sand

Stream-laid sand and gravel

Beach

Younger

Sedimentary facies FIGURE 8.15

SEDIMENTARY FACIES

If you examine a vertical sequence of exposed sedimentary rocks, you may notice differences as you move upward from one bed to the next. You may note changes in sedimentary rock type, color, fossil content, and thickness. The differences indicate that the environmental conditions in that location changed over time. If you trace a single bed laterally for a few kilometers, you may also notice changes, indicating that at any given time during the deposition of the sediment, conditions differed from one place to another.

The changes in the character of sediment from one environment to another are referred to as changes of **sedimentary facies** (pronounced *fay-sheez*). One facies may be distinguished from another by differences in grain size, grain shape, stratification, color, chemical composition, depositional structure, or fossils. Adjacent facies can merge into each other either gradually or abruptly (see FIGURE 8.15). For example, coarse gravel and sand on a beach may gradually pass into finer sand, silt, and clay on the floor of the sea or a lake. Coarse, boulder-like glacial sediment, on the other hand, may end abruptly at the margin of a glacier.

By studying the relationships among different sedimentary facies and using these characteristics to identify original depositional settings, we can reconstruct a picture of the environmental conditions that prevailed in a region during past geologic times (see Case Study).

CONCEPT CHECK STOP

How do fossils tell us about the climate in a particular location millions of years ago?

What kind of sedimentary deposits would you find at the foot of a mountain range? On a beach? At the bottom of the ocean?

Where in a stream bed would you expect to find each of the following types of sediment—gravel, sand, and mud?

What is the most characteristic property of wind-transported sediments?

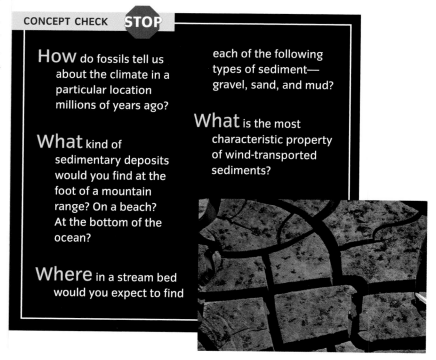

Sedimentary Facies and the History of Humankind

Once upon a time, 3.3 million years ago, a little girl wandered near a stream in Africa. Around her were open grasslands and shaded woodlands, home to elephants, hippos, rhinoceroses, and antelopes. Though the tiny three-year-old walked upright on two feet, she had strong muscles and gorilla-like shoulders and could climb a nearby tree to escape from predators, take shelter, or forage for fruit.

Today we know this part of the world as Ethiopia. The rocky hills, south of the Awash River near a place called Dikika, are now dry, hot, and seemingly barren (**FIGURE A**). But the tiny set of bones that emerged from these rocks is the oldest and most complete fossil of a human-like (*hominid*) child ever found (**FIGURE B**). The paleontologists who found the girl's remains named her Selam—"peace" in the Amharic language. The strata of the Hadar Formation that held her bones reveal much about how Selam lived—and died.

The facies reveal that Selam and others of her species, *Australopithecus afarensis* (including the famous fossil "Lucy" found not far away in 1974), lived in an environment of rapidly fluctuating, ephemeral lakes and fast-moving streams, in a deltaic depositional system. Flooding was a common occurrence in this constantly shifting terrain. Perhaps the little girl died by falling into a swiftly flowing stream—

A Here, members of the team sift loose sediment, looking for small fossils, Sedimentart strata of the Ethiopian badlands are in the background.

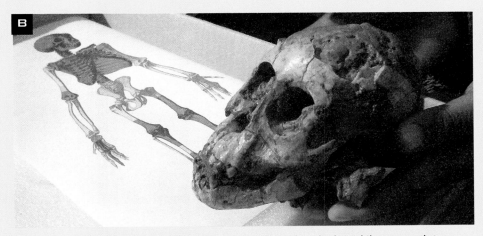

B Dr. Zeresenay Alemseged, the paleontologist who heads the Dikika research team, holds the baby's fossilized skull, which fits easily into the palm of his hand.

her bones were found encased in river channel deposits of gravel and sand, with clear evidence of rapid sedimentation. Her small body would have been buried quickly by sediment-laden water, protected from predators and the elements for millennia.

NATIONAL GEOGRAPHIC

How Plate Tectonics Affects Sedimentation

LEARNING OBJECTIVES

Identify plate-tectonic features that are favorable for the accumulation of sediments.

Explain how the sediments in a rift valley or passive continental margin differ from those in a collision zone or a subduction zone.

Clastic sediment is far more abundant than chemical or biogenic sediment. The locations where clastic sediment is commonly formed and deposited are largely controlled by plate tectonics. Clastic sediment originates from the rock and mineral debris produced by weathering and erosion of continental masses. The sediment accumulates in low-lying areas—troughs, trenches, and basins of various types. FIGURE 8.16 illustrates the plate-tectonic settings where clastic sediment is most likely to accumulate. (Chemical and biogenic sediments are not as strongly influenced by plate tectonics, although the locations of ocean basins, where many of these sediments occur, are controlled by plate tectonics.)

DIVERGENT PLATE BOUNDARIES— RIFT VALLEYS

Low-lying **rift valleys** are formed when a continent splits apart because of tensional forces. The East African Rift is an example of a rift that is still active in pulling apart a continent; the Red Sea is a young rift where a new ocean is already forming; and the Atlantic Ocean is a mature (i.e., well-developed) rift. A rift valley may eventually become a continental margin bordering an ocean, or the tensional forces may stop at some point and the valley would then become a failed rift, such as the Newark basin of New Jersey, or the 1.1-billion-year-old rift that underlies the sedimentary rocks of the North American midwest.

■ **rift valley** A linear, fault-bounded valley along a divergent plate boundary or spreading center.

Both the East African Rift and the ancient Newark Basin hold deep wedges of immature clastic sediments deposited by streams. In the Red Sea, marine sediments now cover the clastic sediments that were deposited before the Red Sea was wide enough for the sea to enter and the evaporites that formed while the sea water was still shallow. On the Atlantic Ocean margin of North America, sediment that has eroded from the adjacent continent has accumulated over many millions of years to a thickness of over 14 kilometers (see Figure 8.16A). Most of the strata deposited on the shelf of a passive continental margin are mature, shallow-water marine sediments. The continental shelf slowly subsides as accumulation takes place, and the pile of sediment grows thicker and thicker. Lithification occurs as sediments are buried progressively deeper in the pile.

CONVERGENT PLATE BOUNDARIES— COLLISIONAL-TYPE

A variety of low-lying regions are found within and along the edges of high mountain ranges. Many of these are structural basins or troughs caused by faulting and folding of rocks associated with the process of mountain building. Coarse, immature (i.e., angular and poorly sorted) stream sediments eroded from a rising mountain range will accumulate in these low-lying areas as vast thicknesses of conglomerate, gravel, and sand (see Figure 8.16B). In the Himalaya, for example, thick sequences of conglomerate and coarse sandstone flank the southern edge of the range, while the finest-grained sediment has been transported all the way to the sea by the many streams that flow to the Indian Ocean.

CONVERGENT PLATE BOUNDARIES— SUBDUCTION-TYPE

The lowest-lying points on Earth's surface are the long, deep trenches that form on the ocean floor in subduction zones (see Figure 8.16C). Subduction zones are

Clastic sediment accumulates in low-lying areas, which occur in specific locations that are strongly controlled by plate tectonics.

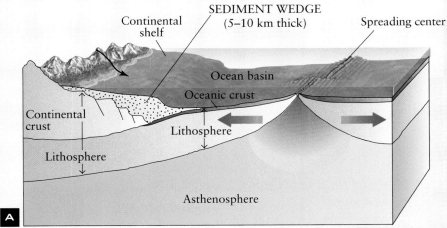

A Thick sedimentary wedges accumulate in rift valleys and along passive continental margins that are formed when continental crust rifts and a new ocean basin opens.

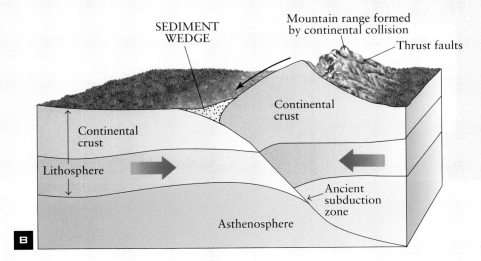

B Sediments accumulate in structural basins along the edges of mountain ranges thrust up by continental collisions.

C Sediments are shed from continents into deep-sea trenches above subduction zones. The sediment forms a wedge that is compressed and crushed as the oceanic plate subducts under the continental plate. The *back-arc basin* that forms on the over-riding plate, behind the subduction zone, is another location where sediment is likely to accumulate.

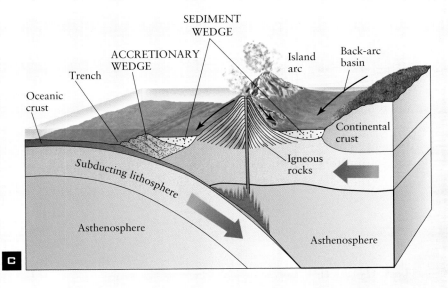

How Plate Tectonics Affects Sedimentation 235

The eastern edge of Taiwan is an active accretionary wedge. **A** This chaotic terrain of rocks broken during accretion is near Taitung, at southern end of the wedge.

often found near continental margins, such as along the western margin of South America and the eastern margins of Asia. The elevation of Earth's surface plunges from the top of the volcanoes on the edge of the continent (a common feature in subduction zones) to the bottom of the trench, all within the lateral distance of a few hundred kilometers. Because erosion proceeds most rapidly where the slopes are steepest, we would expect to find rapid erosion and deposition of sediment here—and we do. Sediment is transported from the continent to the adjacent trench by streams and turbidity currents, and it accumulates in the trench as turbidites.

At some convergent margins, the sediment is scraped off the subducting plate in thick slabs, separated by thrust faults, which pile up like a stack of playing cards on the overriding plate. Adjacent volcanic arcs ensure that a lot of volcanic debris is present in the sediment. The wedge-shaped accumulations of volcaniclastic sediment are called *accretionary wedges* (see **FIGURE 8.17**). For example, the islands of Taiwan and Barbados largely consist of accretionary wedge sediments thrust up above sea level. At other subducting margins, the sediment travels down into the mantle along with the subducting

B Steep gorges can result from rapid uplift during accretion. This is Taroko Gorge, Hualien, at the northern end of the Taiwan accretionary wedge.

plate. The water contained in the sediment is released there, and facilitates wet melting at depths of 100 kilometers and greater. These partial melts fuel the explosive volcanic eruptions that are characteristic of subduction zones.

In some cases, the volcanic arc forms offshore, creating volcanic islands and a *back-arc basin* between the islands and the mainland. Though not as dramatic topographically as the deep ocean trenches, these basins are also places where ocean-floor sediment accumulates.

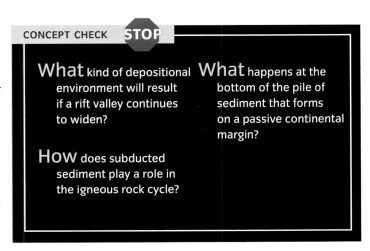

CONCEPT CHECK STOP

What kind of depositional environment will result if a rift valley continues to widen?

What happens at the bottom of the pile of sediment that forms on a passive continental margin?

How does subducted sediment play a role in the igneous rock cycle?

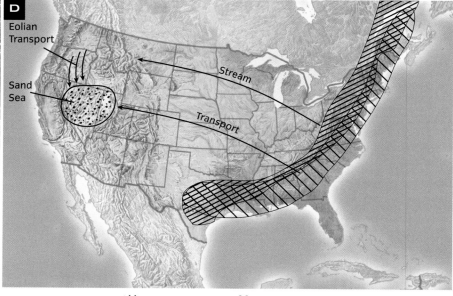

Many of the rock formations in Utah's Zion National Park resemble sand dunes—because that is what they originally were. The imposing cliffs (FIGURE A), the roadside hill (FIGURE B), and the stunningly etched ground seen from above (FIGURE C) were all once part of a vast sea of sand, larger and thicker than the Sahara Desert is today. The dunes lithified into sandstone and subsequently eroded into the formations you see today. The sedimentary formation they belong to, called the Navajo Sandstone, extends over several states in the American Southwest, and attains a thickness of 700 meters in Zion National Park.

But if all of this rock was once sand, that poses a puzzle: Where did all the sand come from? The dune patterns indicate that the prevailing winds came from the north, but no mountain ranges of suitable size or age can be found there. The mountain range we call the Rockies did not yet exist in the Lower Jurassic Period, 190 million years ago, when the Navajo sand was being deposited.

\\\ Appalachians   ⧄ Grenville Mountains

However, the Appalachian Mountains did exist then. Lifted up by a plate tectonic collision that began 500 million years ago, they were once as tall as the Himalaya are today, and extended all the way into what is now Texas. By the Lower Jurassic, the Appalachians had eroded considerably—and geologists now think that it was their sediment, that formed the Navajo Sandstone. Rivers flowing westward carried the erosional sediment toward what is today the center of the continent (FIGURE D). As the climate became arid, winds transported the now dried river sands southward, where they became one of the largest sand seas that has ever existed on Earth.

1 Sediments and Sedimentation

1. Sediments and sedimentary rocks are our best record of how climates and environments have changed throughout geologic history. Geologists use three broad categories to distinguish sediment types—*clastic*, *chemical*, and *biogenic*.

2. **Clastic sediment** consists of fragmented rock and mineral debris produced by weathering, together with broken remains of organisms. The fragments, called *clasts*, are classified on the basis of size. Clasts may become *rounded* and *sorted* during transport by water and wind, but they typically remain unsorted during transport by glaciers or as a result of mass wasting. Volcaniclastic sediments are clastic sediments in which the fragments, also called *pyroclasts*, are volcanic in origin.

3. **Chemical sediment** is formed when substances carried in solution in lake water or seawater are precipitated, generally as a result of evaporation or other processes that concentrate the dissolved substances.

4. **Biogenic sediment** is composed of the accumulated remains of organisms. Plants, animals, and microscopic life forms may all contribute skeletal and/or organic material to sediments.

2 Sedimentary Rocks

1. When sediment is turned into sedimentary rock, the **bedding**, or layered arrangement of the strata, is generally preserved. The presence of bedding and **bedding surfaces**, the boundaries between adjacent strata, indicates that the rock was once sediment. In a *graded bed*, the coarsest clasts are at the bottom, and the finest are at the top. *Cross bedding* is produced by turbulent stream or wind flow, or by ocean waves.

2. **Lithification** is the group of processes that transforms loose sediment into sedimentary rock. Lithification takes place through a variety of changes, including **compaction**, the reduction of pore space as a result of increased pressure; **recrystallization**, the formation of new minerals from old ones; and **cementation**, the process by which substances dissolved in pore water precipitate out and cement grains of sediment together. The low-temperature, low-pressure changes that happen to sediment after **deposition** are called **diagenesis**.

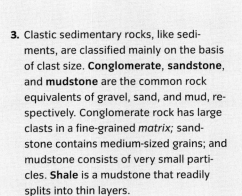

3. Clastic sedimentary rocks, like sediments, are classified mainly on the basis of clast size. **Conglomerate**, **sandstone**, and **mudstone** are the common rock equivalents of gravel, sand, and mud, respectively. Conglomerate rock has large clasts in a fine-grained *matrix;* sandstone contains medium-sized grains; and mudstone consists of very small particles. **Shale** is a mudstone that readily splits into thin layers.

4. Chemical sedimentary rocks formed by evaporation are **evaporites**. **Banded iron formations**, though uncommon, are significant both economically and for understanding the history of our planet. **Limestone**, **peat**, and **coal** are important kinds of biogenic sedimentary rocks. The lithification of peat into coal is known as *coalification*.

3 Depositional Environments

1. Sedimentary rocks often contain shape and color clues to the kind of environment in which they formed. Common sediments on land are stream, lake, glacial, and **eolian sediments** such as *loess*. We can interpret environmental clues, such as *ripple marks*, *mud cracks*, and fossil tracks to deduce past surface conditions.

2. Streams and lakes are common transporters of sediment on land and can form several types of depositional environments, including *alluvial fans* and **deltas**, triangle-shaped deposits that form where streams enter a standing body of water. Eolian sediment that is transported by wind tends to be finer than that moved by other erosional agents.

3. Other depositional environments are located in or near the ocean. Some sediment may be trapped in **estuaries**, semi-enclosed bodies of water along the coast, where fresh water and ocean

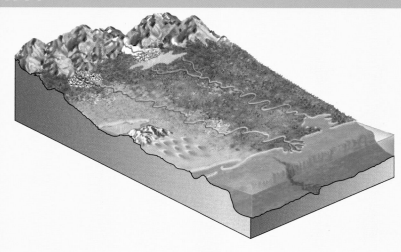

water mix. Beaches tend to comprise finer sediments, as the particles have been repeatedly abraded by surf. Finer sediment is also deposited on the continental shelves and slopes and in the deep sea. *Reefs* are deposits made of the skeletons of marine invertebrates. **Turbidity currents** are underwater landslides that transport sediment from the edge of the continental shelf to the deeper water at the foot of the continental slope. Another type of deposi-

tional environment in the ocean is the seafloor, which commonly hosts two types of deposits: *calcareous ooze* and *siliceous ooze*.

4. A **sedimentary facies** is a single geologic unit consisting of sedimentary rocks with the same composition, deposited at more or less the same time. Geologists learn about the history of a region by studying the way that different sedimentary facies adjoin and grade into each other in space and in time.

4 How Plate Tectonics Affects Sedimentation

1. Clastic sediment, which originates from the rock and mineral debris produced by weathering and erosion, is the most abundant type of sediment. Clastic sediment tends to collect in low-lying areas whose location is strongly affected by plate tectonics.

2. Divergent margins create **rift valleys** and passive continental margins. Colliding tectonic plates create structural basins and valleys associated with mountain ranges as a result of the folding and faulting of rock. Subducting plates form deep oceanic trenches along with volcanic arcs that may have *back-arc basins*. All of these locations are effective traps for sediment.

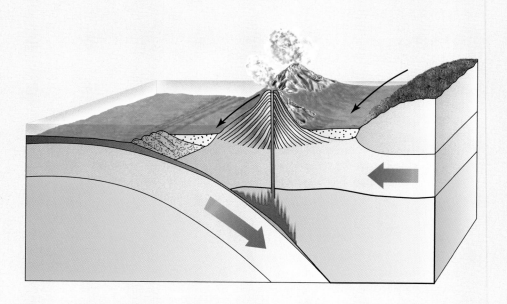

- **deposition** p. 214
- **clastic sediment** p. 214
- **chemical sediment** p. 216
- **biogenic sediment** p. 216
- **lithification** p. 217
- **bedding** p. 217
- **bedding surface** p. 217
- **compaction** p. 219
- **cementation** p. 219
- **recrystallization** p. 219
- **conglomerate** p. 222
- **sandstone** p. 222
- **mudstone** p. 222
- **shale** p. 222
- **evaporite** p. 222
- **banded iron formation** p. 222
- **limestone** p. 222
- **peat** p. 224
- **coal** p. 224
- **delta** p. 228
- **eolian sediment** p. 228
- **estuary** p. 229
- **turbidity current** p. 231
- **rift valley** p. 234

CRITICAL AND CREATIVE THINKING QUESTIONS

1. Estuaries are generally shallow, yet there are thick accumulations of estuarine sediments in the geologic record. What hypothesis can you suggest to explain this?

2. Do any sedimentary rocks outcrop in the area where you live? If so, see if you can recognize the kinds of rocks present and identify the environment in which the sediments were deposited.

3. It is estimated that as much as 25 billion tons of soil are lost through erosion as a result of farming every year. What happens to that soil?

4. Exploration for oil has led to the discovery of up to 14 kilometers of sedimentary rock on the continental shelf of North America. If the oldest of the strata were deposited 150 million years ago during the Jurassic Period, and the youngest are still being deposited today, what is the average rate of deposition?

5. Investigate the formation of graded bedding by filling a large beaker with a half-and-half mixture of water and sediment. The sediment should have a variety of grain sizes, including fine clays, sand, and gravel. Shake up the mixture and let it settle quietly until the water is completely clear. Which grains settle first and which settle last? What does the final sediment deposit look like?

What is happening in this picture ?

This slab of rock from the Cambrian Period was collected in Quebec, Canada. Can you tell the fossil tracks apart from the ripple marks? What kind of animal could have made tracks like these, and what kind of environment did it live in? (No one knows exactly what the animal, called *Climactichnites*, looked like.)

1. ——————————— sediment forms from loose rock and mineral debris produced by weathering and erosion.

    a. Clastic
    b. Biogenic
    c. Chemical

2. These photographs are close-up views of two sediments labeled A and B. Which sediment shows a greater degree of rounding?

    a. Sample A
    b. Sample B
    c. Neither sample is a well-rounded sediment.
    d. Both samples are well-rounded sediments.

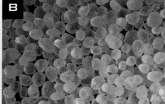

3. Which of the two sediments in the photographs shows a greater degree of sorting?

    a. Sample A
    b. Sample B
    c. Neither sample is a well-sorted sediment.
    d. Both samples are well-sorted sediments.

4. Examine the middle sedimentary bed depicted in this photograph. Which statement best describes the bed?

    a. cross-bedded
    b. glacial
    c. graded
    d. eolian

5. Three separate processes can lead to the lithification of sediments. During ——————————, the weight of accumulating sediment reduces pores space and forces out water from a sediment. —————————— occurs when ions dissolved in solution precipitate out, forming minerals that hold the grains together. Pressure can lead to ——————————, which causes less stable minerals to rearrange crystals into more stable forms.

    a. compaction; Cementation; recrystallization
    b. compaction; Recrystallization; cementation
    c. dehydration; Recrystallization; cementation
    d. dehydration; Cementation; recrystallization

6. Which of the three rock samples in these photographs is <u>not</u> a sedimentary rock?

    a. Sample A
    b. Sample B
    c. Sample C
    d. All of the rock samples displayed are sedimentary rocks.

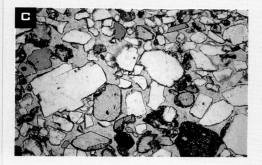

7. Which of the following sedimentary rocks is typically composed entirely of calcite?

   a. sandstone
   b. limestone
   c. shale
   d. conglomerate
   e. evaporite

8. Which of the following is a medium-grained sedimentary rock of clastic origin?

   a. sandstone
   b. limestone
   c. shale
   d. conglomerate
   e. evaporite

9. Which of the following rocks is most likely to have formed in a desert playa?

   a. sandstone
   b. limestone
   c. shale
   d. conglomerate
   e. evaporite

10. If mud cracks are found on a bedding surface of sedimentary rock, what can we deduce about the environmental conditions during deposition of the original sediment?

   a. Deposition occurred during a strong wind storm.
   b. The environment was subject to strong currents.
   c. The environment was moist and then dried out.
   d. Deposition occurred as a result of glaciation.

11. In which of the sedimentary environments listed below, would you expect to find siliceous and calcareous oozes?

   a. shallow marine environment
   b. stream environment
   c. deep marine environment
   d. estuarine reason

12. Given the vertical association of depositional facies presented in this block diagram, what can we deduce about sea level over time?

   a. Sea level was rising during deposition
   b. Sea level was falling during deposition
   c. Sea level was stable during deposition

13. In what type of tectonic setting would you expect to find voluminous amounts of arc-derived volcaniclastic sediments?

   a. Divergent margin–rift valley
   b. Convergent margin–subduction zone
   c. Convergent margin–continental collsion
   d. Passive continental margin

14. Accretionary wedges composed of sediment scraped off the oceanic crust are typical of what type of tectonic setting?

   a. divergent margin–rift valley
   b. convergent margin–subduction zone
   c. convergent margin–continental collsion
   d. passive continental margin

15. In what type of tectonic setting would you expect to find sediments in structural basins adjacent to mountain ranges?

   a. divergent margin–rift valley
   b. convergent margin–subduction zone
   c. convergent margin–continental collision
   d. passive continental margin

Folds, Faults, and Geologic Maps

9

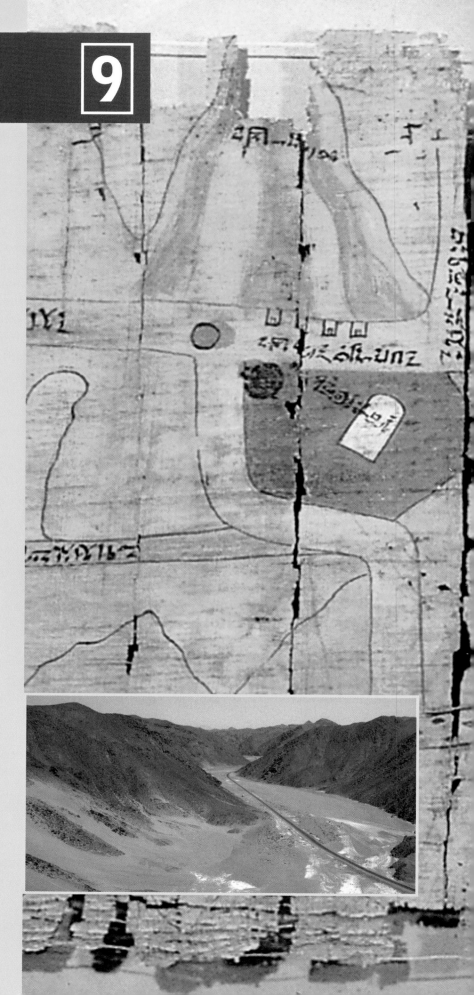

This 3100-year-old map may have been drawn by Amennakht, the chief scribe in the Village of the Craftsmen in ancient Egypt during the reign of Ramesses IV, around 1150 B.C. It is by far the earliest known example of a geologic map.

The papyrus, which now resides in the Museum of Egypt in Turin, Italy, shows topographic and geologic features along a 15-kilometer stretch of the dry stream valley in eastern Egypt known as Wadi Hammamat (*inset photo*). The mountains are represented as stylized conical forms on both sides of the streambed. The colors correspond in a general way to the actual appearance of the rocks in the mountains—dark gray metamorphic rocks to the southwest, and pink granites and rhyolites to the northeast. The map recorded the location of economically valuable deposits, particularly the quarries where the Egyptians mined bekhen stone (a desirable kind of sandstone) for their monuments and buildings. Many bekhen stones remain in the valley today and bear inscriptions that tell of the emperors' great quarrying expeditions.

In this chapter, you will learn how geologists draw similar maps today, and how these maps are used to indicate locations and types of rocks, folds, and faults in the underlying strata. Just as in Rameses' time, these features help geologists identify sites with valuable resources. They also help us understand the geologic history of an area, including the tectonic forces of stretching and compression to which it has been subjected.

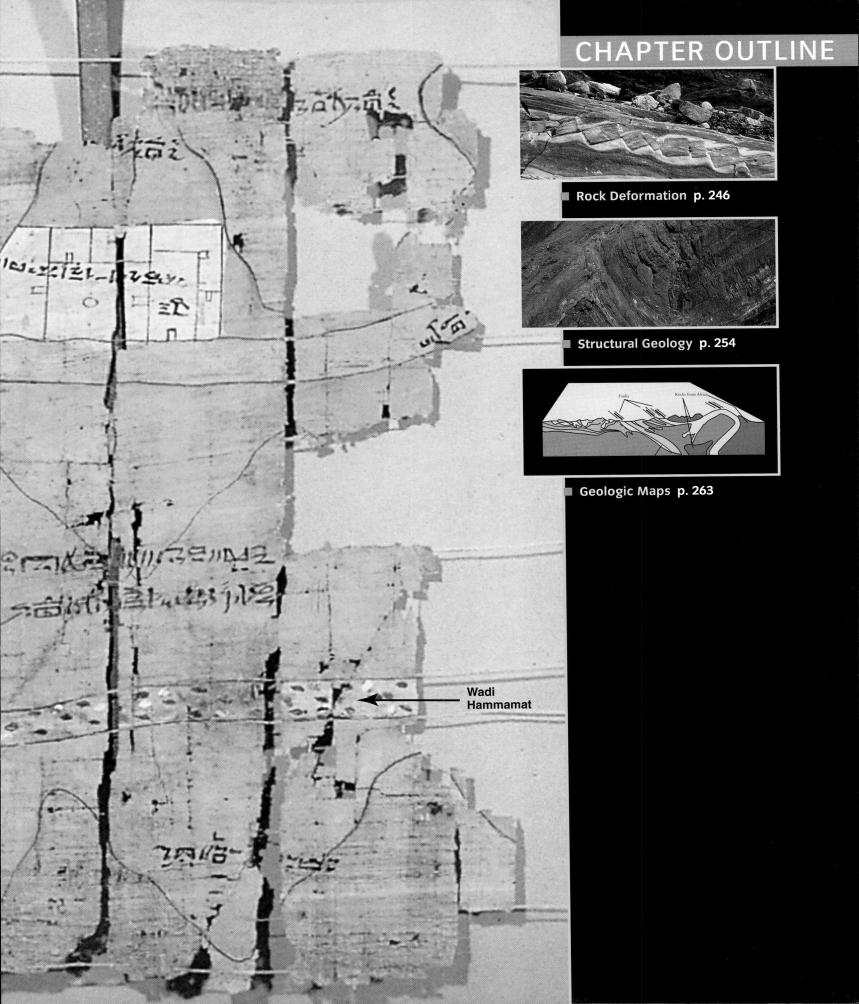

Wadi
Hammamat

Rock Deformation

A s you learned in Chapter 5, lithospheric plates are constantly moving around, colliding with one another, and interacting along their margins. Rocks are recycled into the mantle at subduction zones, and new crust is created at divergent margins (see Chapter 4). But what happens at convergent margins in collision zones? Here, enormous forces deform huge masses of continental crust and uplift them into great mountain ranges. We usually can't actually see the rocks being twisted and bent by tectonic forces, because this happens very slowly and deep underground. However, the aftermath is easily visible in locations where the deformed rock has been uplifted by tectonic forces and exposed by erosion. In **FIGURE 9.1**, for instance, the originally flat and horizontal beds of sedimentary rock have been folded and tilted into a spectacular zigzag. To understand how folding like this occurs, we rely both on laboratory measurements and on field studies of deformed rock.

STRESS AND STRAIN

In discussing rock deformation, we use the word **stress** rather than **pressure**. These two words are related in meaning but different in connotation. Both are defined as the force acting on a surface per unit area. The term *pressure*, as used in geology, implies that the forces on a body of rock are essentially uniform in all directions.

stress The force acting on a surface, per unit area, which may be greater in certain directions than in others.

pressure A particular kind of stress in which the forces acting on a body are the same in all directions.

The core of a former mountain FIGURE 9.1

These tightly folded rocks on South Georgia Island (near Antarctica, off the southern tip of South America) bear witness to the intense geologic stresses that bent them into a zigzag shape. Such intense deformation is typical for rocks that once formed the core of a great mountain chain.

Three types of stress FIGURE 9.2

The shape of a cube of rock changes, depending on the type of stress applied to it. The arrows indicate tensional, compressional, and shear stress. Rocks that are subjected to differential stress—stress that is stronger in one direction than another—typically respond by changing their shape, as shown by these blocks.

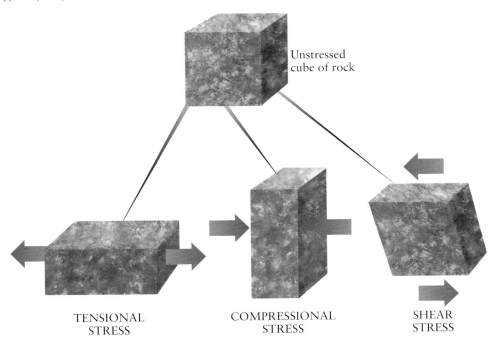

Unstressed cube of rock

TENSIONAL STRESS   COMPRESSIONAL STRESS   SHEAR STRESS

tension A stress that acts in a direction perpendicular to and *away from* a surface.

compression A stress that acts in a direction perpendicular to and *toward* a surface.

Sometimes this is also called *uniform stress* or *confining stress.* These are appropriate terms to describe, for instance, the stress on a small body immersed in a liquid such as water or magma.

Rocks, however, are solids; unlike liquids and gases, solids can resist different pressures in different directions at the same time. For this reason, stress is a more versatile term for discussing rock deformation, because it does not imply that the forces are necessarily the same in all directions. To be even more precise, we sometimes use the term *differential stress* when the force is greater from one direction than from another. The stresses that cause rock to change shape are differential.

They can be classified into three different kinds, as illustrated in FIGURE 9.2: **tension, compression,** and **shear**.

In response to stress, a rock will experience **strain**. Uniform stress causes a change in volume only, while differential stress may cause a change in shape. For example, if a rock is subjected to uniform stress by being buried deep in Earth, its volume will decrease; that is, the rock will be compressed. If the spaces (or *pores*) between the grains become smaller as water is expelled from them, or if the minerals in the rock are transformed into more compact structures, the volume change may be relatively large.

shear A stress that acts in a direction *parallel* to a surface.

strain A change in shape or volume of a rock in response to stress.

BRITTLE AND DUCTILE DEFORMATION

elastic deformation
A temporary change in shape or volume from which the material rebounds after the deforming stress is removed.

The way a rock responds to differential stress depends not only on the amount and kind of stress but also on the nature of the rock itself. For example, a rock may stretch like a metal spring and then return to its original shape when the stress is removed. Such a nonpermanent change is called **elastic deformation**. For most solids, including rocks, there is a degree of stress—called the *elastic limit*—beyond which the material is permanently deformed. If the rock is subjected to more stress than this, it will not return to its original size and shape when the stress is removed (see **FIGURE 9.3**).

When rocks are stretched past their elastic limit, they can deform in two different ways. **Ductile deformation**, also called *plastic deformation*, is one type of permanent deformation in a rock (or other solid) that has been

ductile deformation
A permanent but gradual change in shape or volume of a material, caused by flowing or bending.

Visualizing
Deformation of rock

Elastic limit FIGURE 9.3

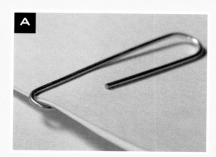

A A paper clip that is holding only three sheets of paper returns to its original shape after the papers are removed.

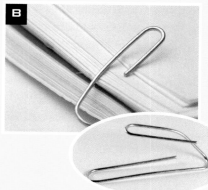

B The same paper clip becomes permanently bent when used to hold together a fifty-page document (inset). Its elastic limit is somewhere between three and fifty pages.

Glass deformation FIGURE 9.4

A At room temperature, glass deforms in a brittle manner, as shown by this light bulb shattering when it hits the floor.

B However, if the glass is heated slowly over a flame, it can bend and flow in a ductile manner, as in this glassmaker's studio.

stressed beyond its elastic limit. Alternatively, the rock may undergo **brittle deformation**.

A brittle material deforms by fracturing, whereas a ductile material deforms by changing its shape. Drop a piece of chalk on the floor and it will break. Drop a piece of play dough, and it will bend or flatten instead of breaking. Under the conditions of room temperature and atmospheric pressure, chalk is brittle, and play dough is ductile. Similarly, some rocks behave in a brittle manner, and others behave in a ductile manner.

However, a rock that is brittle in one set of conditions may be ductile in different conditions (see **FIGURE 9.4**).

The difference between these two kinds of deformation is illustrated in **FIGURE 9.5**. The main factors that affect how a rock deforms are temperature, confining pressure, rate of deformation, and composition of the rock. Let's look briefly at each of these.

Temperature   As Figure 9.4 shows, glass is brittle at room temperature but becomes ductile at a higher temperature. Rocks are like glass in this respect; they are brittle at the surface, but they become ductile deeper inside the planet, where temperatures are higher.

Ductile and brittle deformation
FIGURE 9.5

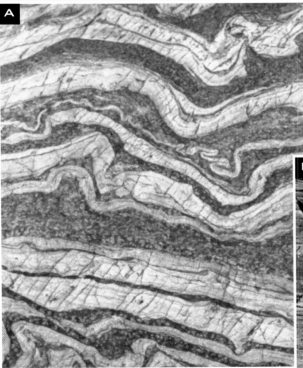

◀ A These rocks responded to stress by folding and flowing, a ductile deformation.

B This rock, fractured under stress, a ▼ brittle deformation.

Confining pressure
The effect of (uniform) confining pressure on deformation is not familiar from our everyday experience. High confining pressure reduces the brittleness of rocks because it hinders the formation of fractures. FIGURE 9.6 shows the results of a series of experiments that demonstrates this effect. Rocks near Earth's surface, where confining pressure is low, exhibit brittle behavior and develop many fractures. At great depth, however, where confining pressure is high, rocks tend to be ductile and deform by flowing or bending.

Rate of deformation
The rate at which stress is applied to a solid is another important factor in determining how a material will deform. If you take a hammer and suddenly whack a piece of ice, it will fracture. But if stress is applied to the ice little by little over a long period, it will sag, bend, and behave in a ductile manner. The same is true of rocks. If stress is applied quickly, the rock may behave in a brittle manner, but if small stresses are applied over a very long period, the same rock may behave in a ductile manner. The term *strain rate* refers to the rate at which a rock is forced to change its shape or volume. The lower the strain rate, the greater the tendency for ductile deformation to occur (see FIGURE 9.7).

To summarize, low temperature, low confining pressure, and high strain rates tend to enhance the brittle behavior of rocks. Low-temperature and low-pressure conditions are characteristic of Earth's crust, especially the upper crust. As a result, fracturing is common in upper-crustal rocks. High temperature, high confining pressure, and low strain rates, which are characteristic of the deeper crust and the mantle, reduce the brittle properties of rocks and enhance their ductile properties. The depth below which ductile properties predominate is referred to as the *brittle-ductile transition*. Fractures are uncommon deep in the crust and in the mantle, because rocks at great depths (below about 10–15 km) tend to behave in a ductile manner.

Composition
The composition of a material determines the exact point at which its brittle-ductile transition will occur. For example, both chalk and play dough are brittle at 50 degrees below zero. When warmed to room temperature, however, the play dough behaves in a ductile manner, while the chalk is still

Rock deformation under pressure FIGURE 9.6

These cylinders show the results of a series of experiments on the effects of confining pressure on rocks.

A This is an undeformed cylinder of rock.

B This cylinder was subjected to high confining pressure (uniform in all directions) and, at the same time, compression from above. It deformed in a ductile manner, becoming shorter and fatter.

C An identical cylinder was subjected to the same amount of compression from above, but this time with a lower confining pressure. It deformed in a brittle manner, with many large fractures.

A At very slow strain rates, Silly Putty flows like a liquid. Starting from an apparently solid ball shape, it changes to the shape of a liquid drop over a few hours.

B At medium strain rates, such as when stretched by human hands, it deforms in a ductile manner.

C At high strain rates, it deforms elastically. (Bouncing is an elastic effect, caused by the material's tendency to return to its original shape.)

D At extremely high strain rates, such as when shot by a bullet, Silly Putty shatters like glass.

Silly Putty FIGURE 9.7

The effects of different strain rates can be observed with Silly Putty.

brittle. This is because they have different chemical compositions, and therefore different properties. The same is true of rocks and their mineral constituents. Some minerals—notably quartz, garnet, and olivine—are strong and brittle, as are the rocks that contain these minerals (such as sandstone and granite). For the most part, quartz-bearing rocks control the depth of the brittle-ductile transition. Other minerals—notably mica, calcite, and gypsum—are more often ductile under natural conditions. Thus, the rocks that contain them (such as limestone, marble, shale, and slate) also tend to deform in a ductile manner.

Water is another component that enhances the ductile properties of rock. It reduces the friction between mineral grains and dissolves material at points of high stress, which permits the material to move to places where the stress is lower. Trace amounts of water can also enter strong minerals such as quartz and olivine and significantly weaken them by a process called *hydrolytic weakening*.

WHERE ROCK DEFORMATION OCCURS

In which tectonic environments may the different types of stress—tension, compression, and shear stress—be expected to occur? Tensional stress involves a pulling-apart motion; it is characteristic of environments in which divergent motion is occurring, such as a continental rift or mid-ocean ridge. Shear stress occurs where lithospheric plates are sliding past one another, such as along the boundary between the Pacific Ocean plate and the North American plate. Compression is characteristic of environments in which convergent motion is occurring, such as subduction zones and continent-continent collisions.

craton A region of continental crust that has remained tectonically stable for a very long time.

The deformation of rocks in continent-continent collisions provides some of the best evidence that Earth has been tectonically active for billions of years. As discussed in Chapter 4, the conclusive evidence for plate tectonics—seafloor spreading—occurs in oceanic crust. However, all of today's oceanic crust is relatively young—less than 200 million years old. Therefore, evidence for earlier plate tectonics comes from continents and how they are put together (see FIGURE 9.8), including the placement of **cratons** and **orogens**.

orogen An elongated region of crust that has been deformed and metamorphosed by a continental collision.

Some orogens, such as the Alps and the Himalaya, are still active today. Others, such as the Appalachian Mountains, are inactive and now highly eroded. Through radiometric dating, some orogens have been found to be as old as 4.0 billion years.

Cratons and orogens FIGURE 9.8

Even though all of North America lies on one lithospheric plate today, a closer look at the rocks shows that it has been assembled like a jigsaw puzzle from older parts. The core of a continent is formed of cratons, which represent the very oldest tectonic units. Between the cratons lie elongated regions called orogens. These contain highly deformed rock strata and lots of metamorphic rock (see Chapter 10)—both signs that these were once sites of intense mountain-building activity.

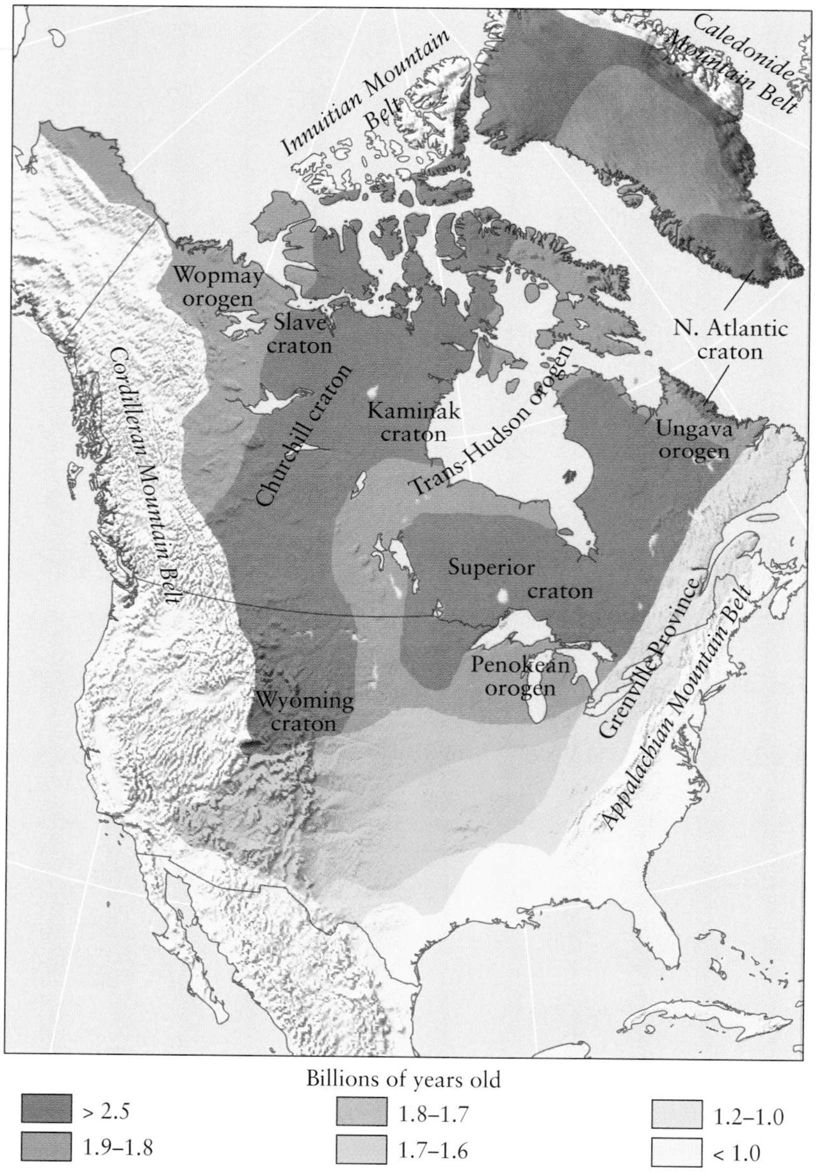

Adjusting the height of continents FIGURE 9.9

The rigid lithosphere "floats" on the ductile asthenosphere.

A In a collision of two cratons, the lithosphere thickens—both upward and downward—and the lithosphere-asthenosphere boundary is pushed down, creating a mountain belt (an orogen) with a deep root.

Compressional tectonic forces create mountain range.

Lithosphere

Asthenosphere

Craton   Orogen   Craton

A

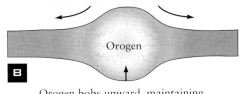

Erosion wears down the mountains.

Orogen

B

Orogen bobs upward, maintaining isostasy and preserving mountains even after the tectonic forces that created them are gone.

B As erosion slowly reduces the height of the mountains in the orogen, the root slowly rises to compensate for the removal of mass. This continuing adjustment to maintain buoyancy, like a cork finding its level in water, is called *isostasy*.

isostasy The flotational balance of the lithosphere on the asthenosphere.

Orogens like the Appalachians can still display some topographic relief even after hundreds of millions of years of erosion, because of a phenomenon called **isostasy** (see FIGURE 9.9). Isostasy is the process in which a floating object automatically adjusts to a position of equilibrium with the medium in which it is floating. From everyday experience you know that a block of wood floating in water will sink if a weight is put on top of it and will bob up if the weight is removed. The same thing happens with icebergs floating in the ocean. Icebergs are partially visible above the ocean surface, but every iceberg has a deep root that keeps it floating in balance. As the exposed top of an iceberg melts, the entire block of ice will bob upward to maintain this flotational balance.

Isostasy plays an important role in the shaping of Earth's topography, particularly in orogens. Continental collisions cause the entire continental crust to thicken; as a result, beneath every great mountain range is a root of thickened continental crust that dips down into the asthenosphere, similar to the root of an iceberg. This root keeps the entire orogen in flotational balance with respect to the underlying hot, weak asthenosphere. As erosion weathers away the top of the mountain range, the root bobs upward, like the root of a melting iceberg. But unlike an iceberg, which floats in seawater, the orogen is solid rock that floats in the hot, ductile asthenosphere, which makes the process extremely slow. The Appalachian Mountains are still adjusting isostatically after more than 300 million years of weathering.

CONCEPT CHECK STOP

Why is a rock buried deep underground more likely to undergo ductile deformation than brittle deformation?

How would slate and granite respond differently to stress?

Which kind of stress tends to stretch rocks? Which kind of stress shortens and thickens rocks?

Why do orogens preserve a record of ancient deformations long after you might expect them to have eroded away?

Structural Geology

he concepts of stress and strain play a central role in **structural geology**. Structural geologists attempt to decipher the geologic history of a region by identifying and mapping deformational features in rocks. These features are also important from a practical perspective. For example, faults control the locations of certain types of ore deposits. Other rock structures can affect slope stability, influence the flow of groundwater, or trap oil and natural gas deep underground (see **FIGURE 9.10**).

Many of these practical applications, such as oil prospecting or mapping the flow of groundwater, require geologists to make inferences about structures that lie underground, hidden from view. Most of their information comes from observations made at the surface, with additional information from drilling and other methods of subsurface study. A systematic program of measurements and mapping can give structural geologists a very good idea of what might lie beneath the surface.

structural geology The study of stress and strain, the processes that cause them, and the deformation and rock structures that result from them.

www.wiley.com/college/murck

STRIKE AND DIP

The *principle of original horizontality* (see Chapter 3) tells us that sedimentary strata are horizontal when they are first deposited. Where such rocks are tilted, we generally can assume that deformation has occurred. To describe this deformation, the geologist starts by

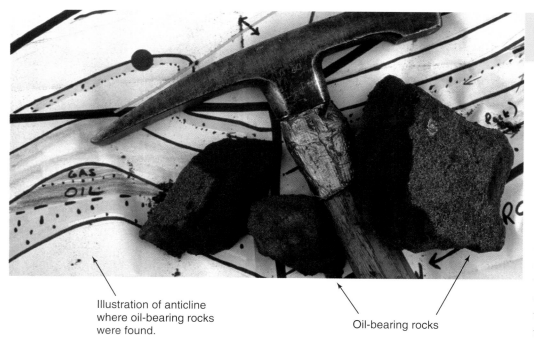

Illustration of anticline where oil-bearing rocks were found.

Oil-bearing rocks

Structural geology and oil FIGURE 9.10

The rocks in this photograph, which contain oil, came from an exploratory oil well site on the North Slope of Alaska. The diagram underneath the rock pick shows a typical rock formation that harbors oil. A layer of porous rock (shown in yellow) forms an arch-shaped fold with a layer of impervious rock such as shale overlying it. Oil and natural gas, because they are less dense than the water in the pore spaces of the rock, gradually rise through the porous rock until they are trapped at the top of the arch.

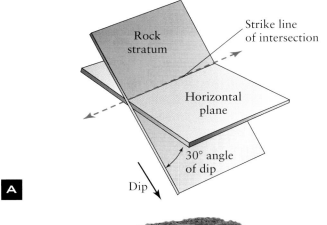

Rock stratum

Strike line of intersection

Horizontal plane

30° angle of dip

Dip

A

Geologists use *strike* and *dip* to describe the orientation of a tilted layer of rock.

A The strike is formed by the intersection of a rock layer and a (sometimes imaginary) level plane. The dip is the angle of tilt of the rock stratum, measured from the horizontal plane.

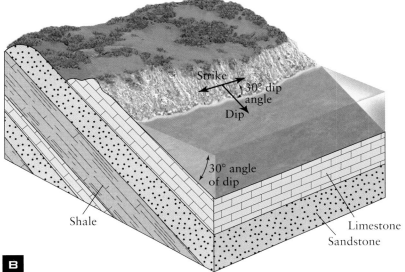

Strike

30° dip angle

Dip

30° angle of dip

Shale

Limestone

Sandstone

B

B In this drawing, the top of the water provides a suitable horizontal plane. The shoreline (the intersection of the water with the rock surface) indicates the strike. The dip is the angle at which the rock layers are tilted. In this drawing, they are tilted 30° from the horizontal. The patterns shown in the diagram are commonly used to distinguish different rock types: The brick-like pattern represents limestone, the tiny dots represent sandstone, and the dashes represent shale.

■ **strike** The compass orientation of the line of intersection between a horizontal plane and a planar feature, such as a rock layer or fault.

■ **dip** The angle between a tilted surface and a horizontal plane.

measuring the orientation of the tilted rock layer. This information is given by the **strike**. It is a basic fact of geometry that a sloping plane—in this case, the tilted rock layer—intersects a horizontal plane to form a line (see **FIGURE 9.11A**). The strike of the rock layer is the compass direction of that line (measured in degrees north, south, east, or west).

We need another measurement to fully describe the orientation of a tilted rock layer. **Dip** is measured as an angle downward from the horizontal plane in degrees, using an instrument similar to a protractor. Dip is always measured in a direction perpendicular to strike. In Figure 9.11, the rock strata are tilted at 30°. If they were dipping more shallowly, the angle would be smaller. If they were dipping more steeply, the angle would be greater. The maximum possible dip is 90°, which represents a layer that has been tipped so that it is completely vertical. When rock layers are tilted even further than the vertical orientation, we say they are *overturned*.

It is important to remember that strike and dip describe the orientation and slope of the rock layer, specifically its *bedding surface*, not the ground itself. Strike and dip are also used to describe the orientation of other types of planar structural features, including faults.

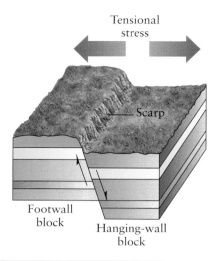

Tensional stress

Scarp

Footwall block

Hanging-wall block

Normal fault FIGURE 9.12

When the crust is stretched by tension, normal faults occur. The block on the overhanging part of the fault (called the *hanging-wall block*) moves down relative to the block underneath the fault (called the *footwall block*). The motion may expose a cliff-like landform at the surface called a *scarp*.

Horst and graben FIGURE 9.13

A Normal faults often occur in pairs. In a graben, the two faults dip towards each other and the block between them drops down. In a horst, the faults dip away from each other and the block between them rises. Although the relative motion of the blocks is nearly vertical, keep in mind that it is the *horizontal* stretching of the crust, due to tensional stress, that causes it.

FAULTS AND FRACTURES

Fractures, or cracks, are characteristic of brittle rock deformation. Fractures occur in all sizes. Some are so tiny that you would need a microscope to see where an individual mineral grain has cracked. Cracks this small are sometimes called *microfractures*. A *fault*, as defined in Chapter 4, is a fracture in rock along which movement has occurred along the fracture surface. Some faults are small, only meters long, but others are very large. For example, the East African Rift Valley is a huge system of roughly parallel faults that extends for more than 6000 kilometers through the countries of East Africa.

There are many types of faults, caused by different kinds of stress. Tensional (or *extensional*) stress, stress that stretches or pulls apart the crust, causes **normal faults** (see FIGURE 9.12).

> ■ **normal fault**
> A fault in which the block of rock above the fault surface moves downward relative to the block below.

B The Great Rift Valley in Kenya is under tensional stress because it is a divergent plate margin that is stretching the crust beneath the African and Somali plates. A series of horsts and grabens can be seen in this photograph.

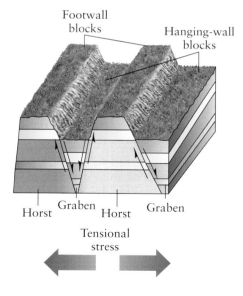

Footwall blocks

Hanging-wall blocks

Horst Graben Horst Graben

Tensional stress

A

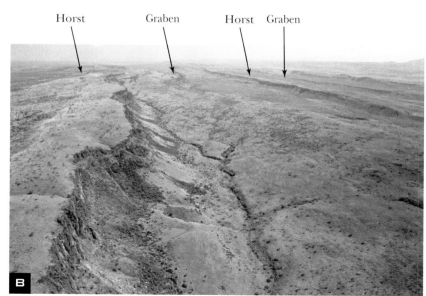

Horst   Graben   Horst Graben

B

REVERSE FAULT

Compressional stress

Lake

Footwall block

Hanging-wall block

A

THRUST FAULT

Compressional stress

Symbols used on a map to indicate a thrust fault

Footwall block

Hanging-wall block

B

Reverse and thrust faults FIGURE 9.14

A In a reverse fault, compressional stress pushes the hanging-wall block up and over the footwall block. The direction of movement along the fault is opposite to that on a normal fault.

B A reverse fault with a very shallow dip is called a thrust fault. Geologists indicate this with a row of triangles pointing toward the hanging-wall block.

Normal faults sometimes occur in pairs, as shown in FIGURE 9.13, creating a distinctive pairing of up-lifted and down-dropped blocks called *horsts* and *grabens*. For example, the region lying between the Sierra Nevada and the Rocky Mountains, known as the Basin and Range Province, is made up of alternating horsts and grabens.

Compressional stress is responsible for **reverse faults** and **thrust faults** (see FIGURE 9.14). In reverse faults, the hanging-wall block is pushed over the footwall block, shortening and thickening the crust. The reverse fault in Figure 9.14A dips steeply. When a reverse fault dips more shallowly, less than 45°; it is called a thrust fault (see Figure 9.14B). Thrust faults are common in mountain chains along convergent plate boundaries.

In large thrust faults, the hanging-wall block may move thousands of meters, coming to rest on top of much younger rock in the footwall block. This seems to contradict the principle of superposition. However, that principle says that in any *undisturbed* sequence of strata, the younger strata are deposited on top. A thrust fault represents a disturbance of the original stratigraphic sequence.

reverse fault
A fault in which the block on top of the fault surface moves up and over the block on the bottom.

thrust fault A reverse fault with a shallow angle of dip.

Strike-slip faults Shear stress typically creates **strike-slip faults**, along which two adjacent blocks are displaced horizontally relative to one another. The name comes from the fact that the blocks slip in a direction parallel to the strike of the fault (not the strike of the rock layers), as shown in FIGURE 9.15.

strike-slip fault
A fault in which the direction of the movement is mostly horizontal and parallel to the strike of the fault.

Strike-slip fault FIGURE 9.15

In a strike-slip fault, the movement is mostly horizontal and parallel to the strike of the fault.

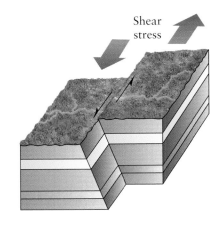

Shear stress

The San Andreas Fault system FIGURE 9.16

The San Andreas Fault is actually part of a complicated system of faults along which the Pacific Plate is moving in a northwesterly direction relative to the North American Plate. This map shows the major faults in the vicinity of San Francisco along which motion has occurred in the last 10,000 years.

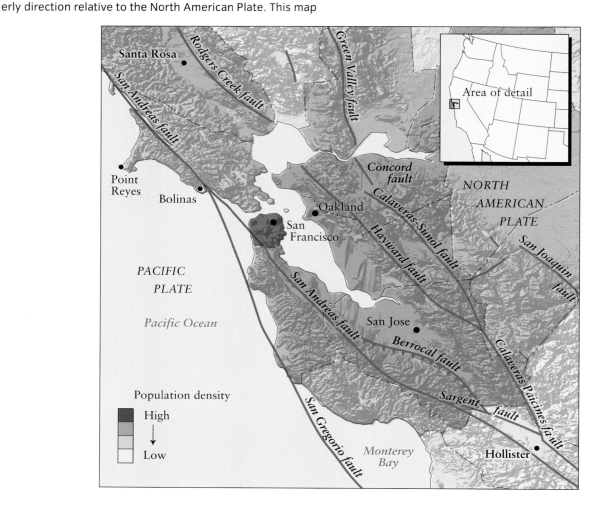

One strike-slip fault is so famous that almost everyone has heard of it: the San Andreas Fault in California. Along this fault, the Pacific Plate is moving toward the northwest relative to the North American Plate (see **FIGURE 9.16**). The word *relative* is very important here. In fact, both the Pacific Plate and the North American Plate are moving in a roughly northwesterly direction, but the Pacific Plate is moving more quickly, about 10 cm/yr, like a fast runner overtaking a slower one. Over the last 15 million years, the Pacific Plate may have moved more than 600 km northwest relative to the North American Plate, which is moving at about 5 cm/yr.

Strike-slip faults can be described according to the direction of relative horizontal motion as follows: To an observer standing on either block, the movement of the other block is *left-lateral* (also called *sinistral*) if it has moved to the left, and *right-lateral* (or *dextral*) if it has moved to the right. The San Andreas Fault is right-lateral, because to an observer standing on the Pacific Plate, the North American Plate appears to be moving to the right (that is, toward the southeast). Note that it does not matter which plate you stand on; if you stand on the North American Plate, the Pacific Plate will still appear to be moving toward your right (that is, toward the northwest), so the San Andreas is still a right-lateral fault.

FOLDS

fold A bend or warp in a layered rock.

anticline A fold in the form of an arch, with the rock strata convex upward and the older rocks in the core.

When rocks deform in a ductile manner, they bend and flow, creating **folds**. A fold may be a broad, gentle warping over many hundreds of kilometers, a tight flexing of microscopic size, or anything in between.

The simplest type of fold is a *monocline*, a local steepening in otherwise uniformly dipping strata (see What a Geologist Sees). An easy way to visualize a monocline is to lay a book on a table and drape a handkerchief over one side of the book. So draped, the handkerchief forms a monocline. However, most folds are more complex than monoclines. In fact, a monocline that is visible at the surface may be just one part of a larger and more complicated fold that is partially hidden underground. Folds are often combinations or variations of two basic types: **anticlines** and **synclines**. If you push the edge of a carpet with your foot, it will form a series of anticlinal and synclinal folds.

syncline A fold in the form of a trough, with the rock strata concave upward and the younger rocks in the core.

A

Monocline

FIGURE A shows a monocline in southern Utah that interrupts the generally flat-lying sedimentary strata of the Colorado Plateau. Note how the strata at the top of the photo become steeper as they descend underground.

In **FIGURE B**, a schematic drawing of the area in the photo, you can see that the strata become nearly vertical on the right-hand side of the monocline. The land surface cuts across the rock strata on the right, but it lies parallel to the rock strata on the left.

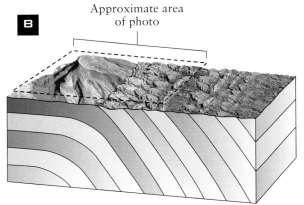

B

Approximate area of photo

What a Geologist Sees

Several measurements are needed to describe the geometry and orientation of a fold. Imagine a plane dividing the fold in half, as symmetrically as possible (see FIGURE 9.17). This is the *axial plane* of the fold. The two halves of the fold, on either side of the axial plane, are the *flanks* or *limbs*. Axial planes can be vertical, as in Figure 9.17B, or they can be tilted. The axial planes of simple folds are planar, but in a complicated fold the axial plane can be a curved surface.

Notice that the axial planes in Figure 9.17B connect the most strongly curved parts of each rock layer—in other words, the axial plane passes through the crests of the rock layers in the anticline and the troughs of the rock layers in the syncline. The line where the axial plane intersects the fold is called the *axis* or *hinge*. The axis of a simple fold may be horizontal (as in the anticline on the left in Figure 9.17B), but some fold axes are tilted. A fold in which the axis is not horizontal is said to be *plunging* (like the anticline on the right in Figure 9.17B).

Maps are two-dimensional representations of three-dimensional geologic structures, so we must use lines to represent axial planes. The projection of a fold axis or axial plane onto a horizontal plane is called the *axial trace*. To fully represent the geometry of a fold, the map needs to show the following elements: (1) a line representing the axis and the axial plane (that is, the axial trace); (2) a symbol to indicate the type of fold (anticlines are illustrated by two arrows pointing away from the axial trace, synclines are illustrated by two arrows pointing toward the axial trace); (3) an arrow to show the direction in which the axis is plunging; and (4) a number (measured in degrees from horizontal) to show the plunge angle. FIGURE 9.18 gives an example of how these symbols are used on a map.

Simple folds FIGURE 9.17

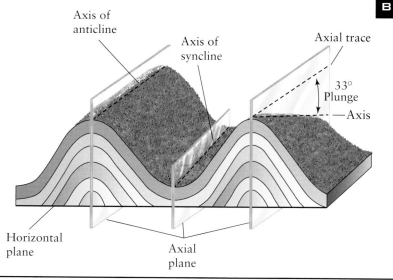

A On the left in this photo of the Old Red Sandstone in Wales is an anticline, an arch-shaped fold in which the layers of rock are concave down, with the oldest rocks underneath. On the right is a syncline, a trough-like fold with the youngest rocks on top.

B The axial plane separates the two limbs of a fold. The fold axis corresponds to the line of maximum curvature in the folded rock layers—usually along the crest of the arch in an anticline, or the deepest part of the trough in a syncline. The axial trace is the horizontal projection of the axis.

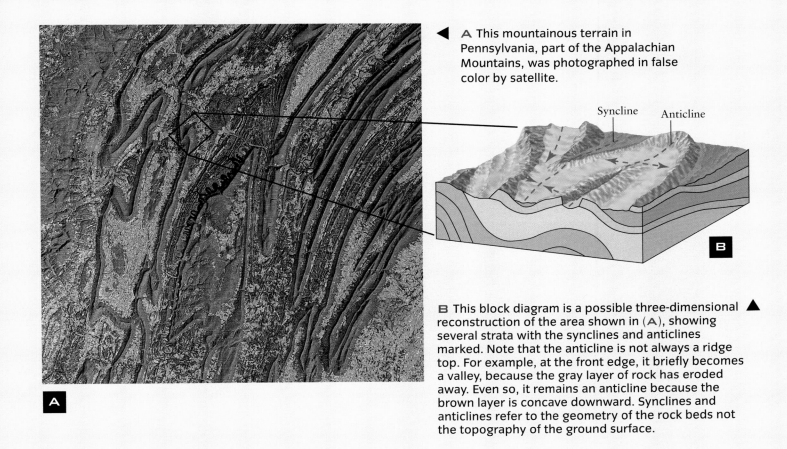

◀ **A** This mountainous terrain in Pennsylvania, part of the Appalachian Mountains, was photographed in false color by satellite.

Syncline   Anticline

B

B This block diagram is a possible three-dimensional ▲ reconstruction of the area shown in (**A**), showing several strata with the synclines and anticlines marked. Note that the anticline is not always a ridge top. For example, at the front edge, it briefly becomes a valley, because the gray layer of rock has eroded away. Even so, it remains an anticline because the brown layer is concave downward. Synclines and anticlines refer to the geometry of the rock beds not the topography of the ground surface.

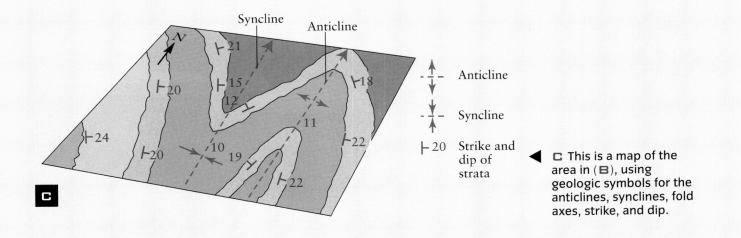

Syncline   Anticline

C

Anticline

Syncline

├20 Strike and dip of strata

◀ **C** This is a map of the area in (**B**), using geologic symbols for the anticlines, synclines, fold axes, strike, and dip.

Structural Geology   261

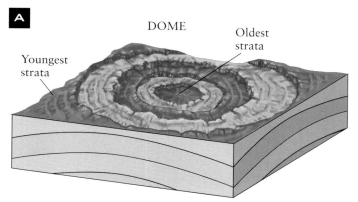

A Upwarping of the crust causes the formation of domes. When domes erode, they expose older rocks at their centers.

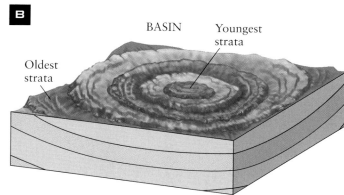

B Downwarping of the crust causes the formation of basins. In a basin, the youngest rocks will be at the center and the oldest rocks around the outer edge.

C This is a dome in northern Flinders Range, South Australia. The strata are late Proterozoic in age.

Domes and basins FIGURE 9.19

Note that synclines do not always form valleys, and anticlines do not always form ridges. Sometimes an anticline may expose a rock stratum that is more susceptible to erosion than the surrounding layers; in that case, the anticline will erode away quickly and form a valley. In general, the contours of the land do not always follow the bedding surfaces. It is the shape of the bedding surfaces that matters most for geology, because they record the long-term deformational history of the rock.

Folds can also take more complicated forms than the ones we have discussed. Sometimes a large area of crust undergoes upwarping or downwarping, which forms broad, gentle folds. Upwarping of strata forms *domes*, while downwarping forms large, bowl-like *basins* (see FIGURE 9.19). Folds can also be *asymmetrical*, with one limb dipping more steeply than the

other, or even *overturned*, with one limb tilted so far over that it is upside down. Folds that are so strongly overturned that they are almost lying flat are called *recumbent*. This is another way in which rock deformation can change the normal sequence of strata.

Geologic Maps

Geologists cannot see all the structural details of deformed rocks in a given area; soil, water, vegetation, and buildings cover much of the evidence. They must gather geologic information from *outcrops*—places where bedrock is exposed at the surface (see **FIGURE 9.20**). At outcrops, geologists will note the type of rock present, the orientation of the layers, and the presence of structural features such as fractures or folds. They will extrapolate from these data—perhaps using additional sources of information, such as drilling or seismic studies—to determine what lies beneath the soil, vegetation, water, and buildings in the areas between the outcrops. The final result is a **geologic map**.

The oldest known geologic map—possibly the first one ever made—was drawn on a papyrus scroll in ancient Egypt (see the chapter introduction). Evidently, it was used to show the locations of certain rock types for quarrying purposes. Today, geologic maps still help geologists interpret the geologic history of an area. Geologists from mining companies, oil companies, engineering firms, environmental agencies, consulting firms, and government agencies refer to such maps regularly.

> **geologic map**
> A map that shows the locations, kinds, and orientations of rock units, and structural features such as faults and folds.

Visualizing

Making a geologic map FIGURE 9.20

A This block diagram shows a landscape with tilted rock strata. Much of the rock is covered at the surface by grass and soil, but there are some outcrops where bedrock is exposed.

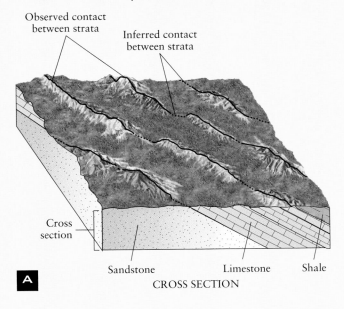

Observed contact between strata

Inferred contact between strata

Cross section

Sandstone   Limestone   Shale
CROSS SECTION

A

B GEOLOGIC MAP

N

50°
51°
48°
46°

X   Y

B The geologist has transferred the information about rock types and boundaries (*geologic contacts*) between rocks onto a map of the area. In areas where the bedrock is covered, the geologist makes an educated guess about the types of rock and the location of boundaries. This educated guess can be used to construct a vertical profile of the strata, shown as the front panel in (A); it is called a *geologic cross section*.

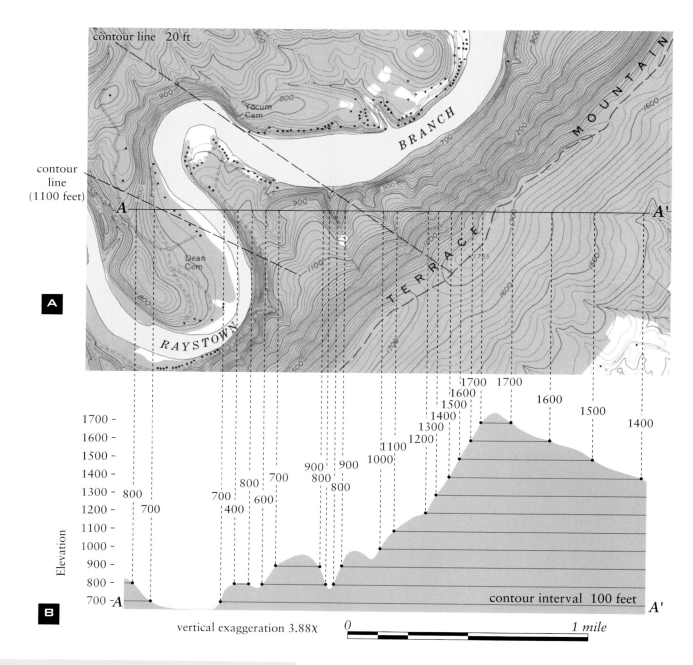

contour line 20 ft

contour line (1100 feet)

A

BRANCH

MOUNTAIN

TERRACE

RAYSTOWN

Yocum Cem

Dean Cem

755

B

Elevation

1700 –
1600 –
1500 –
1400 –
1300 –
1200 –
1100 –
1000 –
900 –
800 –
700 –

800
700
700
400
800
600
700
900
800
800
900
1000
1100
1200
1300
1400
1500
1600
1700
1700
1600
1500
1400

A — A'

contour interval 100 feet

vertical exaggeration 3.88X

0   1 mile

Creating a topographic profile FIGURE 9.21

A Topographic maps contain information about the ups and downs of the land. Contour lines indicate points of equal height; for example, every point on the line labeled "1100" is 1100 feet (335 m) above sea level. B You can use the height information from a topographic map to plot a *topographic profile* of the landscape. First, pick a line along which you would like to construct the topographic profile (such as the line labeled A – A' on the map). Next, set up your topographic profile: use the line A – A' as the horizontal axis, and draw and label a vertical axis to represent height or elevation. Directly below each point where the A – A' crosses a contour line on the map, place a dot at the appropriate height on the profile. Finally, connect the dots with a smooth curve. This curve represents the topography of the mountain. (We have exaggerated the height in order to make the differences in elevation stand out.)

MAKING AND INTERPRETING GEOLOGIC MAPS

topographic map A map that shows the shape of the ground surface, as well as the location and elevation of surface features, usually by means of contour lines.

Commonly, a **topographic map** is used as the base for a geologic map (see FIGURE 9.21). You may have used a topographic map if you have ever gone backpacking or hiking. An important aspect of topography is *relief*, the difference between the lowest and highest elevations in the area. A mountainous region has *high relief*, whereas a flat plain has *low relief*. Topographic maps use *contour lines*, lines of equal elevation, to portray topography.

However, topography is not geology. A geologist also needs to know about the rocks that form the topography, as well as the rocks beneath the surface. A *formation* is a unit of rock that can be distinguished from units above and below it on the basis of rock type and recognizable boundaries, or *geologic contacts*, with other rock units.

The colors of the rock units do not reveal whether the underlying rock strata are vertical or horizontal or tilted in one direction or another. This information can be indicated by symbols showing the locations and orientations of folds, faults, and other geologic features. The finished product looks like the map in FIGURE 9.22. The colors and symbols most commonly used to portray rock formations on maps are shown in FIGURE 9.23 on the next page.

Interpreting geologic maps   FIGURE 9.22

This is part of a real geologic map, showing the Canmore Quadrangle in Alberta, Canada. Note the colors, which indicate the age and type of rock, and the contour lines, which show the elevation. In the inset, a magnified part of the map, you can see the symbols for thrust faults, strike and dip, synclines, and anticlines.

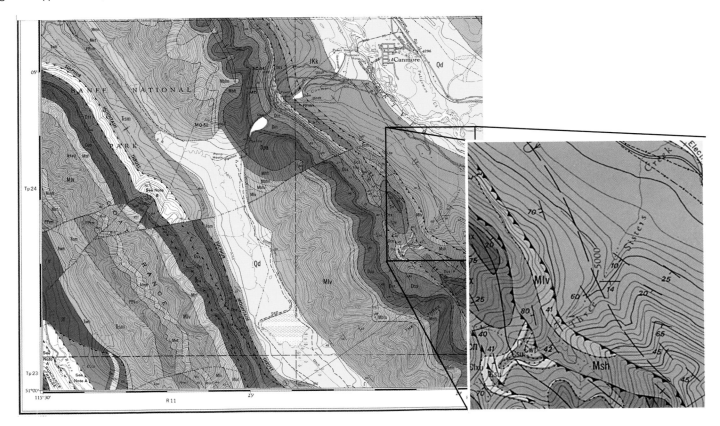

Symbols commonly used on geologic maps FIGURE 9.23

SYMBOL   EXPLANATION

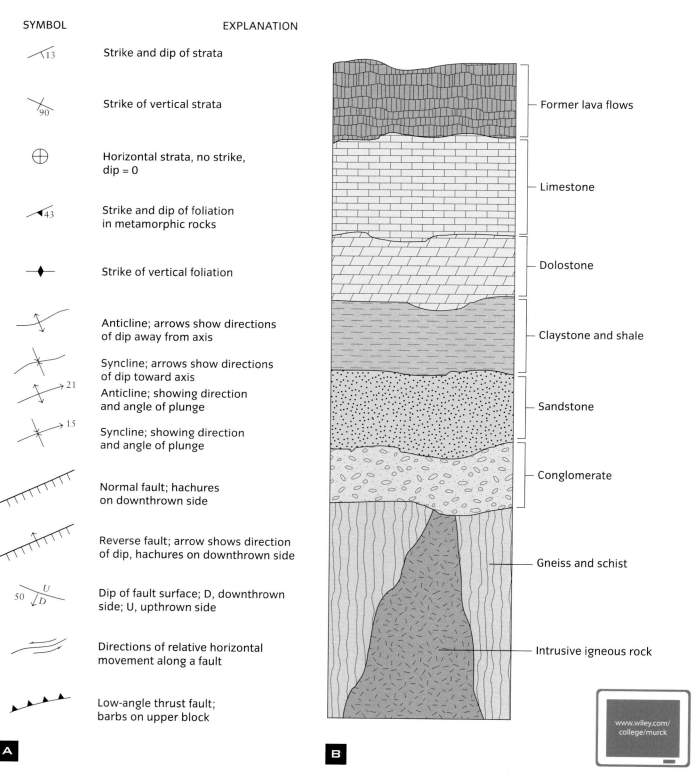

Strike and dip of strata

Strike of vertical strata

Horizontal strata, no strike, dip = 0

Strike and dip of foliation in metamorphic rocks

Strike of vertical foliation

Anticline; arrows show directions of dip away from axis

Syncline; arrows show directions of dip toward axis

Anticline; showing direction and angle of plunge

Syncline; showing direction and angle of plunge

Normal fault; hachures on downthrown side

Reverse fault; arrow shows direction of dip, hachures on downthrown side

Dip of fault surface; D, downthrown side; U, upthrown side

Directions of relative horizontal movement along a fault

Low-angle thrust fault; barbs on upper block

Former lava flows

Limestone

Dolostone

Claystone and shale

Sandstone

Conglomerate

Gneiss and schist

Intrusive igneous rock

www.wiley.com/college/murck

A

B

Representative patterns commonly, but not universally, used to show various kinds of rock in geologic maps and cross sections

Underneath the Alps FIGURE 9.24

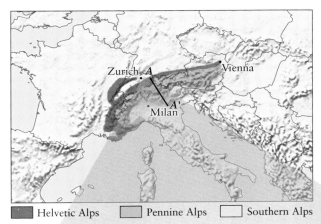

◀ A The map shows part of the Alps, a great mountain chain that borders Italy, France, and Switzerland, which was formed by crustal deformation and compressive forces. The points labeled A and A' are the beginning and end of the cross-section shown below, in (B).

▼ B The cross section A – A' reveals the intense folding and faulting that have resulted in the mountainous landscape of the Alps. Features you should be able to spot in the cross section include several faults, an enormous anticline-like structure (called a *nappe*), and the remnants of a subduction zone.

■ Helvetic Alps	■ Pennine Alps	□ Southern Alps

A

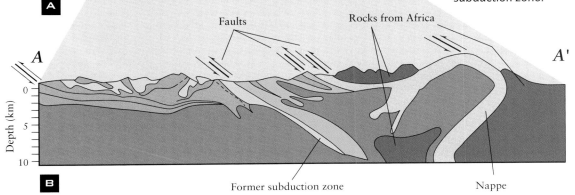

B

GEOLOGIC CROSS SECTIONS

A geologic map shows the locations of all rock outcrops on the ground surface, and the orientations of layering, structures, and other geologic features. It also shows the geologist's educated guess as to what lies under the soil, vegetation, and buildings between the outcrops. A geologic map can be used to make inferences about what happens to the rock layers just under the ground. Do they bend completely around and come back to the surface? Do they level out and become flat? Do they grade into a different type of rock? To answer these questions, we must try to visualize the area in three dimensions, even though we may only have information concerning the rocks at the surface.

Geologists do this three-dimensional visualization by constructing a

■ **geologic cross section** A diagram that shows geologic features that occur underground.

geologic cross section like the one shown in FIGURE 9.24. This is a bit like a topographic map to which geologists add his or her best guess as to how the strata fold and fault underneath the ground. As you can see, the layering can become very complicated indeed.

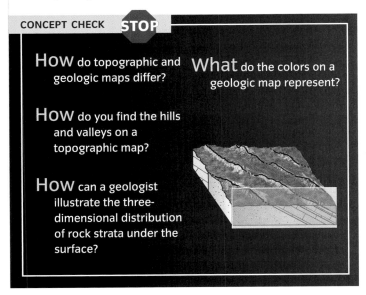

CONCEPT CHECK STOP

How do topographic and geologic maps differ?

How do you find the hills and valleys on a topographic map?

How can a geologist illustrate the three-dimensional distribution of rock strata under the surface?

What do the colors on a geologic map represent?

Amazing Places: The Canadian Rockies

Just to the north of Glacier National Park, the Canadian Rockies in western Alberta and eastern British Columbia offer many beautiful views of folding and faulting.

FIGURE A This dramatic fold at Mount Kidd marks the end of the Lewis Thrust Fault into folded rock.

FIGURE B Visible from the Kananaskis Highway west of Calgary, a peak in the Opal Range, known locally as Opal Mountain, is a combination of folded and faulted strata.

FIGURE C The Canadian Rockies form what geologists call a fold-thrust belt. A good way to visualize the relationship between the folds and the faults is to imagine a snowplow pushing a layer of snow. The snow compresses horizontally and thickens vertically. In the process, a series of thrust faults form in front of the plow, and the snow is highly folded between the faults.

A

B

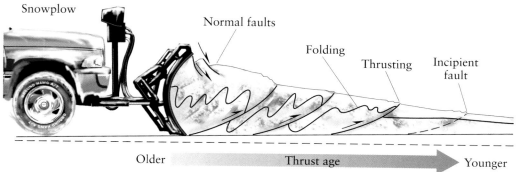

C

Snowplow

Normal faults

Folding

Thrusting

Incipient fault

Older   Thrust age   Younger

NATIONAL GEOGRAPHIC

CHAPTER SUMMARY

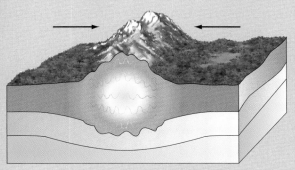

1 Rock Deformation

1. In response to **stress**, a rock may undergo **strain**; that is, it may change its shape, volume, or both. A nonpermanent change is **elastic deformation**. A permanent change that involves folding or flowing is **ductile** (or *plastic*) deformation. A permanent change that involves fracturing is **brittle deformation**.

2. Rock deformation results from **pressure** or stress placed on rocks by the movements and interactions of lithospheric plates.

3. Stress can be *uniform* (the same in all directions) or *differential* (stronger in one direction than in another). **Compression** results from forces that squeeze a rock. **Tension** results from forces that stretch a rock or pull it apart. **Shear** stress causes a body of rock to be twisted and to change shape.

4. Low temperature, low confining pressure, and high rate of strain enhance the brittle properties of rocks. Failure by fracture is common in upper-crustal rocks, where temperature and pressure are low. High temperature, high confining pressure, and low rate of strain, which are characteristic of the deeper crust and the mantle, enhance the ductile behavior of rocks. The composition of a rock also influences its deformational properties.

5. Many deformed rocks can be found in **orogens** where former tectonic plates collided. Today's continents are complicated assemblages of orogens and **cratons**—regions of ancient crust eroded to near sealevel. Orogens are collision zones between cratons and they remain subject to vertical movement long after the collisions that produced them have ended, because of **isostasy**.

2 Structural Geology

1. **Structural geology** deals with the processes and structures associated with stress and strain, such as faults, fractures, and **folds**.

2. The **strike** of a stratum, a fault, or other geologic surface is the orientation of the line marking the intersection of the surface with a horizontal plane. The **dip** is the angle between the tilted surface and a horizontal plane, measured down from the horizontal in degrees. Any planar features in rocks can be described using these two measurements.

3. **Normal faults**, in which the hanging-wall block moves down relative to the footwall block, are caused by tensional (pull-apart) stress. In **reverse faults**, caused by compressional (squeezing) stress, the hanging-wall block moves up and over the footwall block. Shallowly dipping reverse faults are called **thrust faults**. In **strike-slip faults**, caused by shear stress, the movement is mainly horizontal and parallel to the strike of the fault.

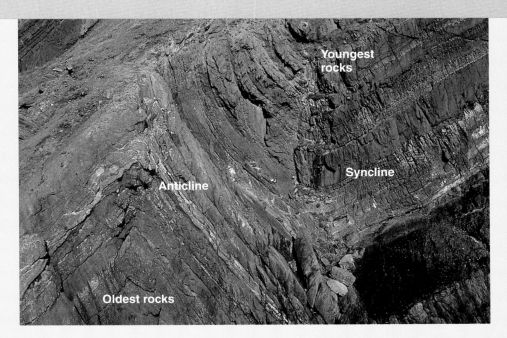

4. An **anticline** is an upward fold in the form of an arch, with the oldest rocks in the core of the arch. A **syncline** is a downward fold in the form of a trough, with the youngest rocks in the concave "valley" of the trough. Many folds are combinations or variations of these two basic types. To fully portray the geometry of a fold, we must describe the fold axis and (if necessary) the direction in which the axis is plunging, as well as the dip of the plunging axis.

CHAPTER SUMMARY

3 Geologic Maps

1. Geologists use **geologic maps** to portray rock units and structures found at Earth's surface. It is common to use a **topographic map** as the base for a geologic map. Topographic maps give information about the *relief*, or "ups and downs" of the land surface, using *contour lines* to connect points of equal elevation.

2. Geologic maps provide information about the distribution, locations, and orientations of rock *formations* and structures at the surface, using standardized symbols, colors, and fill textures to portray various geologic features.

3. Sometimes it is important to be able to visualize an area in three dimensions, in order to understand what happens to the geologic structures in the subsurface, and to interpret the geologic history of the area. This is done by constructing a **geologic cross section**. Geologic cross sections are based partly

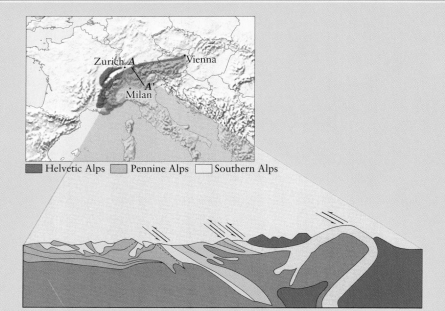

Helvetic Alps   Pennine Alps   Southern Alps

on the surface geology, as revealed by geologic mapping, and partly on other sources of information about underground geology, such as drilling or seismic studies.

KEY TERMS

- **stress** p. 246
- **pressure** p. 246
- **tension** p. 247
- **compression** p. 247
- **shear** p. 247
- **strain** p. 247
- **elastic deformation** p. 248
- **ductile deformation** p. 248
- **brittle deformation** p. 249

- **craton** p. 252
- **orogen** p. 252
- **isostasy** p. 253
- **structural geology** p. 254
- **strike** p. 255
- **dip** p. 255
- **normal fault** p. 256
- **reverse fault** p. 257
- **thrust fault** p. 257

- **strike-slip fault** p. 257
- **fold** p. 259
- **anticline** p. 259
- **syncline** p. 259
- **geologic map** p. 263
- **topographic map** p. 265
- **geologic cross section** p. 267

1. Find real examples of plate boundaries along which each of the following types of stress predominates: (a) compression, (b) tension, and (c) shearing. Try to find different examples than those used in the text.

2. What might happen to a rubber ball or a banana or Silly Putty when they are frozen in liquid nitrogen? Can you find other common materials or objects that make a transition from ductile to brittle (or vice versa) at different temperatures, pressures, or strain rates?

3. To construct geologic cross sections, geologists use the information on topographic maps, often supplemented with information about the subsurface provided by remote study techniques. What are some of the techniques geologists use to study rocks beneath Earth's surface? (Hint: Look back at Chapter 5.)

4. In this chapter we mentioned the role of structural geology in determining the locations of mineral ores, oil, and natural gas. Try to find out about some other practical uses and applications of structural geology, geologic maps, and cross sections.

5. Imagine that you are taking a field course in the Alps. From geologic cross sections, you expect to find some very large folds (such as the one illustrated on the right of Figure 9.24) where the rock layers have been bent over backwards. What kinds of evidence could you look for in the rocks that would help you determine whether a particular sequence of rocks is right-side up or upside down? In other words, how would you determine whether a trough-shaped fold was a right-side up syncline, or an upside down (i.e., overturned) anticline? Discuss this question with fellow students or your instructor.

What is happening in this picture ?

Chief Mountain, in Glacier National Park in Montana, is a huge chunk of Precambrian rock that lies on top of a layer of much younger Cretaceous shale and sandstone (underneath the arrow). How is such a formation possible? The arrow points to a fault. Can you guess what kind of fault it is?

1. _____ is a type of stress that acts in a direction perpendicular to and away from a fault surface.

   a. Shear stress
   b. Compression
   c. Tension

2. _____ is a type of stress that acts in a direction parallel to and along a fault surface.

   a. Shear stress
   b. Compression
   c. Tension

3. In _____ rocks will bend as long as stress is applied to the crust but resume their original shape if the stress is released.

   a. elastic deformation
   b. brittle deformation
   c. ductile deformation

4. What type of crustal deformation is depicted in this photograph?

   a. elastic deformation
   b. brittle deformation
   c. ductile deformation

5. The structure depicted in the photograph could have formed as a result of stress, under conditions that have any combination of _____

   a. high confining pressure, low temperature, or high strain rate.
   b. low confining pressure, low temperature, or high strain rate.
   c. high confining pressure, low temperature, or high strain rate.
   d. high confining pressure, high temperature, or low strain rate.

6. A(n) _____ is an elongate region of crust that has been deformed and metamophosed by a continetal collision. By contrast a(n) _____ is a region of continental crust that has remained undeformed for a very long time.

   a. syncline; anticline
   b. anticline; syncline
   c. orogen; craton
   d. craton; orogen

7. On this block diagram of tilted strata, draw symbols to show the direction of strike and dip and indicate the 30° angle of dip.

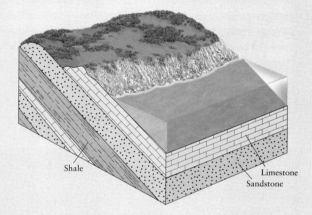

Shale
Limestone
Sandstone

8. This diagram shows a faulted block of Earth's crust. What type of fault is depicted in the diagram?

   a. normal fault
   b. thrust or reverse fault
   c. strike-slip fault

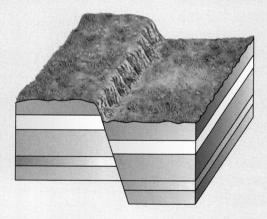

9. The fault depicted in the diagram in question 8 must have formed in response to what kind of stress?

   a. tension
   b. compression
   c. shear

10. A _____ fault is a product of compression of Earth's crust, whereas a _____ fault is a product of tension. _____ faults result from shear stress.

   a. strike-slip; normal; Thrust (reverse)
   b. normal; thrust (reverse); Strike-slip
   c. thrust (reverse); strike-slip; Normal
   d. thrust (reverse); normal; Strike-slip
   e. strike-slip; thrust (reverse); Normal

11. Label this illustration with the following terms:

   anticline   axial plane
   syncline   plunge
   axial trace

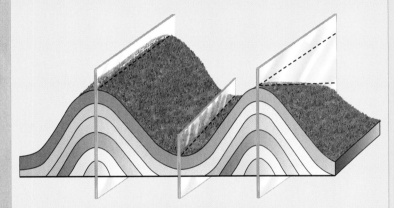

12. A(n) _____ is a local steepening in otherwise uniformly dipping strata.

   a. anticline
   b. syncline
   c. monocline
   d. dome
   e. basin

13. A _____ would allow a geologist to locate the steepest cliff face in a region.

   a. geologic map
   b. topographic map
   c. geologic cross section

14. A _____ can be used to show the subsurface structure of an orogen.

   a. geologic map
   b. topographic map
   c. geologic cross section

15. The areal distribution of rocks and structures in an area is best depicted on a _____ .

   a. geologic map
   b. topographic map
   c. geologic cross section

Metamorphism: New Rocks from Old 10

To carve the Pietà, a statue portraying Mary holding the lifeless body of Jesus, Michelangelo Buonarroti needed a block of perfect marble two meters wide. The year was 1498; the place was Rome. After Michelangelo had resigned himself to a visit to the quarries in distant Carrara, word reached him that a block had been cut on order and shipped as far as Rome but not paid for. It would go no further. As Irving Stone writes in *The Agony and the Ecstasy*:

> *He watched the rays of the rising sun strike the block and make it transparent as pink alabaster, with not a hole or hollow or crack or knot in all its massive white weight. It tested out perfect against the hammer, against water, its crystals soft and compacted with fine graining. His Pietà had come home.*

White marble starts out as marine sediment composed of fragments of shells, most of them made of calcite. The sedimentary rock that forms from this sediment as a result of diagenesis is limestone. When subjected to further heat and pressure, the calcite recrystallizes, all traces of the shells disappear, and the limestone becomes a crystalline white marble. This transformation is called metamorphism. In this chapter you will learn about the processes of change during metamorphism and the many kinds of metamorphic rock and where they are found.

Which is more remarkable: the metamorphism from calcareous sediment to pristine marble or the transformation from a raw quarried marble block to a finished sculpture? We are amazed by them both.

What Is Metamorphism?

Perhaps the best everyday analogy for **metamorphism** is cooking—a process that takes, for instance, flour, salt, sugar, yeast, and water and transforms them into a completely new substance called bread, or turns raw meat into a medium-rare steak (see FIGURE 10.1). Like cooking, and like diagenesis (see Chapter 8), metamorphism involves heat and pressure, but the temperatures and pressures involved are considerably higher. Some of the same physical processes that occur in diagenesis, such as compaction and recrystallization, continue during metamorphism but in a more intense way. In metamorphism these processes are accompanied by chemical reactions that change the rock's mineral assemblage and texture, and sometimes its chemical composition as well. Metamorphism also takes place at greater depths than diagenesis—typically between 5 and 40 kilometers, which is most of the extent of the thickness of the crust.

> **metamorphism**
> The mineralogical, textural, chemical, and structural changes that occur in rocks as a result of exposure to elevated temperatures and/or pressures.

There is one important kind of change that does *not* take place during metamorphism: Rocks do not melt. This is one of the things that make metamorphic rocks so interesting. They may have been squeezed, stretched, heated, and altered in complex ways (like the gneiss in Figure 10.1B), but they have remained solid. Solids, unlike liquids and gases, preserve clues to the events that changed them. For example, if you throw a stone into a pond, the splash and resulting ripple soon disappear. Throw a stone at a window, however, and the result is permanently cracked glass. Metamorphic rocks preserve a record of all the heatings, stretchings, and grindings that have happened to them throughout geologic history. When we encounter an area with many metamorphic rocks, we can be certain that it has been geologically busy. Decoding the record of that activity is a challenge, but the rewards are high because it can tell us about long-ago collisions of continents, the break-up of ancient supercontinents, and whether rich mineral deposits might be present as a result of that activity.

THE LIMITS OF METAMORPHISM

The two most important factors in metamorphism are heat and pressure. The heat that causes metamorphic reactions is Earth's internal heat. We know from drilling deep gas and oil wells, and from deep gold mines, that temperature in the continental crust increases with depth at a rate of about 30°C/km. At a depth of about 5 kilometers, the temperature is about 150°C. This temperature represents the dividing line between the processes of diagenesis (Chapter 8), which change sediments into sedimentary rocks (below 150°C), and the processes of metamorphism (above 150°C). Geologists have settled on 150°C as the "official" boundary between diagenesis and metamorphism, but the actual transition from diagenesis into metamorphism is more gradational; some rocks begin to show distinct signs of metamorphic changes at 150°C, but others do not.

A

◄ **A** In the thermal metamorphism of a steak, the outer, brown layers of the meat have been subjected to higher temperatures than the inner, red layers. They have therefore undergone a higher degree of metamorphism from the raw state, including a change of texture and color. The outermost layer is in the well-done zone; moving in, the lighter-colored meat is in the medium rare zone; and in the center, the reddest colored meat is in the rare zone.

B

B High temperatures and pressures deep in the crust long ago converted the ingredients in this mix of minerals (dark-colored clay and light-colored quartz) into a new metamorphic rock called gneiss. This rock outcropping was photographed near Isua, Greenland. ►

The upper temperature limit of metamorphism is about 800°C. Partial melting above this temperature marks the transition from metamorphic processes to magmatic processes. As discussed in Chapter 6, the onset of melting can occur at a variety of temperatures depending on the pressure, the composition of the rock, and the amount of fluid present, particularly water and carbon dioxide. Thus, 800°C is only a generalization of the temperature at which partial melting typically begins in the crust.

Just as pressure influences the temperature at which melting begins, it also influences the kind of metamorphic changes that occur at temperatures below the onset of melting. The air pressure at sea level on Earth's surface is defined to be one *atmosphere*—about half the pressure of the air inside an automobile tire. However, rocks are heavier than air. In the crust, pressure increases with depth at a rate of about 300 atmospheres (atm) per kilometer, or about 30 megapascals (MPa) per kilometer. Therefore, the pressure 5 kilometers below the surface is 1500 times greater than the surface atmospheric pressure—that is, 1500 atm, or 150 MPa. This is the depth at which both temperature and pressure are high enough for recrystallization and the growth of new minerals to start.

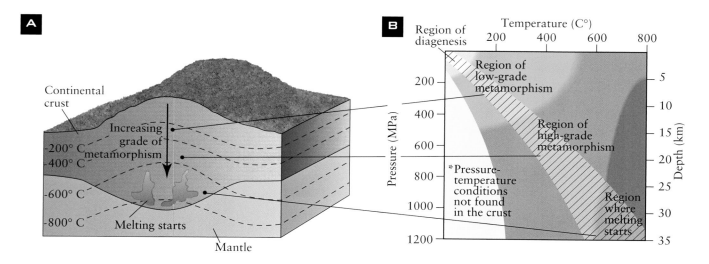

Temperature and pressure conditions for metamorphism FIGURE 10.2

A This sketch shows schematically where in the crust metamorphism and melting occur.

B Four colored bands illustrate the temperature and pressure conditions for diagenesis, low-grade and high-grade metamorphism, and melting. Note that the vertical scale is given in units of both pressure and depth beneath the surface. The shaded diagonal band indicates the pressures and temperatures most commonly found in continental crust.

FIGURE 10.2 shows the range of temperatures and pressures over which metamorphism occurs. In Figure 10.2B, pressure is represented on the vertical axis, increasing with depth. Temperature is represented on the horizontal axis, increasing from left to right. The pressure and temperature conditions under which sediments are formed and changed into sedimentary rocks during diagenesis occupy the upper left-hand corner of the diagram. Below this, and up to a depth of about 15 kilometers, where the pressure reaches 400 MPa (4000 atm) and the temperature is about 400°C, metamorphism occurs but is not very intense. In these **low-grade** metamorphic rocks, the minerals have changed but the appearance is still similar to that of sedimentary rocks. At higher temperatures and pressures, extending to the onset of melting, **high-grade** metamorphic rocks are formed. In the lower right-hand corner of the diagram, representing the hottest temperatures and highest pressures, rocks may begin to melt and would no longer be considered metamorphic.

■ **low-grade**

Rocks that are metamorphosed under temperature and pressure conditions up to 400°C and 400 MPa.

■ **high-grade**

Rocks that are metamorphosed under temperature and pressure conditions higher than about 400°C and 400 MPa.

FACTORS INFLUENCING METAMORPHISM

As in cooking, the end product of metamorphism is controlled by the ingredients (the composition of the rock) and the temperature. However, metamorphism is also influenced by pressure, the duration and rate at which high pressures and temperatures are applied, and the presence or absence of fluids. The directionality of pressure also plays an important role in metamorphism. As explained in Chapter 9, geologists refer to pressures that are greater in one direction than another as *stress*. Many metamorphic rocks clearly show the effects of directional stress. Let's look at all of these factors.

Temperature and pressure   When rocks are heated, some of the original minerals recrystallize but don't change composition; others are involved in chemical reactions that form new minerals. FIGURE 10.3 shows the effect of increasing temperature and pressure

From shale to gneiss FIGURE 10.3

As shale is subjected to higher and higher temperatures and pressures, it develops into a sequence of metamorphic rocks that have different mineral assemblages. The photos in the diagram were all taken under a microscope, with a 3-mm field of view.

www.wiley.com/college/murck

Diagenesis

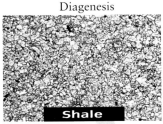
Shale

Shale
(sedimentary)

1. Shale is a sedimentary rock made of clay particles and quartz grains.

Increasing temperature and pressure (metamorphism)

Low-Grade

Slate

Slate
(metamorphic)

2. Low-grade slate develops from shale. It contains quartz, chlorite, muscovite mica, and feldspars, but no clay.

Medium-Grade

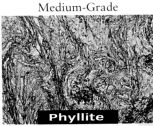
Phyllite

Phyllite
(metamorphic)

3. As temperature continues to increase, chlorite disappears and is replaced by biotite mica.

High-Grade

Gneiss

4. At even higher temperatures schist or gneiss is formed. Minerals like garnet, kyanite, and sillimanite appear.

Schist (mica-rich)
Gneiss (mica-poor)
(metamorphic)

Clay

Chlorite

Muscovite (white) mica

Biotite (dark) mica

Garnet

Kyanite*

Sillimanite*

Feldspar

Quartz

*Sillimanite and kyanite both have the same formula— Al_2SiO_5—but different crystal structures.

on shale, a common sedimentary rock. Notice that as the grade of metamorphism increases, completely new minerals appear—and, in some cases, disappear again at higher grades. Thus, identifying the component minerals is a good way to tell these metamorphic rocks apart.

Pore fluids

Another factor that affects metamorphism is the presence of open spaces, called *pores*. The same term is used for all the tiny open spaces in a rock or sediment, such as those between the grains in a sedimentary rock, as well as small fractures in igneous and metamorphic rocks. Most pores are filled either by a watery fluid or by a gaseous fluid, such as carbon dioxide. The watery or *aqueous* fluid is never just pure water—it has small amounts of gases and salts dissolved in it, along with traces of all the mineral constituents present in the enclosing rock. At high temperatures, the pore fluid is likely to be entirely gaseous. Regardless of their specific characteristics, pore fluids play a vital role in metamorphism.

Pore fluids enhance metamorphism in two ways. The presence of pore fluid permits material to dissolve from one place, move quickly via the fluid, and be precipitated in another place. In this way, pore fluids speed up recrystallization, and generally lead to rocks with relatively large mineral grains. Pore fluids also enhance metamorphism by acting as a reservoir during the growth of new minerals. When the temperature and pressure of a rock undergoing metamorphism change, so does the composition of the pore fluid. Some of the dissolved constituents move from the fluid to the new minerals growing in the metamorphic rock. Other constituents move in the other direction, from the minerals to the fluid. Pore fluids speed up chemical reactions in the same way that water in a stew pot speeds up the cooking of a tough piece of meat. Because of this, metamorphism proceeds rapidly when pore fluids are present, but when pore fluids are absent, or present in tiny amounts, metamorphic reactions occur very slowly.

As pressure increases and metamorphism proceeds, the amount of pore space decreases and the pore fluid is slowly driven out of the rock. The escaping pore fluid carries with it small amounts of dissolved minerals. As the fluid flows through fractures in the rock, some of the dissolved minerals may precipitate, creating *veins* (see FIGURE 1 0.4). Veins are quite common in metamorphic rock.

The presence of pore fluid also greatly influences the onset of rock melting. The effect of fluid on rock is to lower the melting temperature of the rock. The upper temperature limit of metamorphism therefore depends, in part, on the amount of pore fluid present. When a tiny amount of fluid is present, only a small amount of melting occurs and the magma remains trapped in small pockets in the metamorphic rock.

Quartz vein FIGURE 1 0.4

This sample of gneiss contains many quartz *veins* (white) that criss-cross the rock. The quartz precipitated out of pore fluid that was expelled from the rock during metamorphism.

Effects of uniform and differential stress FIGURE 10.5

The two rocks in these photos have similar mineral assemblages but look very different because of their different stress histories.

A This granite consists of quartz (glassy), feldspar (white), and biotite (dark), which crystallized from magma (a liquid) under conditions of uniform stress. Note that the biotite grains are randomly oriented.

B This gneiss, a high-grade metamorphic rock, contains the same minerals as the granite, but they developed entirely in the solid state and under differential stress. The biotite grains are aligned, giving the rock a pronounced layered texture.

When the rock cools, so do the pockets of magma. The result is a composite rock, part igneous and part metamorphic, called **migmatite**. When abundant pore fluid is present and large volumes of magma develop, the magma rises and intrudes into overlying metamorphic rock. In such cases we observe batholiths of granitic rock (see Chapter 6) closely associated with large volumes of metamorphic rock. This igneous-metamorphic rock association occurs along subduction and collision margins of tectonic plates.

Stress   As you learned in Chapter 9, rocks that are buried deep underground typically experience nonuniform forces or differential stresses that have a strong effect on their appearance (see FIGURE 10.5).

> **foliation** A planar arrangement of textural features in a metamorphic rock, which give the rock a layered or finely banded appearance.

Differential stress causes rock undergoing metamorphism to develop a distinctive layered or planar texture called **foliation**. Figure 10.5A is a nonfoliated rock; 10.5B is a foliated rock. Foliation is particularly evident when minerals belonging to the *mica* family are present. Micas are minerals in which the silicate anions link together (polymerize) in flat sheets (see Chapter 2 to review the structure of mica and other minerals mentioned below). Under differential stress, micas grow so that the sheets are perpendicular to the direction of maximum stress (see FIGURE 10.6).

Foliation under a microscope FIGURE 10.6

This image was created by gluing a rock chip to a glass microscope slide and grinding the chip down until it was thin enough to let light pass through, then viewing it in polarized light. The sample is a foliated metamorphic rock consisting mostly of muscovite (a mica). The direction of maximum stress is indicated by the arrows.

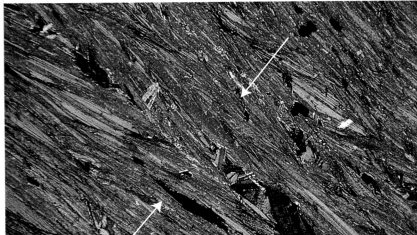

Low-grade metamorphic rocks tend to be so fine-grained that the new mineral grains can be seen only under a microscope. Foliation in these rocks produces a distinctive style of fracture, called **slaty cleavage**. The orientation of the cleavage planes says a great deal about the conditions under which metamorphism occurred, as explained in FIGURE 10.7.

Slaty cleavage is similar to cleavage in a mineral (see Chapter 2) but with a very important difference. A mineral grain is a single crystal, and cleavage occurs because the bonds between atoms in the crystal are weaker in some directions than others. Cleavage in a rock, in contrast, involves a great many crystal lattices, and it happens because the crystals themselves are roughly parallel. The flat cleavage planes that result make slate an excellent material for roofing and paving tiles, blackboards, billiard tables, and other uses that require a durable natural material with a flat surface (see Case Study).

Visualizing

Slaty cleavage FIGURE 10.7

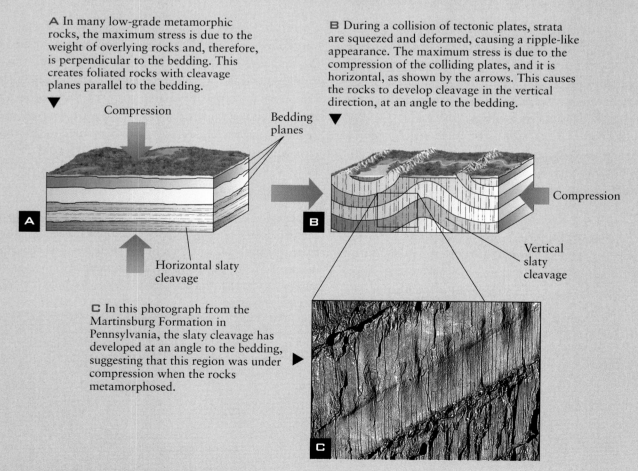

A In many low-grade metamorphic rocks, the maximum stress is due to the weight of overlying rocks and, therefore, is perpendicular to the bedding. This creates foliated rocks with cleavage planes parallel to the bedding.

Compression

Bedding planes

Horizontal slaty cleavage

B During a collision of tectonic plates, strata are squeezed and deformed, causing a ripple-like appearance. The maximum stress is due to the compression of the colliding plates, and it is horizontal, as shown by the arrows. This causes the rocks to develop cleavage in the vertical direction, at an angle to the bedding.

Compression

Vertical slaty cleavage

C In this photograph from the Martinsburg Formation in Pennsylvania, the slaty cleavage has developed at an angle to the bedding, suggesting that this region was under compression when the rocks metamorphosed.

Metamorphism and Billiards

The modern game of billiards, and its close relative, pool, are played on tables made from metamorphic rock. Early billiard tables resembled modern ones in that they were rectangular, surrounded by cushions, and fitted with six pockets. However, they were made of wood, which had a tendency to vibrate and thus

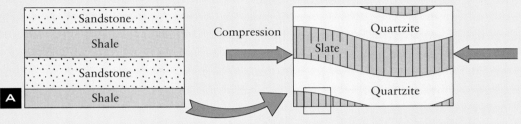

Metamorphosis converts shale to slate.

Slate comes from the low-grade metamorphism of shale (see FIGURE A). In North America a belt of slate runs along the eastern seaboard from Georgia to Maine. Thus, the collision of continents that created the supercontinent of Pangaea 400 million years ago also created the rocks (see FIGURE B) that enabled these men (see FIGURE C) to play billiards in the shadow of the Kunlun Mountains, in the far west of China.

interfere with the accuracy of the players. In addition, the wood warped after only a few years. Around 1825, at the height of the Industrial Revolution, an Englishman named John Thurston invented the slate bed. Manufacturers of billiard tables discovered that if slate were used for the table (with the slaty cleavage surface forming to the table top), they could ensure a surface that was very smooth and free of vibrations. The slate table was also much sturdier and less prone to wear and tear.

schistosity

Foliation in coarse-grained metamorphic rocks.

Another kind of foliation, **schistosity**, forms under conditions of high-grade metamorphism, which causes mineral grains to grow large enough to be seen with the naked eye. This kind of foliation differs from slaty cleavage primarily in the size of the mineral grains. Another difference is that schistose rocks generally break along wavy or distorted surfaces, while slaty cleavage is strictly planar.

Duration and rate of metamorphism Chemical reactions can occur rapidly or slowly. Some reactions, such as the burning of natural gas to produce carbon dioxide and water, happen so fast that they can cause an explosion if not handled carefully. At the other end of the scale are reactions that take thousands or even millions of years to complete. Most of the reactions that happen during metamorphism are of the latter kind.

Despite the slowness of most metamorphic reactions, scientists have been able to use laboratory experimentation to demonstrate that high temperatures, high pressures or stresses, abundant pore fluids, and long reaction times produce large mineral grains. Coarse-grained rocks—those with mineral grains the size of a thumbnail or larger—probably form under long-sustained metamorphic conditions (possibly over millions of years) with high temperatures, high stresses, and abundant pore fluids. On the other hand, fine-grained rocks—those with mineral grains the size of a pin head or smaller—were either produced under conditions involving lower temperatures and lower stresses or formed under conditions where pore fluids were scarce or reaction times were short.

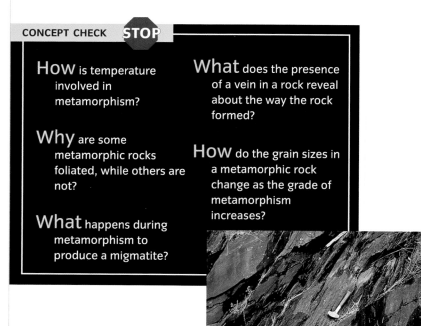

CONCEPT CHECK **STOP**

How is temperature involved in metamorphism?

Why are some metamorphic rocks foliated, while others are not?

What happens during metamorphism to produce a migmatite?

What does the presence of a vein in a rock reveal about the way the rock formed?

How do the grain sizes in a metamorphic rock change as the grade of metamorphism increases?

Metamorphic Rocks

LEARNING OBJECTIVES

Describe the common metamorphic products of shale.

Identify the common metamorphic products of basalt.

Explain how basalt can be metamorphosed in different ways.

Identify several common foliated and nonfoliated metamorphic rocks.

The names of metamorphic rocks are mostly based on their textures and mineral assemblages. The most widely used names apply to metamorphic rocks derived from the sedimentary rocks shale, sandstone, and limestone, and from the igneous rock basalt. This is because shale, sandstone, and limestone are the most abundant of the sedimentary rocks, and basalt is the most abundant of the igneous rocks. We will first describe the metamorphic rocks that always have some degree of foliation; then we will move on to describe metamorphic rocks that tend to occur without foliation.

ROCKS WITH FOLIATION

As we mentioned earlier, the metamorphic products of shale form a sequence of rocks, beginning with slate, a low-grade product. As the metamorphism proceeds to higher grades, the rock looks less and less like its parent rock, shale, and develops larger and larger mineral grains. In **FIGURE 10.8** we describe each of these rocks in greater detail.

There are some subtleties of terminology in the sequence of rocks deriving from shale. The names **slate** and **phyllite** describe textures. They are usually used without adding mineral names as adjectives, because their mineral assemblages are not easy to see. On the other hand, the mineral grains in **schist** and **gneiss** are large enough to be identified, so geologists usually add the mineral assemblage to the name, as in quartz-plagioclase-biotite-garnet gneiss.

Visualizing

Metamorphic rocks derived from shale FIGURE 10.8

SHALE
A sedimentary rock

A The source rock, *shale*, consists primarily of clay minerals and quartz.

SLATE
A low-grade metamorphic rock

B Under conditions of low-grade metamorphism, muscovite and/or chlorite forms from the clay minerals and pore fluids, and the rock becomes a *slate*. The tiny new mineral grains (too small to be seen without a microscope) produce slaty cleavage.

PHYLLITE
A medium-grade metamorphic rock

C Continued low-grade metamorphism of a slate produces larger grains of mica and a changing mineral assemblage. In a *phyllite* the grains of mica are just large enough to be visible.

D Still further metamorphism of phyllite leads to *schist*. Though the most obvious change is the size of the grains, there are also changes in the mineral assemblage. At high grades of metamorphism, minerals may begin to segregate into separate bands. In *gneiss* (pronounced "nice"), the dark bands contain mica, and the light bands consist of quartz or feldspar.

SCHIST
A mica-rich high-grade metamorphic rock with pronounced *schistosity*.

GNEISS
A mica-poor high-grade metamorphic rock with a coarsely banded (*gneissose*) texture.

Metamorphic Rocks   285

When basalt is subjected to metamorphism under conditions in which pore water can enter the rock, distinctive mineral assemblages develop.

Increasing grade of metamorphism →

Increasing pressure (without increasing temperature)

A Low-grade metamorphism produces a rock called greenschist, so named because of the presence of a green, mica-like mineral called chlorite. Like phyllite, greenschists are fine-grained with pronounced foliation.

B Under high-grade metamorphism, the chlorite is replaced by amphibole, and the resulting high-grade rock is an amphibolite. Foliation is present in amphibolites but is not pronounced because micas are usually absent.

C Under conditions of very high stress, as in subduction zones, the low-grade metamorphic product of basalt is blueschist, which owes its color to a mineral called glaucophane.

D At even higher pressures and moderate temperatures, blueschist is replace by the high-grade rock eclogite, which contains such minerals as jadeite and garnet.

Metamorphic rocks don't always develop from sedimentary rocks; igneous rocks also can undergo metamorphism. The names given to metamorphosed igneous rocks differ from those used for the metamorphic rocks that develop from shale and other sedimentary rocks. Basalt—the most abundant igneous rock—is a very common parent rock. Basalt produces several distinct mineral assemblages, depending on the metamorphic conditions to which it is subjected (see FIGURE 1 0.9).

Under the conditions of high stress and low to moderate temperature that occur in subduction zones, the metamorphic rocks *blueschist* and *eclogite* are produced. Eclogite is an important source for such semiprecious gemstones as garnet and jadeite (see What a Geologist Sees). Under moderate- to high-temperature conditions, particularly where pore fluids are available, the metamorphic rocks *greenschist* and *amphibolite* develop from igneous rocks of the same basaltic composition. Metamorphic rocks made from basalt also are less foliated than the series of rocks formed by the metamorphism of shale, because they do not contain as much mica. Amphibole—the mineral that typically defines the foliation in amphibolites—is not as elongate as the mica minerals.

Two Kinds of Jade

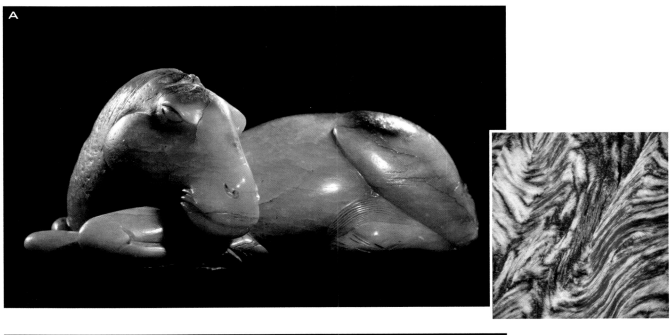

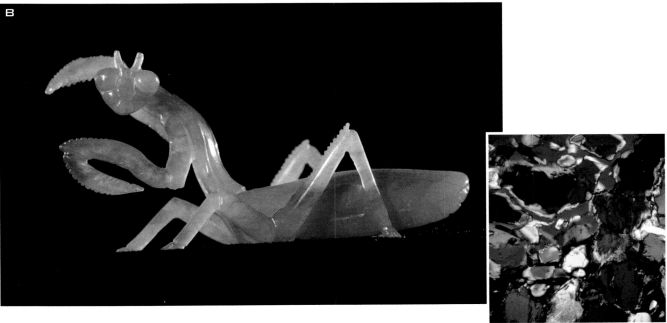

Jade is a tough, compact rock, commonly green or white, that takes a high polish and is strong enough to be carved in intricate ways. Rocks with densely packed, small grains of either of two different minerals, nephrite or jadeite, are the most common forms of jade. Nephrite (**FIGURE A**) is an amphibole formed by metamorphism of impure limestone. The inset photo is a microscope view of tightly packed amphibole fibers, which make nephrite a desirable carving medium. Jadeite (**FIGURE B**) is a pyroxene formed by metamorphic transformation of albite in subduction zones. The inset photo shows the tight packing of the individual small grains of jadeite in a sample of jade. Chinese artists referred to jadeite as "kingfisher stone" when they started receiving it through trade with Burma in the 1700s, because it was more vividly colored than the nephrite they were used to. Nephrite and jadeite are only semiprecious stones; the value of jade lies mainly in the exquisite workmanship of jade carvings.

Nonfoliated metamorphic rocks FIGURE 10.10

A *Marble* is a metamorphic rock composed mainly of calcite. Pure marble is snow white. The pink color of this sample (from Tate, Georgia) and the attractive patterns (called "marbling") that can be seen in the background of the chapter-opening photograph, come from impurities in the marble.

B Quartz-rich sandstone becomes *quartzite* when its pore spaces are filled with silica and the entire mass recrystallizes as a result of metamorphism. The specimen shown here is from Minnesota.

ROCKS WITHOUT FOLIATION

quartzite The product of metamorphism formed by recrystallization of sandstone.

Two kinds of sedimentary rock consist almost entirely of a single mineral species and therefore are said to be *monomineralic*. The first is sandstone, a clastic sedimentary rock that is usually dominated by quartz grains. The second is limestone, a chemical sedimentary rock whose only essential constituent is calcite. Neither of these rocks contains the ingredients (mainly aluminum, potassium, and iron) necessary to form micas and other minerals that might impart foliation to a metamorphic rock.

marble The product of metamorphism formed by recrystallization of limestone.

As a result, **quartzite** and **marble**—the metamorphic rocks derived from quartz-rich sandstone and limestone, respectively—usually lack foliation (see FIGURE 10.10). (In contrast, those ingredients *are* present in sedimentary rocks such as shale, where they occur in clay minerals; therefore, the metamorphism of shale *does* lead to the formation of micas, resulting in a foliated texture.) Under a microscope, the individual mineral grains in quartzite and marble tend to be uniform in size, closely packed together, and show few or no signs of their sedimentary origin, such as bedding planes or fossils.

CONCEPT CHECK STOP

What is schistosity, and how does it differ from slaty cleavage?

What is the rock sequence that develops from shale under increasingly higher grades of metamorphism?

Why do some metamorphic rocks have weaker foliation than others?

What are the two common nonfoliated metamorphic rocks, and what are their parent rocks?

Metamorphic Processes

LEARNING OBJECTIVES

Identify two types of physical changes that occur in rocks during metamorphism.

Describe contact, burial, and regional metamorphism.

Identify the tectonic settings where the different types of metamorphism are likely to occur.

Define metasomatism and explain how it differs from metamorphism.

T he processes that cause changes in texture and mineral assemblages in metamorphic rocks are *mechanical deformation* and *chemical recrystallization*. Mechanical deformation includes grinding, crushing, bending, and fracturing. FIGURE 10.11B, which shows flattened pebbles in a conglomerate, is an example of mechanical deformation. Chemical recrystallization includes changes in mineral composition, growth of new minerals, recrystallization of old minerals, and changes in the amount of pore fluid due to chemical reactions that occur when a rock is heated and squeezed. Metamorphism typically involves *both* mechanical deformation and chemical recrystallization processes, but their relative importance varies greatly.

TYPES OF METAMORPHISM

We can distinguish among several kinds of metamorphism on the basis of the dominant processes (that is, the relative importance of mechanical deformation and chemical recrystallization) and the tectonic environments in which they occur.

Mechanical deformation FIGURE 10.11

A This undeformed conglomerate contains pebbles rounded from stream transport.

B In the deformed conglomerate differential pressure has squeezed the once-rounded pebbles to such an extent that they are now flat.

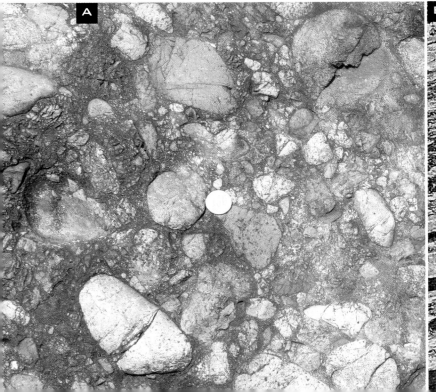

contact metamorphism
Metamorphism that occurs when rocks are heated and chemically changed adjacent to an intruded body of hot magma.

Contact metamorphism

Where hot magma intrudes into cooler rocks, high temperatures cause chemical recrystallization in surrounding rocks. The magma itself may also release pore fluids, chiefly heated water, as it cools. These fluids in turn accelerate the growth of new minerals. **Contact metamorphism** is primarily temperature-driven; mechanical deformation—and therefore foliation—are minor.

An igneous intrusion is commonly surrounded by a zone or *aureole* of contact metamorphism (see **FIGURE 10.12**). If the rock is relatively impervious, such as shale, or if the magma contains little water, then the aureole may extend only a few centimeters. However, a large intrusion contains more heat energy than a small one and may release a large volume of pore fluid as it cools. When intrusions are very large (the size of stocks and batholiths), and when they intrude into highly reactive and relatively pervious rock, such as limestone, the effects of contact metamorphism may extend for hundreds of meters. Unlike other kinds of metamorphism, contact metamorphism is not associated with any particular tectonic setting; it may occur wherever magma intrusions occur.

Burial metamorphism

The first stage of metamorphism to occur in a sedimentary rock after diagenesis is **burial metamorphism**. The metamorphic processes caused by burial begin at about 150°C, when a sedimentary rock has been buried at a depth of about 5 kilometers. The maximum stress exerted during burial metamorphism tends to be vertical; foliation, if present, is thus parallel to bedding. Burial metamorphism requires a relatively thick layer of overlying rock or sediment, so it is usually observed in deep sedimentary basins such as those found along passive margins or fore-arc basins (see Chapter 8). Because the temperature of the rock in these basins seldom exceeds 300°C,

burial metamorphism
Metamorphism that occurs after diagenesis, as a result of the burial of sediments in deep sedimentary basins.

burial metamorphism tends to be low-grade. As temperature and pressure increase beyond this point, or as sediment on a continental margin is deformed during a tectonic collision, burial metamorphism grades into regional metamorphism.

Contact metamorphism FIGURE 10.12

A layer of limestone undergoes contact metamorphism when a granite magma intrudes into it. The intrusion is surrounded by an *aureole* of altered rock, with several distinct zones. In the outermost zone, farthest from the heat of the granite intrusion, the limestone has been metamorphosed to marble. Inside the marble is a zone where chemicals dissolved in the pore fluid have reacted with the limestone to form chlorite and serpentine. Closest to the magma lies a zone of high-grade garnet and pyroxene.

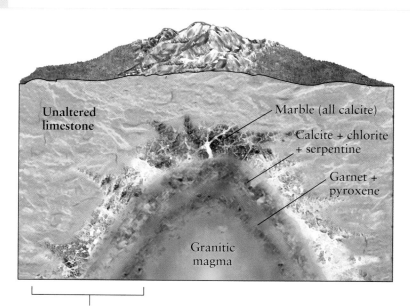

Unaltered limestone

Marble (all calcite)

Calcite + chlorite + serpentine

Garnet + pyroxene

Granitic magma

Aureole of metamorphic rock

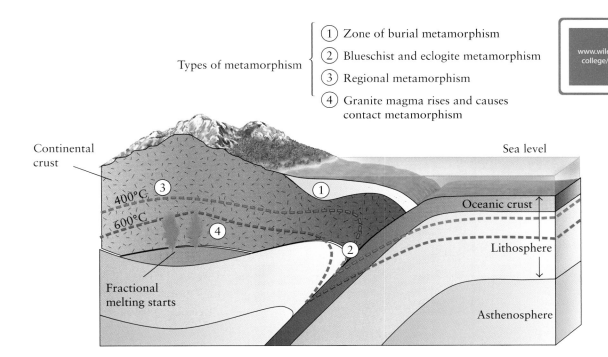

Types of metamorphism

① Zone of burial metamorphism

② Blueschist and eclogite metamorphism

③ Regional metamorphism

④ Granite magma rises and causes contact metamorphism

www.wiley.com/college/murck

Tectonics and metamorphism FIGURE 10.13

The theory of plate tectonics provides a unified view of burial (1), regional (3), and contact metamorphism (4). Each style of metamorphism occurs in its own distinct region of the crust, and the location of these regions is largely a result of plate tectonics. Blueschist and eclogite metamorphism (2), which involves high pressures but relatively cool temperatures, occurs in subduction zones. The dashed lines are *isotherms* (lines denoting equal temperature).

Regional metamorphism

The most common kinds of metamorphic rocks of the continental crust quartzite—slate, phyllite, schist, and gneiss—are found in areas extending over tens of thousands of square kilometers. They are formed by a process called **regional metamorphism**, which occurs only at convergent plate margins. Regionally metamorphosed rocks are found in mountain ranges and in the eroded remnants of former mountain ranges (see FIGURE 10.13). They are formed as a result of subduction or by collisions between masses of continental crust. Rocks at convergent margins are subjected to intense differential stresses; the foliation that is characteristic of regionally metamorphosed rocks is a consequence of such stresses, although, as discussed above, the full development of foliation depends on the composition of the rock.

Regionally metamorphosed rocks can also be found in places where no plate collisions are occurring today. We described these regions (called *orogens*) in Chapter 9. These are regions where plates have collided in the past, and their fragments have been assembled into the lithospheric plates we see today. Orogens provide some of our best evidence that plate tectonics and regional metamorphism have been active on Earth for billions of years.

A special kind of regional metamorphism occurs at subduction margins (Figure 10.13). In region 2 of Figure 10.13, you can see the isotherms bend down steeply, indicating that these rocks are much cooler than the surrounding crust. This phenomenon occurs because the solid, cold oceanic crust dives rapidly (by geologic standards) into the hot, weak asthenosphere.

regional metamorphism

Metamorphism of an extensive area of the crust, associated with plate convergence, collision, and subduction.

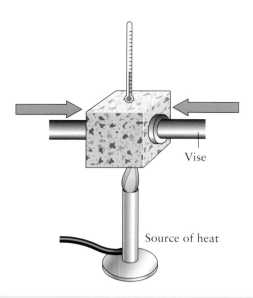

When a block of rock is squeezed, the entire block "feels" the stress immediately. When the rock is heated on one side, however, it takes a long time for the other side to get hot. A thermometer on the side away from the heat source will still register a low temperature after a fairly extended period of heating.

The pressure on the subducting rock rises quickly, but the temperature of the rock cannot rise at the same rate. This conclusion follows from a basic property that can be observed in a laboratory: Rock transmits pressure immediately but conducts heat slowly (see FIGURE 10.14). Therefore, in a subduction zone, metamorphism occurs at very high pressures but low to moderate temperatures. This produces a distinctive series of rocks, including blueschist and (at higher grades) eclogite.

Other types of metamorphism
Occasionally metamorphism occurs in other geologic settings where high temperatures and/or elevated pressure or stress exist, even if only briefly or sporadically. These include fault zones, meteorite impact craters, and even the sites of large fires. As geologists have become more aware of the role of meteorite impacts in our planet's geologic history, the unique kinds of metamorphism produced by these rare events have become an important diagnostic tool for detecting highly eroded ancient impact sites (see FIGURE 10.15).

A

Instantaneous metamorphism FIGURE 10.15

◀ A Nördlingen Cathedral in Germany sits in the center of an ancient impact crater. The cathedral was constructed of local rocks that were metamorphosed by the heat and high pressure associated with the meteorite impact.

▼ B This is a microscopic view of shocked quartz of the type found in Nördlingen Cathedral building stone—the lines in the quartz record a kind of very high-pressure deformation that is found only at impact craters and atomic bomb blast sites. The stones of the cathedral also contain tiny diamonds, which ordinarily form deep underground under high pressure but were formed here instantaneously by the blast wave from the exploding meteorite.

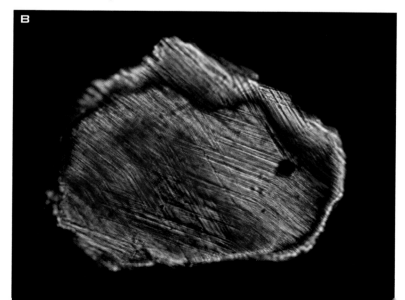

B

METASOMATISM

Metamorphic processes may dramatically change the appearance of a rock, and may cause its chemical constituents to move and crystallize into new mineral assemblages. However, they have little effect on the *overall* chemical composition of the rock. This is one reason why geologists can tell the identity and composition of metamorphic parent rocks.

There is an important exception to this general rule. Pore fluids, chiefly water and carbon dioxide, can be squeezed into or out of a body of rock during metamorphism. Most metamorphic environments have a small fluid-to-rock ratio, which means that the amount of fluid present compared to the amount of rock, is less than about 1:10. That is enough fluid to facilitate metamorphism but not enough to dissolve much of the rock and thus change its composition noticeably.

In a few circumstances, though, large fluid-to-rock ratios of 10:1 or even 100:1 can occur. One example is a large, open rock fracture through which a lot of fluid flows. The rocks adjoining the fracture can be drastically altered by the addition of new material, the removal of material into the fluid solution, or both. The term **metasomatism**, which comes from the Greek words for "change" and "body," is applied in such cases. Note that *metamorphism* refers to a change of form (*morph*, from the Greek word for "shape" or "form"), often accompanied by a redistribution of chemical constituents. *Metasomatism*, on the other hand, refers to a change in the chemical composition of the rock as a whole.

> **metasomatism**
> The process whereby the chemical composition of a rock is altered by the addition or removal of material by solution in fluids.

FIGURE 10.16 is a photograph of a contact metamorphic rock that was originally a limestone. Without the addition of new material, the limestone would have become a marble, dominated by calcite. Through metasomatism, however, it gained the constituents needed to crystallize garnet and a pyroxene called diopside, in addition to calcite. Metasomatic fluids may also carry valuable minerals in solution, which can lead to the formation of ore deposits, a topic discussed in Chapter 15.

Metasomatism FIGURE 10.16

Metasomatism of a limestone produced this colorful rock from the Gaspe Peninsula in Quebec. The white is calcite, the red is garnet, and the dark brown is pyroxene. The sample is 2.5 x 1.5 centimeters.

CONCEPT CHECK **STOP**

How can an underground magma body cause metamorphism in the surrounding rocks?

What distinguishes burial metamorphism from regional metamorphism?

How does regional metamorphism in a subduction zone differ from regional metamorphism in a collision zone?

What rare kind of very high-pressure metamorphism happens extremely rapidly?

What process changes the chemical composition of a rock, rather than just its texture or mineral assemblage?

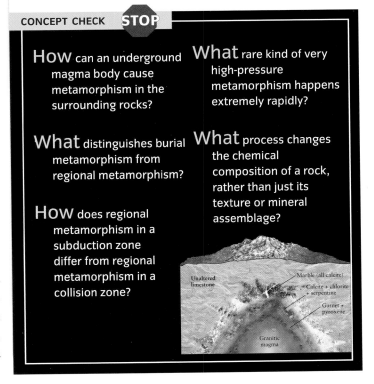

Metamorphic Facies

LEARNING OBJECTIVES

Define metamorphic zones and metamorphic facies.

Explain how metamorphic facies correlate with the temperature and pressure the rock is exposed to.

Much of the early research on metamorphism and metamorphic rocks was done in the Scottish Highlands in the late 1800s. Geologists discovered that mineral assemblages varied a great deal from place to place but that the overall chemical composition was essentially the same as that of shale. The rocks thus differed in terms of the grade of metamorphism they had been subjected to and the mineral assemblages they had developed but not in the composition of their source rock. The geologists identified characteristic *index minerals* that marked the appearance of each new mineral assemblage in a progression from low-grade to higher-grade metamorphic rocks. They drew lines, called *isograds* (lines of equal grade), connecting the map locations where a given index mineral first appeared (see FIGURE 10.17). Rocks lying between one isograd and the next would thus have similar mineral assemblages, indicating that they formed under similar metamorphic conditions. This method of mapping, first used in Scotland a century ago, has been successfully tested on different types of rocks around the world.

The metamorphic rocks in Scotland all came from source rock of similar composition (shale), that was exposed to a range of metamorphic conditions. It is also possible to group together metamorphic rocks of *differing* compositions that have all been exposed to the *same* temperature and stress conditions; such a grouping is called a **metamorphic facies**. Each of

> **metamorphic facies** The set of metamorphic mineral assemblages that form in rocks of different compositions under similar temperature and stress conditions.

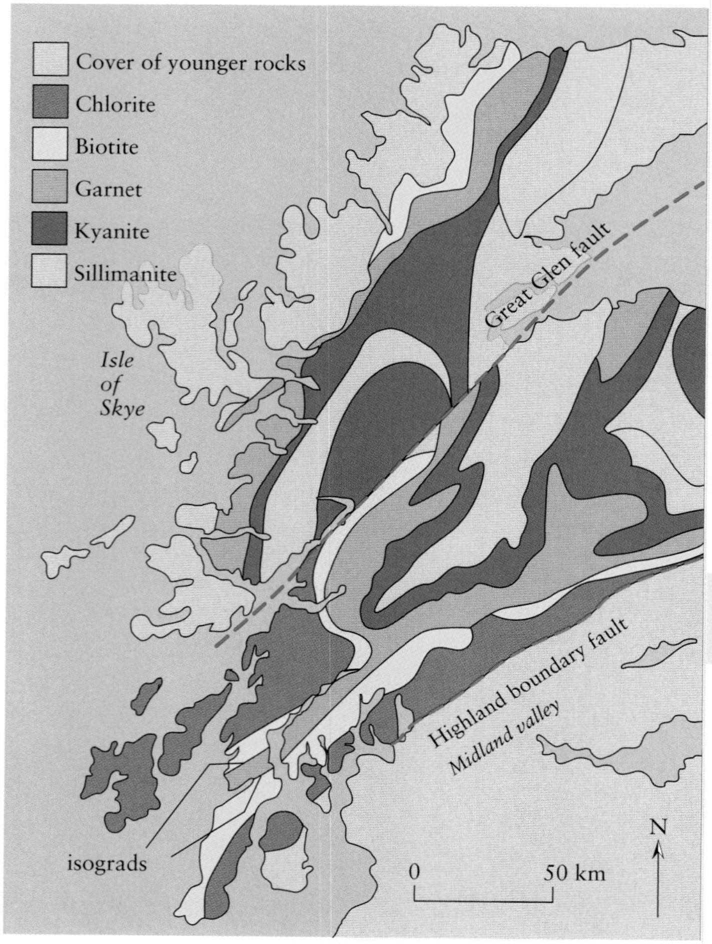

Regional metamorphism in Scotland
FIGURE 10.17

This map shows isograds in a body of regionally metamorphosed rocks in Scotland. The isograds mark the first appearance of various index minerals. Rocks between the isograds are said to be in a particular *metamorphic zone*. For example, a rock lying between the biotite and garnet isograds would contain the index mineral biotite, but would not yet have reached the metamorphic conditions in which garnet would appear; it would be said to be in the "biotite zone."

Map legend:
- Cover of younger rocks
- Chlorite
- Biotite
- Garnet
- Kyanite
- Sillimanite

Great Glen fault
Isle of Skye
Highland boundary fault
Midland valley
isograds
N
0   50 km

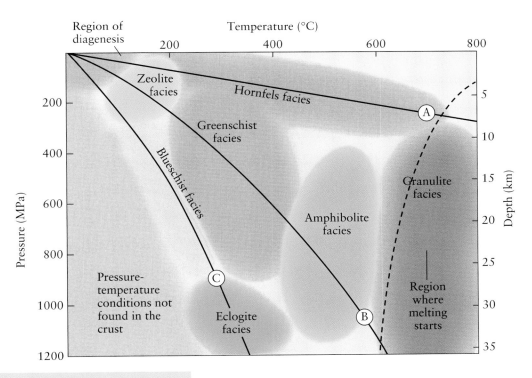

Metamorphic facies FIGURE 10.18

This graph shows the regions of pressure and temperature that characterize the different metamorphic facies. The zeolite facies is characteristic of burial metamorphism, and the hornfels facies is characteristic of contact metamorphism. Blueschist and eclogite facies are typical for subduction zones. Greenschist, amphibolite, and granulite facies occur in the re- gional metamorphism of continental crust thickened by plate collisions. The lines indicate how pressure and temperature change under conditions of (A) contact metamorphism, (B) regional metamorphism in continental plate collisions, and (C) regional metamorphism in subduction zones.

the metamorphic facies has been named, as you can see in FIGURE 10.18. The names typically reflect some obvious feature, such as the color of a common rock (blueschist and greenschist); the presence of a distinctive mineral (zeolite and amphibole); an un- usual type of rock (eclogite); or a distinctive texture (granulite). Each facies is representative of a specific set of metamorphic conditions. For example, the blueschist facies represents relatively high-pressure, low-temperature metamorphism, while the hornfels facies represents relatively high-temperature, low- pressure metamorphism (i.e., contact metamorphism).

The pressure-temperature diagram in Figure 10.18 shows the characteristic pressure and tempera- ture ranges for each metamorphic facies and how they relate to specific tectonic settings. For any given composition of source rock, it is possible to prepare a pressure–temperature grid like the one in Figure 10.18 that shows the specific mineral assemblages character- istic of each metamorphic facies.

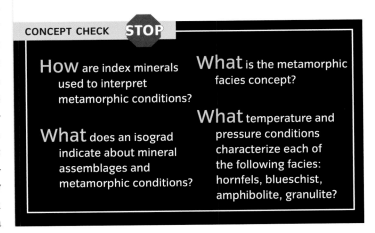

CONCEPT CHECK STOP

How are index minerals used to interpret metamorphic conditions?

What does an isograd indicate about mineral assemblages and metamorphic conditions?

What is the metamorphic facies concept?

What temperature and pressure conditions characterize each of the following facies: hornfels, blueschist, amphibolite, granulite?

GUATEMALA

Global Locator

Jade, in either of its two common forms, jadeite or nephrite, has long been treasured in cultures around the world. Not only is it visually beautiful, but it is exceedingly tough, which makes it suitable for cutting tools and ideal for intricate carving, because fine details will not break.

The world's largest jadeite deposits are in Burma and Central America. The Burmese deposit has been known for centuries. Until recently, however, the Central American deposits posed a mystery. From roughly 1200 to 400 BC, the first known civilization in Central America, the Olmec, lived in the region that is today occupied by eastern Mexico and Guatemala. They are known, among other things, for beautiful jade carvings such as the axe shown here (see FIGURE A). However, geologists and

A

All of this changed in 1998, however, when Hurricane Mitch eroded the slopes on the *other* side of the Motagua River and exposed more extensive jadeite deposits. In FIGURE B, a geologist stands on an outcrop of the jadeite-bearing rock. Subsequent explorations have revealed that the jade-bearing area is six times larger than geologists previously knew. The newly discovered jadeite is a very close match to the Olmec jade, with the same translucent blue-green color and white inclusions (see FIGURE C). Mystery solved!

B

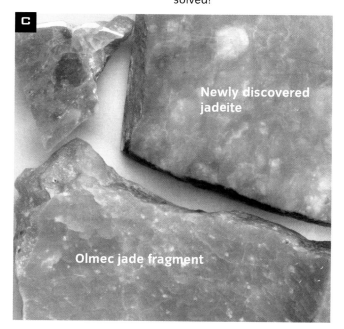

C

Newly discovered jadeite

Olmec jade fragment

archaeologists did not know where the Olmec jade had come from. They did know about a band of jadeite deposits along the northern bank of the Motagua River in Guatemala, where local miners called *jaderos* made a good living extracting jadeite. However, this jadeite is inferior to the Olmec jade, which is translucent and has a distinctive blue-green color; this small band of deposits could not have been the source for the Olmec jade.

NATIONAL GEOGRAPHIC

CHAPTER SUMMARY

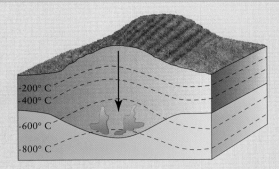

1 What Is Metamorphism?

1. New rock textures and new mineral assemblages develop when rocks are subjected to elevated temperatures and stresses. **Metamorphism** is a term that describes all such processes that occur at higher temperatures and pressures than *diagenesis* (Chapter 8) but without melting the rock.

2. The transition from diagenesis to metamorphism is gradual, but the dividing line between the two is customarily taken to be a temperature of around 150°C, which typically occurs at a depth of about 5 kilometers. The temperature and pressure are both high enough at that depth to initiate the formation of new minerals. Between 5 kilometers and 15 kilometers beneath the surface, rocks are subjected to **low-grade** metamorphism. The region of **high-grade** metamorphism lies from 15 kilometers to the depth at which melting occurs, often between 30 and 40 kilometers. The kind of metamorphism that occurs depends on both temperature and pressure, as well as composition.

3. *Pore* fluids enhance metamorphism by permitting material to dissolve, move around, and be precipitated somewhere else in the rock. They also speed up some chemical reactions. Pore fluids are driven out of rocks as metamorphism progresses. *Veins* in metamorphic rocks mark the passageways through which pore fluids once flowed. At the high-temperature limit of metamorphism, pore fluids can reduce the melting point of rock to the point where small pockets of magma form. The resulting rock, with small pockets of igneous rock surrounded by metamorphic rock, is called a *migmatite*.

4. Differential stress during metamorphism produces a distinctive texture known as **foliation**, marked by parallel cleavage planes and plates formed by crystals that are all aligned in the same direction. It is particularly noticeable in rocks containing minerals of the *mica* family. Foliation developed in low-grade metamorphic rocks is termed **slaty cleavage**; in coarser-grained, higher-grade metamorphic rocks it is called **schistosity**. The orientation of the foliation is perpendicular to the direction of maximum stress. Under conditions of low-grade metamorphism, the maximum stress is usually downward, and the slaty cleavage is horizontal and parallel to bedding. Under conditions of compression, produced by plate tectonics, the maximum stress is horizontal and the slaty cleavage is vertical.

5. High temperatures, high pressures, abundant pore fluid, and long reaction times tend to produce metamorphic rocks with large mineral grains. Metamorphic rocks containing small mineral grains usually form at lower temperatures and pressures but can also form if pore fluid is scarce or reaction time is short.

2 Metamorphic Rocks

1. The names of metamorphic rocks are based partly on texture and partly on composition. Foliated metamorphic rocks derived from shales are **slate**, **phyllite**, **schist**, and **gneiss**, in order of increasing metamorphic grade. The first two terms refer to textures of fine-grained rocks. In schist and gneiss, where the mineral grains are large enough to see, it is customary to identify the rock further by its mineral composition, as in "biotite schist" or "quart-biotite-amphibole gneiss." Gneiss is characterized by a coarsely foliated, or *gneissose,* texture, with alternating bands of mica-rich and mica-poor material, giving the rock a banded appearance.

2. Foliated rocks can also derive from igneous rocks, the most common of which is basalt. One series of rocks derived from basalts are *greenschists* (a low-grade metamorphic rock) and *amphibolites* (a high-grade metamorphic rock). Under conditions of high pressure but moderate temperature, which occur in subduction zones, a different series of metamorphic rocks forms from basalts. These are *blueschists* (low-grade) and *eclogites* (high-grade).

3. Two common kinds of metamorphic rocks have no foliation. These are **marble**, a metamorphic product of limestone, and **quartzite**, a metamorphic product of sandstone. These tend to be *monomineralic,* or nearly so: The main mineral present in marble is calcite, and the main mineral present in quartzite is quartz.

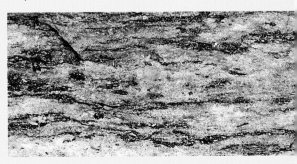

3 Metamorphic Processes

1. The processes that are primarily responsible for changes in texture and mineral assemblages in metamorphic rocks are *mechanical deformation* and *chemical recrystallization*. Typically both sets of processes are involved in metamorphism, but their relative importance depends on the specific pressure and temperature conditions of metamorphism.

2. Contact metamorphism is caused by the intrusion of hot magma or fluids (primarily water) into cooler rocks. Because contact metamorphism is caused by high temperatures, the main process involved is chemical recrystallization, and there is little mechanical deformation of the rock.

3. Sedimentary rocks that are subjected to sufficiently high temperatures and pressures due solely to the weight of overlying strata undergo **burial metamorphism**. Burial metamorphism is a low-grade metamorphic process. If the rocks are foliated, the foliation is usually parallel to the bedding planes.

4. Regional metamorphism occurs under the conditions of differential stress and elevated temperatures that result from collisions between tectonic plates. A special kind of regional metamorphism occurs at subduction zones, where pressures are high but temperatures are moderate, owing to the very rapid burial of the rock.

5. Metasomatism occurs when a large volume of fluid flows into or out of a rock. It is a process that changes the overall chemical composition of the rock, by moving chemical components into or out of the rock rather than just altering the texture or mineral assemblage.

4 Metamorphic Facies

1. For a given rock composition, the assemblages of minerals that are formed under a given set of temperature and stress conditions are the same regardless of where in the world the metamorphism happens.

2. Rocks of different chemical compositions that are metamorphosed under the same temperature and stress conditions are said to belong to the same **metamorphic facies**, even though they will develop different mineral assemblages.

3. Each metamorphic facies is associated with a particular tectonic setting. Blueschist facies and eclogite facies conditions occur in the high-pressure environment characteristic of subduction zones. Greenschist, amphibolite, and granulite facies conditions occur along convergent margins where continental masses collide. The zeolite facies is characteristic of burial metamorphism, and the hornfels facies is characteristic of contact metamorphism.

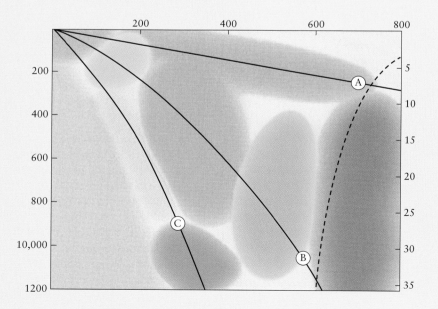

KEY TERMS

CRITICAL AND CREATIVE THINKING QUESTIONS

1. Briefly describe how pressure and temperature might change over time in rocks being subjected to contact metamorphism, burial metamorphism, and subduction-related regional metamorphism.

2. Suppose a large meteorite struck a pile of clastic sedimentary rocks. How would pressure and temperature change in the rocks involved? Would the effects be the same for rocks on the Moon as for rocks in a sedimentary basin on Earth?

3. Examining the texture of a rock is an important "rule-of-thumb" for geologists, because the texture can reveal whether the rock has been metamorphosed, and under what conditions. Explain why. Would this rule work for both foliated and nonfoliated rocks? Why (or why not)?

4. Compare the concept of metamorphic facies to that of sedimentary facies (see Chapter 8). In what ways are they similar? In what ways are they different?

5. Let's say that you have bought a property in a hilly area, and you are going to build a new house on the property. You like the idea of building on a hillslope with a view, but you are concerned about safety issues. One of the hills on the property is composed of gneiss, another is slate, and a third is a marble in which there are caves. Which one of these hillslopes do you think would provide the most stable location for your home, and why? Try sketching the hillslope as a way of helping you think about the different rock types and how they might be distributed.

What is happening in this picture ?

This roof in Ireland was made from a commonly occurring planar metamorphic rock. What kind of rock do you think it is? What properties of this rock would make it very suitable for use as a roof tiling material? Do you think this rock is used in modern roofing? Why, or why not?

1. How does metamorphism differ from diagenesis and lithification?

   a. Diagenesis and lithification occur at higher temperatures and pressures.
   b. Diagenesis and lithification occur at lower temperatures and pressures.
   c. Metamorphism requires melting of pre-existing rock.
   d. Diagenesis and lithification do not cause any significant changes in the rock or sediment.

2. Metamorphism will occur under relatively _____ conditions.

   a. high temperature and high pressure
   b. high temperature and low pressure
   c. low temperature and high pressure
   d. All of the above are correct.

3. For this illustration, locate and label the areas on the temperature and pressure diagram that correspond to regions of Earth's crust listed below.

   region of diagenesis   region of high grade metamorphism

   region where melting starts   region of low grade metamorphism

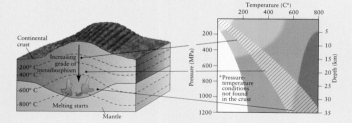

4. Pore fluids enhance metamorphism by _____.

   a. permitting material to dissolve
   b. permitting chemical components to move around and be precipitated somewhere else
   c. speeding up some chemical reactions
   d. All of the above is correct.

5. _____ is the planar alignment of minerals due to differential stress during metamorphism.

   a. Ionization
   b. Metasomatism
   c. Foliation
   d. Diagenesis

6. Foliated metamorphic rocks derived from shales are _____ , in order of increasing metamorphic grade.

   a. phyllite, slate, and schist or gneiss
   b. phyllite, gneiss, and schist or slate
   c. slate, phyllite, and schist or gneiss
   d. slate, gneiss, and phyllite or schist

7. Basalt can be metamorphosed to form a series of foliated rocks ranging from greenschists (low-grade) to amphibolites (high-grade). It is also possible for a different series of metamorphic rocks to form, ranging from blueschists (low-grade) to eclogites (high-grade). Under what conditions does this latter series form?

   a. low pressure and moderate temperature
   b. low pressure but high temperature
   c. high pressure but moderate temperature
   d. high pressure and high temperature

8. The two rock samples shown in these photographs are nonfoliated metamorphic rocks. Sample A is composed of calcite and Sample B was formed through the metamorphism of sandstone. Which of the following correctly identifies the two samples?

   a. Sample A is quartzite and Sample B is schist.
   b. Sample A is quartzite and Sample B is marble.
   c. Sample A is marble and Sample B is quartzite.
   d. Sample A is marble and Sample B is schist.

9. Metamorphism involves two main groups of processes, _____ and _____ .

   a. fractional melting; chemical recrystallization
   b. fractional melting; fractional crystallization
   c. fractional melting; mechanical deformation
   d. mechanical deformation; chemical recrystallization

10. _____ occurs over an extensive area of the crust and is associated with plate convergence, collision, and subduction.

   a. Regional metamorphism
   b. Burial metamorphism
   c. Contact metamorphism
   d. Diagenesis

11. _____ occurs when rocks are heated and chemically changed adjacent to an intruded body of hot magma.

   a. Regional metamorphism
   b. Burial metamorphism
   c. Contact metamorphism
   d. Diagenesis

12. On this illustration, label the areas designated 1 through 4 from the following choices:

- regional metamorphism
- blueschist- and eclogite-facies metamorphism
- burial metamorphism
- zone where fractional melting starts, granite magma rises, and contact metamorphism occurs

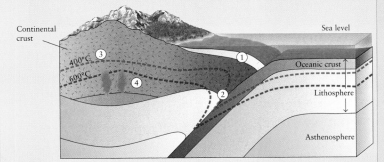

13. Metasomatism is a process by which the chemical composition of rock is altered. What is the mechanism of this chemical alteration?

   a. addition or removal of material by solution in fluids
   b. loss of chemical material during fractional melting
   c. loss of chemical material during fractional crystallization
   d. None of the above is correct.

14. Label this illustration with the proper metamorphic facies.

zeolite facies	granulite facies
blueschist facies	greenschist facies
amphibolite facies	hornfels facies
eclogite facies	

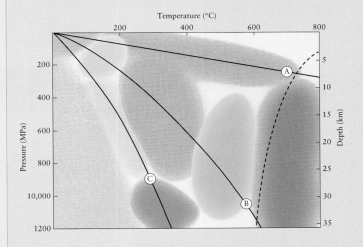

15. In the illustration in Question 14, the three lines labeled A, B, and C represent temperature and pressure pathways for different types of metamorphism. Which of these three pathways describes the suite of metamorphic facies that you would expect to see in an area of regional metamorphism due to continental collision?

   A.
   B.
   C.

Water on and under the Ground

Often our prosperity depends on managing water: tapping it with wells, storing it behind dams, holding it behind levees, moving it from one place to another. However, our efforts to control it sometimes exacerbate the problem. This paradox became painfully evident when Hurricane Katrina struck Louisiana on August 29, 2005.

The subdivision shown here *(inset)*, in St. Bernard Parish, was constructed on sand pumped from the Mississippi River and was protected from flooding by a levee. But the levee was not sufficient to withstand Katrina's storm surge. Ten days after the storm, when the large photo was taken, the development was still under almost two meters of water.

In New Orleans, three separate levee breaches flooded large parts of the city that lay below sea level, but the levee failures were only part of the story. New Orleans wasn't always below sea level. The French founded the city in 1718 on natural levees formed by deposition of coarse sediment as flood waters spilled out of the channelway. Over the years, people raised the levees still higher to protect the town from flooding; however, this prevented the river from depositing new sediment to build up its flood plain. In addition, pumping of water from the flood plain caused the soil to become compacted. Over time, this combination of factors raised the river level and lowered the land surface under the city, with drastic consequences.

In this chapter, we will explore the ways in which fresh water on the surface and underground affect and are affected by both geology and human activity.

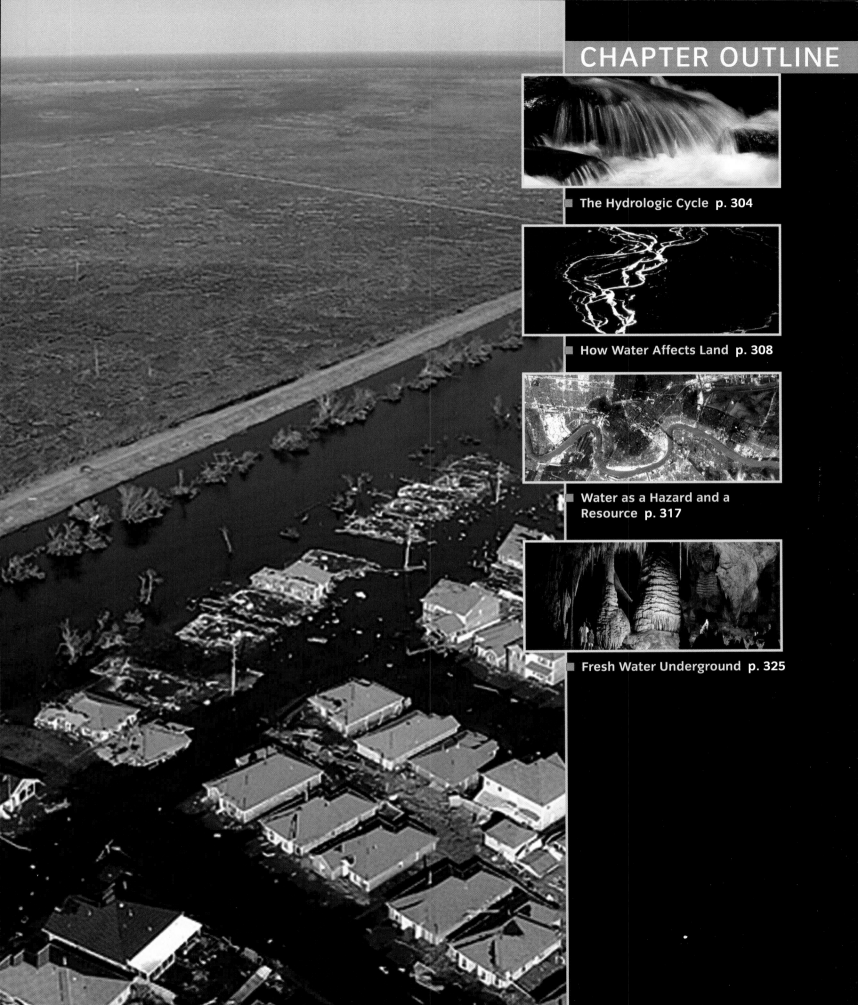

The Hydrologic Cycle

There are four great reservoirs that make up the Earth system (see Chapter 1): the lithosphere, hydrosphere, atmosphere, and biosphere. Water moves—and helps move materials—among all four spheres. Water vapor is an important part of the atmosphere. Water is a constituent of many common minerals (such as micas and clays) in the lithosphere, where it is tightly bonded in their crystal structures. And, of course, water is a fundamental component of living things in the biosphere. The **hydrologic cycle**, also called the **water cycle**, describes how water moves among these four reservoirs (see **FIGURE 11.1**), and the scientific study of water is called **hydrology**.

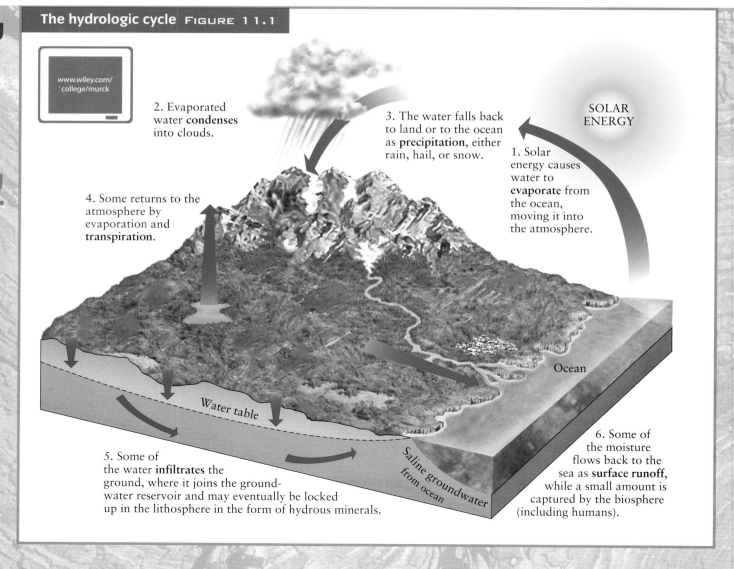

The hydrologic cycle FIGURE 11.1

www.wiley.com/college/murck

SOLAR ENERGY

1. Solar energy causes water to **evaporate** from the ocean, moving it into the atmosphere.

2. Evaporated water **condenses** into clouds.

3. The water falls back to land or to the ocean as **precipitation**, either rain, hail, or snow.

4. Some returns to the atmosphere by evaporation and **transpiration**.

5. Some of the water **infiltrates** the ground, where it joins the groundwater reservoir and may eventually be locked up in the lithosphere in the form of hydrous minerals.

6. Some of the moisture flows back to the sea as **surface runoff**, while a small amount is captured by the biosphere (including humans).

Water table

Saline groundwater from ocean

Ocean

The hydrologic cycle as part of the Earth system FIGURE 11.2

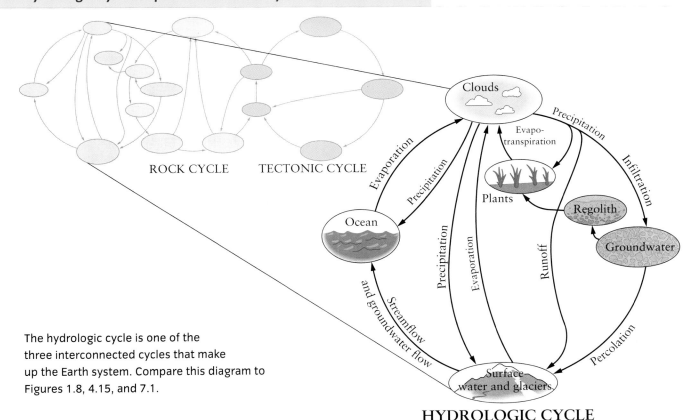

The hydrologic cycle is one of the three interconnected cycles that make up the Earth system. Compare this diagram to Figures 1.8, 4.15, and 7.1.

HYDROLOGIC CYCLE

■ **evaporation** The process by which water changes from a liquid into a vapor.

■ **transpiration** The process by which water taken up by plants passes directly into the atmosphere.

■ **condensation** The process by which water changes from a vapor into a liquid or a solid.

Water moves through the hydrologic cycle along numerous pathways and processes. These include **evaporation** and **transpiration**, both of which are powered by energy from the Sun. Depending on local conditions of temperature, pressure, and humidity, some of the water vapor in the atmosphere will undergo **condensation**, changing to liquid or solid form and falling back to the land or ocean as rain, snow, or hail via the process of **precipitation**. Some of this precipitation becomes **surface runoff**, whereas some trickles directly into the ground via **infiltration**.

WATER IN THE EARTH SYSTEM

The schematic representation of the hydrologic cycle (see FIGURE 11.2) will look familiar to you if you have read Chapters 1, 4, and 7, which introduced the tectonic cycle and the rock cycle. The water cycle is linked to both of these other cycles. Like them, it is a *closed cycle* of *open systems*. Because it is a closed cycle, the total amount of water is fixed. However, all of the local reservoirs within the cycle, such as rivers and trees, are free to gain or lose water. They sometimes do so quite dramatically, as during a flood or drought.

■ **precipitation** The process by which water that has condensed in the atmosphere falls back to the surface as rain, snow, or hail.

■ **surface runoff** Precipitation that drains over the land or in stream channels.

■ **infiltration** The process by which water works its way into the ground through small openings in the soil.

The world's water resources (in proportion):
97.5 liters saltwater (A)

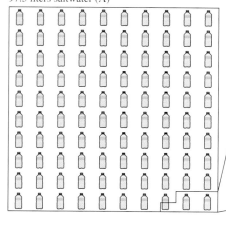

1.85 liters frozen water (B)

0.64 liters groundwater (C)

10 milliliters surface water (D)*
(= 2 teaspoons)

*includes water in biosphere and atmosphere

The vast majority of Earth's water is either salty (**A**), frozen (**B**), or underground (**C**). The most visible everyday sources of fresh water, such as rivers, lakes, and the atmosphere (**D**), together comprise less than a hundredth of a percent of Earth's water budget.

Living on the water's edge FIGURE 11.4

Many of the world's great cities are built on riverbanks or coastlines. St. Louis has long benefited from its proximity to the Mississippi River. In the foreground, a towboat pushes barges upstream; in the background, the Mississippi Queen riverboat brings tourists to the Gateway Arch.

Unlike the other two cycles, the water cycle is easily observable to us. We can measure the amount of global precipitation; using satellite monitoring, we can even measure the amount of evaporation. With these measurements, along with the overall mass balance of the water cycle, geologists can roughly deduce how much water is exchanged along each of the pathways shown in Figure 11.2.

It is also instructive to compare the sizes of each of the reservoirs in the water cycle (see FIGURE 11.3). The largest reservoir—by far—is the world's oceans, which hold 97.5 percent of Earth's water. Thus, the vast majority of Earth's water is saline (salty), not fresh. This has important consequences for humans, because we depend on fresh water as a resource for drinking, agriculture, and industrial use.

The other numbers in Figure 11.3 may surprise you. Most of Earth's fresh water (almost 74%) is locked up in polar ice sheets, where it is almost inaccessible to humans. The vast majority of the unfrozen fresh water (98.5%) lies underground. Historically, human settlement has concentrated along lakes or rivers, where the other 1.5 percent of the world's fresh water—namely, surface water—is readily available (see FIGURE 11.4). But in most rural areas and quite a few urban areas as well, underground water sources are much more plentiful.

There is a correlation between the size of a reservoir and the *residence time*, the average length of time spent by a water molecule in that reservoir. Residence times in the large-volume reservoirs, such as the oceans and ice sheets, are several thousands of years. The time spent by water in the groundwater system may amount to hundreds of years or more. In small-volume reservoirs, the residence time of water is much shorter—weeks in streams and rivers, days in the atmosphere, and hours in some living organisms.

Although water is continuously cycling from one reservoir to another, the total volume of water in each reservoir is approximately constant over short time intervals. However, the volume of water in each reservoir can change dramatically over longer intervals. During glacial ages, for example, vast quantities of water evaporate from the ocean and are precipitated on land as snow. The snow slowly accumulates to build ice sheets that are thousands of meters thick and cover vast areas. At such times, the amount of water removed from the ocean is so large that the global sea level can fall by many meters, and the expanded glaciers increase the ice-covered area of Earth.

CONCEPT CHECK STOP

What are the major reservoirs in the hydrologic cycle?

How does water move among these reservoirs?

How does the residence time of water correlate to the size of a reservoir?

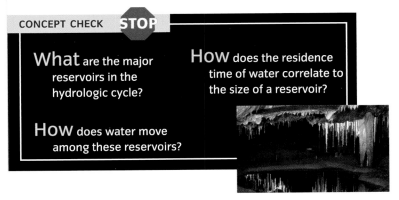

How Water Affects Land

I f you stand outside during a heavy rain, you can see that, initially, water tends to move downhill in a process called *overland flow* (or *sheet flow*, because the flowing water often takes the form of a thin, broad sheet). After traveling a short distance, overland flow begins to be concentrated into well-defined channels, thereby becoming *streamflow*. Overland flow and streamflow together constitute surface *runoff*, one of the pathways in the water cycle. Let's look more closely at streams, streamflow, and their interactions with the land.

STREAMS AND STREAMFLOW

stream A body of water that flows downslope along a clearly defined natural passageway.

channel The clearly defined natural passageway through which a stream flows.

gradient The steepness of a stream channel.

Every **stream** or **river** has a **channel**. Several factors affect the shape of a channel and the types of landforms it creates. The most important are the channel's **gradient**, **discharge**, and **load**.

Gradient, discharge, and load are interrelated. For example, if the gradient of a stream becomes steeper along a particular stretch of channel, the velocity of the flow is likely to increase as well. If the velocity is high, a greater load can be carried. If the discharge increases, it means that the channel must handle more water in a given period; as a result, both the velocity of flow and the depth of the water in the stream will increase. (Note that this is the first of two different meanings of "discharge" used in this chapter.) In some cases, the channel itself will increase in width and depth as the water scours the banks and the bottom. This scouring, in turn, adds sediment to the load. When the velocity of the water eventually decreases, the sediment settles out, filling in the channel and allowing it to return to its original size.

discharge (1) The amount of water passing by a point on the channel's bank during a unit of time.

load The suspended and dissolved sediment carried by a stream.

Types of channels

Streams create landforms through two processes: *erosion* (see Chapter 7) and *deposition* (see Chapter 8). Both processes go on throughout a stream's existence and along its length, but one or the other may predominate at a particular location or during a particular time, depending on a variety of factors.

The gradient, discharge, and load are all important determinants of a channel's size and shape. This means that topography (which determines gradient), climate (which determines the amount of precipitation), and the general character of the landscape through which the stream flows are all important factors, too. The geologic characteristics of the underlying rocks are also important. For example, a stream may suddenly bend or its gradient may increase when it passes from erosion-resistant rock into rock that is easily eroded. Because these factors interact in different ways, no two channels are exactly alike. Nevertheless, we can classify them into three broad categories: *straight*, *meandering*, and *braided* (see **FIGURE 11.5**).

Unlike engineered aqueducts and canals, natural streams are never really straight from start to finish. Straight channels may occur over short distances, particularly in *upstream* areas (that is, near the *headwaters* or source of the stream), where the gradient is high and the

A Straight channels, such as this stream that drains a glacier in Alaska, usually occur only in relatively short stretches. They are often found where streams have a high gradient (near the stream's headwaters), and they generally have a classic V-shaped valley shaped by mass wasting and overland flow.

Oxbow lake

Main channel

Cut bank

Point bars

B Rio Itaquai, a tributary of the Amazon River in Brazil, shows several features typical of a meandering river: a low gradient, light-colored sandy bars on the inside edge of the bends, and nearby oxbow lakes marking previous watercourses that have been cut off.

C This stream flows north across the Arctic National Wildlife Refuge in Alaska. It has the classic profile of a braided stream, with a low gradient and a large and variable load of sediment—produced in this case by the action of glaciers.

In the background of this photograph, taken several months after the Mount St. Helens eruption, you can see the braided form of the Toutle River. In the foreground, engineers have built an artificial levee to straighten the river's course and control erosion.

Braided flow created by high bed load upstream

Muddy water due to high rate of erosion upstream

Straight channel flow

NATIONAL GEOGRAPHIC

channel deeply incised. Even in a "straight" channel, close examination will show that the deepest part of the channel oscillates from side to side.

Meandering streams tend to develop where the stream gradient is low, typically in the lower or *downstream* parts of a stream system (close to the *mouth*, where the stream empties into another surface water body.) The erosion in a meandering stream concentrates along the sides of the channel rather than the bottom. As water sweeps around a bend, it flows more rapidly along the outer bank, undercutting and steepening it to form a *cut bank*. Meanwhile, along the inner side of each meander, where the water is shallow and velocity is low, sediment accumulates to form a *point bar*, as shown in Figure 11.5B. Thus, meanders slowly change shape and shift their position along a valley as the stream erodes material from one bank and deposits sediment on the other. Sometimes the water finds a shorter route downstream, bypassing a meander by cutting across the narrow part of the loop. As sediment is deposited along the banks of the new channel route, the former meander is cut off and converted into a curved *oxbow lake*.

Braided channels arise when a stream's ability to move its sediment load varies over time. At times of high flow, a stream can carry more sediment. If the discharge decreases but the load does not, the stream deposits the excess sediment in its own channel as bars or islands. These variations in flow over time cause the channel to repeatedly divide and reunite, as shown in Figure 11.5C. Braided patterns tend to form in streams with a highly variable discharge and large load of coarse sediment. Braided patterns can form in streams with discharge that varies seasonally (e.g., following rapid snowmelt) and in streams with easily eroded banks.

As an extreme example of dramatic changes in a river channel, Washington's Toutle River changed almost overnight from a meandering stream to a braided stream in 1980, when Mount St. Helens erupted (see **FIGURE 11.6**). The debris avalanche that resulted from the eruption deposited huge amounts of loose and easily eroded pyroclasts and volcanic ash into the Toutle Valley, dramatically increasing the river's load. In addition, there were no longer any living trees to hold the old riverbanks in place. Changes in an ancient river's flow patterns (as revealed in sedimentary strata) can tell geologists about the geologic history of a region.

STREAM DEPOSITS

Stream deposits form along channel margins, valley floors, mountain fronts, and at the stream's mouth where it opens into the ocean or a lake. These are all places where the stream loses energy, and therefore its ability to carry a load. A point bar is one example of a stream deposit. Others include *floodplains*, *alluvial fans*, and *deltas*.

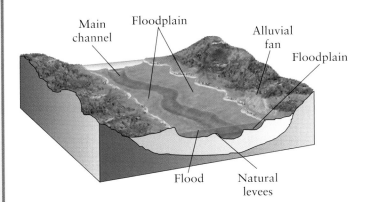

A Some of the landforms created by sediment in stream valleys are floodplains, natural levees, and alluvial fans.

B Where this stream emerges from the mountains into Death Valley, California, it abruptly slows and deposits its sediment load. This has created a symmetrical alluvial fan, covered by a braided system of channels. The stream was dry at the time the photograph was taken.

C Where the Nile River empties into the Mediterranean Sea, the sediment it deposits has formed a fan-shaped delta that supports the green vegetation seen in this satellite image. Its triangular shape, similar to the Greek letter delta (Δ), gave rise to the term "delta" that we use today.

floodplain The relatively flat valley floor adjacent to a stream channel, which is inundated when the stream overflows its banks.

When a stream rises during a flood, water overflows the banks and inundates the **floodplain** (see FIGURE 11.7A). As sediment-laden water flows out of the channel, its depth, velocity, and turbulence decrease abruptly at the margins of the channel. This results in sudden, rapid deposition of the coarser part of the load along the margins, which builds up a broad, low ridge of **alluvium** atop each bank, called a *natural levee*. Farther away, the finer particles settle out in the quiet water covering the valley. This creates the broad, flat, fertile land that is typical of floodplains.

alluvium Unconsolidated sediment deposited in a recent geologic time by a stream.

How Water Affects Land   311

Another kind of alluvial structure develops where a stream draining a steep upland region suddenly emerges onto the floor of a much broader lowland valley. The stream will slow down and lose some of its ability to carry sediment. It will deposit the coarser part of its load (the part that can no longer be transported) in an *alluvial fan* (see FIGURE 11.7B). Such fans are typical of semi-arid conditions where vegetation is sparse and infrequent rainfall creates streams that are heavily laden with sediment.

A similar situation occurs when a stream flows into a standing body of water, such as an ocean or lake. It quickly loses velocity and fans out, dropping its sediment load (heaviest particles first and the finer particles farther seaward). Over time, the sediment builds up a deposit called a **delta** (so named because of its shape, which is commonly triangular and resembles the Greek letter delta "Δ"). Most of the world's great rivers, including the Nile, Ganges-Brahmaputra, Huang He, Amazon, and Mississippi, have built massive deltas (see FIGURE 11.7C).

Prominent deltas do not form in places where strong wave, current, or tidal action redistributes sediment as quickly as it reaches the coast. However, if the rate of sediment supply exceeds the rate of coastal erosion, a delta will form. The converse holds, too: If the rate of deposition slows down, the delta will disappear (see FIGURE 11.8).

Disappearing coastline FIGURE 11.8

The taming of the Mississippi River with levees and channels has slowed the deposition of new sediment in its delta. As a result, the entire Mississippi Delta has been shrinking. Marshes give way to open water, ponds turn into lakes, and barrier islands shrink. The top image shows the extent of the delta in 1839, the middle image shows its extent in 1993, and the bottom image shows a projection of its extent in 2090. The degradation of the delta is at least partly responsible for the vulnerability of New Orleans to hurricane damage.

DECONSTRUCTING A COAST

1839

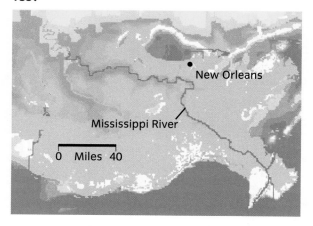

1993 ☐ Land loss on coast

2090 *projection*

The Mississippi River drainage basin FIGURE 11.9

The drainage basin of the Mississippi River encompasses most of the midwestern United States and extends into southern Canada. In this diagram the widths of the rivers are exaggerated to represent the discharge in cubic meters per second.

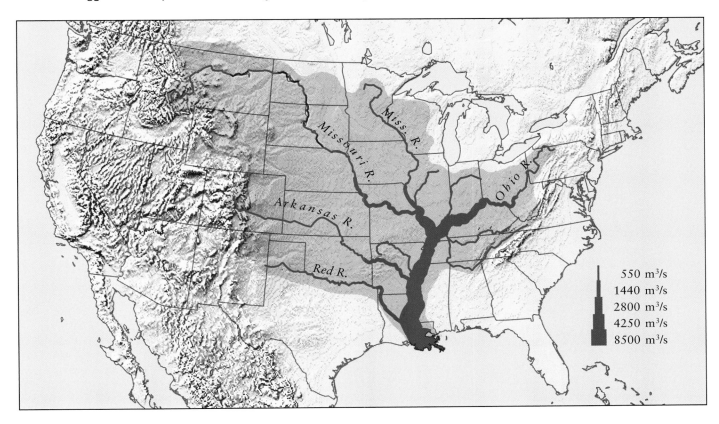

550 m³/s
1440 m³/s
2800 m³/s
4250 m³/s
8500 m³/s

LARGE-SCALE TOPOGRAPHY OF STREAM SYSTEMS

Streams are governed by a simple principle: Water flows downhill. Rainwater that falls on the land surface will move from higher to lower elevations under the influence of gravity. A stream's headwater region is the area of relatively higher elevation from which streams have their source. Small, high-gradient *tributary streams* carry water downslope from the headwater region, combining their flow to form a larger stream. The gradient gradually decreases toward the low-lying region of the stream's mouth.

Every stream is surrounded by its **drainage basin** (sometimes called a *catchment* or a *watershed*). Drainage basins range in size from less than a square kilometer to areas the size of subcontinents. In general, the greater a stream's annual discharge, the larger its drainage basin. The vast drainage basin of the Mississippi River encompasses more than 40 percent of the total area of the contiguous United States (see **FIGURE 11.9**). From an environmental perspective, a drainage basin is a more natural geographic entity than a country or a state, because

drainage basin The total area from which water flows into a stream.

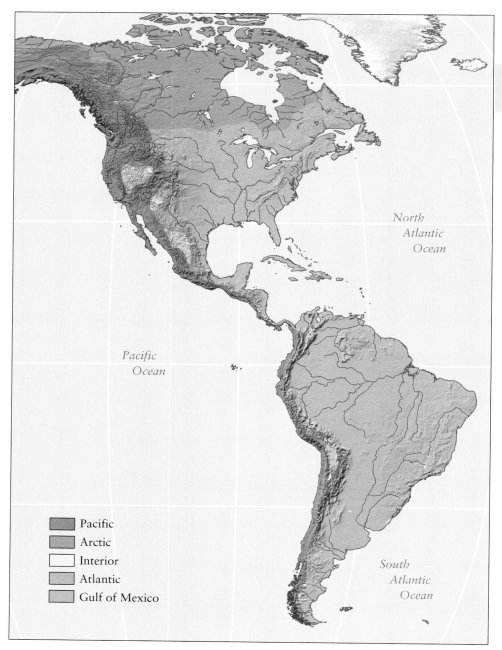

This map of North and South America shows the location of continental drainage basins and divides.

North Atlantic Ocean

Pacific Ocean

- ■ Pacific
- ■ Arctic
- □ Interior
- ■ Atlantic
- ■ Gulf of Mexico

South Atlantic Ocean

issues of water supply, pollution, or wildlife management that affect one part of a watershed are likely to affect it all. (See What a Geologist Sees.)

If you have driven across North America, you may have seen highway signs marking the *continental divide*. This **divide** separates streams that drain toward one side of the continent from streams that drain toward the other side. The continental divide of western North America lies along the length of the Rocky Mountains (see FIGURE 11.10). In

> ■ **divide** A topographic high that separates adjacent drainage basins.

general, any two adjacent watersheds are separated by a divide, even if they ultimately flow into the same ocean.

LAKES

Lakes are standing bodies of water that have open surfaces, in direct contact with the atmosphere. Water enters lakes from streams, overland flow, and groundwater, and exits either by evaporation or by flowing through an outlet. All freshwater lakes have outlets; however, some saline lakes lack an outlet and therefore lose water only through evaporation, which inevitably

Drainage Basins

A The drainage basins of many rivers are hard to see from a satellite because they are covered by vegetation. However, the basin of this river, Wadi Al Masilah in South Yemen, adjacent to the Rubh-al-Khali, a desert, is very easy to spot. We have outlined the boundary—that is, the divide—that surrounds and defines the drainage basin of one tributary here; note that this basin would itself be a part of the drainage basin of the larger river at the top of the photograph.

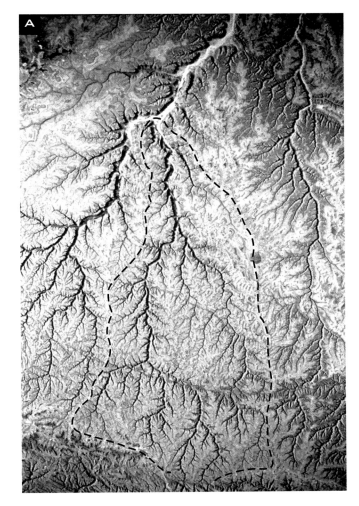

B Watershed boundaries are more important than political boundaries for management of the environment and water resources. This can create transboundary water management challenges. Shown below is a world map in which the "countries" are the world's 106 largest drainage basins. (Blank regions are drained by smaller rivers or are too cold or dry for regular water flow.) Some of the river names are shown.

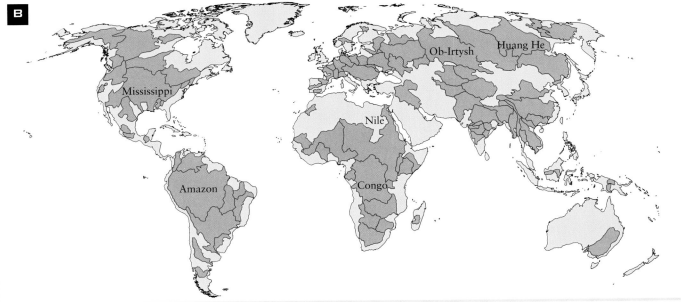

A dying lake FIGURE 11.11

This lake in Florida has turned green and mucky-looking because of the growth of algae stimulated by excessive nutrients (probably from sewage or fertilizer). Eventually, the algae will use up all the oxygen in the water, making it impossible for other life to survive in the lake. This is called *eutrophication*.

leads to a buildup of salt. The Great Salt Lake in Utah is an important example of an inland, saline lake.

Lakes are important to us as sources of fresh water and food, as well as hydroelectric power. The majority of lakes in the United States have been altered, and in some cases, even created, by humans. Sometimes the changes are deliberate, as when a *reservoir* lake is created by a dam, or a wetland is drained to make it accessible for human development. In other cases the human effects are inadvertent. For example, runoff of sewage or fertilizer into a lake can cause *eutrophication*, which can kill most of the life in the lake (see **FIGURE 11.11**). Eutrophication can occur as a natural part of the process of swamp formation.

Lakes can form as a result of several different geologic processes. Crustal faulting creates many large, deep lakes. Lava flows often form a dam in a river valley, causing water to back up as a lake. Landslides suddenly create lakes by blocking valleys. Throughout formerly glaciated regions of North America and Europe, plains of glacial sand and gravel contain natural pits and hollows left by the melting of stagnant ice masses that were buried in the sand and gravel deposits. Kettle lakes, a common landform in New England, form when these pits fill with water.

An important characteristic of lakes is that they are short-lived features on the geologic time scale. They disappear by one of two processes, or a combination of both. First, lakes that have stream outlets will be gradually drained as the outlets are eroded to lower levels. Second, lakes accumulate inorganic sediment carried by streams entering the lake, and organic matter produced by plants within the lake. Eventually, they fill up, forming a boggy *wetland* with little or no free water surface.

In arid climates, many lakebeds are either dry or only intermittently filled with shallow water. Streams bring dissolved salts to these *ephemeral* lakes. Since evaporation removes only pure water, the salts remain behind and salinity levels increase. Eventually the salts may be precipitated as solid evaporites, some of which have economic value.

CONCEPT CHECK **STOP**

What do geologists mean by the term "stream"?

What factors influence stream behavior?

How do the three main types of stream channels develop?

What are the major types of stream deposits?

Why are lakes geologically short-lived?

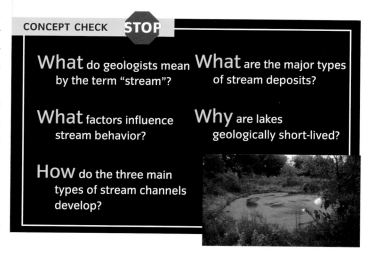

Water as a Hazard and a Resource

LEARNING OBJECTIVES

Describe how floods occur and what factors may make them worse.

Define recurrence interval and show how it is used to predict floods.

Explain why flood prevention efforts sometimes have the opposite effect.

Explain why interbasin transfer of water can be a flawed solution to water scarcity.

T hough water is a vital resource, it can also be a dangerous force. Uneven distribution of rainfall through the year causes some water bodies to dry up, but others rise and overflow their banks, creating hazardous circumstances for people who live in the area.

FLOODS

Under natural circumstances, all water bodies undergo changes in the volume of water they hold or transport. From time to time, when the discharge or water level becomes too much to handle, the water body will **flood** (see FIGURE 11.12).

> ■ **flood** An event in which a water body overflows its banks.

Mississippi River flood FIGURE 11.12

A pair of satellite images shows the region where the Missouri River joins the Mississippi River at St. Louis, Missouri. **A** This photo shows a dry summer with low flows (July 1988). **B** This photo shows the same region in July 1993. Weeks of rain hundreds of kilometers away caused the rivers to overflow their levees. Numerous towns, along with 44,000 square kilometers of farmland in nine states, were flooded by an estimated 3 cubic kilometers of floodwater.

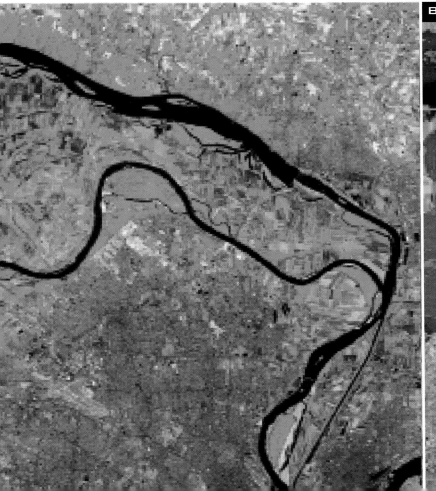

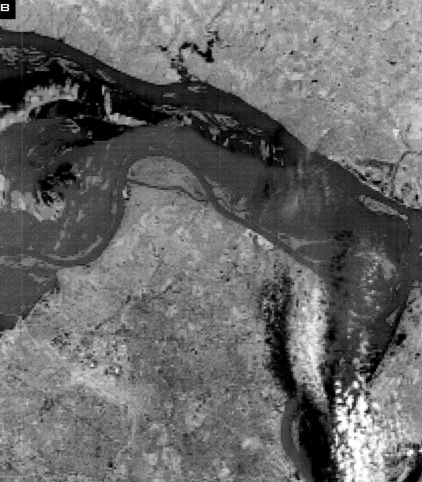

During a stream flood, the "extra" water flowing in the channel, contributed by excess precipitation, is called *storm runoff*. The peak discharge of a flood usually comes well after the rains that produced it. Geologists record the development of the flood with a *hydrograph*, which shows the stream's discharge as a function of time. FIGURE 11.13 shows an example in which a passing storm generated a brief interval of intense rainfall. As the runoff moved into the stream channel, the discharge quickly rose. The *crest* of the resulting flood—the time when the peak flow passed the *hydrologic station* where the measurements were made—occurred about two hours after the storm. It took another eight hours for the flood runoff to pass through the channel and for the discharge to return to its normal level.

A hydrograph of stream discharge FIGURE 11.13

This diagram illustrates the hydrograph of a stream after a brief, intense storm.

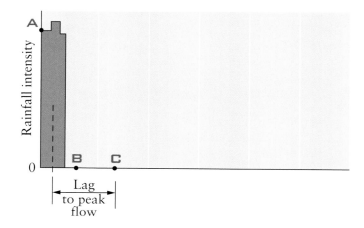

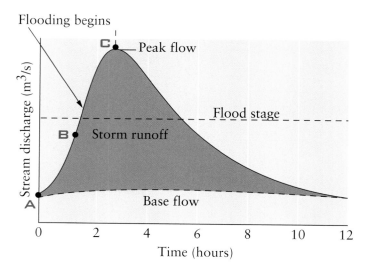

*Base flow is the "normal" flow of water in a stream, contributed by groundwater.

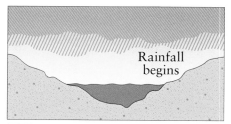

A Onset of storm (0 hours): The peak discharge is delayed as the runoff collects and runs down the stream channel.

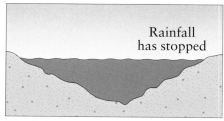

B 1 hour: One hour after the cloudburst, the stream can still contain the increased volume.

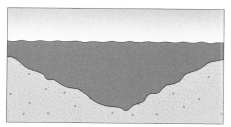

C 2 hours: After two hours, the stream reaches its peak flow and cannot be contained by its banks any more.

17th Street Canal breached on east side

Lake Pontchartrain

City Park

Downtown New Orleans

Mississippi River

Coastal flood FIGURE 11.14

As Hurricane Katrina moved toward New Orleans from the Gulf of Mexico in August of 2005, high waters breached the city's protective levees. In this satellite view, a week after the hurricane, dark areas are flooded, and Lake Pontchartrain, from which much of the water came, is at the top. The 17th Street Canal, where one of the levee breaks occurred, is clearly visible as the western edge of the flooded region.

Lakes also flood, as do oceanic coastal zones. In the case of coastal flooding, it may be the inflow of water from the ocean, rather than the runoff of water from the land, that does most of the damage. The *storm surge* associated with Hurricane Katrina in 2005 temporarily raised the water level near New Orleans by 6 meters or more and breached the levees even before the main force of the storm hit the city (see FIGURE 11.14).

Though both coastal flooding and stream flooding often take people by surprise, geologists view them as normal and inevitable events. The geologic record shows that floods have been occurring throughout Earth's history, for as long as there has been a hydrosphere. Even though flooding is a natural geologic process, it can quickly become a human catastrophe when it affects population centers. The Huang He in China, sometimes called the Yellow River because of its heavy load of yellowish-brown silt, has a long history of catastrophic floods. In 1887, the river inundated 130,000 square kilometers and swept away many villages in the heavily populated floodplain. In 1931, another Huang He flood killed a staggering 3.7 million people.

Yet these same floods help replenish the soil in the floodplain, which explains why people keep moving back to the area.

Human activity sometimes increases the chance that flooding will occur (instead of decreasing it). Urban development can exacerbate the problem of flooding in a variety of ways. Urban construction on compressible sediments, often accompanied by withdrawal of groundwater, can lead to **subsidence** (a drop in the ground surface), which increases the danger of flooding (as it did in New Orleans). The impermeable ground cover associated with urbanization can add substantially to surface runoff in urban areas. Storm sewers can contribute to flooding because they allow the runoff from paved areas to reach the river channel more quickly. Floods in urbanized basins often have higher peak discharges and reach their peaks more quickly than floods in undeveloped basins. This quicker and higher crest means that people living in a flood-prone area must be able to move very quickly, or be able to predict when a flood might strike.

Flood prediction and prevention

Because floods can be so damaging, predicting and preparing for them is essential (see FIGURE 11.15). To do this, the frequency of past floods of different sizes is plotted on a graph, producing a *flood-frequency curve*. The average time interval between two floods of the same magnitude is called the *recurrence interval*. For example, a "10-year flood" has a recurrence interval of 10 years, which means that there is a 1-in-10 (or 10%) chance that such a flood will occur in any given year. A flood with an even greater discharge having a recurrence interval of 50 years would be termed a "50-year flood" for this particular stream, and have a 1-in-50 (2%) chance of occurring in any given year. Regional planners should (and do) keep these intervals in mind when planning development on or near a flood plain.

Predicting floods FIGURE 11.15

The graph below shows the frequency of floods of different sizes on the Skykomish River at Gold Bar, Washington. A flood with a discharge of 1750 cubic meters per second has a recurrence interval of 10 years, and hence a 1-in-10 chance of occurring in any given year.

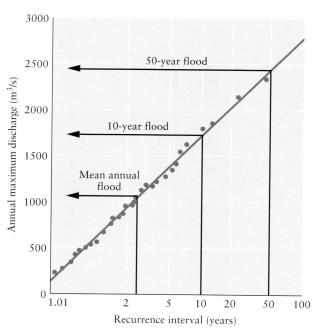

The landscape shows the normal, nonflood discharge for a hypothetical stream in darkest blue. Lighter shades of blue correspond to 2-year, 10-year, 50-year, and 100-year floods for this stream.

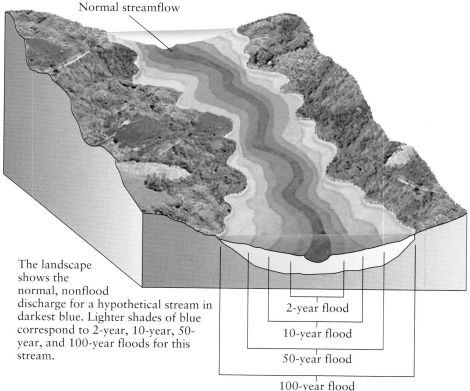

Another aspect of flood prediction is the real-time monitoring of storms and water levels. Hydrologists can combine information about the weather with their knowledge of a river basin's geology and topography to forecast the peak height of a flood and the time when the crest will pass a particular location. Such forecasts, which are often made with the aid of computer models and *Geographic Information Systems (GIS)*, can be very useful for planning evacuation or defensive measures.

Understandably, many people throughout history have been unsatisfied with simply predicting floods and have attempted to prevent them. River channels are often modified or "engineered" for the purpose of flood control and protection as well as to increase access to floodplain lands, facilitate transport, enhance drainage, and control erosion. The modifications usually consist of some combination of widening, deepening, straightening, clearing, or lining of the natural channel. All of these approaches are collectively called **channelization**.

Like dams, channelization projects can contribute to the economic well-being of a community, but at a price. Channel modifications interfere with natural habitats and ecosystems. The aesthetic value of the river can be degraded, and water pollution aggravated. Projects sometimes control flooding in the immediate area but contribute to more intense flooding downstream. Perhaps most importantly, any modification of a channel's course or cross-section renders invalid the hydrologic data collected there in the past. During the Mississippi River floods of 1973 and 1993, experts could not account for water levels that were higher than predicted by the historical data; the likely cause was extensive upstream modifications of the river channel by humans.

SURFACE WATER RESOURCES

A reliable water supply is critical—not only for human survival and health, but also for the role it plays in industry, agriculture, and other economic activities (see FIGURE 11.16 on pages 322–323). Twenty-six countries worldwide, with a total population of almost 250 million people, are today designated as *water-scarce*. The lack of water in these countries places serious constraints on agricultural production, economic development, health, and environmental protection.

Globally, crop irrigation accounts for about 73 percent of the demand for water, industry for about 21 percent, and domestic use for the remaining 6 percent, though the proportions vary from one region to another. Demand in each of these sectors has more than quadrupled since 1950. Population growth is partly responsible for the increasing demand, but improvements in standards of living around the world have also contributed to the large increase in water use per capita over the past few decades. The total amount of water being withdrawn (that is, diverted from rivers, lakes, and groundwater) for worldwide human use is now about eight times the annual streamflow of the Mississippi River.

Sometimes, because of population growth and development, regions with the greatest demand for water do not have an abundant and readily available supply of surface water. For this reason, surface water is often transferred from one drainage basin to another, sometimes over long distances. Besides raising political issues related to water rights, such *interbasin transfer* can have negative environmental impacts (see Case Study on page 324).

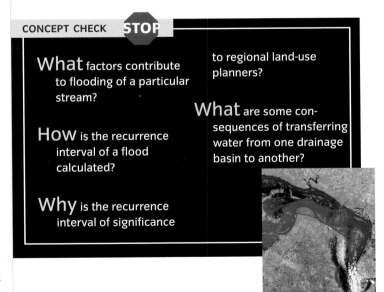

CONCEPT CHECK STOP

What factors contribute to flooding of a particular stream?

How is the recurrence interval of a flood calculated?

Why is the recurrence interval of significance

to regional land-use planners?

What are some consequences of transferring water from one drainage basin to another?

It's as vital to life as air. Yet fresh water is one of the rarest resources on Earth. Only 2.5% of Earth's water is fresh, and of that the usable portion for humans is less than 1% of all fresh water, or 0.01% of all water on Earth. Water is constantly recycling through Earth's hydrologic cycle. But population growth and pollution are combining to make less and less available per person per year, while global climate change adds new uncertainty. Efficiency, conservation, and technology can help ensure that the water you absorb today will still be usable and clean hundreds of years from now.

▶ **PRIMARY WATERSHEDS AND CRITICAL AREAS**
Watersheds are Earth's rain barrels. They collect precipitation and filter it as they channel it to streams, rivers, lakes, and aquifers. Many watersheds are stressed by human activities.

The map at right shows sensitive areas worldwide. The Amazon River, the world's largest watershed, is still relatively pristine, but Brazil plans on building dozens of dams, some to power aluminium smelters. In Africa and Asia, lack of access to water and water-related diseases are the main problems. In Europe and the Middle East, overuse, pollution, and disagreement over diverting water are the major challenges. Hope rests in better planning and community-scale projects.

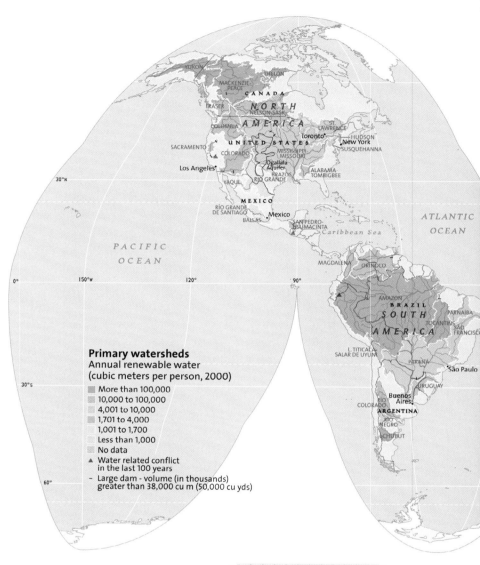

Primary watersheds
Annual renewable water
(cubic meters per person, 2000)

- More than 100,000
- 10,000 to 100,000
- 4,001 to 10,000
- 1,701 to 4,000
- 1,001 to 1,700
- Less than 1,000
- No data
- ▲ Water related conflict in the last 100 years
- – Large dam - volume (in thousands) greater than 38,000 cu m (50,000 cu yds)

Percent of total population using improved drinking water sources, 2000

- More than 90%
- 76% - 90%
- 51% - 75%
- 25% - 50%
- Less than 25%
- No data available

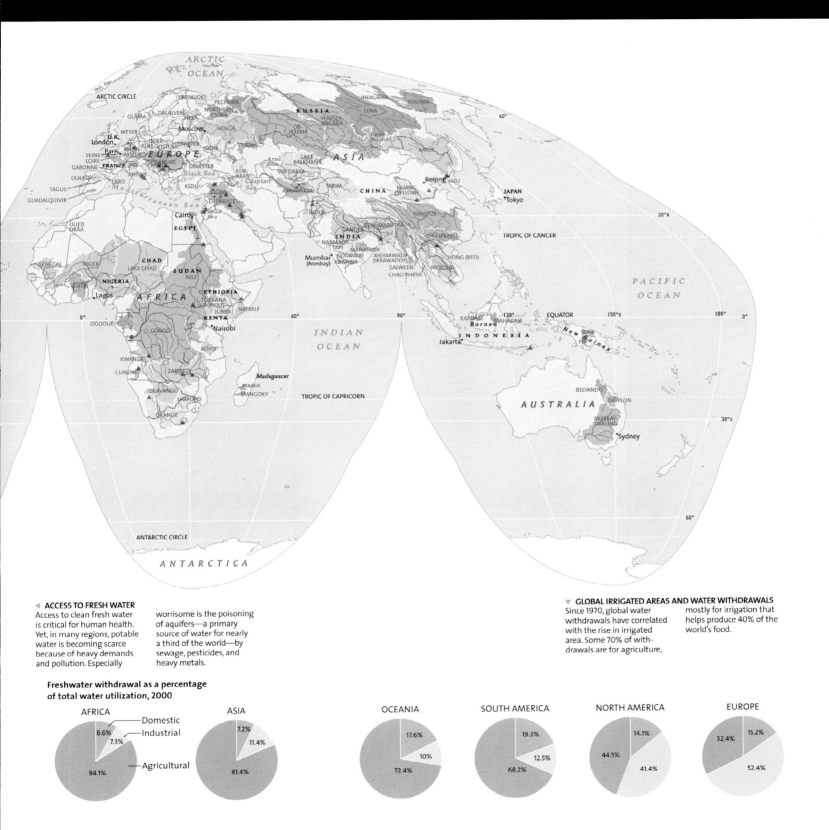

◀ **ACCESS TO FRESH WATER**
Access to clean fresh water is critical for human health. Yet, in many regions, potable water is becoming scarce because of heavy demands and pollution. Especially worrisome is the poisoning of aquifers—a primary source of water for nearly a third of the world—by sewage, pesticides, and heavy metals.

▼ **GLOBAL IRRIGATED AREAS AND WATER WITHDRAWALS**
Since 1970, global water withdrawals have correlated with the rise in irrigated area. Some 70% of withdrawals are for agriculture, mostly for irrigation that helps produce 40% of the world's food.

Freshwater withdrawal as a percentage of total water utilization, 2000

AFRICA
- Domestic 8.6%
- Industrial 7.3%
- Agricultural 84.1%

ASIA
- 7.2%
- 11.4%
- 81.4%

OCEANIA
- 17.6%
- 10%
- 72.4%

SOUTH AMERICA
- 19.3%
- 12.5%
- 68.2%

NORTH AMERICA
- 14.1%
- 44.5%
- 41.4%

EUROPE
- 15.2%
- 32.4%
- 52.4%

Mono Lake

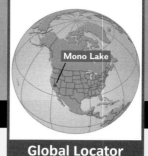

California's Mono Lake, in the Sierra Nevada Mountains, has been the site of a collision between human water needs and the needs of a unique habitat and its wildlife.

These calcium carbonate spires (see FIGURE A) were once underwater. But in 1941, the Los Angeles Aqueduct began diverting water from four of the six streams that empty into Mono Lake. With insufficient input to make up for its evaporation losses, the lake shrank to half its original volume and doubled in salinity. Migratory birds, such as the nation's second-largest colony of California gulls, were placed in jeopardy. Their food supply (brine shrimp) was dying because of the high salinity of the water, and the islands on which they nested were in danger of being connected to the mainland because of the retreating water, making the birds vulnerable to predators.

By the late 1980s the lake and its ecosystem were nearing collapse. In 1994, after a court battle between the City of Los Angeles and environmental groups, both sides agreed to a plan to raise the water level by 5 meters and compensate Los Angeles for the loss of part of its water supply.

Water levels in the lake have risen about halfway to the target set by the agreement, and the lake is expected to reach its target level in 15 to 20 years. Dams

that once diverted streams into the aqueduct (see FIGURE B) have been reengineered to do the opposite: They maintain a steady flow into Mono Lake and divert water into the aqueduct only if there is an overflow. For the California gulls (see FIGURE C), the change may have come just in time.

Fresh Water Underground

LEARNING OBJECTIVES

Define water table.

Explain how porosity and permeability of rocks affect the motion of groundwater.

Identify two types of aquifers.

Explain why some wells require pumping, while others flow unaided.

Describe how subsidence relates to groundwater.

Describe how a cave forms.

L ess than 1 percent of all water in the hydrosphere cycle is **groundwater**. Although this sounds small, the volume of groundwater is 40 times greater than that of all the water in freshwater lakes and streams. Water can be found everywhere beneath the land surface, even beneath parched deserts. About half of it is near the surface, no more than 750 meters below ground. At greater depths, the pressure exerted by overlying rocks reduces the pore space, making it difficult for water to flow freely. In addition, water that occurs at great depths tends to be *briny*, or rich in dissolved mineral salts, and not well suited for human use. Therefore, from a practical perspective, we can think of groundwater as the water found between the land surface and a depth of about 750 meters, even though an equally large amount of water is present at greater depths.

groundwater Subsurface water contained in pore spaces in regolith and bedrock.

water table The top surface of the saturated zone.

THE WATER TABLE

Much of what we know about groundwater has been learned from the accumulated experience of generations of people who have dug or drilled millions of wells. This experience tells us that a hole penetrating the ground ordinarily first encounters a zone in which the spaces between the grains in regolith or bedrock are filled mainly with air, although the material may be moist to the touch. This is the *zone of aeration*, also known as the *vadose zone* (see FIGURE 11.17).

After passing through the zone of aeration, the hole reaches the **water table** and enters the *saturated*

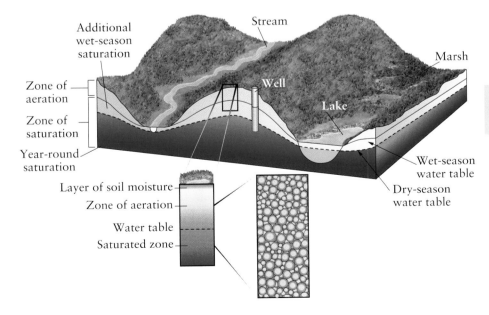

Additional wet-season saturation
Stream
Marsh
Zone of aeration
Well
Lake
Zone of saturation
Year-round saturation
Layer of soil moisture
Zone of aeration
Water table
Saturated zone
Wet-season water table
Dry-season water table

Water under the ground
FIGURE 11.17

A well first passes through the zone of aeration, where pores in the soil are filled with both air and water. Eventually, it reaches the water table, where the pore spaces are completely filled with water. Underground water exists everywhere, and surface water occurs wherever the ground intersects the water table.

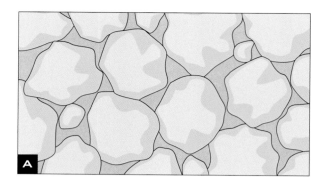

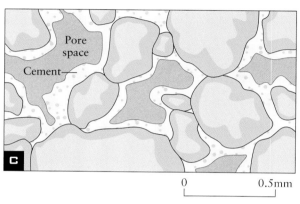

Pore
space

Cement

0   0.5mm

Porosity in sediments and rocks

FIGURE 11.18

In these examples, all of the pore spaces are filled with water, as they would be in the saturated zone. **A** The porosity is about 30% in this sediment with particles of uniform size. **B** This sediment, in which fine grains fill the space between larger grains, has a lower porosity, around 15%. **C** In sedimentary rock, the porosity may be reduced by cement that binds the grains together and fills the pores.

zone, also known as the *phreatic zone*, in which all openings are filled with water. The water table is high beneath hills and low beneath valleys. This may seem surprising, because the surface of a glass of water or a lake is always level. But water underground flows very slowly and is strongly influenced by surface topography. If all rainfall were to cease, the water table would slowly flatten. Seepage of water into the ground would diminish and then stop entirely, and streams would dry up as the water table fell. During droughts, the depression of the water table is evident from the drying up of springs, streambeds, and wells. Repeated rainfall, which soaks the ground with fresh supplies of water, maintains the water table at a normal level and keeps surface water bodies replenished.

Whether it is deep or shallow, the water table marks the upper limit of readily usable groundwater. For this reason, a major aim of groundwater specialists and well drillers is to determine the depth and shape of the water table. To do this they must first understand how groundwater moves and what forces control its distribution underground.

HOW GROUNDWATER MOVES

Most groundwater is in motion. Unlike the swift flow of rivers, however, which is measured in kilometers per hour, the movement of groundwater is so slow that it is measured in centimeters per day or meters per year. The reason is simple: Whereas the water of a stream flows through an open channel, groundwater must move through small, constricted passages. Therefore, the rate of groundwater flow is dependent on the nature of the rock or sediment through which the water moves, especially its porosity and permeability.

Porosity and permeability

Porosity determines the amount of fluid a sediment or rock can contain. The porosity of sediment is affected by the size and shape of the particles and the compactness of their arrangement (see FIGURE 11.18A and B). The porosity

porosity The percentage of the total volume of a body of rock or regolith that consists of open spaces (pores).

permeability
A measure of how easily a solid allows fluids to pass through it.

percolation The process by which groundwater seeps downward and flows under the influence of gravity.

of a sedimentary rock is also affected by the extent to which the pores have been filled with cement (see **FIGURE 11.18C**). Plutonic igneous rocks and metamorphic rocks, which consist of many closely interlocked crystals, generally have lower porosities than sediment and sedimentary rocks do. However, joints and fractures may increase their porosity.

A rock with low porosity is likely also to have low **permeability**. However, high porosity does not necessarily mean high permeability, because both the sizes and the continuity of the pores (that is, the extent to which the pores are interconnected) influence the ability of fluids to flow through the material.

Percolation After water from a rain shower soaks into the ground, or infiltrates, some of it evaporates, while some is taken up by plants. The remaining water continues to **percolate** under the influence of gravity until it reaches the water table. The "perc test" that must be carried out when a new septic system is being installed is a measure of percolation. The movement of

groundwater in the saturated zone is similar to the flow of water that occurs when you gently squeeze a water-soaked sponge. Water moves slowly through very small pores along threadlike paths. The water flows from areas where the water table is high toward areas where it is lower. In other words, it generally flows toward surface streams or lakes (see **FIGURE 11.19**). Some of the flow paths turn upward and enter the stream or lake from beneath, seemingly defying gravity. This upward flow occurs because groundwater is under greater pressure beneath a hill than beneath a stream or lake. Because water tends to flow toward points where pressure is low, it flows toward bodies of water at the surface.

Recharge and discharge
Recharge of groundwater occurs when rainfall and snowmelt infiltrate the ground and percolate downward to the saturated zone

recharge Replenishment of groundwater.

discharge (2) The process by which subsurface water leaves the saturated zone and becomes surface water.

(see Figure 11.19). The water then moves slowly along its flow path toward zones where **discharge** occurs. (Note that earlier in the chapter we used the term "discharge" for a somewhat different concept—the flow

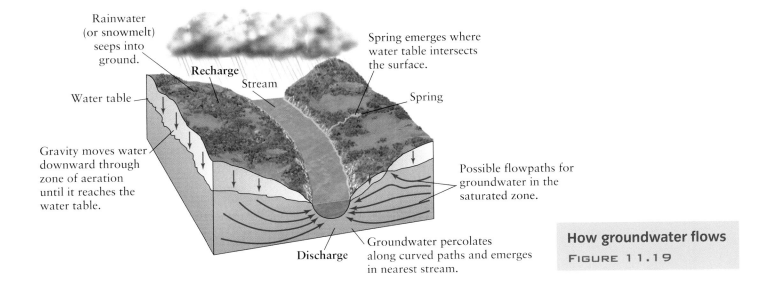

Rainwater (or snowmelt) seeps into ground.

Spring emerges where water table intersects the surface.

Recharge

Stream

Spring

Water table

Gravity moves water downward through zone of aeration until it reaches the water table.

Possible flowpaths for groundwater in the saturated zone.

Discharge

Groundwater percolates along curved paths and emerges in nearest stream.

How groundwater flows
FIGURE 11.19

spring Where the water table intersects the land surface, allowing groundwater to flow out.

of water along a stream channel.) In discharge zones, subsurface water either flows out onto the ground surface as a **spring** or joins bodies of water such as streams, lakes, ponds, swamps, or the ocean. The discharge of groundwater is what maintains the base flow of a stream. Pumping groundwater from a well also creates a point of discharge. The amount of time water takes to move through the ground to a discharge area depends on distance and rate of flow. It may take as little as a few days or as long as thousands of years.

WHERE GROUNDWATER IS STORED

When we wish to find a reliable supply of groundwater, we search for an **aquifer** (Latin for "*water carrier*"). An aquifer is not a body of water—it is a body of rock or regolith that is water-saturated, as well as porous and permeable. Gravel and sand generally make good

aquifers; many sandstones are also good aquifers, as are fractured or cavernous limestone and granite. An aquifer in which the water is free to rise to its natural level is called an *unconfined aquifer* (see FIGURE 11.20). In a well drilled into an unconfined aquifer, the water will rise to the level of the surrounding water table. To bring it to the surface one would need a pump or a bucket.

A *confined aquifer* is overlain by impermeable rock units, called *confining layers*, or **aquicludes** (see Figures 11.20 and FIGURE 11.21). Common aquicludes are shale and clay layers, as well as unfractured, nonporous, impermeable bedrock. The water in a confined aquifer is held in place by the overlying impermeable unit, and its recharge zone may be many kilometers away at a higher elevation. If a well is drilled into the aquifer, the high water pressure due to the elevation of the recharge zone will cause the

aquifer A body of rock or regolith that is water-saturated, porous, and permeable.

aquiclude A layer of impermeable rock.

Aquifiers, confined and unconfined FIGURE 11.20

An unconfined aquifer is open to the atmosphere through pores in the rock and soil above the aquifer. In contrast, the water in a confined aquifer is trapped between impermeable rock layers.

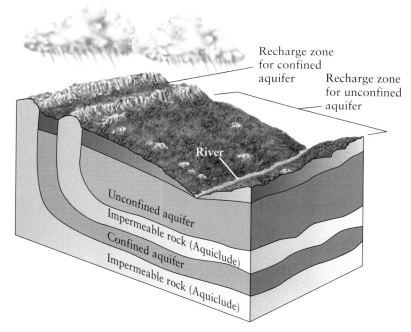

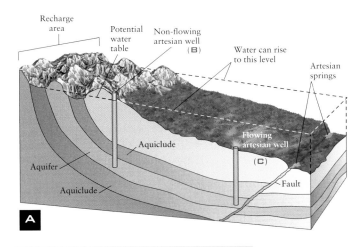

A

Artesian water FIGURE 11.21

A This diagram illustrates the geologic conditions that produce artesian wells or springs. Note that the water enters the aquifer at a higher elevation than the well. As a result, the water flows or gushes out of the well with positive water pressure.

B This gushing artesian well in Texas drew on the same aquifer that provides drinking water to San Antonio.

water to rise or even flow out of the well without having to be pumped. This is called an *artesian well*. A fault can also serve as a natural conduit for artesian water.

A change in permeability of the rocks at ground level often gives rise to a spring, which is a natural analogue of an artesian well. Such a change may be due to the presence of an aquiclude (see FIGURE 11.22) or it may happen along the trace of a fault (see Figure 11.21A).

What causes springs? FIGURE 11.22

A This spring in the Grand Canyon is fed by water from the porous Redwall and Muav Limestones. These cavernous limestones are the water source for many springs. The impermeable shale unit beneath them, the aquiclude, is the Bright Angel Shale. In **B** Water flows from a spring in a limestone aquifer underlain by an impermeable shale aquiclude.

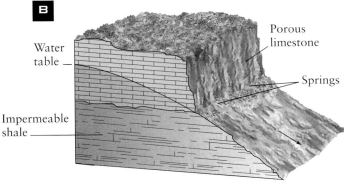

GROUNDWATER DEPLETION AND CONTAMINATION

A well will supply water if it is deep enough to intersect the water table. As shown in FIGURE 11.23, a shallow well may become dry during periods when the water table is low, whereas a deeper well may yield water throughout the year. When water is pumped from a well, a *cone of depression* (a cone-shaped dip in the water table) will form around the well. In most small domestic wells, the cone of depression is hardly discernible. Wells pumped for irrigation and industrial uses, however, sometimes withdraw so much water that the cone may become very wide and steep and can lower the water levels in surrounding wells. When large cones of depression from pumped wells overlap, the result is regional depression of the water table.

If the rate of withdrawal of groundwater regularly exceeds the rate of natural recharge, the volume of stored water steadily decreases; this is called *groundwater mining*. It may take hundreds or even thousands of years for a depleted aquifer to be replenished. The results of excessive withdrawal include lowering of the water table; drying up of springs and streams; *compaction* of the aquifer; and *subsidence*, a decline in land surface elevation. Sometimes it is possible to recharge an aquifer by pumping water into it. In other cases the effects of depletion may be permanent. When an aquifer suffers compaction—that is, when its mineral grains collapse on one another because the pore water that held them apart has been removed—it is permanently damaged and may never be able to hold as much water as it originally held.

Wells: year-round and seasonal FIGURE 11.23

Seasonal changes affect the height of the water table. A well will only produce year-round if it extends into the year-round zone of saturation.

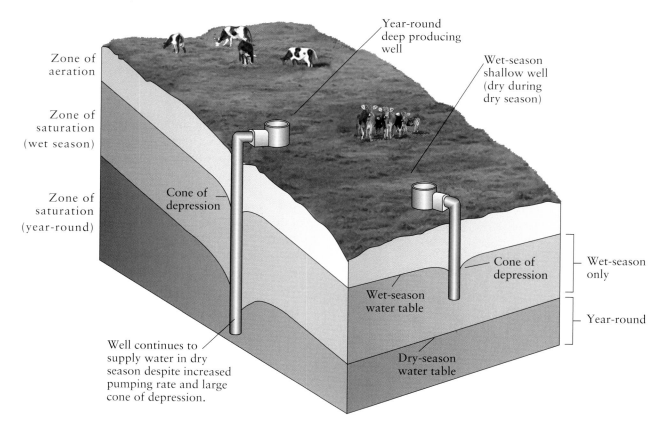

Subsidence FIGURE 11.24

In Galveston, Texas, these homes had to be abandoned because pumping of groundwater lowered the level of the land, allowing seawater to flow in from the Gulf of Mexico. The snapshot in the foreground shows the same neighborhood, high and dry, in 1968.

Urban development can increase the rate of depletion of groundwater, not only by increasing the demand for water but also by increasing the amount of impermeable ground cover in the area. When roads, parking lots, buildings, and sidewalks cover a recharge area, the rate of infiltration and groundwater recharge can be substantially reduced. Subsidence caused by withdrawal of groundwater (see FIGURE 11.24) is a problem in many urban areas, such as Mexico City, Bangkok, and Venice. The weight of buildings also contributes to the compaction of compressible sediments.

Laws and policies relating to water rights are very complicated, and the application to groundwater is even more complicated than for surface water. Because groundwater is hidden from view, it is difficult to monitor its flow and regulate its use. If you drill a well into an aquifer underlying your property, are you entitled to withdraw as much water as you need from that well? Should you withdraw water only for your own purposes, or should you be permitted to withdraw the water and sell it elsewhere? What happens if withdrawing the groundwater depletes the aquifer and your neigh-

bor's well runs dry? Similar problems arise when a landowner's actions cause an aquifer to become contaminated. In some jurisdictions it is technically legal to contaminate groundwater; it only becomes illegal when the contaminated groundwater flows into a neighboring property. To make matters worse, because of the slow movement of groundwater, those responsible for the problem may be long gone before their actions begin to affect their neighbors.

Many of the types and sources of contaminants that affect surface water also cause groundwater contamination. Because of its hidden nature, however, groundwater contamination can be much more difficult to detect, control, and clean up than surface water contamination. The most common source of water pollution in wells and springs is untreated sewage. Agricultural pesticides and fertilizers, significant sources of surface water pollution, are also common contaminants of groundwater. Harmful chemicals leaking from waste disposal facilities can also infiltrate into groundwater reservoirs and contaminate them. Probably the most serious groundwater contamination problem in North America is caused by leaking

▲

A Newtown Creek runs for about 5 kilometers through New York City, where it forms part of the border between Queens (at right) and Brooklyn (at left).

B The surface of the creek is constantly fouled with oil, from an underground spill that took place more than half a century ago. According to a Coast Guard study, in 1948 approximately 65 million liters of oil leaked from a refinery owned by the Standard Oil Company of New York (later renamed Mobil). (For comparison, this is more than the amount of oil spilled in the much more famous *Exxon Valdez* accident in 1989.) The successor company, ExxonMobil Company, has cleaned up part of the mess, but most of the spill continues to contaminate the groundwater and ooze to the surface in Newtown Creek.

The forgotten oil spill FIGURE 11.25

underground storage tanks at gas stations, refineries, and other industrial settings (see FIGURE 11.25).

WHEN GROUNDWATER DISSOLVES ROCKS

In regions underlain by rocks that are highly susceptible to chemical weathering (see Chapter 7), groundwater creates extensive systems of underground caverns. In such areas, a distinctive landscape forms on the surface. *Karst topography*, named for the Karst region of the former Yugoslavia (now Slovenia), is characterized by many *sinkholes* (small, closed basins) and disrupted drainage patterns (see FIGURE 11.26). Streams disappear into the ground and join the groundwater. Large springs form downstream, where the water table intersects the land surface, or along favorable routes of escape, such as faults. Karst is most typical of regions underlain by soluble carbonate rocks (limestone and dolostone), although it can also occur in regions with extensive evaporite (salt) deposits. In carbonate terrains with karst topography, the rate of dissolution is faster than the average rate of erosion of surface materials by streams and mass wasting.

Karst FIGURE 11.26

These sinkholes in Florida are part of a karst terrain. Most of Missouri and large parts of Kentucky, Tennessee, Texas, and New Mexico also have carbonate karst landscapes. In Arecibo, Puerto Rico, the world's largest radio telescope was built in a sinkhole, taking advantage of its naturally circular shape.

CAVES AND SINKHOLES

Caves and **caverns** are formed when circulating groundwater at or below the water table dissolves carbonate rock. The process begins with dissolution along interconnected fractures and bedding planes. A cave passage then develops along the most favorable flow route.

> **cave** Underground open space; a cavern is a system of connected caves.

The development of a continuous passage by slowly moving groundwater may take up to 10,000 years, and the further enlargement of the passage by more rapidly flowing groundwater needed to create a fully developed cave system may take an additional 10,000 to 1 million years. Finally, the cave may become accessible to humans after the water table drops below the floor of at least some of the chambers.

In the parts of the cave that lie above the water table (in the vadose zone), groundwater continues to percolate downward, dripping from the ceiling to the floor. Calcium carbonate dissolved in the water precipitates out of solution and builds up beautiful icicle-like decorations on the cave walls, ceilings and floor. These include *stalactites* (hanging from the ceiling) and *stalagmites* (projecting upward from the floor), as well as columns, draperies, and flowstones. In the saturated zone, below the water table, water movement is largely horizontal, creating tubular passages.

Caves are dissolution cavities that are closed to the surface or have only a small opening. In contrast, a *sinkhole* is a dissolution cavity that is open to the sky. Some sinkholes are formed very abruptly when the roofs of caves collapse. Most sinkholes, though, develop much more slowly and less catastrophically, simply growing wider over time as the carbonate bedrock slowly dissolves.

CONCEPT CHECK STOP

What is the geologic difference between an unconfined and a confined aquifer?

What is a water table? What lies above and below the water table?

How can surface water become groundwater, and vice versa?

What are some of the problems associated with the use of groundwater resources?

How and why do caves and sinkholes develop in certain regions?

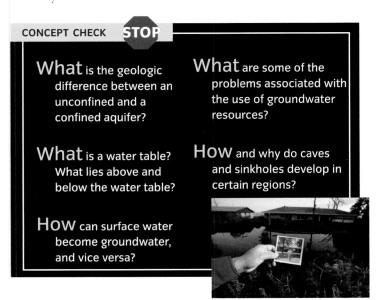

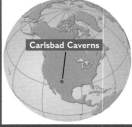
Amazing Places: Lechuguilla Cave

In 1986, cavers (also known as *spelunkers*) discovered the deepest known cave in the United States, called Lechuguilla Cave, in Carlsbad Caverns National Park. Its entrance had been known for decades, but it had been considered a dead-end until cavers dug through the floor to a huge network of passages on the other side.

Lechuguilla Cave has now been explored to a depth of 475 meters, and has almost 160 kilometers of mapped passages. It is as spectacular as it is deep, but is closed to the public to preserve its unusual formations.

▲ A A "bush" made of fragile aragonite pokes out of a stalagmite made of calcite.

B "Soda straws" reach down from the ceiling in this small chamber. Water flows down through the center of the straw. If water starts flowing down the outside, it will build a stalactite. ▶

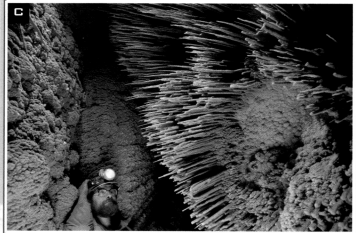

◀ C The origin of this rare formation, called "pool fingers," is a mystery, perhaps related to bacterial activity. The "fingers" crystallized in a pool of water and were left behind when the water retreated.

D Gypsum crystals provide a clue to this cave's unusual history. Unlike most limestone caves, which are formed by carbonic acid in rainwater, Lechuguilla formed from the bottom up. Hydrogen sulfide from lower-lying oil deposits percolated up into the groundwater and formed sulfuric acid, which dissolved away the rock. As the water table dropped, gypsum deposits precipitated out of the acidic water. ▶

1 The Hydrologic Cycle

1. The **water cycle**, or **hydrologic cycle**, describes the movement of water from one reservoir to another in the hydrosphere. The ocean is the largest reservoir, followed by the polar ice sheets. The largest reservoir of unfrozen fresh water is groundwater. Surface water bodies, the atmosphere, and the biosphere are much smaller water reservoirs.

2. The pathways or processes by which water moves from one reservoir to another include **evaporation**, **condensation**, **precipitation**, **transpiration**, **infiltration**, and **surface runoff**.

3. Like the tectonic and rock cycles, the hydrologic cycle is a *closed cycle* of *open systems*. Because it is closed, the global hydrologic cycle maintains a mass balance.

2 How Water Affects Land

1. **Streams** and **rivers** flow downslope along a clearly defined natural passageway, the **channel**. Interrelated factors that influence the behavior of a stream include the **gradient**, **discharge**, **load**, and velocity of the water.

2. Streams create landforms through *erosion* and *deposition*. *Straight*, *braided*, and *meandering* channels and *oxbow lakes* are erosional landforms. *Point bars*, *alluvial fans*, *deltas*, and **floodplains** are depositional landforms made of recently deposited sediment, or **alluvium**.

3. Every stream is surrounded by its **drainage basin**, the total area from which water flows into the stream. The topographic "high" that separates adjacent drainage basins is a divide. Streams originate in a *headwater* region, where they collect water from a large number of smaller *tributary streams*. Water exits a stream system from the stream's *mouth*, which may empty either into a larger river, a lake, or an ocean.

4. *Lakes* form in topographic basins created by faults, glacial debris, landslides, and lava flows, as well as in areas of poor drainage. Lakes are usually short-lived geologic features, because they tend to disappear by erosion of the outlet or by silt deposition.

3 Water as a Hazard and a Resource

1. A **flood** occurs when a stream's discharge becomes so great that it exceeds the capacity of the channel, causing the stream to overflow its banks. The risk of floods is increased by *subsidence*, and flooding can be exacerbated by human activity.

2. Prediction of flooding is based on analysis of the frequency of occurrence of past events and on real-time monitoring of storms using a *hydrograph*. River channel modifications made for purposes of flood control and protection, as well as for navigation and other purposes, are collectively known as *channelization*.

3. Agriculture is by far the largest consumer of fresh water. Surface water can be transported from one drainage basin to another, but only at the risk of long-term environmental changes in the affected basins.

CHAPTER SUMMARY

4 Fresh Water Underground

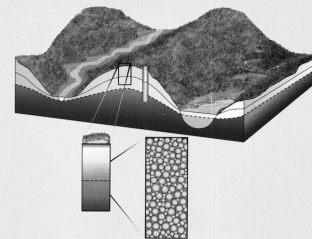

1. **Groundwater** is subsurface water contained in spaces within bedrock and regolith. In the *zone of aeration (unsaturated* or *vadose zone)*, water is present, but does not completely saturate the ground. In the *phreatic* or *saturated zone*, all openings are filled with water. The top of the saturated zone is the **water table**.

2. The rate of groundwater flow is dependent on the characteristics of the rock or sediment through which the water must move. **Porosity** is the percentage of the total volume of a body of rock or regolith that consists of open spaces. **Permeability** is a measure of how easily a solid allows fluids to pass through it. Groundwater in the saturated zone moves slowly by **percolation** through very small pores from areas where the water table is high to where it is lower.

3. Groundwater **recharge** occurs when rainfall and snowmelt infiltrate and percolate downward to the saturated zone. **Discharge** occurs where subsurface water leaves the saturated zone and becomes surface water in a stream, lake, or **spring**.

4. An **aquifer** is a body of permeable rock or regolith in the zone of saturation. An *unconfined aquifer* is in contact with the atmosphere through the pore spaces of overlying rock or regolith, while a *confined aquifer* lies between layers of impermeable rock, called **aquicludes**. The water pressure in a confined aquifer may be high enough to force the water partway or all the way to the surface when a well is drilled into it. Such a well is called *artesian*. Excessive withdrawal from an aquifer can cause the water table to drop, springs and streams to dry up, and regolith to compact and subside.

5. Many of the types and sources of contaminants that affect surface water also cause groundwater contamination. These include untreated sewage, agricultural pesticides and fertilizers, and leaks or spills of chemicals from commercial and industrial sites.

6. **Caves**, *caverns*, *sinkholes*, and *karst topography* are formed when rocks—most commonly carbonate rocks—are dissolved by circulating groundwater, creating underground cavities. *Stalactites* and *stalagmites* are built up from calcium carbonate precipitated from percolating groundwater.

KEY TERMS

- **evaporation** p. 305
- **transpiration** p. 305
- **condensation** p. 305
- **precipitation** p. 305
- **surface runoff** p. 305
- **infiltration** p. 305
- **stream** p. 308
- **channel** p. 308
- **gradient** p. 308

- **discharge** (1) p. 308
- **load** p. 308
- **floodplain** p. 311
- **alluvium** p. 311
- **drainage basin** p. 313
- **divide** p. 314
- **flood** p. 317
- **groundwater** p. 325
- **water table** p. 325

- **porosity** p. 326
- **permeability** p. 327
- **percolation** p. 327
- **recharge** p. 327
- **discharge** (2) p. 327
- **spring** p. 328
- **aquifer** p. 328
- **aquiclude** p. 328
- **cave** p. 333

1. List as many ways as you can in which we depend on the availability of fresh water in our daily lives.

2. Investigate the ways in which water shortages, floods, or other processes associated with water (such as erosion and deposition of sediment) have affected human history. How did societies respond, and what effects did those responses have?

3. The concept of residence time is very important in geology. It applies not only to water but also to any substance that moves from one reservoir to another in the Earth system. What other substances, besides water, move around in the Earth system? Why would we want to monitor these substances and keep track of their residence times?

4. What evidence has been gathered for water-shaped landscapes on Mars or elsewhere in the solar system? How are these landforms similar to or different from their analogues on Earth?

5. Where does your community obtain its water supply? Is it a groundwater or a surface freshwater source? Is either the quantity or the quality of the water threatened?

6. Visit a stream before and after an intense rainfall. Observe the gradient, discharge, load, and velocity of the streamflow. What changes do you notice?

What is happening in this picture ?

This 100-meter-wide sinkhole opened up one day in Winter Park, Florida, and grew to the point where it eventually swallowed up part of a house, six commercial buildings, and the municipal swimming pool. What could have caused this to happen? What kind of rock is prone to this sort of collapse?

1. On this illustration, label each stage of the hydrologic cycle (1 through 6) using the following terms:

   precipitation   cloud formation through condensation
   surface runoff   surface evaporation and transpiration
   infiltration   evaporation from oceans

2. The illustration above depicts Earth's hydrologic cycle. Which reservoir holds most of the fresh water resources of the planet?

   a. the oceans
   b. ice sheets
   c. lakes
   d. groundwater

3. The _____ of a stream will have a large influence over channel development and the evolution of associated landforms.

   a. discharge
   b. gradient
   c. sediment load
   d. All of the above statements are correct.

4. _____ will form in streams when there is a low slope gradient and large and variable sediment load.

   a. Straight channels
   b. Meandering channels
   c. Braided channels

5. Lakes are ephemeral features that disappear by _____.

   a. the accumulation of inorganic sediment carried in by streams
   b. the accumulation of organic matter produced by plants within the lake
   c. gradually being drained by stream outlets eroding to lower levels
   d. All of the above statements are correct.

6. On this illustration, label each of the landforms in a stream valley using the following terms:

   flood   alluvial fan
   floodplain   natural levees
   main channel

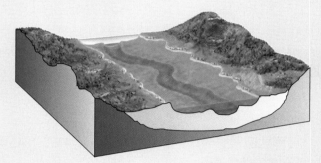

7. A(n) _____ is the total area from which water flows into a stream.

   a. alluvial fan
   b. braided channel
   c. water cycle
   d. drainage basin

8. Urban development can lead to increased risk of flooding by _____.

   a. compressing underlying sediments causing subsidence
   b. increasing surface runoff
   c. by channeling runoff more quickly to rivers through storm drains
   d. All of the above are correct.

9. There is a _____ chance of a "50-year flood" occurring in any given year.

   a. 1 percent
   b. 2 percent
   c. 5 percent
   d. 10 percent
   e. 50 percent

10. Though the proportions vary from one region to another, globally what accounts for the greatest demand for water?

    a. industry
    b. domestic use
    c. crop irrigation

11. What factors control the porosity of an aquifer?

    a. size of the particles in the material
    b. shapes and uniformity of the material
    c. amount of cementing agent in the pore space of the material
    d. All of the above are correct.

12. A(n) _____ is an underground reservoir of water that that is overlain by impermeable rock units.

    a. aquiclude
    b. aquifer
    c. confined aquifer
    d. unconfined aquifer

13. In a well drilled into a(n) _____, the water will rise to the level of the surrounding water table.

    a. aquiclude
    b. aquifer
    c. confined aquifer
    d. unconfined aquifer

14. Placing a well in a position where Earth's surface is below the water table will result in a well that_____.

    a. requires pumping to produce water
    b. does not require pumping to produce

15. On this illustration, label each groundwater feature using the following terms:

    dry-season water table   additional wet season saturation
    wet-season water table   year-round saturation
    zone of saturation

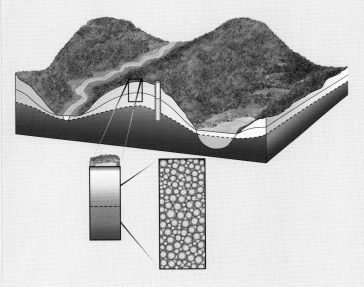

The Oceans and the Atmosphere

On September 26, 1991, eight humans were sealed inside an airtight greenhouse in the Arizona desert north of Tucson. The structure was grandiosely named "Biosphere 2," taking its name after "Biosphere 1"—Earth. The goal of the experiment was to design a closed system that could support human life, as a possible dry run for future bases on the Moon or Mars. The greenhouse covered 16.4 hectares and contained seven different Earth-like environments, including a simulated "ocean" (inset) and jungle.

Although the eight "Biospherians" did survive two years inside the enclosure, the experiment was widely viewed as a failure. Not only was the crew unable to produce enough food (despite the rich organic soil inside), they also began to run out of oxygen: The percentage of oxygen in the air dropped from 21 to 14 percent, comparable to living at an altitude of more than 6000 meters.

Project scientists concluded that they had miscalculated how much oxygen would be used by microbes in the soil. As bacteria feed on organic matter, they consume oxygen and release carbon dioxide. Ironically, a second unanticipated event actually worked in the Biospherians' favor: As the cement in the concrete walls "cured," it absorbed the excess carbon dioxide.

Biosphere 2 is now nothing more than a tourist destination and an object lesson in the difficulty of maintaining such a complex closed system. There is still only one true biosphere. Every day, Earth's plants, animals, atmosphere, oceans, and land environments interact in a complex manner that humans have not yet learned to duplicate.

The Oceans

LEARNING OBJECTIVES

Explain why seawater is salty and what processes add or remove salt.

Identify the ocean's three layers and describe the differences between them.

Describe the overall pattern of surface and deep-water currents.

Explain how the ocean affects the world's climates.

In Chapter 11 we introduced the hydrologic cycle and described the events in this cycle that take place on land. If we stopped there, we would be left with a rather distorted view of the hydrologic cycle as a whole. The oceans are by far Earth's largest reservoir of water. The atmosphere, a much smaller reservoir than either the oceans or fresh water on land, also plays a critical role in redistributing Earth's water. Indeed, it is best to think of the oceans and the atmosphere as a linked system, although we will discuss them one at a time. Atmospheric circulation drives ocean waves and currents, and in turn the ocean regulates the temperature and humidity of the lower atmosphere. It is also the conveyor that transfers water vapor from the oceans to the land.

OCEAN BASINS

Water has been present on Earth's surface since its earliest days. In fact, we do not have any geologic record of a time before water. The oldest rock so far discovered on Earth (about 4.0 billion years old) is gneiss that was once sedimentary strata. These strata were deposited in water and are similar to strata being deposited today. We can be reasonably certain, as a result, that the ocean formed sometime between 4.56 billion years ago (when Earth was formed) and 4.0 billion years ago. Where the water in the oceans came from is still an open question (see **FIGURE 12.1**).

Most of the water on our planet is contained in four huge interconnected ocean basins. The Pacific, Atlantic, and Indian oceans are connected with the Southern Ocean, the body of water that encircles Antarctica. (The Arctic Ocean is considered an extension of the North Atlantic.) Collectively, these four bodies of water, together with a number of smaller ones such as the Mediterranean Sea, Hudson Bay, and the Persian Gulf, are often referred to as the *world ocean* and cover 71 percent of Earth's surface.

THE COMPOSITION OF SEAWATER

As you know if you have ever swum in the ocean, seawater contains a lot of salt, which makes it not only unpalatable but also dangerous for human consumption. (If you tried living on salt water, you would actually become dehydrated, as your body would try to rid itself of the excessive amounts of ions such as sodium and potassium.)

The **salinity** of seawater ranges between 3.3 and 3.7 percent. This may seem quite small, but it is about 70 times the salinity of tap water. Not surprisingly, when seawater is evaporated, more than three-quarters of the dissolved matter is sodium chloride—that is, table salt. Seawater contains most of the other natural elements as well, but many of them in such low concentrations that only extremely sensitive analytical instruments can detect them.

> **salinity** A measure of the salt content of a solution.

The elements dissolved in seawater come from several sources. Chemical weathering of rock releases soluble materials such as salts of sodium, potassium, and sulfur. The soluble compounds are leached out and become part of the dissolved load in river water flowing to the sea. Volcanic eruptions, both on land and beneath the sea, also contribute soluble compounds via volcanic gases and hot springs. The "black smoker" vents found at spreading centers (see Chapter 4) add dissolved

Where did the water come from? FIGURE 12.1

Scientists do not yet agree on where Earth got its water.

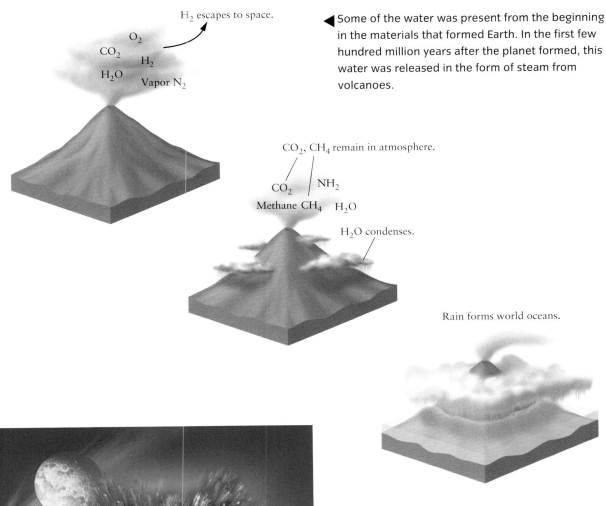

H_2 escapes to space.

O_2
CO_2
H_2
H_2O Vapor N_2

◀ Some of the water was present from the beginning in the materials that formed Earth. In the first few hundred million years after the planet formed, this water was released in the form of steam from volcanoes.

CO_2, CH_4 remain in atmosphere.

CO_2 NH_2
Methane CH_4 H_2O

H_2O condenses.

Rain forms world oceans.

◀ B Additional water may have been added to Earth's chemical mix by comets and meteorites coming in from the outer regions of the solar system, after the planet had already formed.

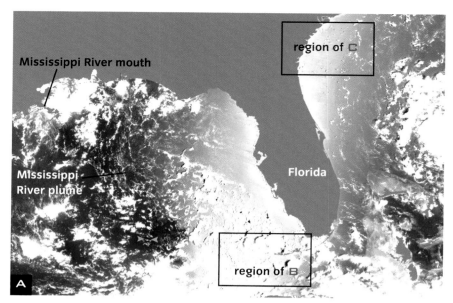

July 30, 2004

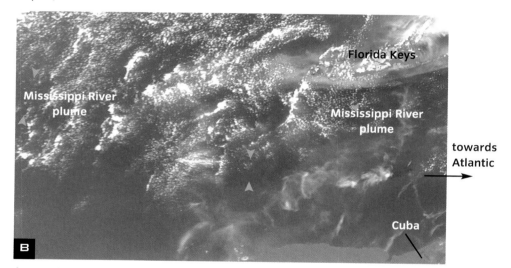

August 4, 2004

August 9, 2004

A river runs through it FIGURE 12.2

These satellite photos, taken in 2004, show a "river" (or, as geologists call it, a plume) of less salty water running from the mouth of the Mississippi River through (A) the Gulf of Mexico and (B) the Florida Straits and into (C) the Atlantic Ocean, before it finally mixes sufficiently with the ocean water that it cannot be seen any more. The river water appears brown in the photos. Shipboard measurements showed the plume was 50 kilometers wide and 10 to 20 meters deep. It had lower salinity, higher temperature, and double the chlorophyll concentration of the other water in the Florida Straits. Scientists estimate that 23 percent of the Mississippi River's discharge flows directly to the Atlantic Ocean in this manner.

minerals from the hot rocks in the midocean rift. Two other processes, evaporation of surface water and freezing of seawater, tend to make seawater saltier because they remove fresh water while leaving the salts behind.

Why, then, doesn't the sea continue to become saltier and saltier? The answer is that it is an open system and constantly receives fresh water from *precipitation* and *river flow* (see FIGURE 12.2). Also, aquatic plants and animals withdraw some elements, such as silicon, calcium, and phosphorus, to build their shells or skeletons. Other elements precipitate out in mineral form and settle to the sea bottom. All of these processes balance each other, keeping the composition of seawater essentially unchanged. This is the problem that John Joly ran into when he tried to use the salinity of seawater to estimate the age of Earth (Chapter 3).

LAYERS IN THE OCEAN

The salinity and temperature of seawater combine to control its density: Cold salty water will sink, whereas warm, less salty water will rise. Figure 12.2 shows this very clearly; the warm and less salty Mississippi River water stays on top of the Gulf of Mexico water for hundreds of kilometers.

Ocean scientists have discovered three major layers, or zones, in the ocean, in which the density of the water differs. The differences are caused by changes in both temperature and salinity, with temperature being the major factor (see FIGURE 12.3).

OCEAN CURRENTS

In 1492, when Christopher Columbus set sail across the Atlantic Ocean in search of China, he took an indirect route. On the outward voyage he sailed southwest toward the Canary Islands and then west on a course that carried him to the Caribbean Islands, where he first sighted land. In choosing this course, he was following the path of the prevailing winds and surface ocean currents, instead of fighting the westerly winds and currents at 40° N latitude. On the return trip, he sailed a more northerly route in

Structure of the ocean FIGURE 12.3

The three major density zones in the ocean: the surface layer, the transitional zone (the thermocline), and the deep zone.

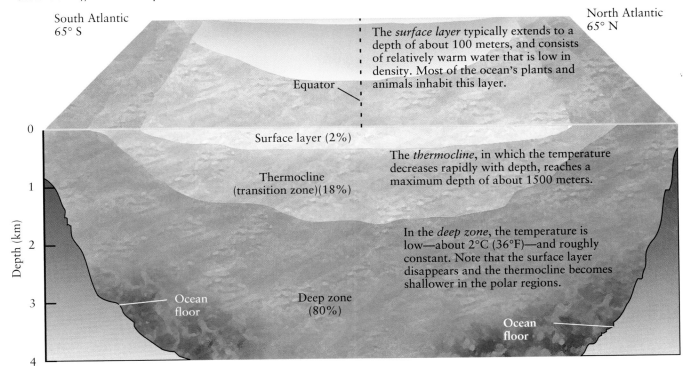

South Atlantic
65° S

North Atlantic
65° N

Equator

The *surface layer* typically extends to a depth of about 100 meters, and consists of relatively warm water that is low in density. Most of the ocean's plants and animals inhabit this layer.

Surface layer (2%)

Thermocline (transition zone)(18%)

The *thermocline*, in which the temperature decreases rapidly with depth, reaches a maximum depth of about 1500 meters.

In the *deep zone*, the temperature is low—about 2°C (36°F)—and roughly constant. Note that the surface layer disappears and the thermocline becomes shallower in the polar regions.

Ocean floor

Deep zone (80%)

Ocean floor

Depth (km)

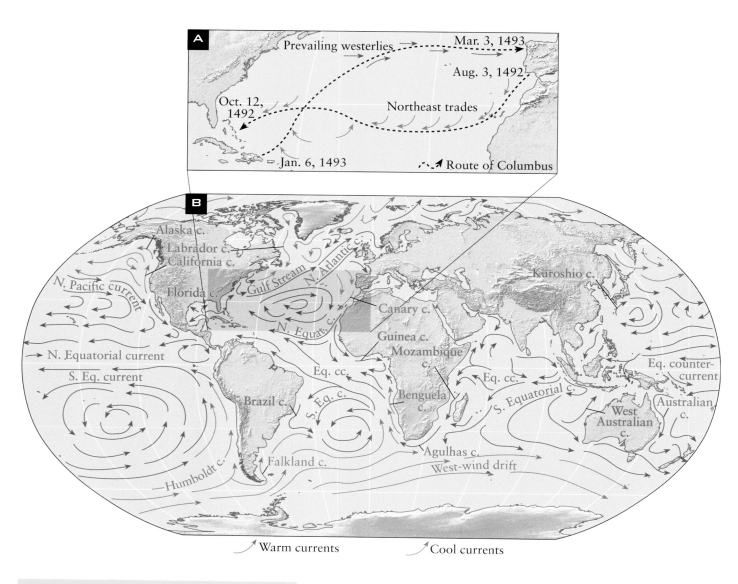

A

Prevailing westerlies → → → Mar. 3, 1493

Aug. 3, 1492

Oct. 12, 1492

Northeast trades

Jan. 6, 1493

Route of Columbus

B

Alaska c.
Labrador c.
California c.
N. Pacific current
Florida c.
N. Equatorial current
S. Eq. current
Brazil c.
Humboldt c.
Falkland c.

Gulf Stream
N. Atlantic c.
N. Equat. c.
Eq. cc.
S. Eq. c.

Canary c.
Guinea c.
Mozambique c.
Benguela c.
Agulhas c.
West-wind drift

Kuroshio c.

Eq. cc.
S. Equatorial c.
Eq. counter-current
West Australian c.
Australian c.

Warm currents   Cool currents

Surface ocean currents FIGURE 12.4

Sailors have long been aware of the prevailing currents in the ocean, and many of these currents even have names. Christopher Columbus used the North Atlantic current to good advantage in 1492. Surface ocean currents in the Northern Hemisphere tend to curve to the right (clockwise), whereas in the Southern Hemisphere they curve to the left (counterclockwise)

order to take advantage of the westerly winds at that latitude. His route is shown by the dotted black arrows in the top portion of FIGURE 12.4.

Surface ocean currents, like those Columbus followed, are set in motion by the prevailing winds. For this reason, they are only about 50 to 100 meters deep, and are confined to the surface zone.

The pattern of water flow in the deep ocean is quite different and needs to be pictured in three di-

mensions (see FIGURE 12.5). Differences in temperature and salinity—which change the water's density—drive this circulation pattern, and the current directions are controlled by the Coriolis effect (discussed later in the chapter). Water off the coast of Greenland is cold and salty (because it has had fresh water removed from it to form sea ice), and therefore dense; it sinks to form a large mass of water called the *North Atlantic Deep Water*. This mass of cold, salty, dense water moves and mixes in the South Atlantic with another mass of cold, salty, dense water, known as the *Antarctic Bottom Water*. The current of combined water masses flows across the seafloor and northward into the Pacific. Eventually, in the northern Pacific, the cold water wells up, becoming shallower and warmer. The shallower, warmer surface currents that result eventually flow back into the Atlantic and towards the north, where they are largely responsible for maintaining a mild climate in Europe. The *Gulf Stream* is part of this northward flow. By the time the water reaches the North Atlantic, it begins to cool and sink; the process is continous. Because of its continuous, repetitive nature, this circuit has been called "the great ocean conveyor belt." Its scientific

The ocean conveyor belt FIGURE 12.5

A As in a conveyor belt, deep ocean currents flow in the opposite direction from surface currents. Vertical motions drive the conveyor: Cold, salty water near Antarctica and the Arctic Ocean sinks and eventually flows northward into the Pacific Ocean where it wells up, eventually becoming warmer and fresher. It takes about 1000 years for water to complete one circuit of this global flow pattern.

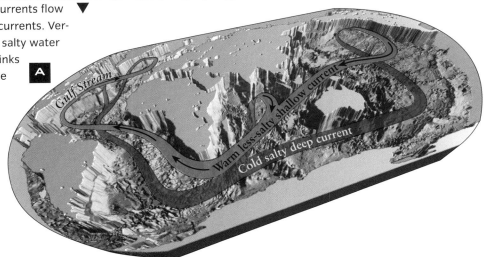

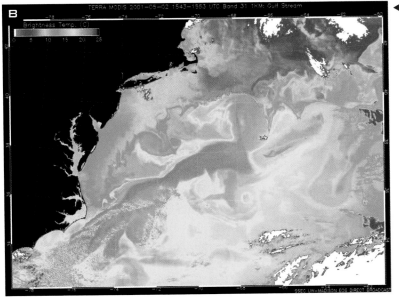

B The Gulf Stream is a well-known regional link in the ocean conveyor belt. In this satellite image, the colors represent the surface temperature of the water, with red water being warmer (20°C and above). The Gulf Stream flows northeast along the Carolina coast (lower left) and then out to sea. Black areas are land (the East Coast of the United States), and white areas are clouds, which blocked the collection of data by the satellite.

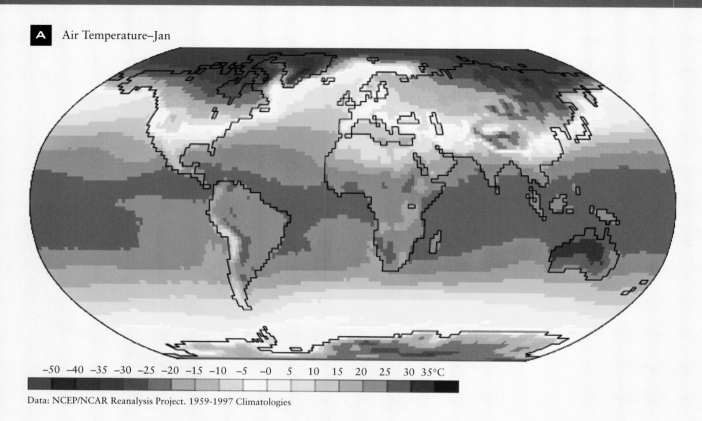

A Air Temperature–Jan

−50 −40 −35 −30 −25 −20 −15 −10 −5 −0 5 10 15 20 25 30 35°C

Data: NCEP/NCAR Reanalysis Project. 1959-1997 Climatologies

These maps show the average daily air temperature in (**A**) January and (**B**) July over both land and sea. Notice the much broader range of temperatures over land in both of the maps. Also, notice that there is a pronounced seasonal change in land temperatures, whereas in the world's oceans there is little difference between summer and winter (especially in the tropics).

name is the *thermohaline circulation*, because it is driven by variations in ocean temperature (*"thermo"*), and salinity (*"haline"*).

HOW OCEANS REGULATE CLIMATE

The ocean differs from the land in the amount of heat it can store. When the Sun's rays strike the land, only the top meter or so is warmed—below the top few meters, rock temperatures are controlled by the geothermal gradient and at any given depth are roughly constant year round. In the ocean, mixing by currents and waves means the Sun's heat affects much more than just the surface layers.

Both the ocean and the atmosphere are also affected by water's *latent heat*—the energy that the ocean absorbs when water evaporates and that which it releases when it freezes. When water evaporates from the ocean, it carries some of the Sun's heat energy with it into the atmosphere, leaving the ocean water slightly cooler. The reverse happens when water freezes; it releases its latent heat and maintains the temperature of the water around it. Both of these processes tend to reduce the amount of variation in ocean temperatures (see FIGURE 12.6). The ocean, in turn, moderates the climate of coastal regions on land. Along the Pacific coast of Washington and British Columbia, for example, winter air temperatures seldom drop to freezing, whereas inland temperatures plunge to -30°C or lower.

Air Temperature–Jul

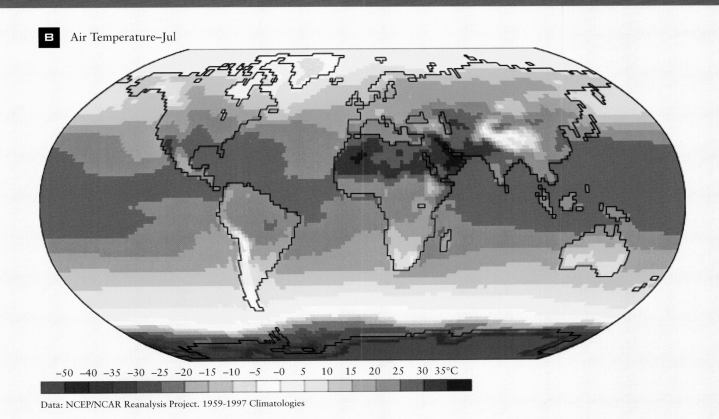

−50 −40 −35 −30 −25 −20 −15 −10 −5 −0 5 10 15 20 25 30 35°C

Data: NCEP/NCAR Reanalysis Project. 1959-1997 Climatologies

One of the most important natural influences on climate is the ocean's "conveyor belt" or thermohaline circulation. Because the conveyor belt redistributes heat from the tropics to the poles, any weakening of the circulation would be expected to increase the temperature contrast between the two. The cause and effect may work the other way, too: During the last ice age, the polar ice caps may have cut off the "conveyor belt," which would have made the glaciation longer-lasting and be more severe. As always, the parts of the Earth system, the hydrosphere and atmosphere, interact with each other and both of them interact with the land surface, controlling the formation of ice sheets, the weathering of rock, and the distribution of plants and animals.

CONCEPT CHECK STOP

What are two possible origins of Earth's water?

What are the layers of the ocean?

How do surface currents differ from deep-ocean currents?

What is the ocean's "conveyor belt" circulation and how does it affect climate?

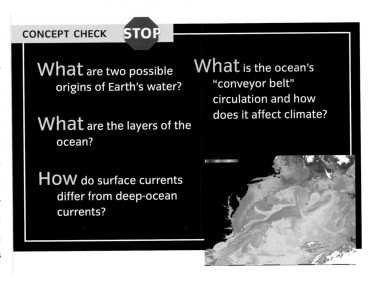

Where the Ocean Meets Land

LEARNING OBJECTIVES

Explain why worldwide sea levels do not always stay the same.

Explain what causes tides.

Describe how waves and longshore currents transport sediment.

Identify three common types of shorelines.

The majority of the world's population lives within 100 kilometers of a coastline. This reflects our dependence on the oceans as well as the richness of resources in coastal zones. However, the concentration of large numbers of people in coastal areas means that the coastal environment must absorb the impacts of a wide range of human activities. It also means that human vulnerability to hazards can be particularly high in coastal zones; for example, infrequent events such as large storms can cause major loss of life and damage to property.

If you visit almost any coast line on two occasions a year apart, you will see changes. Sometimes the changes are small, but often they are substantial. Large sand dunes may have shifted. Sand may have built up behind barriers or been eroded away. Steep sections of coastline may have collapsed. Channels may have broken through from the sea to lagoons on the landward side, where there were no channels before. The energy driving these continual changes comes from storms and the waves that accompany them.

CHANGES IN SEA LEVEL

Water plays a powerful role in shaping coastlines. Before describing the landforms that result from the water's action, we will describe the processes themselves, which operate on very different time scales. On a very long time scale, over many thousands of years, the sea level can rise or fall by hundreds of meters. On a shorter time scale—twice a day—tides produce changes in water level that can amount to several meters in certain places. On the shortest time scale, ocean waves constantly stir up the sediment along the coast, and, especially during storm season, waves batter the coastline and can produce dramatic effects.

Global volume changes One of the most important factors in determining the long-term position of the shoreline is the volume of the ocean. This in turn is strongly tied to global climate change. When the climate warms, water stored on continents in glaciers and ice caps melts and returns to the sea. This causes a worldwide rise in sea level. Conversely, during cold climatic periods, glaciers and ice caps expand, and water is withdrawn from the ocean and stored on land (in the form of ice); this results in a drop in sea level. Currently, global warming is suspected to be responsible for a worldwide rise in sea level of approximately 2.4 mm/yr. Although such a trend is nearly imperceptible on the scale of a human lifetime, over geologic time it may account for great change in the position of a shoreline (see **FIGURE 12.7**).

Tides From a human perspective, a much more obvious change in sea level occurs due to the effect of the **tides**. The gravitational attraction of the Moon makes ocean water bulge toward the Moon on the side of Earth nearest the Moon. There is

tides A regular, daily cycle of rising and falling sea level that results from the gravitational action of the Moon, the Sun, and Earth.

also a bulge away from the Moon on the opposite side, whose origin is the *inertial force* (the same force that pulls the string of a yo-yo tight if you swing it around your finger). The Sun also affects the tides, but because it is so much farther away than the Moon, its tide-producing force is only about half as strong.

To visualize how tides work, consider the tidal bulges oriented with their maximum height lying along a line running through the center of Earth and the

The Bering land bridge FIGURE 12.7

A About 21,000 years ago, during the most recent ice age, sea levels were about 120 meters lower than they are today. This had important geologic ramifications, including the emergence of a "land bridge" between Asia and North America. Because there is no definitive evidence of humans in America before that time, it has been suggested that humans migrated across the "land bridge" into the Western Hemisphere. The green regions are above sea level.

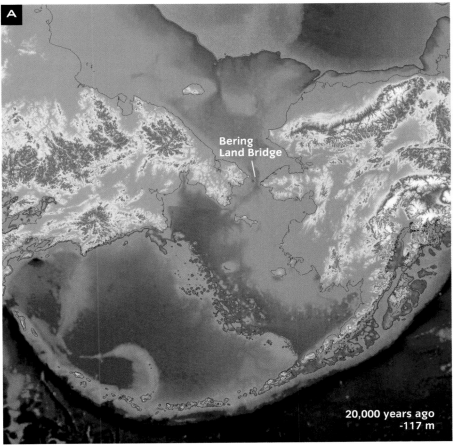

B Today Asia and America are separated by an 85-kilometer wide stretch of sea called the Bering Strait.

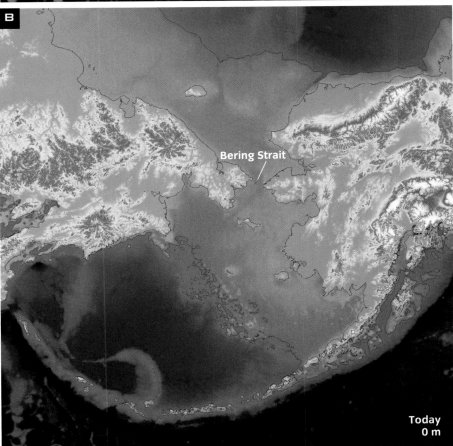

center of the Moon, as shown in What a Geologist Sees. Whereas Earth rotates around its axis, the tidal bulges remain stationary opposite the Moon. Thus, any given coastline will move eastward through both tidal bulges each day. Every time a landmass encounters a tidal bulge, the water level along the coast rises. As Earth rotates, the coast passes through the highest point of the tidal bulge (high tide) and the water level begins to fall. Along most coastlines two high tides and two low tides are observed each day.

In the open sea the effect of tides is small. However, the shape of a coastline can greatly influence the tidal *run-up* height, the highest elevation reached by the incoming water. Narrow openings into bays, rivers, estuaries, and straits can amplify normal tidal fluctuations. At the Bay of Fundy in Nova Scotia, a *tidal range* (the difference between high and low tide) of up to 16 meters has been reported (see What a Geologist Sees). The bay is very long and narrow, which causes the incoming tide to rush in, forming a steep-fronted wall of water called a *tidal bore* that can move faster than a person in average physical condition can run.

In most places, though, the tides act too slowly to have much geologic impact. Perhaps their most im-

High-tide water level

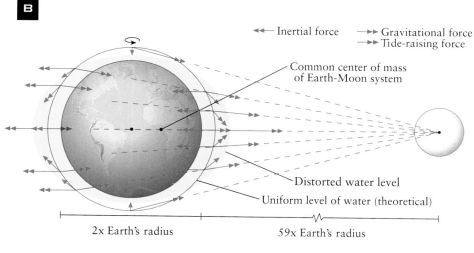

Inertial force → Gravitational force → Tide-raising force

Common center of mass of Earth-Moon system

Distorted water level

Uniform level of water (theoretical)

2x Earth's radius

59x Earth's radius

What Causes Tides?

The tidal range in the Bay of Fundy in eastern Canada is one of the greatest in the world (see **FIGURE A**). These fishing boats are grounded at low tide. At high tide, the water will rise up to the level where the posts in the photograph change color.

The Moon's gravitational attraction and the inertia of the rotating Earth-Moon system combine to produce tides. As shown in this greatly exaggerated sketch (see **FIGURE B**), the two forces stretch Earth into an oblong shape, with bulges directed toward and away from the Moon. The bulges remain essentially stationary while Earth rotates through them, creating two high tides and two low tides per day.

In addition, local topography strongly affects both the timing and size of the tides in any particular place. The tides in the Bay of Fundy are so high because the frequency of the tides closely matches the natural period of oscillation of water in the long, thin bay. It is like sloshing the water in a bathtub—if you move back and forth at just the right frequency, the water will overflow. Other places in the world have almost no tides, or even once-a-day instead of twice-a-day tides, all depending on the local conditions.

portant effect is global in scale. The tides gradually slow down Earth's rotation, causing the length of the day to increase. In the early Phanerozoic Eon, the day was about 22 hours long and there were nearly 400 days in a year.

WAVES

Like surface ocean currents, ocean waves receive energy from winds. The size of the largest wave depends on how fast, how far, and how long the wind blows across the water surface. A gentle breeze blowing across a bay may ripple the water or form low waves less than a meter high. In contrast, storm waves whipped up by intense winds over several days may tower over ships unfortunate enough to be caught in them (see **FIGURE 12.8**). The waves generated by a storm can travel great distances. For example, surfers in California keep an eye on the weather report for storms far away in the tropical Pacific, because they know that a good storm creates a "swell" that will arrive at their shores several days later.

Rogue waves FIGURE 12.8

A

◀

A In a storm, the highest waves can wash over the deck of a ship.

B Sailors have told tales for centuries about "rogue waves" as high as a 10-story building, which are big enough to capsize a ship. For years scientists did not believe such stories. However, evidence such as this radar image, taken from a satellite in 1996, has removed all doubt. In the center of the white bar, you can see a dark trough that goes down more than 10 meters below sea level, next to a bright peak that reaches more than 15 meters above sea level. A ship caught in the trough would have seen a wall of water more than 25 meters high (inset). Note that the rogue wave is quite localized; all around it are waves that are less than 10 meters high, from trough to crest.

▶ B

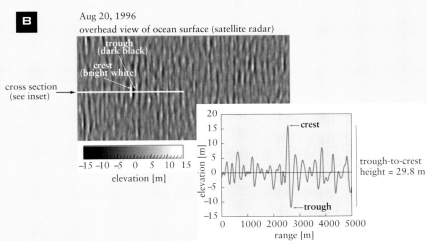

Aug 20, 1996
overhead view of ocean surface (satellite radar)

trough (dark black)

crest (bright white)

cross section (see inset)

−15 −10 −5 0 5 10 15
elevation [m]

crest

trough-to-crest height = 29.8 m

trough

How waves change near the shore FIGURE 12.9

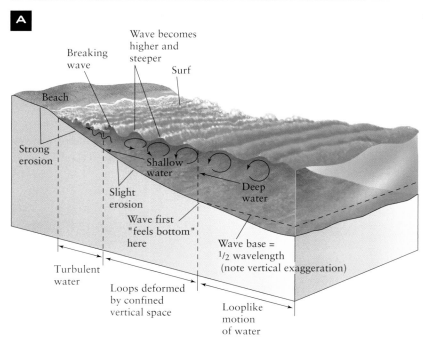

A

Wave becomes higher and steeper

Breaking wave

Surf

Beach

Strong erosion

Shallow water

Slight erosion

Deep water

Wave first "feels bottom" here

Wave base = ¹/₂ wavelength (note vertical exaggeration)

Turbulent water

Loops deformed by confined vertical space

Looplike motion of water

◀ **A** Waves change form as they travel through shallow water. In deep water, the water moves in a nearly circular loop as the wave passes. Near shore, the loop flattens as the wave begins to "sense" the bottom. Finally, the wave breaks, and the water's motion changes to the familiar in-and-out action of the surf.

B

B Orienting the board for the best ride, a surfer skims the inside of a breaking wave off Maui. Surfing is possible only in shallow water, because waves do not break in deep water. ▶

Wave action along coastlines

As a wave approaches the shore, it undergoes a rapid transformation. In deep water, a buoy or a parcel of water makes circular loops as the waves pass by. But in shallower water, the circular loops become flatter (see FIGURE 12.9A). Where water depth becomes less than half a wavelength, the increasingly shallow sea floor interferes with wave motion and distorts the wave's shape. The wave height increases, and the wavelength decreases. Now the front of the wave is in shallower water

and is also steeper than the rear. Eventually, the front becomes too steep to support the advancing wave. As the rear part continues to move forward, the wave collapses, or *breaks* (see FIGURE 12.9B).

When a wave breaks, the motion of the water instantly becomes turbulent, like that of a swift river. **Surf** is found between the line of

surf The "broken," turbulent water found between a line of breakers and the shore.

breakers and the shore, forming an area known as the *surf zone*. Each wave finally dashes against the rocks or rushes up a sloping beach until its energy is expended; then the water flows back toward the open sea. Water that has piled up against the shore returns seaward in an irregular and complex way, partly as a broad sheet along the bottom and partly in localized narrow channels known as *rip currents*. These can sweep unwary swimmers out to sea. However, experienced swimmers know that the rip current is not wide. By swimming sideways to the current, rather than against it, they can get out of trouble.

A wave approaching a coast generally does not encounter the bottom simultaneously all along its length. When a segment of a wave touches the seafloor, that part of the wave slows down. Gradually the trend of a wave becomes realigned to parallel the contours of the seafloor. This process, known as *wave refraction*, causes a series of waves approaching a shoreline at an angle to change their direction of movement. (Note that the word "refraction," in this context, means "bending"—just as it does in the context of seismic wave or light wave refraction.)

Erosion and transport of sediment by waves

Surf is a powerful erosive force because it possesses most of the original energy of the waves that created it. Wave erosion takes place not only at sea level but also below sea level and—especially during storms—above sea level. In the surf zone, rock particles are worn down, becoming smoother, rounder, and smaller. At the same time, through continuous rubbing and grinding with these particles, the surf scours and deepens the bottom. Onshore, the surf acts like a saw cutting horizontally into the land. Its energy is eventually consumed in turbulence, in friction at the bottom, and in the movement of sediment thrown up from the bottom. Two processes, one on land and one underwater, transport the sediment: the *longshore current* and *beach drift*.

Most waves reach the shore at an oblique angle (see FIGURE 12.10A). Part of the force of an incoming wave is oriented perpendicular to the shore; this produces the crashing surf. Another component of the wave motion is oriented parallel to the shore. The parallel component sets up a **longshore current**. Whereas surf erodes sediment at the shore, the longshore current moves the sediment along in the surf zone.

> ■ **longshore current** A current within the surf zone that flows parallel to the coast.

Longshore current and beach drift FIGURE 12.10

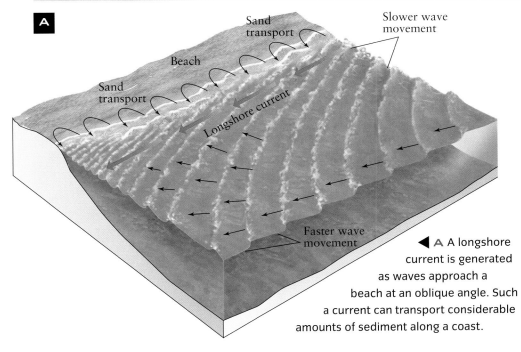

A A longshore current is generated as waves approach a beach at an oblique angle. Such a current can transport considerable amounts of sediment along a coast.

B Surf swashes obliquely onto a Brazilian beach and forms a series of arc-shaped cusps. A grain of sand moves along a zigzag path, as successive waves wash up on the shore at an angle and retreat back downslope in a direction nearly perpendicular to the shoreline. ▼

beach drift The movement of particles along a beach as they are driven up and down the beach slope by wave action.

Meanwhile, on the exposed beach, incoming waves produce an irregular pattern of water movement along the shore. Because waves generally strike the beach at an angle, the *swash* (uprushing water) of each wave travels obliquely up the beach before gravity pulls the water down (see FIGURE 12.10B). This zigzag movement of water carries sand and pebbles first up, then down the beach slope. Successive movements of this type gradually transport sediment along the shore, causing **beach drift**. The greater the angle of waves to the shore, the greater will be the rate of drift. Marked pebbles have been observed to drift along a beach at a rate of more than 800 meters per day.

The processes of erosion and deposition along a coast are anything but steady. A single winter storm can erode cliffs or beaches more than a full year's worth of ordinary surf. The balance between erosion and deposition can change with the seasons. Along parts of the Pacific coast of North America, winter storm surf tends to carry away fine sediment, and the beach becomes narrow and steep. In calm summer weather, fine sediment drifts in and the beach assumes a gentler profile.

Human actions can also shift the balance between erosion and deposition. When a seawall interrupts the flow of sediment along the coastline, the beach builds up on one side of the seawall but erodes away on the other side, because the longshore current no longer brings in enough material to replenish the eroded sediments (see FIGURE 12.11).

Disturbing the shoreline FIGURE 12.11

In Ocean City, Maryland (right), the construction of a jetty has caused sand to accumulate and make the beach wider. Just south of Ocean City, Assateague Island in Virginia (left) has been deprived of the sand that the longshore current would have deposited on its shore. Thus the beach on Assateague is retreating inland.

Ocean City

Direction of longshore current

Jetty

Sand accumulates on the up-current side of the jetty

Assateague

pocket beach

upper bench

headland

lower bench

Wave-cut benches FIGURE 12.12

The coast of New Zealand at Tongue Point has two terraces or *benches*. They are both former sea floor, and were elevated above sea level by two stages of tectonic uplift. In this photo-graph you can see two other common features of rocky coasts: *headlands* that jut out into the sea, and *pocket beaches* that form in the bays between the headlands.

SHORELINES AND COASTAL LANDFORMS

The end result of the constant interplay between erosional and depositional forces along coastlines is a wide variety of shorelines and coastal land-forms. Their forms depend on the geologic processes at work, the susceptibility to erosion of coastal rocks, and the length of time these processes have been operating. Changes in sea level can also influence the development of coastal features. Many coastal land formations show clear signs of different sea levels at different times in the past (see FIGURE 12.12).

Despite the variability of coasts and shore-lines, three basic types are most common: the *rocky (cliffed) coast*, the *lowland beach and barrier island coast*, and the *coral reef*. Each one has a particular set of erosional and depositional landforms.

Rocky coasts The most common type of coast, comprising about 80 percent of ocean coasts world-wide, is a rocky or cliffed coast. Seen in profile, the usual elements of a cliffed coast are a vertical **wave-cut cliff** and a horizontal *wave-cut bench* at its base, both products of erosion. As the upper part of the cliff is undermined, it collapses and the resulting debris is redistributed by waves. An undercut cliff that has not yet collapsed may have a well-developed notch at its base. The bench may be covered by sand or the bedrock may be exposed, especially at low tide.

wave-cut cliff A coastal cliff cut by wave action at the base of a rocky coast.

If the coast has been uplifted by tectonics, a wave-cut bench and its sediment cover can be lifted up out of the water and become a *marine terrace*. In some lo-cations, you can see two or more terraces ascending out of the ocean like a giant staircase (see Figure 12.12).

Expensive follies FIGURE 12.13

A Houses dangle on the edge of a cliff in Pacifica, California, after winter storms caused a section of cliff to collapse. Note the rubble on the beach and the scar on the cliff where the debris slide occurred. B Typically this kind of slope failure occurs because waves have undercut the base of the cliff. Unfortunately, this scenario will continue to repeat itself as long as people choose home sites based on the desire for "million-dollar views" rather than on sound geologic reasoning.

Area of collapse

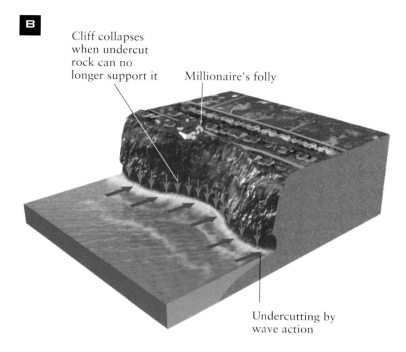

Cliff collapses when undercut rock can no longer support it

Millionaire's folly

Undercutting by wave action

The rocky character of cliffed coasts may be misleading to people looking for a dramatic home site. They see a rocky cliff as a sign of permanence and stability, whereas in fact quite the reverse is the case. Shorelines with cliffs are susceptible to frequent landslides and rock falls as erosion eats away at the base of the cliff. Roads, buildings, and other structures built too close to such cliffs can be damaged or destroyed when sliding occurs (see FIGURE 12.13).

Beaches and barrier islands   Beaches are a striking feature of many coasts. Most people think of a beach as the sand surface above the water along a shore. Actually, a **beach** also includes sediment in the surf zone, which is underwater and therefore continually in motion. At low tide, when a large part of a beach is exposed, onshore winds may blow beach sand inland to form belts of coastal dunes. A landform commonly associated with beaches is the **barrier island** (see FIGURE 12.14). Sand dunes are typically the highest topographical points on a barrier island. Barrier islands are found along most lowland coasts. For example, the Atlantic and Gulf coasts of the United States consist mainly of a series of barrier beaches ranging from 15 to 30 kilometers in length and 1.5 to 5 kilometers in width, located 3 to 30 kilometers offshore. Examples include Coney Island, New York, and the long chain of islands known as the Outer Banks of North Carolina.

Barrier islands are topographically low, so they are very susceptible to flooding. During a major storm, surf washes across the low places and erodes them,

beach Wave-washed sediment along a coast.

barrier island A long, narrow, sandy island lying offshore and parallel to a lowland coast.

A This aerial view of the Outer Banks in North Carolina shows a series of barrier islands. The action of wind and waves constantly pushes the sediment on the island toward the mainland (left in this photo).

www.wiley.com/college/murck

tidal inlet

B Salishan Spit in Oregon, shown here, is attached to the mainland at one end while the other end terminates at a tidal inlet. ▶

C Spits and barrier islands often create a sheltered lagoon on the landward side. This lagoon is found on Cayo Costa Island in Florida.

Where the Ocean Meets Land   359

cutting *tidal inlets* that may remain open permanently. At such times fine sediment is washed between the barrier island and the mainland. Because of this deposition and erosion of sediment, the length and shape of barrier islands is always changing. Unfortunately, the ever-changing nature of barrier island coasts conflicts with our desire to erect permanent buildings on them. Property owners often protect their properties with artificial seawalls, which, though they may serve as a local fix, only hasten the erosion or deposition processes elsewhere on the coast.

Barrier island beaches typically exhibit depositional landforms such as *spits* (elongated ridges of sand or gravel that project from land into the open water of an embayment along the coast), *tombolos* (spit-like ridges of sand and gravel that join an island to the mainland), and *bay barriers*, which may completely close off the mouth of a small bay. A well-known example of a large, complex spit is Cape Cod, Massachusetts.

The elongated bay lying inshore from a barrier island or other low, enclosing strip of land (such as a coral reef) is called a *lagoon* (see Figure 12.14C). Lagoons are commonly fed by *estuaries*, the wide, fan-shaped mouths of rivers in the tidal zone where fresh and salt water meet. Lagoons and estuaries are important habitats for a wide variety of plants, birds, and animals. They also play an important role in the protection of mainland shorelines because they serve as buffers against storm waves. Human activities can also adversely affect these sensitive environments.

> **reef** A hard structure on a shallow ocean floor, usually but not always built by coral.

Coral reefs

Many of the world's tropical coastlines consist of limestone **reefs** built by vast colonies of organisms, principally corals, which secrete calcium carbonate (the main chemical

Coral reef hosts bountiful life FIGURE 12.15

Anthias fish swim through a multicolored coral reef off the coast of Fiji, a Pacific island. Reefs are a vital component of many ocean ecosystems because they provide shelter and nutrients for many kinds of fish.

constituent of limestone) as their skeletal material. Reefs are built up very slowly over thousands of years. Each of the tiny coral animals, called *polyps*, deposits a protective layer of calcium carbonate; over time the layers build up, forming a complex reef structure. *Fringing reefs* form coastlines that closely border the adjacent land. *Barrier reefs* are separated from the land by a lagoon, as in the case of the Great Barrier Reef off Queensland, Australia (see FIGURE 12.15). Reefs are highly productive ecosystems that support a diversity of marine life forms. They also perform an important role in the recycling of nutrients in shallow coastal environments. They provide physical barriers that dissipate the force of waves, protecting the ports, lagoons, and beaches that lie behind them, and are an important aesthetic and economic resource.

Corals require shallow, clear water in which the temperature remains above 18°C but does not exceed 30°C. Reefs therefore are formed only at or close to sea level and are characteristic of warm, low latitudes. They also favor places where there is deep sea upwelling that provides abundant nutrients, normal salinity—not too salty, not too fresh—and an absence of sediment being

deposited from large streams. Because of their very specific requirements, coral reefs are highly susceptible to damage from human activities as well as from natural causes such as tropical storms.

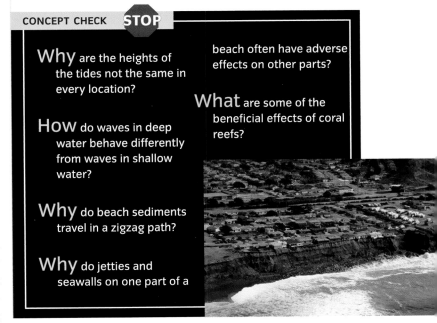

CONCEPT CHECK STOP

Why are the heights of the tides not the same in every location?

How do waves in deep water behave differently from waves in shallow water?

Why do beach sediments travel in a zigzag path?

Why do jetties and seawalls on one part of a beach often have adverse effects on other parts?

What are some of the beneficial effects of coral reefs?

Earth's Atmosphere

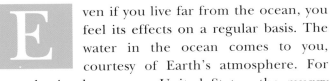

LEARNING OBJECTIVES

Identify the main chemical constituents of Earth's atmosphere.

Identify the four layers of Earth's atmosphere.

Explain how the atmosphere protects Earth from radiation and contributes to the warming of Earth's surface.

Describe how the Sun's heat and Earth's rotation affect the movement of air.

Relate the movement of air masses to climate zones.

Even if you live far from the ocean, you feel its effects on a regular basis. The water in the ocean comes to you, courtesy of Earth's atmosphere. For example, in the eastern United States, the muggy weather and thunderstorms of summertime originate with water that evaporated off the Gulf of Mexico. The atmosphere is the second component in the engine that drives the hydrologic cycle, and the main component responsible for the hydrologic cycle's effects on land.

COMPOSITION OF EARTH'S ATMOSPHERE

Atmosphere is the generic term for the gaseous envelope that surrounds a planet or any other celestial body.

A portrait of Earth's atmosphere FIGURE 12.16

In this spectacular photograph, taken from the International Space Station as it flew over the Pacific Ocean on July 21, 2003, you can clearly see the troposphere (the lower atmosphere) as a thin blue band above the curved horizon. Anvil-shaped clouds in the foreground are thunderstorms. Their unique shape is caused by the fact that warm air in the clouds cannot rise past the tropopause (the boundary between the troposphere and stratosphere).

air A mixture of 78 percent nitrogen, 21 percent oxygen, and trace amounts of other gases, found in Earth's atmosphere.

Air is the gaseous envelope that surrounds one planet in particular, Earth (see FIGURE 12.16).

Air is an invisible, normally odorless mixture of gases and suspended particles. Two components of air are highly variable in concentration: *aerosols* and *water vapor*. Aerosols are liquid droplets or solid particles that are so small that they remain suspended in the air. Water droplets in fogs are liquid aerosols. Examples of solid aerosols are tiny ice crystals, smoke particles from fires, and sea-salt crystals from ocean spray. Water is always present in the air, but in varying amounts expressed by the *humidity*. On a hot, humid day in the tropics, as much as 4 percent of the air by volume may be water vapor. On a crisp, cold day, less than 0.3 percent may be water vapor. (Note that 100 percent relative humidity does not mean the air is 100 percent water vapor! It simply means that the air contains all the water vapor it can carry at the temperature of the air. The temperature at which the relative humidity is 100% is called the dew point.)

What air is made of FIGURE 12.17

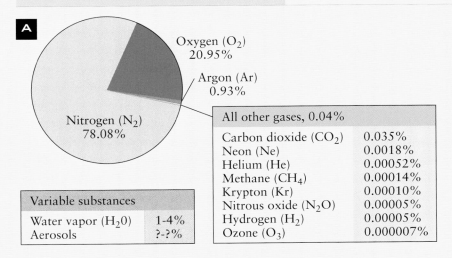

A

Oxygen (O$_2$)
20.95%

Argon (Ar)
0.93%

Nitrogen (N$_2$)
78.08%

All other gases, 0.04%	
Carbon dioxide (CO$_2$)	0.035%
Neon (Ne)	0.0018%
Helium (He)	0.00052%
Methane (CH$_4$)	0.00014%
Krypton (Kr)	0.00010%
Nitrous oxide (N$_2$O)	0.00005%
Hydrogen (H$_2$)	0.00005%
Ozone (O$_3$)	0.000007%

Variable substances	
Water vapor (H$_2$O)	1-4%
Aerosols	?-?%

◄ **A** Air contains two substances whose concentration varies from place to place and day to day: water vapor and aerosols. The rest of the atmosphere consists primarily of nitrogen and oxygen, with a smattering of other gases.

B The specific percentages of each gas can change over ▶ time. This graph shows the atmosphere's carbon dioxide content, measured at the Mauna Loa Observatory in Hawaii since 1958. You can see two kinds of variation in the figure. One is a seasonal fluctuation, as plants grow rapidly and consume carbon dioxide in the spring, and then die and decay in the fall, releasing carbon dioxide back into the atmosphere. The second change is a steady increase in carbon dioxide due to the burning of fossil fuels. The amount of carbon dioxide has increased from 320 ppm (parts per million) to 380 ppm in 50 years. If the trend continues, it will have profound implications for Earth's climate, as we discuss in Chapter 13.

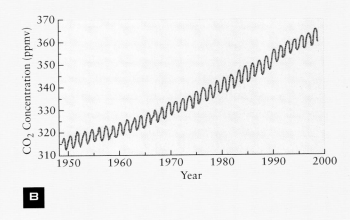

B

Both the water vapor and aerosol contents of air vary widely. For this reason, the relative amounts of the remaining gases are generally reported as if the air were entirely lacking in water vapor and aerosols. When these two components are ignored, the relative proportions of the remaining gases in the air—termed dry air—turn out to be essentially constant. As shown in FIGURE 12.17, three gases—nitrogen (78%), oxygen (21%), and argon (0.93%)—make up 99.96 percent of dry air by volume. The remaining gases (carbon dioxide, 0.035%; neon, 0.0018%; and six others) are present in very small quantities. However, some of these minor gases are profoundly important for life on Earth because they absorb certain wavelengths of sunlight. They act both as a warming blanket and as a shield against deadly ultraviolet radiation.

The oxygen and carbon dioxide in the atmosphere are constantly being removed or replenished by plants, animals, and chemical processes such as the weathering of rock. If anything happens to affect the rate of these processes, the makeup of the atmosphere will change. The designers of Biosphere 2 learned this the hard way, and we may be learning it too (see Figure 12.17B).

Four major layers FIGURE 12.18

Temperature varies with altitude in the atmosphere. In the lowest level, called the *troposphere*, the temperature drops rapidly with increasing altitude. In the next layer, the *stratosphere*, the reverse is true. Two more reversals occur in the *mesosphere* and *thermosphere*. The air in the thermosphere is so tenuous that this zone is defined by rocket scientists (but not by geologists) to be part of "space." NASA awards a space-flight badge to anybody who flies above 80 kilometers.

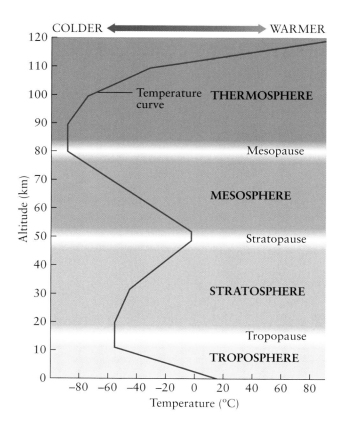

As sunlight passes through the atmosphere, water vapor and some of the minor gases absorb certain wavelengths. Absorption of the Sun's radiation raises the temperature of Earth's surface and of the air itself. Scientists have discovered that the atmosphere is composed of four layers with distinct temperature profiles, each separated by boundaries called *pauses* (see FIGURE 12.18).

The troposphere and greenhouse effect

Most humans, with the exception of military pilots and astronauts, never fly outside of the troposphere. The troposphere contains 80 percent of the actual mass of the atmosphere, including virtually all the water vapor and clouds. The thin blue layer of atmosphere that you saw in Figure 12.16 is mainly the troposphere. This is also where almost all weather-related phenomena originate. Although very little mixing occurs between the troposphere and the stratosphere, the troposphere itself is constantly moving and thoroughly mixed by winds.

The troposphere contains most of the heat-absorbing gases (also called *greenhouse gases*) that play a role in warming Earth's surface. The most important greenhouse gas is water vapor. Another greenhouse gas is carbon dioxide, and still another is methane, the main constituent of natural gas. About 30 percent of the incoming solar energy is simply reflected back into space by clouds and the ocean, but the remaining 70 percent is absorbed by the ocean, atmosphere, land, and biosphere. Eventually, all of this energy is re-emitted in the form of long-wavelength infrared radiation from Earth back into outer space. The greenhouse gases intercept some of the infrared radiation and prevent it from escaping, in just the same way that the glass of a greenhouse does (see FIGURE 12.19). That is why this natural atmospheric process is called the **greenhouse effect**. Without the greenhouse effect, the surface of Earth would be a cold and inhospitable place. However, it is a matter of serious concern that the amount of greenhouse gases in the atmosphere is steadily increasing.

greenhouse effect The process through which long-wavelength (infrared) heat energy is absorbed by gases in the atmosphere, thereby warming Earth's surface.

How a greenhouse works FIGURE 12.19

The glass in a real greenhouse (**A**) and certain "greenhouse gases" in the troposphere (**B**) work in approximately the same way.

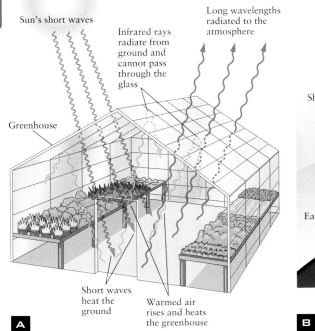

www.wiley.com/college/murck

Sun's short waves

Infrared rays radiate from ground and cannot pass through the glass

Long wavelengths radiated to the atmosphere

Greenhouse

Short waves heat the ground

Warmed air rises and heats the greenhouse

A

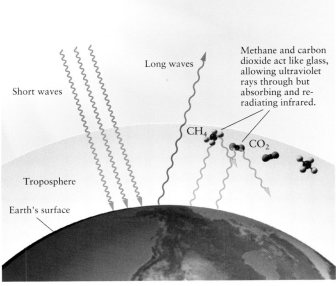

Short waves

Long waves

Methane and carbon dioxide act like glass, allowing ultraviolet rays through but absorbing and re-radiating infrared.

CH_4

CO_2

Troposphere

Earth's surface

B

The stratosphere and the ozone layer

The stratosphere contains 19 percent of the atmosphere's total mass, which means that the troposphere and the stratosphere combined contain 99 percent of the atmosphere. The three outer layers of the atmosphere—the stratosphere, mesosphere, and thermosphere—play very important roles for life on Earth despite their relatively small masses (when compared to the troposphere). Oxygen in various forms absorbs harmful ultraviolet rays coming from the Sun (see FIGURE 12.20). The thermosphere absorbs short ultraviolet wavelengths, the mesosphere absorbs intermediate wavelengths, and the stratosphere absorbs long wavelengths. The gas ozone (O_3) in the stratosphere, in tiny but vital amounts, absorbs the most dangerous of the ultraviolet rays.

A shield against radiation FIGURE 12.20

The ultraviolet radiation coming from the Sun can be harmful or lethal; generally speaking, the shorter the wavelength, the more harmful the radiation. Fortunately, the atmosphere protects us from almost all of these rays because they are absorbed by three kinds of oxygen—O, O_2, and O_3 (ozone).

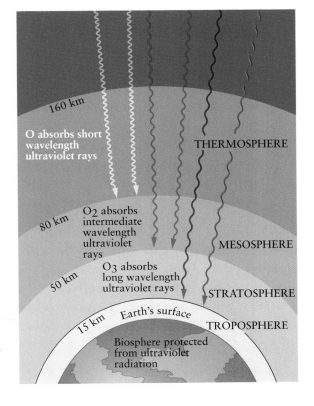

160 km

O absorbs short wavelength ultraviolet rays

THERMOSPHERE

80 km

O_2 absorbs intermediate wavelength ultraviolet rays

MESOSPHERE

50 km

O_3 absorbs long wavelength ultraviolet rays

STRATOSPHERE

15 km Earth's surface

TROPOSPHERE

Biosphere protected from ultraviolet radiation

NATIONAL GEOGRAPHIC

In 1985, scientists discovered that a previously unnoticed gap (blue) in the ozone layer was forming over Antarctica every winter. (A smaller hole also formed over the Arctic Ocean.) For several years the hole grew larger and the depletion at its center grew more severe (purple). However, in recent years it has apparently begun to stabilize.

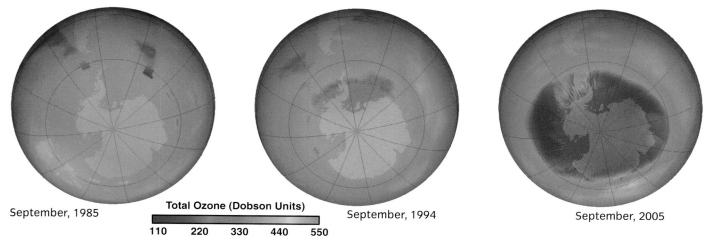

September, 1985

Total Ozone (Dobson Units)

110 220 330 440 550

September, 1994

September, 2005

> **ozone layer** A zone in the stratosphere where ozone is concentrated.

In recent years scientists have become very concerned about the destruction of the **ozone layer** by chemical pollutants (see FIGURE 12.21). Among them are chlorofluorocarbons (CFCs), which were formerly an ingredient in aerosol sprays but were banned by an international treaty in 1996. Though it is still too early to be certain, there are signs that the destruction of ozone is abating. Perhaps, in the future, this will be cited as an example of a successful environmental intervention.

MOVEMENT IN THE ATMOSPHERE

Two things energize the atmosphere: the Sun's heat and Earth's rotation. Because Earth is a sphere, the Sun does not warm every place on Earth equally. In places where the Sun is directly overhead, the incoming rays are perpendicular to the surface and a maximum amount of heat is received per unit area (see FIGURE 12.22). But because of the curvature of Earth's surface, at all other locations the surface is at an angle to the incoming rays. Therefore, they receive less heat per unit of surface area. Winds and ocean currents are the natural processes by which the Earth system redistributes the heat more evenly. Currents move heat from the equator—where the input of solar heat is greatest—toward the poles, where it is least.

Unequal heating of Earth's surface causes *convection currents* in the atmosphere. Heated air near the equator expands, becomes lighter, and rises. Near the top of the troposphere it spreads outward toward the poles. As the upper air travels northward and southward toward the poles, it gradually cools, becomes heavier, and sinks. Upon reaching the surface, this cool air flows back toward the equator, warms up, and rises again, thereby completing a convective cycle. In reality, it is not quite so simple: The global circulation actually organizes itself into three *convection cells*, which interlock like gears (see FIGURE 12.23).

The Coriolis effect If Earth did not rotate, the convection currents would simply flow north and south. But of course Earth does rotate, and its rotation complicates the convection currents in the atmosphere (and in the ocean). The **Coriolis effect** causes anything that moves freely with respect to the rotating Earth

> **Coriolis effect** An effect due to Earth's rotation that causes a freely moving body to veer from a straight path.

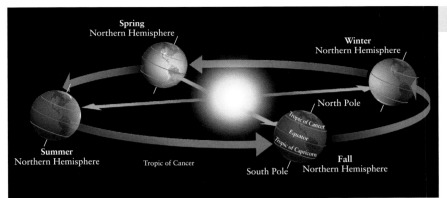

At all times, the point on Earth's surface where the Sun is directly overhead receives the maximum amount of solar energy (about 1366 watts per square meter). At the spring and fall equinoxes (center) the maximum energy falls on the equator. In December (right), because of the 23.5° tilt of Earth's axis of rotation to the plane of its passage around the Sun, solar input is most intense at 23.5° S, the Tropic of Capricorn. In June (left), the incoming solar energy is most intense at 23.5° N, the Tropic of Cancer.

Global atmospheric circulation
Figure 12.23

Huge convection cells transfer heat from equatorial regions, where the input of solar energy is greatest, toward the poles, where the solar input is least. Because Earth is rotating, the flow of air toward the poles and the return flow toward the equator are constantly deflected sideways, creating the circulating air masses you may have seen on weather maps. The convection cells are permanent features of Earth's atmosphere, and therefore have a great influence both on day-to-day weather and on long-term climate.

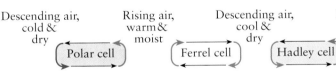

(such as a plane, a missile, or the wind) to veer off course. It is similar to trying to throw a ball to a friend when you are both on a spinning merry-go-round. In the Northern Hemisphere the Coriolis effect causes a moving mass to veer to the right, and in the Southern Hemisphere to the left. The effect strongly influences both the global pattern of wind systems (see Figure 12.23) and the pattern of ocean currents, including the deep-ocean thermohaline "conveyor belt" circulation.

Wind systems   The Coriolis effect breaks up the flow of convective air between the equator and the poles into belts. For example, a large belt or cell of circulating air lies between the equator (0°) and about 30° latitude in both the Northern and Southern

Hemispheres. Warm air rises at the equator, creating a low-pressure zone called the *intertropical convergence zone* (see FIGURE 12.24A). The air rises to the top of the troposphere and begins to flow toward the poles, but it veers off course as a result of the Coriolis effect. By the time it reaches a latitude of 30°, the high-altitude air mass has cooled and started to sink. The descending air flows back across Earth's surface toward the equator. As it flows, the land and sea warm the air so that it eventually becomes warm enough to rise again. The low-latitude cells (from 0° to 30° N and S) created by this circulation pattern are called *Hadley cells*. The prevailing surface winds in Hadley cells are deflected by the Coriolis effect so that they blow northeasterly in the Northern Hemisphere (they flow from the northeast toward the southwest), whereas in the Southern Hemisphere they are southeasterly. These winds are called *trade winds* because their consistent direction and flow carried trade ships across the tropical oceans at a time when winds were the chief source of power (see FIGURE 12.24B.)

A second set of convecting air cells, called *polar cells*, lies over the polar regions. In a polar cell, frigid air flows across the surface away from the pole and toward the equator, slowly warming as it moves. When the polar air has reached about latitude 60° N or S, it has warmed sufficiently to rise convectively high into the troposphere and flow back toward the pole where it

cools and descends again, thereby completing the convection cell. Because of the Coriolis effect, the cold air that flows away from the poles is deflected to the right, giving rise to a wind system called *polar easterlies*.

Between the Hadley cells and the polar cells, a third, but less well-defined set of convection cells is found in the mid-latitudes between about latitude 30° to 60° N and S. In these mid-latitude cells, called *Ferrel cells*, air flows northerly so by the Coriolis effect is deflected to the right, creating a wind that blows from the west. That is why weather systems in the continental United States (and southern Canada) travel from west to east. These circulation patterns will become clearer if you take some time to study Figure 12.23.

HOW THE ATMOSPHERE REGULATES CLIMATE

The global patterns of airflow ultimately control the variety and distribution of Earth's climatic zones. *Climate* is an average of weather patterns over a long period. The processes we think of as *weather*—wind, rain, snow, sunshine, storms, and even floods and droughts—are just temporary local variations when viewed against the more stable, longer-term background of global climate.

Trade winds FIGURE 12.24

◀ **A** The Intertropical Convergence Zone, where the trade winds of the North and South Hemispheres merge, shows up very clearly in this satellite photograph. Warm, rising air near the equator causes ocean water to evaporate and form a nearly perpetual band of storm clouds.

B The continual blowing of the east-to-west trade winds has caused this tree, near the south tip of Hawaii, to lean far to one side. ▶

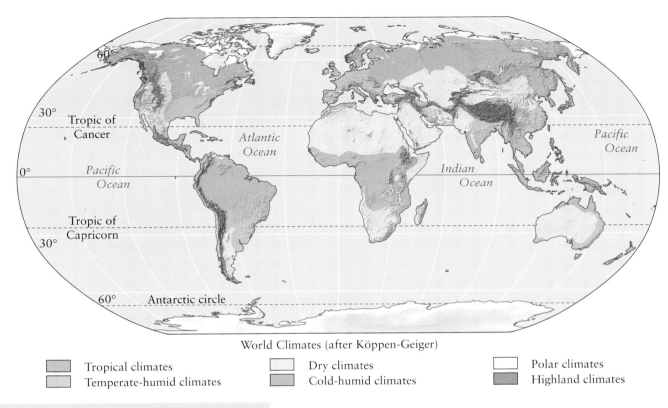

World Climates (after Köppen-Geiger)

▨ Tropical climates	▨ Dry climates	▢ Polar climates
▨ Temperate-humid climates	▨ Cold-humid climates	▨ Highland climates

Global climate zones FIGURE 12.25

In a simplified version of Köppen's climate classification system, there are six basic types of climate: tropical, dry, temperate-humid, cold-humid, polar, and highland. These climate zones are strongly related to the global pattern of atmospheric circulation.

Climatologists use several classification schemes to describe and categorize Earth's climatic zones. In FIGURE 12.25, we illustrate the *Köppen-Geiger climate system*, originally devised in 1918 by the Austrian scientist Vladimir Köppen. This system defines climate zones on the basis of temperature and precipitation. The most striking feature in this map is the close link between climate types and the prevailing wind currents illustrated in Figure 12.23. The world's deserts are mostly concentrated along the 30° N and 30° S parallels, where the barometric pressure is high and the descending air mass is dry. The tropical jungles of Africa and South America lie near the equator, where we expect low barometric pressure and lots of warm, moist air.

In this chapter we have emphasized that the hydrosphere and atmosphere interact in a complex and ever-changing way. We will conclude with a Case Study of a recently discovered phenomenon that highlights this interaction: El Niño.

CONCEPT CHECK STOP

What are the two components of air that vary significantly in concentration from day to day and place to place?

How do greenhouse gases warm Earth's surface?

Why does the ozone hole allow more ultraviolet radiation from the Sun to reach Earth's surface?

Why does the Sun's heat not warm Earth's surface uniformly?

What are trade winds, westerlies, and polar easterlies, and what are their physical causes?

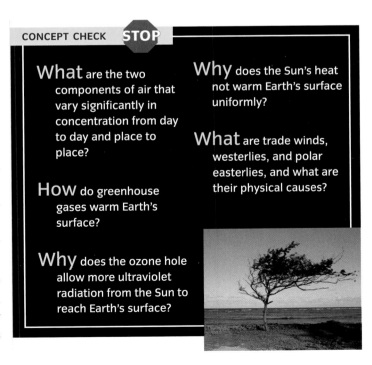

El Niño (The Return of the Child)

The fishing grounds off the coast of Peru are among the richest in the world, sustained by upwelling cold waters filled with nutrients. However, each year around Christmas time a slightly tepid current warms up the coastal waters, reducing the fish population a little and giving the fishermen a rest over the holidays. The fishermen have long called this current "El Niño," a reference in Spanish to the Christ child. At irregular intervals of 2 to 7 years, however, the El Niño current is exceptionally warm and long-lived. Climatologists have transferred the name "El Niño" to this unusually warm current and its associated effects on the climate.

In a normal year (see **FIGURE A**), persistent trade winds blow westward across the tropical Pacific from the zone of upwelling water off Peru. This water warms as it moves westward, forming a large pool of warm water above the thermocline in the western Pacific. The warm water causes the moist maritime air to rise and cool, bringing abundant rainfall to Indonesia.

During an El Niño year (see **FIGURE B**), the trade winds slacken and the pool of warm water moves eastward to the central Pacific. Descending cool, dry air brings drought conditions to Indonesia, while rising moist air above

the warm-water pool greatly increases rainfall in the mid-Pacific. Surface waters in the eastern Pacific become warmer, and downwelling shuts off the supply of deep-water nutrients. This adversely affects the fishery off Peru.

During a strong El Niño year, such as 1997–98, the weather is disrupted over the entire planet.

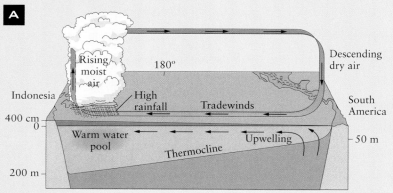

Normal conditions in the tropical Pacific

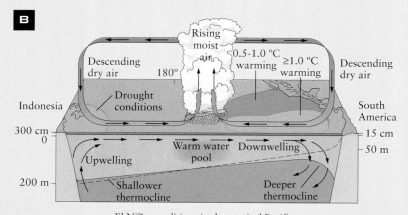

El Niño conditions in the tropical Pacific

NATIONAL GEOGRAPHIC

Global Locator

The world's third-longest coral barrier reef, and the only living coral reef in the continental United States, runs parallel to the Florida Keys. It stretches for 133 kilometers, and in most places is just a short boat ride offshore. Here is a place where you can enjoy a fun vacation and watch geology in a action at the same time. Unlike other animals, whose skeletons become stone only after thousands of years of burial, the skeletons of corals are limestone from day one, and they gradually add to the landmass of the continent.

In the Florida Keys Reef, you can find a variety of coral, such as fire coral, mustard hill coral (see FIGURE A, foreground) and brain coral (see FIGURE A, background). Unfortunately, though, the coral in the Florida Keys has been in declining health for a long time. In FIGURE B, you can see how part of Carysfort Reef progressed from a healthy state in 1975 to a sick one in 1985, and by 1995 was almost completely dead. By 2000, Carysfort Reef had lost 90 percent of its original coral cover.

The causes of coral decline are numerous and not completely understood. Coral polyps can be killed by natural causes, such as hurricanes, but the reef usually recovers from them over the long term. Manmade injuries, such as scarring from boats that run aground on the reef, are much harder for coral to recover from. In recent years, a pervasive problem has been coral bleaching (see FIGURE C). When the water gets too warm, coral polyps expel the symbiotic algae that ordinarily give them their bright colors. The bleached coral is not necessarily dead, but without its algae it will die soon. The white parts of the staghorn coral in this photo are bleached, whereas the brown parts are still healthy (but probably in danger).

NATIONAL GEOGRAPHIC

CHAPTER SUMMARY

1 The Oceans

1. There has been an ocean on this planet for at least 4.0 billion years. The water in the ocean may have condensed from steam produced by primordial volcanic eruptions or some of the water may have been delivered to the planet's surface via cometary impacts.

2. Seawater ranges in **salinity** from 3.3 to 3.7 percent, and varies measurably from place to place. Freezing and evaporation make the water saltier, whereas rain, snow, and *river flow* make it less salty.

3. The water in the ocean forms layers based on density, which is controlled by temperature and salinity. The water in the surface layer, from the surface to a depth of about 100 meters, is relatively warm, and moves in broad, wind-driven currents. The water in the deep layer moves in great, slow, global currents driven by differences in temperature and salinity. Between the surface layer and the deep layer is a transition zone, the thermocline.

2 Where the Ocean Meets Land

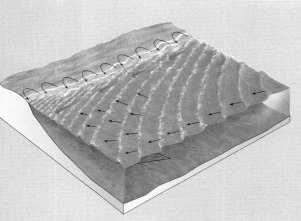

1. The ocean's level varies over a wide range of time scales and for various reasons. Worldwide changes in sea level due to the melting or growth of polar ice sheets take place over thousands of years. Local changes due to the **tides** occur twice a day along most coasts.

2. When a wave moves onto the shore, its motion is distorted, as the shallow bottom interferes with the circular motion of water in the wave. It eventually breaks, creating turbulent **surf**. Much of the resulting erosion and transport of sediment is accomplished by **longshore currents** (in the *surf zone*) and by **beach drift** (on land).

3. Shorelines are highly variable, but three basic types are most common: rocky coasts and **wave-cut cliffs**, lowland **beaches** and **barrier islands**, and coral **reefs**. Shorelines and coastal landforms are shaped by a combination of erosive and depositional processes. Storms and artificial structures often accelerate the process of coastal erosion.

3 Earth's Atmosphere

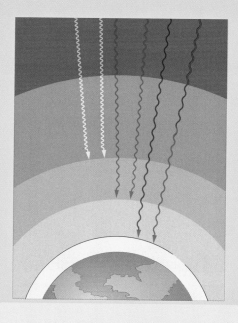

1. The composition of **air**, excluding *water vapor* and *aerosols*, is 99.96 percent by volume nitrogen, oxygen, and argon.

2. Air temperature changes markedly as one moves upward from the surface. As a result, there are four distinct layers in the atmosphere, each with a distinct temperature profile. From the bottom up, they are the *troposphere*, the *stratosphere*, the *mesosphere*, and the *thermosphere*.

3. Most weather-related phenomena occur in the troposphere. The troposphere also contains most of the gases that play a role in Earth's **greenhouse effect**. The **ozone layer**, which protects life on Earth by absorbing harmful incoming ultraviolet radiation, is a concentration of O_3 in the stratosphere.

4. The amount of heat energy from the Sun reaching Earth's surface is greatest near the equator and least near the poles. Winds result from the unequal distribution of energy. *Convection currents* move heat from the equator toward the poles. Earth's rotation produces the **Coriolis effect**, which causes winds to veer to the right in the Northern Hemisphere and to the left in the Southern Hemisphere.

5. *Climate* is the average or "normal" *weather* for a particular location over a long time. The heat-absorbing capacity of the ocean combines with circulation in the atmosphere to control the variety and distribution of Earth's climatic zones.

KEY TERMS

CRITICAL AND CREATIVE THINKING QUESTIONS

1. Compare Earth's atmosphere with what is known about the atmospheres of other planets. How does its composition differ, and why? Why do some planets have dense atmospheres, and others have nearly none? How does its circulation pattern differ? Does Jupiter's Great Red Spot have any analogues on Earth?

2. During the early Tertiary epoch, North and South America were not connected. How might that have changed the global pattern of ocean circulation? Do some library research and suggest some tests for your hypothesis.

3. Visit a shoreline. If possible, visit the same spot on the shoreline on two or more occasions. What kinds of coastal landforms do you observe? What kinds of changes can you notice from one visit to the next?

4. Rocks of the Permian and Triassic age in both North America and Western Europe record the existence of desert conditions. What might have caused the deserts to form?

5. Geologic evidence indicates that the atmosphere of the early Earth was free of oxygen. What constraints might that have had on early life?

What is happening in this picture ?

In this "Calvin and Hobbes" comic strip, Calvin is telling his pet tiger, Hobbes, about his theory that the days are getting colder because the sun is going out. Let's call this Hypothesis 1. His father, on the other hand, says the days are getting colder because Earth is getting farther from the Sun—Hypothesis 2. Another reasonable hypothesis is that Earth's axis is tilted, which makes the sunlight less direct in winter—Hypothesis 3. What scientific tests could you perform to determine why the weather is getting colder? Which explanation do you think is right—Calvin's, his dad's, or Hypothesis 3?

1. Which of the following processes increases the salinity of seawater?

   a. freezing
   b. rain, snow, and river runoff
   c. evaporation
   d. Both a. and c. are correct.

2. Which of the following processes decreases the salinity of seawater?

   a. freezing
   b. rain, snow, and river runoff
   c. evaporation
   d. Both a. and c. are correct.

3. The water in the _____ extends to a depth of 100 m, is relatively warm, and moves in broad, wind-driven currents.

   a. deep zone
   b. surface layer
   c. thermocline
   d. transition zone

4. Deep water currents flow in a/an _____ direction of the currents at the ocean's surface.

   a. parallel
   b. perpendicular
   c. opposite
   d. oblique

5. In coastal regions the ocean can act to _____.

   a. increase average temperatures of the land
   b. decrease average temperatures of the land
   c. induce ice formation
   d. moderate temperatures of the land

6. Which of the following statements about global sea level is true?

   a. Global sea level is a constant over both human and geologic time.
   b. Global sea level can vary tens to hundreds of meters over geologic time.
   c. Global sea level varies only a few meters over geologic time.
   d. Large variations in global sea level can be an effect of climate change.
   e. Both b. and d. are true statements.

7. This illustration depicts the forces involved with the generation of tides on Earth. Label the illustration with the following terms.

distorted water level	common center of mass of Earth-Moon system
inertial force	
gravitational force	uniform level of water (theoretical)
	tide-raising force

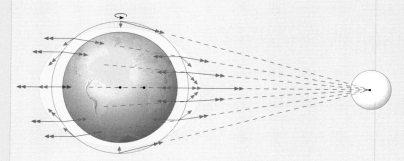

8. Shorelines are highly variable, but three basic types of coastlines are most common. Which of the following is not a common type of coastline?

   a. rocky coasts and wave-cut cliffs
   b. mudflats
   c. lowland beaches and barrier islands
   d. coral reefs

9. This illustration is a block diagram showing the beach and shore environment. Place arrows on the diagram showing the direction of longshore current and sand transport. Be sure to label each of the sets of arrows as being either longshore current or sand transport.

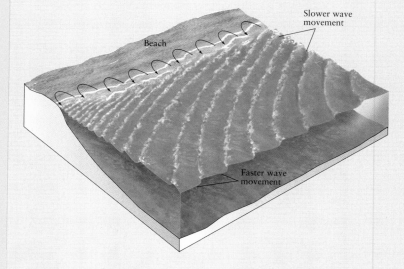

Slower wave movement

Beach

Faster wave movement

10. Which of the following chemical constituents of Earth's atmosphere is the most abundant?

   a. oxygen
   b. carbon dioxide
   c. argon
   d. nitrogen
   e. water vapor

11. This illustration shows a graph of temperature variations vertically through the atmosphere. Label the illustration with the following terms:

   thermosphere   tropopause
   stratosphere   mesopause
   troposphere   stratopause
   mesosphere

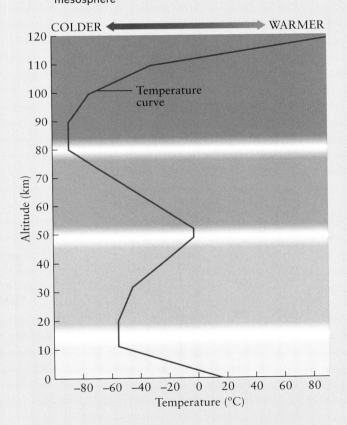

12. How does the atmosphere contribute to the warming of Earth's surface?

   a. by trapping ultraviolet rays in the troposphere
   b. by focusing the Sun's rays through refraction
   c. through evaporation which causes heating of the surface
   d. by absorbing and reradiating infrared radiation

13. The atmosphere can shield us from harmful _____ radiation from the sun as they are absorbed by O, O_2 and O_3 (ozone).

   a. infrared
   b. ultraviolet
   c. gamma
   d. neutron

14. The movement of air masses are a result of _____.

   a. differential heating of the curved Earth by the Sun
   b. the deflection of moving air masses by Earth's rotation
   c. the seasonal melting of ice in the polar regions
   d. Both a. and b are correct.
   e. Both b and c are correct.

15. Subtropical high pressure is created by _____.

   a. descending cool and dry air at the juncture of a Ferrel cell and Hadley cell
   b. rising warm and moist air at the juncture of a Polar cell and Ferrel cell
   c. rising warm and moist air at the juncture of a Ferrel cell and Hadley cell
   d. descending cool and dry air at the juncture of Polar cell and Ferrel cell

Deserts, Glaciers, and Climate Change [13]

The Mandara Lakes in Libya excite the senses. Flat as a mirror, the water reflects a turquoise sky. The palm trees are dwarfed by the dunes of the Sahara Desert that creep up to within a few feet of them. It is easy to see why explorers have always loved the desert: Its vastness and the beauty of its oases make human creations seem trivial by comparison.

Explorers of another stripe, like the mountaineer exulting at the summit of Aurora Peak in Alaska (inset), have always loved the world's icy wildernesses—in this case, the mountaineer is looking directly at Black Rapids Glacier. Almost every description of the Sahara Desert could also apply to this landscape: abstract, beautiful, immense, remote . . . and vulnerable.

For geologists, deserts and glaciers are linked by more than beauty; they are joined by extreme climates. Erosional forces such as ice and wind, which play lesser roles elsewhere, are dominant in these regions.

Deserts and glaciers are important as barometers of climate change. Glaciers and ice caps are retreating, a trend that many climatologists believe will continue throughout the twenty-first century. It is a trend that humans are partially responsible for because of our production of "greenhouse gases." Likewise, deserts grow or shift, in part because of changes in rainfall, and in part because of the way humans manage soil and water resources. Every one of us has good reason to pay attention to deserts and glaciers, even if we never set foot in the Sahara sand or on an Alaskan mountaintop.

376

Deserts and Drylands

LEARNING OBJECTIVES

Identify five different geologic settings for deserts.

Explain how the wind shapes desert landforms through abrasion and deflation.

Describe the structure of a sand dune.

Identify five types of sand dunes.

Explain why human activity sometimes causes desertification, and what can be done about it.

Convection in the atmosphere (see Chapter 12) creates huge belts or cells of rising and falling air masses (see FIGURE 13.1). This results in three global belts of high rainfall and four belts of low rainfall. The high- rainfall belts are regions of *convergence* where huge, moist air masses meet and rise upward. These belts lie in the equatorial region and along the two polar fronts, at approximately 50° N and S latitudes, resulting in the warm-humid (tropical) and cold-humid (polar

The world's deserts FIGURE 13.1

This map shows the distribution of arid and semiarid climates and the major deserts associated with them. Many of the world's great deserts are located where belts of dry air descend along the 30° N and 30° S latitudes. Notice also that re- gions of cold, descending air also surround both of the poles. Despite being covered by ice, the polar regions receive little precipitation and are considered to be frozen deserts—indeed, Antarctica is the world's largest desert.

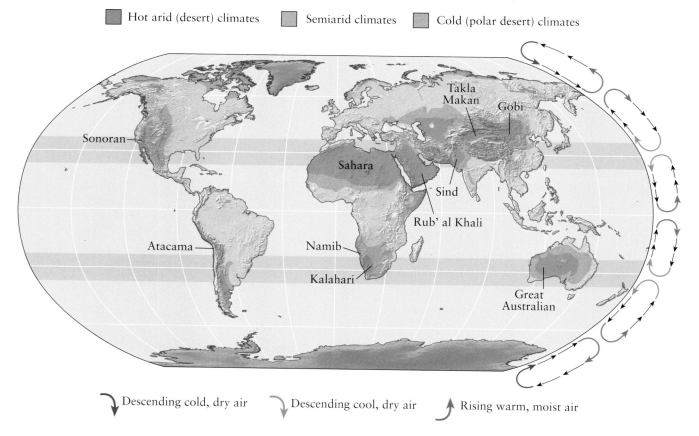

Hot arid (desert) climates   Semiarid climates   Cold (polar desert) climates

Descending cold, dry air   Descending cool, dry air   Rising warm, moist air

front) climate zones. The four global belts of low rainfall are regions of *divergence* where cool, dry air masses move downward and apart. These belts lie in the two polar regions and in the subtropical regions along the 30°N and 30°S latitudes. The dry air masses and resulting low annual precipitation in these regions create two belts of dry climate in the subtropical regions, as well as the dry, cold climates of the polar regions.

TYPES OF DESERTS

The word "desert" literally means a deserted (that is, almost uninhabited) region that is nearly devoid of vegetation. However, in recent years, irrigation has changed the meaning of the word by making many desert regions suitable for agriculture and therefore habitable. As a result, the term **desert** is now defined in terms of the annual precipitation: less than 250 millimeters per year. Desert lands total about 25 percent of the land area of the world outside the polar regions. In addition, there is a smaller percentage of *semiarid* land in which the annual rainfall ranges between 250 and 500 millimeters.

■ **desert** An arid land that receives less than 250 millimeters of rainfall or snow equivalent per year, and is sparsely vegetated unless it is irrigated.

Deserts can be separated into five categories, of which we have already mentioned two. The world's most extensive deserts, *subtropical*, are associated with the two belts of low rainfall near the 30° N and S latitudes. These include the Sahara, Kalahari, and Great Australian deserts. *Polar deserts* receive as little precipitation as subtropical deserts, but because it comes in the form of snow that never gets a chance to melt, these deserts gradually build up a thick ice sheet. The other three categories are less related to global air circulation patterns, and more related to local geography. They are *continental interior deserts*, *rainshadow deserts*, and *coastal deserts* (see **FIGURE 13.2** on the following page).

Later in the chapter we will take a closer look at the unique environment of the polar desert. For now, let's focus on the geologic processes that characterize the world's hot deserts.

WIND EROSION

As we discuss in Chapter 7, wind is an important agent of erosion, transport, and deposition. Processes related to wind—*eolian* processes—are particularly effective in arid and semiarid regions.

Wind-blown sediment Sediment carried by the wind tends to be finer than that moved by water or ice. Because the density of air is far less than that of water, air cannot move as large a particle as water flowing at the same velocity. In most regions, the largest particles that can be lifted in the airstream are grains of sand.

Though they are smaller in size, wind-blown particles are similar to waterborne sediments in the way they travel. The largest grains are transported through **surface creep**. As wind speed increases, smaller grains may be bumped or lifted into the air, where they experience **saltation**. Finer dust-sized particles may be carried aloft to heights of a kilometer or so, where they can travel along in **suspension** as long as the wind keeps blowing. Such particles can be carried all the way across the ocean.

Mechanisms of wind erosion Flowing air erodes the land surface in two ways. The first, **abrasion**, results

■ **surface creep** Sediment transport in which the wind causes particles to roll along the ground.

■ **saltation** Sediment transport in which particles move forward in a series of short jumps along arc-shaped paths.

■ **suspension** Sediment transport in which the wind carries very fine particles over long distances and periods of time.

■ **abrasion** Wind erosion in which airborne particles chip small fragments off rocks that protrude above the surface.

A The Sahara is the greatest of the world's subtropical deserts. Here, a camel caravan crosses the desert in Libya.

B Mongolian nomads transport their belongings through the Altai Mountains, which border on the Gobi desert. This is a typical inland continental desert, which (even though it lies outside the subtropical latitudes) receives little rain because it is so far from the ocean. ▶

◀ **C** Rainshadow deserts form when a mountain range creates a barrier to the flow of moist air, causing a zone of low precipitation to form on the downwind side of the range. Mountains do not completely block air, but they do effectively remove most of its moisture. Death Valley, seen here, lies just east of the Sierra Nevada, the tallest mountain range in the continental United States.

D Baja California, the long narrow peninsula in Mexico just south of its border with the western United States, consists mostly of coastal desert. Coastal deserts occur locally along the western margins of continents, where cold, upwelling seawater cools and stabilizes maritime air flowing onshore, decreasing its ability to form precipitation. ▶

from the impact of wind-driven grains of sand. Abraded rocks acquire distinctive, curved shapes and a surface polish. A bedrock surface or stone that has been abraded and shaped by wind-blown sediment is a *ventifact* ("wind artifact"). When preserved in sedimentary strata, ventifacts can tell geologists the direction of prevailing winds in the past. Other landforms characteristic of desert regions, such as steep-sided but flat-topped buttes, also result, at least in part, from wind erosion of bedrock (see **FIGURE 13.3**).

Abrasion by wind FIGURE 13.3

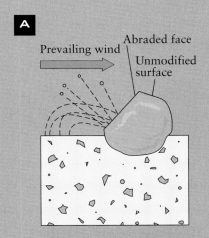

A Wind-blown sand peppers the upwind side of an exposed rock, eventually abrading it to a smooth, inclined surface.

B Ventifacts, each with at least one smooth, abraded surface facing upwind, litter the ground near Lake Vida in Victoria Valley, Antarctica.

C The Finger of God, a pillar-like butte of sandstone in the Kalahari Desert of South Africa, rests precariously on a pyramid of shale. In North America, rock pillars like these are known as *hoodoos*. They form when a resistant stratum lies atop a stratum that is less resistant to weathering. Note the sandstone mesa in the background, which has formed in the same way.

These drawings show how progressive removal of sand from sediment with different-sized particles can lead to the formation of desert pavement.

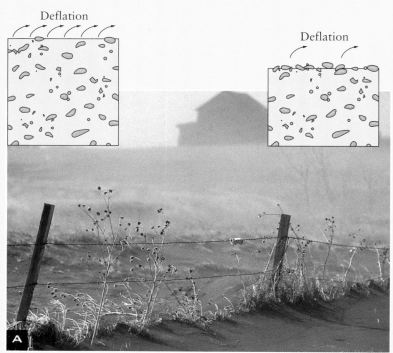

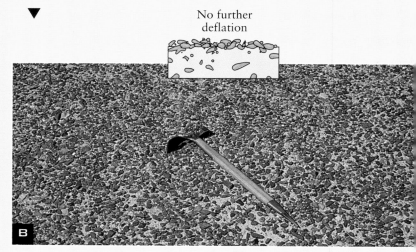

◀ A Deflation is in progress in this plowed field in eastern Colorado. If the deflation continues long enough, and if the soil contains mixed particle sizes, including gravel, the result can be what you see here.

B This photo shows desert pavement on the floor of Searles Valley in California. The gravel is too coarse for the wind to move, and it prevents further deflation.

▼

WIND DEPOSITS AND DESERT LANDFORMS

Any sediment that the wind removes from one place must eventually be deposited somewhere else—sometimes very far away. For example, distinctive reddish dust particles from the Sahara Desert have been identified in the soils of Caribbean islands, in the ice of Alpine glaciers, in deep-sea sediments, and in the tropical rainforests of Brazil. However, the most distinctive and characteristic eolian deposits in the deserts themselves are *dunes*.

■ **deflation** Wind erosion in which loose particles of sand and dust are removed by the wind, leaving coarser particles behind.

The second erosional process is called **deflation** (see FIGURE 13.4A). Deflation on a large scale takes place only where there is little or no vegetation and loose particles are fine enough to be picked up by the wind. It is especially severe in deserts but can occur elsewhere during times of drought when no vegetation or moisture is present to hold soil particles together. Continued deflation sometimes leads to the development of *desert pavement*; most of the fine particles are removed, leaving a continuous pavement-like covering of coarse particles (see FIGURE 13.4B).

Dunes Although little is known about how **dunes** begin, it is likely that they start

■ **dune** A hill or ridge of sand deposited by winds.

where some minor surface irregularity or obstacle distorts the flow of air. On encountering an impediment, the wind sweeps over and around it but leaves a pocket of slower-moving air immediately downwind. In this pocket of low wind velocity, sand grains moving with the wind drop out and begin to form a mound. The mound, in turn, influences the flow of air over and around it and may continue to grow into a dune.

A typical dune is asymmetric, with a gentle windward slope (the side facing toward the wind) and a steep leeward face (the side facing away from the wind). Pushed by the wind, sand moves by surface creep and saltation up the gentle windward slope (see FIGURE 13.5). When it reaches the top, the sand cascades down the steep leeward slope, also called the *slip face*. The slip face is always on the leeward side, so we can tell which way the wind was blowing from the asymmetrical form of the dune. Crisscrossed strata within the dune, called *cross-beds*, are former slip faces.

The sliding sand on the slip face comes to rest at the *angle of repose*, the steepest angle at which loose particles will come to rest. The angle of repose varies for different materials, depending on factors such as the size and angularity of the particles. For dry, medium-sized sand particles it is about 33–34°; the angle is generally steeper for coarser materials such as gravel and gentler for fine materials such as silt.

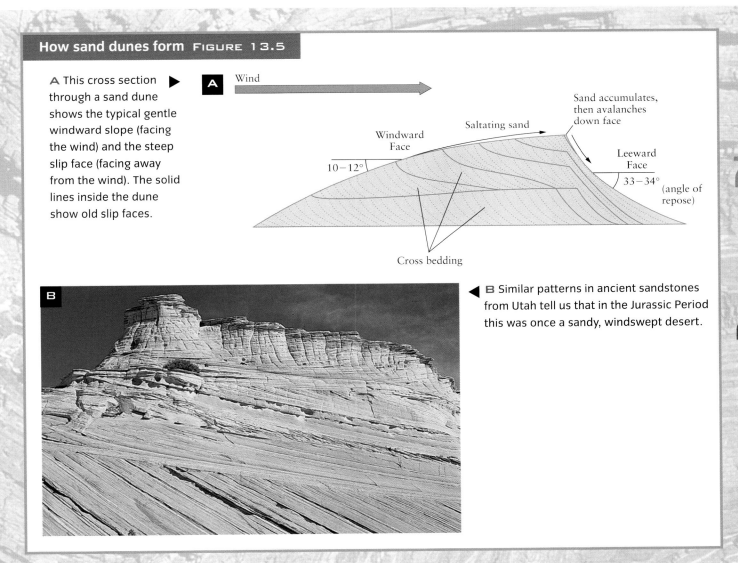

How sand dunes form FIGURE 13.5

A This cross section through a sand dune shows the typical gentle windward slope (facing the wind) and the steep slip face (facing away from the wind). The solid lines inside the dune show old slip faces.

A Wind

Sand accumulates, then avalanches down face

Saltating sand

Windward Face

10−12°

Leeward Face

33−34° (angle of repose)

Cross bedding

B Similar patterns in ancient sandstones from Utah tell us that in the Jurassic Period this was once a sandy, windswept desert.

Process Diagram

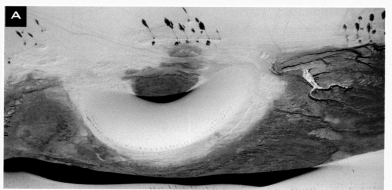

▲ **A** Crescent-shaped dunes called *barchans* are very mobile. These form when the wind blows predominantly in one direction (the "horns" of the crescent point downwind). The dune shown here is invading land adjacent to the Gobi Desert in China. The trees at the top of the photograph have been planted to stop the marching dunes.

▲ **B** When there is a copious sand supply, barchan dunes can merge and form *transverse dunes*, such as these dunes in the Empty Quarter of Saudi Arabia. They are oriented perpendicular to the prevailing winds.

The shape, size, and behavior of a dune depend on the three factors shown in the triangle: sand, wind, and vegetative cover.

Triangle diagram labels: Sand · Wind · Vegetation · • E · Longitudinal dunes · • C · Star dunes · • A Crescent dunes · • B Transverse dunes · Parabolic dunes · No dunes · D •

C If the wind regularly blows in several different directions, it piles the sand up into stationary *star dunes*, such as the ones seen here in the Empty Quarter.

▼

D Coastal regions, where the moist wind off the ocean allows vegetation to grow, form a typical environment for *parabolic dunes*. These are oriented in the opposite direction from barchan dunes: The arms, stabilized by vegetation, point upwind. ▶

▲ **E** *Longitudinal dunes*, like these dunes in Death Valley, run parallel to the prevailing winds. They form in deserts with a meager sand supply. They can also form in areas with bidirectional winds, which first push on one side of the dune and then on the other.

Three factors control the shape and behavior of a dune: the wind conditions, the amount of vegetation cover, and the characteristics and quantity of sand available. These three factors are shown schematically in the triangle in FIGURE 13.6, along with photos of common types of dunes. Clearly, if there is no sand, or if there is plenty of vegetative cover to anchor sediments in place, then dunes will not form. Some types of dunes (see Figure 13.6C) tend to remain stationary. On the other hand, others (see Figure 13.6A) may migrate over long distances, and they can cause severe degradation and loss of agricultural productivity when they invade nondesert lands. This is part of the process of *desertification*.

STREAM EROSION AND DEPOSITION

Contrary to popular belief (and many Hollywood movies), most deserts do not consist of endless expanses of sand dunes. Only a third of the Arabian Peninsula, the sandiest of all dry regions, and only a ninth of the Sahara Desert are covered with sand dunes. The remaining land area is either crossed by systems of stream valleys or covered by alluvial fans and alluvial plains. Running water, therefore, is important in the erosion of deserts, just as it is in rainy regions. However,

instead of acting slowly and steadily, running water in deserts acts suddenly and in short bursts.

Rainfall in a desert region typically occurs during intense downpours that occur during a brief rainy season. The rapid runoff erodes steep-sided canyons called *arroyos* into the landscape (see FIGURE 13.7). These are likely to be dry most of the year, but are subject to *flash floods* during the wet season. When arroyos draining an upland region meet the flat desert floor below, they lose their ability to transport sediment and drop their load of sand and gravel in an *alluvial fan* (Chapter 11). If canyons are closely spaced along the base of a mountain range, the alluvial fans will sometimes coalesce into a broad alluvial apron called a *bajada*.

DESERTIFICATION

In the region south of the Sahara lies a drought-prone belt of dry grassland known as the Sahel. There the annual rainfall is normally only 100 to 300 millimeters (4 to 12 inches), most of it falling during a single brief rainy season. In the early 1970s the Sahel experienced the worst drought of the century. For several years in a row the annual rains failed to appear, causing the adjacent desert to spread southward by as much as 150 kilometers. The drought extended from the Atlantic to the Indian Ocean, and affected a population of at least 20 million.

Flash flood FIGURE 13.7

After a storm, a flash flood thunders down this arroyo on the Navajo reservation in Arizona. As the floodwater subsides, sediments are deposited across the alluvial floor of the canyon.

Desertification FIGURE 13.8

A In the Sahel region of Niger, a herd of goats grazes on pasture at the edge of the desert. As the goats consume the remaining grass and bushes, the dunes of the desert will inevitably advance. ▶

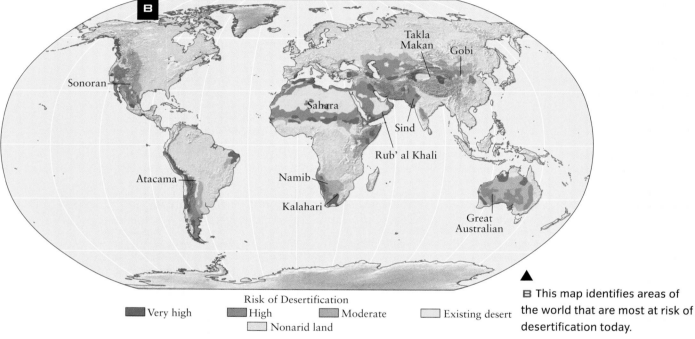

Risk of Desertification
■ Very high ■ High ■ Moderate □ Existing desert
□ Nonarid land

B This map identifies areas of the world that are most at risk of desertification today. ▲

The fringes of deserts naturally migrate back and forth as a result of climate changes, and such was the case in the Sahel. However, the results of the drought were intensified by the fact that between about 1935 and 1970 the human population of the region had doubled and the number of domestic livestock had also increased dramatically. This resulted in severe overgrazing, so the grass cover was devastated by the drought. Millions of people suffered from thirst and starvation. The overgrazing is continuing today (see FIGURE 13.8A), leaving the Sahel at

risk for another devastating famine if the dry weather should ever return.

Desertification can result from natural environmental changes or from human activities, or a mixture of both. The major signs of desertification include lower water tables, higher levels of salt in water and topsoil, reduction in surface water supplies, unusually high rates of soil erosion, and destruction of vegetation. Areas that are most susceptible to desertification, whether by natural or manmade causes, are shown in FIGURE 13.8B. Most are semiarid fringe

> **desertification**
> Invasion of desert conditions into nondesert areas.

lands adjacent to the world's great deserts. In many of these places, humans have lived successfully in a semi-arid environment for centuries; what has changed is an explosion in the size of the population and the adoption of agricultural practices that may not be suited to that part of the world.

One of the best-known examples of desertification occurred in the United States during the mid-1930s when huge dust storms swept across the Great Plains and drove many farm families off their land. John Steinbeck in his award-winning novel, *The Grapes of Wrath*, described this resettlement, the largest forced migration in the country's history. The southern plains came to be called the "Dust Bowl," and historians refer to that period as the "Dust Bowl years" (see **FIGURE 13.9**).

The Dust Bowl had both natural and human causes. Like the Sahel famine, it was triggered by a multiyear drought. However, the effects of the drought were exacerbated by decades of poor land use practices. The grasses that originally grew on the prairies protected the rich topsoil from wind erosion. However, settlers gradually replaced these tall grasses with plowed fields and seasonal grain crops, which left the ground

The Dust Bowl FIGURE 13.9

On April 18, 1935, a massive dust storm closes in on Stratford, Texas. Within a few minutes, the town would be enveloped in pitch darkness, and it would be impossible even to see the house in the foreground.

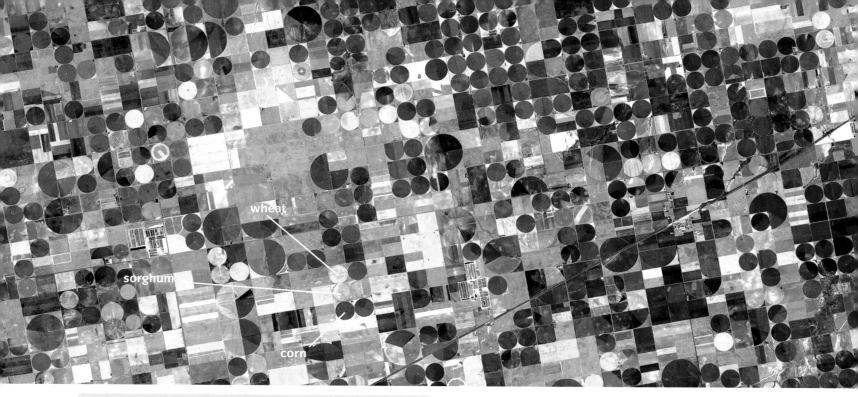

Preventing the next "Dust Bowl" FIGURE 13.10

One of the biggest changes in agriculture in the Great Plains since the 1930s is the extensive use of groundwater for irrigation. These fields in Kansas use central-pivot irrigation, a method that minimizes evaporative loss of water and gives the fields a distinctive circular shape. In June, when this satellite photo was taken, wheat fields are bright yellow. Corn fields, in dark green, are growing vigorously, and the sorghum crop, light green, is just starting to come up. Irrigation helps stabilize the topsoil and thus prevents another "Dust Bowl," but at a cost: It is slowly depleting the High Plains Aquifer.

bare and vulnerable for part of the year. Today, improved farming and irrigation practices have greatly reduced the risk of similar catastrophes (see FIGURE 13.10).

How can desertification be halted or even reversed? The answer lies largely in understanding the geologic principles involved and in the application of measures designed to reestablish a natural balance in the affected areas. Soil management techniques are widely known and readily available; they simply need to be applied more aggressively. These techniques include crop rotation, terracing of steep slopes, and reforestation of vulnerable lands. Geologists can identify and map soils that are unsuitable for agriculture. Land-use planners can eliminate the incentives to exploit arid and semiarid lands beyond their capacity. The preservation of productive lands is essential to maintaining the world's food production capacity at the level needed for an increasing population.

CONCEPT CHECK STOP

Why is a desert more likely to develop in the subtropics than in the tropics?

How do subtropical deserts differ from rainshadow deserts, coastal deserts, and continental interior deserts? Discuss location, causes of aridity, and dune patterns.

Why is water more important than wind in sculpting many desert landscapes?

What are the main causes of desertification?

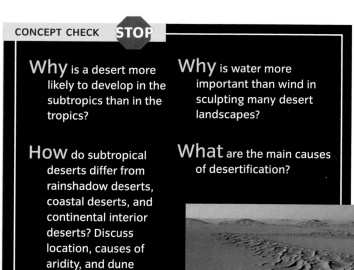

Glaciers and Ice Sheets

LEARNING OBJECTIVES

Distinguish between several different kinds of glaciers and ice formations.

Explain how temperate and polar glaciers differ.

Describe how ice in a glacier changes form, accumulates, ablates, and moves.

Identify several kinds of landforms created by glacial sediments.

Desertification can be an expression of climatic change. Natural processes involving changes in both precipitation and temperature have caused the Sahara Desert to advance and retreat many times over the past 10,000 years, independent of recent human activities in the region.

Climate changes in the hot, sandy deserts and the polar regions are interconnected, and another climate battle is played out in the vast deserts of the polar ice sheets. The expansion and shrinking of glaciers and ice sheets, both in the polar regions and in more temperate alpine settings, is an expression of the complex interplay between temperature and precipitation in the global climate system (Chapter 12). The existence of glaciers and ice sheets is linked to the interaction of several parts of the Earth system: tectonic forces that produce high, mountainous areas; the ocean, a source of moisture; and the atmosphere, which delivers the moisture to the land in the form of snow. We now turn our attention to the great polar deserts and other parts of the **cryosphere**.

> **cryosphere** The perennially frozen part of the hydrosphere.

COMPONENTS OF THE CRYOSPHERE

Annual snowfall is generally very low in polar regions because the air is too cold to hold much moisture. The small amount of snow that does fall doesn't usually melt, because summer temperatures stay very low. In areas where more snow falls each winter than melts during the following summer, the covering of snow gradually grows thicker. As the snow accumulates, its increasing weight causes the snow at the bottom to compact into a solid mass of ice. When the accumulating snow and ice become so thick that the pull of gravity causes the frozen mass to move, a **glacier** is born. The five main types of glaciers are illustrated in **FIGURE 13.11** on pages 390–391.

> **glacier** A semi-permanent or perennially frozen body of ice, consisting largely of recrystallized snow, that moves under the pull of gravity.

Glaciers are cold because they consist primarily of ice and snow. However, scientists have found by drilling holes through glaciers that interior temperatures are not all the same. Some glaciers are warmer than others, and the difference influences the behavior and movement of the ice. In one kind of glacier, the ice is near its melting point throughout the interior. These glaciers, called *temperate glaciers*, form in low and middle latitudes. Meltwater and ice can exist together at equilibrium in temperate glaciers. At high latitudes and altitudes, where the mean annual temperature is below freezing, the temperature in a glacier remains low and little or no seasonal melting occurs. Such a cold glacier is commonly called a *polar glacier*.

Two kinds of ice caps that occur at high latitudes are worth special mention. An *ice sheet* is the largest type of glacier on Earth, a continent-sized mass of ice that covers all or nearly all the land within its margins. At present, ice sheets are found only in Greenland and Antarctica, though they have been much more extensive in the past. They contain 95 percent of the world's glacial ice (and 70 percent of

Glaciers and ice caps FIGURE 13.11

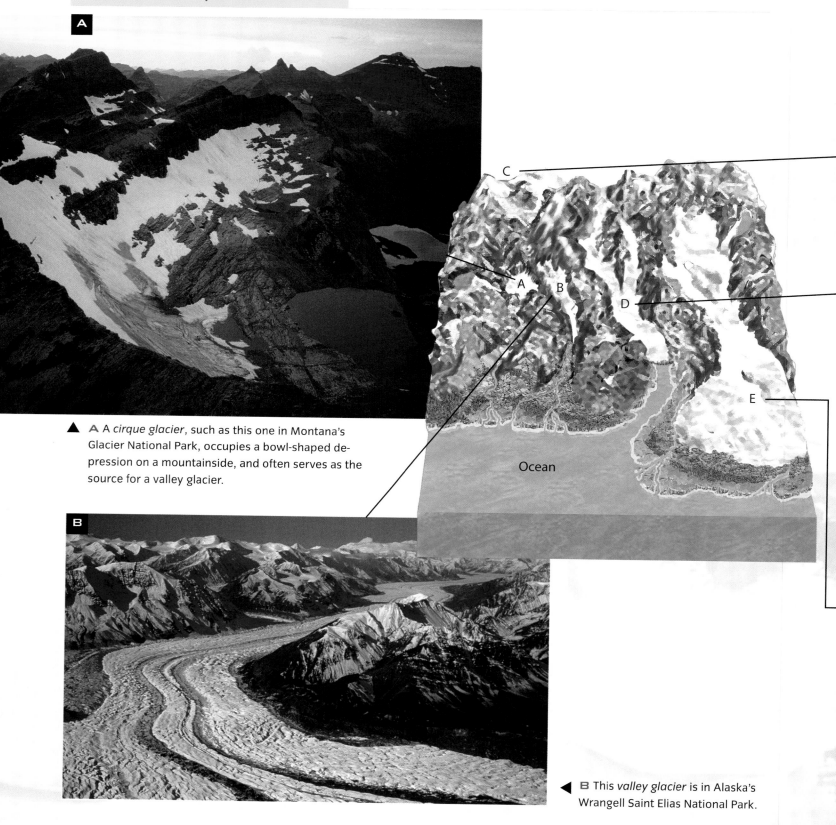

A

▲ **A** A *cirque glacier*, such as this one in Montana's Glacier National Park, occupies a bowl-shaped depression on a mountainside, and often serves as the source for a valley glacier.

Ocean

B

◄ **B** This *valley glacier* is in Alaska's Wrangell Saint Elias National Park.

C An *ice cap* covers a mountaintop (or low-lying land in the polar regions) completely and usually displays a radial flow pattern. In this aerial photograph, the Greenland Ice Cap surrounds the Nunatak Mountains.

◄ **D** When a glacial valley is partly filled by an arm of the sea, the valley is called a *fjord* and the glacier is a *fjord glacier*. Such glaciers often give rise to icebergs that break off and float away.

E When a glacier flows all the way out of ▶ the mountains and onto the surrounding lowlands, it is called a *piedmont glacier*. The Columbia Glacier in Alaska starts as a valley glacier and then spreads out as a piedmont glacier.

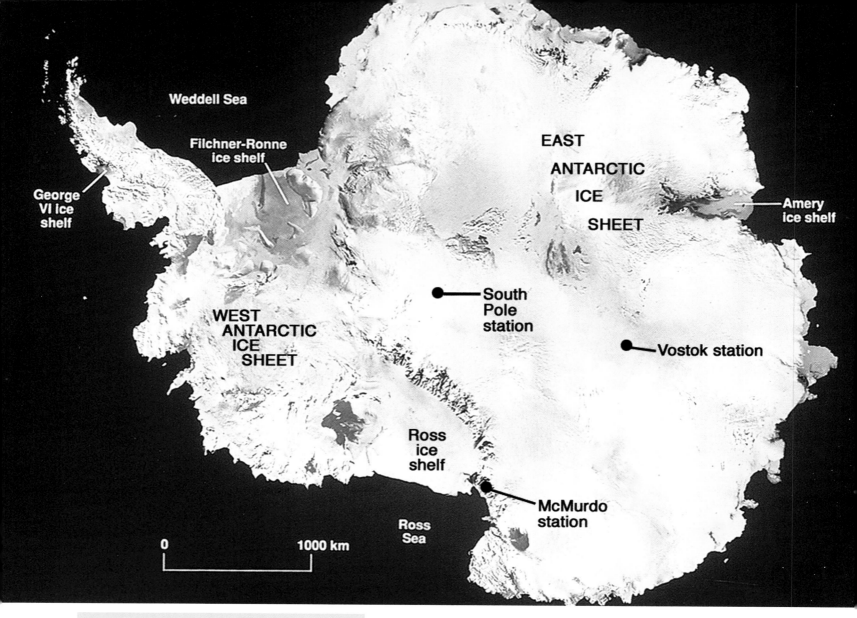

Welcome to Antarctica FIGURE 13.12

The East Antarctic Ice Sheet covers most of the continent of Antarctica, whereas the West Antarctic Ice Sheet overlies a volcanic island arc and the surrounding seafloor. In this satellite image you can also see four ice shelves that occupy large bays. Glaciers that flow down from the mainland feed these shelves.

the world's fresh water). They are so thick and heavy that some of the land underneath Antarctica has actually been pushed below sea level (see FIGURE 13.12).

Ice shelves are thick sheets of floating ice hundreds of meters thick that adjoin glaciers on land (see Figure 13.12). They are constantly replenished by land-based glaciers, but also lose ice when large pieces called *icebergs* break off from them.

Finally, *sea ice* never touches land at all but forms by the direct freezing of seawater. Antarctica is surrounded by sea ice, which in wintertime roughly doubles the apparent area of the continent. Most of the Arctic Ocean is covered by sea ice year-round.

EVER-CHANGING GLACIERS

Although we define a glacier as a semipermanent or perennially frozen body of ice, glaciers are constantly changing in several ways. For example, the snow that falls on the surface of glaciers gradually changes to ice. Glaciers also shrink and grow in response to seasonal changes in temperature and precipitation. The ice in a glacier moves, slowly but surely, under the influence of gravity, and changes in climatic conditions cause the margins of glaciers to advance or retreat. Let's take a closer look at some of these changes.

How glaciers form Newly fallen snow is very porous and easily penetrated by air. The presence of air in the pore spaces allows the delicate points of each snowflake to sublimate (change from solid to vapor without melting). The resulting water vapor crystallizes in tiny spaces in the snowflakes, eventually filling them. In this way the ice crystals in the snow pack slowly become smaller, rounder, and denser, until the pore spaces between them disappear (see FIGURE 13.13). Snow that survives for a year or more becomes more compact as it is buried by successive snowfalls. As the years go by, the snow gradually becomes denser and denser until it is no longer penetrable by air and becomes *glacier ice*. This process may take decades in the case of temperate glaciers to millennia in the case of polar ice sheets.

Further changes take place as the glacial ice is buried deeper and deeper. As snowfall adds to the glacier's thickness, the increasing pressure causes the small grains of glacier ice to grow. This increase in size is similar to what happens when a fine-grained rock recrystallizes as a result of metamorphism in Earth's crust (Chapter 10). Ice is, in fact, a mineral (Chapter 2), and therefore glacier ice is technically a rock. However, the properties of this rock are very different from any other naturally occurring rock, because of its very low melting temperature and its unusually low density. Ice floats in water, because it is only nine-tenths as dense as water.

How glaciers grow and shrink The mass of a glacier constantly changes as the weather varies from season to season and, over time, as local and global climates change. In a way, a glacier is like a checking account. Instead of being measured in terms of money, the balance of a glacier's account is measured in terms of the amount of snow deposited, mainly through snowfall in the winter, and the amount of

From snow to ice FIGURE 13.13

As a new snowflake is slowly converted into a granule of ice, it loses its delicate points, through evaporation and recrystallization, and becomes much more compact.

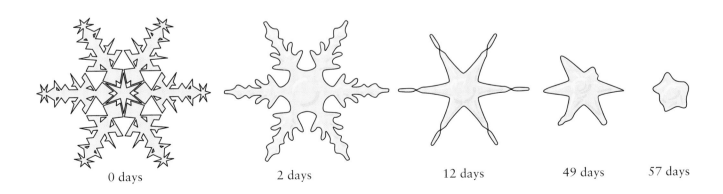

| 0 days | 2 days | 12 days | 49 days | 57 days |

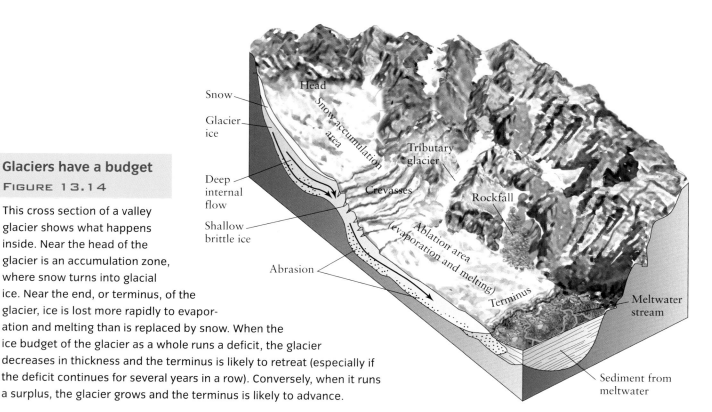

Glaciers have a budget
FIGURE 13.14

This cross section of a valley glacier shows what happens inside. Near the head of the glacier is an accumulation zone, where snow turns into glacial ice. Near the end, or terminus, of the glacier, ice is lost more rapidly to evaporation and melting than is replaced by snow. When the ice budget of the glacier as a whole runs a deficit, the glacier decreases in thickness and the terminus is likely to retreat (especially if the deficit continues for several years in a row). Conversely, when it runs a surplus, the glacier grows and the terminus is likely to advance.

snow (and ice) withdrawn, mainly through melting during the summer. The additions are collectively called **accumulation** and the losses **ablation** (see FIGURE 13.14). The total added to the account at the end of a year—the difference between accumulation and ablation—is a measure of the glacier's *mass balance*. The account may have a surplus (a positive balance) or a deficit (a negative balance), or it may hold the same amount at the end of the year as it did at the beginning.

How glaciers move
Part of the definition of a glacier is that it moves because of the pull of gravity. How can we detect this movement? One method is to carefully measure the position of a boulder on its surface relative to a fixed point beyond the glacier's edge. If you measure the boulder's position again a year later, you will find that it has moved "downstream," usually by several meters. Actually, it is the ice that has moved, carrying the boulder along.

Measurements of velocity show that the ice in the central part of the glacier moves faster than the ice at the sides, and the uppermost layer moves faster than the lower layers. This is similar to what happens to water flowing in a stream (Chapter 7). In most glaciers, flow velocities range from a few centimeters to a few meters a day. It may take hundreds of years for an ice crystal that fell as a snowflake at the head of a glacier to reach the terminus and melt. The glacial ice moves in two basic ways: by internal flow and by basal sliding across the underlying rock or sediment.

Internal flow
As the weight of overlying snow and ice in a glacier increases, individual ice crystals are subjected to higher and higher stress. Under this stress, ice crystals deep within the glacier creep along internal crystal planes (see FIGURE 13.15). As the compacted, frozen mass moves, the crystal axes of the individual ice crystals are forced into the same

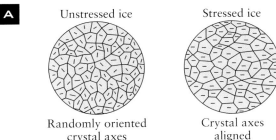

A

Unstressed ice   Stressed ice

Randomly oriented   Crystal axes
crystal axes   aligned

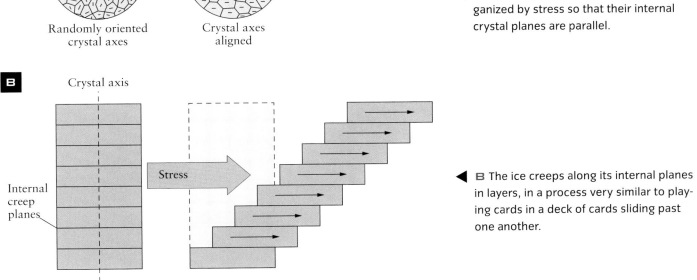

B

Crystal axis

Stress

Internal creep planes

Deforming ice FIGURE 13.15

Ice crystals deep within a glacier move by internal creep.

◀ **A** Randomly oriented ice crystals are reorganized by stress so that their internal crystal planes are parallel.

◀ **B** The ice creeps along its internal planes in layers, in a process very similar to playing cards in a deck of cards sliding past one another.

orientation and end up with their internal crystal planes oriented in the same direction.

In contrast to the deep parts of a glacier, where ice flows by internal creep, the surface portion has relatively little weight on it and is brittle. When a glacier passes over a change in slope, such as a cliff, the surface ice cracks as tension pulls it apart. When the crack opens up, it forms a *crevasse*, a deep, gaping fissure in the upper surface of a glacier (see FIGURE 13.16). Thus, ice moves in a glacier through a combination of ductile deformation at depth and brittle deformation at the surface—a pattern not at all dissimilar to other rocks (Chapter 9).

Crevasses FIGURE 13.16

Deep fissures in the ice, called crevasses, open up as a result of stresses in the brittle surface layer of a glacier. The glacier flows in a direction perpendicular to the crevasse.

Glacial flow

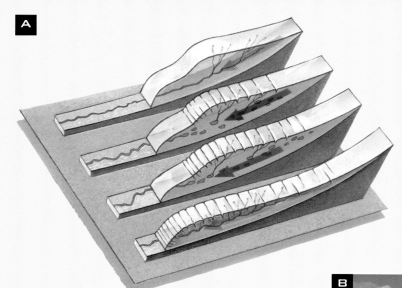

A

A According to one theory, a glacial surge gets started when water at the base of the glacier gets blocked from flowing out. The buildup of pressure lubricates the base and allows the glacier to flow very rapidly, until the water finds a way out again.

◄

B In 1986, Hubbard Glacier in Alaska surged across the mouth of Russell Fjord, damming it up and creating a freshwater lake. This picture was taken soon after the dam broke and reopened the fjord. ►

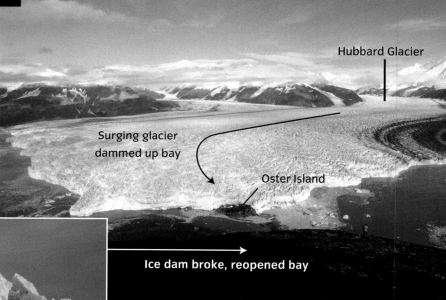

B

Hubbard Glacier

Surging glacier dammed up bay

Oster Island

Ice dam broke, reopened bay

C

◄ **C** Here is a view of the ice dam before it broke. The helicopter hovering in front should give you an idea of the size of this wall of ice.

Basal sliding Sometimes ice at the bottom of a glacier slides across its *bed* (the rock or sediment on which the glacier rests). This is called *basal sliding*. In temperate glaciers, meltwater at the base can act as a lubricant. Basal sliding may account for up to 90 percent of total observed movement in such a glacier, with the remaining 10 percent being internal flow. By contrast, polar glaciers are so cold that they are frozen to their bed; they seldom move by basal sliding, so all movement is by internal flow.

On infrequent occasions, a glacier seems to go berserk (see FIGURE 13.17). Ice in one part of the glacier begins to move rapidly downslope, producing a chaos of crevasses and broken pinnacles. Rates of movement have been observed that are up to 100 times those of ordinary glaciers. These episodes are called *surges*, and their causes are not fully understood. Geologists believe that, in many cases at least, a buildup of water pressure at the base of the glacier reduces friction and permits very rapid basal sliding. The surge stops when the water finds an exit.

The Glacial Landscape As glaciers move, they change the landscape by eroding and scraping away material as well as by transporting and depositing material at their ends and along their margins. In changing the surface of the land over which it moves, a glacier acts like a file, a plow, and a sled. As a file, it rasps away firm rock. As a plow, it scrapes up weathered rock and soil and plucks out blocks of bedrock. As a sled, it carries away the load of sediment acquired by plowing and filing, along with rock debris that falls onto it from adjacent slopes. Let's look more closely at the landforms that result from these processes.

Glacial erosion

The base of a glacier is studded with rock fragments of various sizes that are all carried along with the moving ice. When basal sliding occurs, small fragments of rock embedded in the basal ice scrape away at the underlying bedrock and produce long, nearly parallel scratches called *glacial striations*. Larger particles gouge out deeper *glacial grooves* (see FIGURE 13.18A on the following page). Because glacial striations and grooves are aligned parallel to the direction of ice flow, they help geologists reconstruct the flow paths of former glaciers.

Mountain glaciers produce a variety of distinctive landforms. Bowl-shaped *cirques* are found at a glacier's head. Two cirques on opposite sides of a mountain can meet to form a sharp-crested ridge called an *arête*. Cirques developing on all sides of a mountain may carve its peak into a prominent *horn* (like Mount Everest, seen in FIGURE 13.18B on the following page).

When glacial ice moves downward from a cirque, it scours a valley channel with a distinctive U-shaped cross section and a floor that usually lies well below the level of tributary valleys (see FIGURE 13.18C on the following page). Continental ice sheets can gouge the bedrock to form lakes; some examples of very large glacially formed lakes are the Great Lakes, Lake Winnipeg, and Great Bear Lake. Large glaciers and ice sheets are more effective agents of erosion, and carve deeper valleys and lakes, than small tributary glaciers. At the intersection of a smaller and larger glacier, there will usually be an abrupt change in elevation of the valley floor due to the different depths of erosion.

Glacial deposition

Like streams, glaciers carry a load of sediment particles of various sizes. Unlike a stream, however, a glacier can carry part of its load at its sides and even on its surface. A glacier can carry very large rocks and small fragments side by side. When deposited by a glacier, the load of mixed rocky fragments (called glacial **till**) are not sorted, rounded, or stratified the way stream deposits usually are. In most cases the boulders and rock fragments in a till are different from the underlying bedrock (see FIGURE 13.19A on page 399).

The boulders, rock fragments, and other sediment carried by the glacier may be deposited along its margins or at its terminus. These form ridges called **moraines**; specifically, *lateral moraines* form along the edges and a *terminal moraine* forms at the terminus (see FIGURE 13.19B on page 399), and *recessional* or *end moraines* form as a glacier

> **till** A heterogeneous mixture of crushed rock, sand, pebbles, cobbles, and boulders deposited by a glacier.

> **moraine** A ridge or pile of debris that has been, or is being, transported by a glacier.

Glacial sculpting FIGURE 13.18

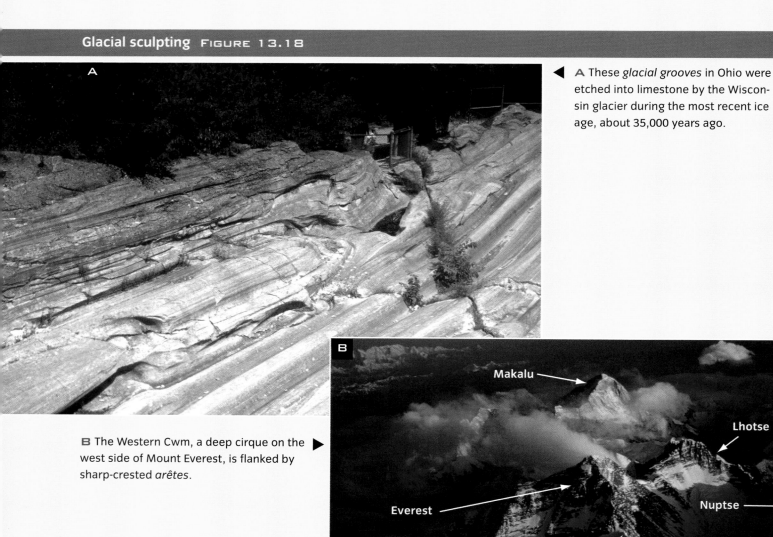

A These *glacial grooves* in Ohio were etched into limestone by the Wisconsin glacier during the most recent ice age, about 35,000 years ago.

B The Western Cwm, a deep cirque on the west side of Mount Everest, is flanked by sharp-crested *arêtes*.

Makalu

Lhotse

Everest

Nuptse

Western Cwm

arête

C The gorgeous Lauterbrunnen Valley in Switzerland has the classic U-shape of a glacial valley. The glacier that formed it no longer exists.

A Glacial till can sometimes include very large boulders, such as these boulders in Yellowstone National Park. When they are different from the bedrock, such boulders are called *erratics*.

▲ B This *terminal moraine* near Mount Robson in British Columbia marks the farthest advance of the glacier at left in the nineteenth and twentieth centuries.

Medial moraine

▲ C The dark stripes running down the center of Kaskawulsh Glacier, in the Yukon, are a *medial moraine*.

▲

D The curving ridge of sand and gravel in this photo is an *esker* in Kettle-Moraine State Park in Wisconsin.

melts and recedes. If two glaciers converge, they may trap lateral moraines between them, forming a ridge of material that rides along the middle of the ice stream, called a *medial moraine* (see FIGURE 13.19C on the previous page). Geologists have used the locations of glacial moraines in the United States and Canada to determine how far the glacial ice cover extended over North America during the last ice age.

The sinuous deposit shown in FIGURE 13.19D (on the previous page) may look perplexing at first: It seems like an upside-down streambed embossed upon the landscape. What could create such a feature? We have already mentioned that the bottoms of some temperate glaciers contain meltwater. This water may actually form a stream that tunnels through the glacier. (These streams can sometimes be seen emerging from the terminus of an active glacier.) Like any other stream, it deposits sediment. If the glacier subsequently retreats, that sediment is left behind in a raised bed called an *esker*, like the one shown.

The retreat of a glacier can leave behind a terrain full of pits and pockmarks, due to abandoned blocks of ice embedded in the glacial debris. These subsequently melt and the depressions left behind are called *kettles*. Many kettles fill with water to form *kettle ponds* and *kettle lakes*. One famous example of a small kettle pond is Walden Pond, immortalized by the writer Henry David Thoreau.

Periglacial landforms The areas near glacial ice also have a distinctive set of landforms, resulting from

Periglacial Landforms

A An ice wedge forms when water seeps into an open crack in the ground and freezes.

B In summer the crack opens or partially melts, allowing more water to enter. In winter the ice freezes again. The ice wedge continues to grow as the melting and refreezing cycle repeats itself hundreds of times. Such wedges can grow as wide as 3 meters and as deep as 30 meters.

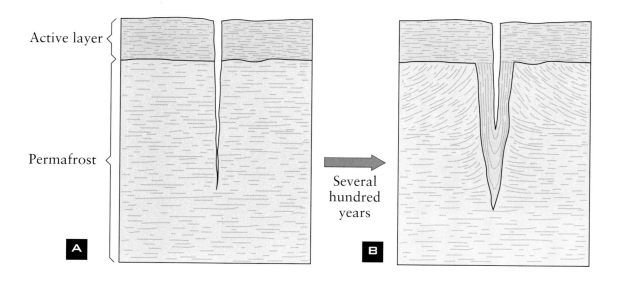

Active layer

Permafrost

Several hundred years

A

B

intense frost action and a large annual range in temperatures. The most common type of environment in present-day *periglacial* regions is *tundra*, a treeless landscape with long winters, very short summers, poorly developed soils, and low, scrubby vegetation. Tundra regions often lie on top of a layer of **permafrost**. During the short summer, the ice melts only in a thin layer near the surface, called the *active layer*. The freeze-thaw cycle produces characteristic geologic formations called *ice wedges* and *patterned ground* (see What a Geologist Sees).

> **permafrost**
> Ground that is perennially below the freezing point of water.

CONCEPT CHECK STOP

How do the glaciers of temperate regions differ from those of polar regions?

What has to happen (in terms of glacial budget) for a glacier to advance?

How does glacial erosion differ between mountain glaciers and continental ice sheets?

What is a moraine? What is significant about a terminal moraine?

What a Geologist Sees

C In Beacon Valley, in Antarctica, ice wedges have grown and interconnected to form what geologists call *patterned ground*.

Glaciers and Ice Sheets   401

Global Climate Change

laciers and polar regions have taken on a special importance in recent years because they are the first and most sensitive indicators of *global warming*, a profound change in the world's climate that most scientists now believe to be real and likely to continue throughout the twenty-first century. You have probably read about global warming in newspapers and magazines. Because

Past climate change FIGURE 13.20

This graph shows an estimate of global temperatures, based on deep-ocean sediments, over (A) the last 60 million years, (B) the last 10 million years, and (C) the last 150,000 years.

A Global Ocean Temperatures, 60 million years ago to present
At the beginning of the Cenozoic Era, Earth's surface was largely free of ice. Sea levels were higher (note that southeastern North America was underwater), and seawater could circulate freely between the Atlantic and Pacific.

B Global Ocean Temperatures, 10 million years ago to present
As plate motions moved the major landmasses near their present locations, temperatures fell and glaciers appeared at the poles. In the last 800,000 years (the blue band in the temperature graph), the climate has fluctuated eight times between ice ages and warm interglacial periods.

C Global Ocean Temperatures, 150 thousand years ago to present
At the peak of the last ice age 18,000 years ago, glaciers blanketed most of North America. Earth is now warmer than it has been at any time in the last 100,000 years, and roughly at the same temperature as it was in the last interglacial period 120,000 years ago.

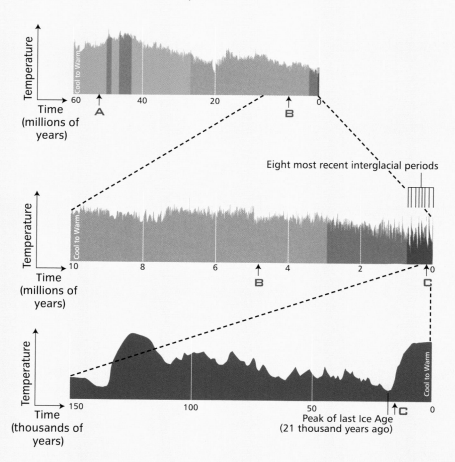

it is a politically and emotionally fraught issue, as well as a scientifically complex one, we believe it is important to emphasize the difference between what we *know* is happening and what we *think* is happening—a distinction that is often lost in the doomsday accounts one sees on television or reads in the press.

CLIMATE CHANGE IN THE PAST

Earth's climate system is complex, comprising multiple interacting parts and subsystems in a state of dynamic equilibrium. Climate is therefore subject to influence from a wide variety of processes, both natural and human. All of these processes, working together, cause cli-

mate to vary cyclically on a number of different time scales. Before we can understand the human role in climate change, present or future, we must first try to understand how natural Earth processes determine climate and control climatic variations. To do this, we must look back in time and study how Earth's climate has changed over geologic history.

What we know In FIGURE 13.20 you can see the trend in worldwide temperatures over the last 60 million years, roughly since the beginning of the Cenozoic Era. This figure highlights three very important points. First, climate change is the norm, and not the exception, on our planet. No matter how much global warming occurs in the next century, it is unlikely to

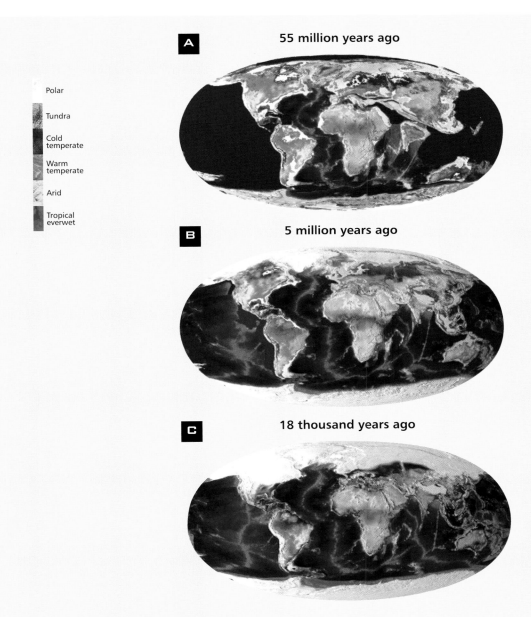

A 55 million years ago

Polar

Tundra

Cold temperate

Warm temperate

Arid

Tropical everwet

B 5 million years ago

C 18 thousand years ago

show up as more than a barely perceptible blip on this graph. Second, the overall trend in temperature in the Cenozoic Era has been downward. The current global warming episode will do nothing to change that. Third, the overall trend of cooling has led to a period in the last 800,000 years with pronounced alternations, each about 100,000 years, of ice ages or **glaciations**, followed by warm *interglacial* periods. In all, more than 20 ice ages have occurred over the past 2 million years. We are currently near the peak of a warm interglacial cycle. In 50,000 years or so Earth will probably experience another ice age—and the current global warming trend is unlikely to delay or hasten it. At most, it may create a sort of "super-interglacial period."

How we know it

How can scientists reconstruct Earth's climate and surface temperature from 20,000 years ago, or even 1 million or 100 million years ago? There are numerous techniques, which make up the subject of *paleoclimatology*.

As described earlier in this chapter, an important piece of evidence is the extent of glacial deposits such as terminal moraines. Layered sediments also provide climatic information. For example, paleontologists infer past climates from assemblages of fossil plants and animals. Fossilized pollen spores in old bogs and lake bottom sediments have been particularly useful for reconstructing past climatic changes on a fine scale. Sedimentologists and stratigraphers study ancient soil horizons, called *paleosols*, which represent former land surfaces and provide information about climate and weather at the time they formed. The minerals present in paleosols allow geochemists to determine the chemical composition of the ambient air and water. Ice cores (see What a Geologist Sees) preserve small bubbles of ancient atmosphere. They tell us the exact chemical composition of the atmosphere up to 160,000 years ago.

Deep-sea sediments provide some of the best evidence we have of past climatic changes. When we ex-amine seafloor sediments obtained by drilling, we find microscopic fossils that record shifts in the animal and plant populations, including warm interglacial forms and cold glacial forms. The proportion of these forms at any time gives us an indication of the average ocean temperature. The temperature graph in Figure 13.20 was derived from the isotopes of oxygen in the fossils. The ratio of oxygen-16 (the lighter isotope of oxygen) to oxygen-18 is also a measure of the temperature of the seawater in which the microscopic creatures lived, although other factors are involved as well.

As you can see, the evidence of past temperature fluctuations and glaciations comes from multiple sources. No single source is definitive.

Causes of climate change

It is clear that the climate has changed dramatically in the past and will continue to change. But the reasons for climatic change are somewhat murkier, and this makes it hard for us to predict *how* it will change and at what rate.

Several mechanisms cause natural climatic changes. Some of these are geographic changes resulting from tectonism. For example, in the Cenozoic Era the Isthmus of Panama formed, joining North and South America. This land bridge severed the connection between the Atlantic and Pacific Oceans, altered oceanic and atmospheric circulation, and thus had a major impact on global climate.

Astronomical factors are also believed to affect Earth's climate (see **FIGURE 13.21** on page 406). The *eccentricity* (departure from circularity) of Earth's orbit, the *tilt* of the planet's axis of rotation, and the *precession* (wobbling) of the axis all affect how much solar radiation reaches Earth's surface, and at what times of year. All three of these factors change gradually over time, but the mechanisms are well understood and the changes are highly predictable. These variations, called *Milankovitch cycles*, correspond reasonably well to the periods of past glaciations. The combined effect of tilt and precession is roughly correlated with 20,000- to 40,000-year interglacial cycles, and variations in eccentricity may contribute to cycles that last 100,000 years.

All of these influences on climate are tempered by the *greenhouse effect* (Chapter 12). Water vapor, carbon dioxide, and methane store heat that radiates from

Ice Cores

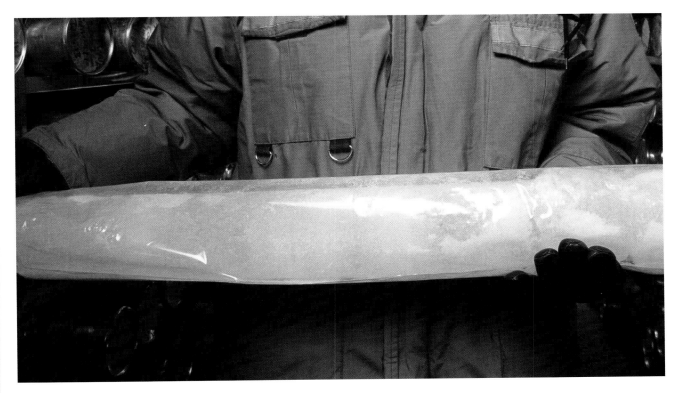

▲ **A** This ice core was extracted from the Quelccaya ice cap in Peru, which is rapidly retreating.

B As the ice in a glacier recrystallizes, ice fills up the pore spaces between the crystals. But some tiny bubbles of air fail to escape, and are locked permanently inside the ice. When the ice is melted under controlled conditions in the laboratory, the chemical composition of the "fossil air" can be measured. Scientists are particularly interested in the concentration of carbon dioxide, a greenhouse gas. Samples from Antarctica and Greenland indicate that the atmosphere contained far less carbon dioxide during glacial ages than during interglacial periods. But the rapid increase of atmospheric carbon dioxide in our time is unprecedented in the ice-core record, and implies that something unusual is happening. ▶

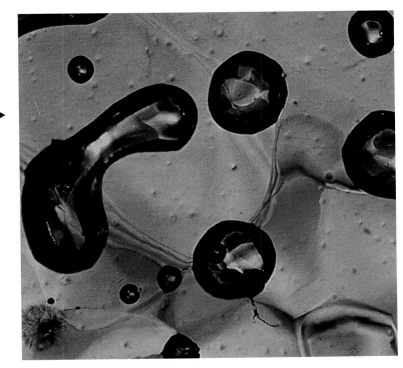

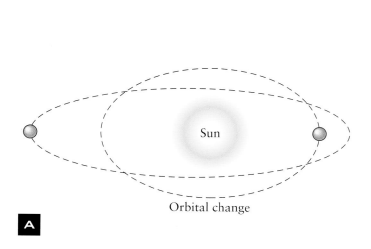

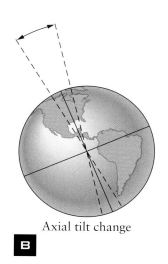

Milankovitch cycles FIGURE 13.21

Three kinds of orbital change affect Earth's climate.

A Earth's orbit becomes more and less elongated over a period of 100,000 years.

B Earth's axis changes its tilt over a period of 41,000 years.

C Earth's axis wobbles in a circle once every 26,000 years.

When these three factors are added together, they greatly affect how much sunlight reaches Earth, and where, at any given time.

Earth's surface and radiate it back downward. Without this natural greenhouse warming by the atmosphere, the average temperature on the surface would be much cooler.

PRESENT-DAY CHANGES

As we have seen, climatic fluctuations are a normal part of the functioning of the atmosphere and climate system. However, within the past two centuries a new player has been added to the system: an industrialized human society that is now capable of significantly affecting climate not only locally but on a worldwide basis.

What we know   The first indisputable evidence of an **anthropogenic** effect on climate was the steady rise in carbon dioxide levels observed over the last 50 years at

> ■ **anthropogenic**
> Produced by human activities.

Mauna Loa Observatory in Hawaii (Chapter 12). Not only are carbon dioxide levels rising, they are rising much more rapidly than at any time in the last 100,000 years. Something altogether new is happening, and it is difficult to escape the conclusion that the new ingredient is the human consumption of fossil fuels. Humans pump roughly 8 billion tons of carbon, most of it as carbon dioxide, into the atmosphere per year—which is actually more than enough to explain the observed increase.

Logic and past climate records suggest that an increase in greenhouse gases should be accompanied by an increase in temperature. The actual evidence is becoming more conclusive year by year. Worldwide temperature records show an increase of about 0.6°C in the last century, and the five warmest years on record (through 2005) were 1998, 2002, 2003, 2004, and 2005. However, even if there is an upward trend, it is not necessarily all due to human activities. Natural climatic fluctuations of this size and duration have certainly happened in the past; we can see the evidence in the growth rings of trees and coral. Such variations may be caused by changes in cloud cover, solar brightness, ocean circulation, or other, unknown factors.

Whatever its cause (or causes), the temperature increase has been greatest in the Arctic, and it is beginning to have real and observable effects. These include the retreat of glaciers, the calving of large icebergs from ice shelves, and the shrinking of some animal habitats and expansion of others (see FIGURE 13.22).

What we think In recent years, climatologists have developed *general circulation models* of the climate that attempt to link processes in the atmosphere, the

Observed effects of climate change FIGURE 13.22

▲ A The Rhône Glacier, source of the Rhone River, used to terminate at this spot in Valais, Switzerland, a century ago. Now the terminus is barely visible in the distance.

▲ B The pack ice in the Beaufort Sea, off the North Slope of Alaska, now breaks up weeks earlier in the spring than it once did.

◄ C On the West Antarctic Peninsula, the rise in temperatures has meant an influx of gentoo penguins, which prefer warmer subarctic temperatures, and a sharp decline in the numbers of Adelie penguins (shown here).

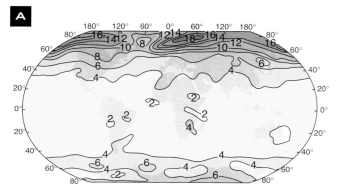

A Temperatures in winter would increase by 10°C or more through most of the Arctic. (Lines labeled "4" show where the projected temperature increase would be 4°C, "6" shows where the increase would be 6°C, and so on.)

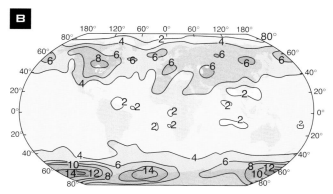

B Temperatures in June, July, and August would increase by 6°C or more in most of Antarctica.

Temperature rise if carbon dioxide doubles FIGURE 13.23

A computer model forecasts the changes in surface air temperature at different times of the year (in °C) that would result from a doubling of carbon dioxide.

hydrosphere, and the biosphere. Many of the linkages in the climate system are still poorly understood and therefore difficult to include in a model. For instance, computer models do not yet adequately portray the dynamics of ocean circulation or cloud formation, two of the most important elements of the climate system. Despite their limitations, these models have been successful in simulating the general character of present-day climates and have greatly improved weather forecasting. This success encourages us to use the models to obtain a general picture of future climate change.

General circulation models allow us to incorporate different assumptions about the anthropogenic factor—that is, the burning of fossil fuels. The models differ in various details, but they all predict that if greenhouse gas levels stay where they are today—almost certainly an optimistic assumption—then we can expect a global temperature increase of 0.5° to 1.5°C over a 50-year period, roughly consistent with what has already happened.

However, if fossil-fuel consumption continues to grow at an increasing rate, atmospheric carbon dioxide is projected to double by 2100, if not sooner. If this proves to be the case, global climate models predict that average global temperatures will rise between 1.5° and 4.5°C. The rise will not be uniform all over the globe, but will continue to be much greater in polar regions (see FIGURE 13.23).

The indirect effects of such a warming are more difficult to quantify, but they include the rising of sea levels due to the melting of glacial ice; increased intensity of severe hurricanes and typhoons, which feed off warm seawater; and significant changes in animal and plant populations due to death or migration. These effects are illustrated in FIGURE 13.24.

CONCEPT CHECK **STOP**

How does the current global warming trend compare with previous climatic fluctuations?

When did glaciations or ice ages begin, and roughly how many have there been?

What methods do scientists use to study past climates?

What do astronomical cycles have to do with Earth's climate?

Why do many experts now believe that Earth is getting warmer and that humans are at least partially responsible?

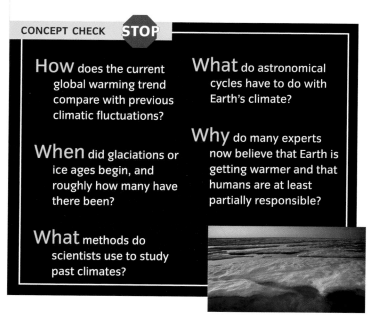

A

◀ **A** Male, the capital of the Maldives, could find itself underwater if sea levels rise. The Maldives, an island nation in the Indian Ocean, are built on coral reefs that grow over time, but cannot possibly grow fast enough to keep up with a rapid change in sea level.

B

▲ **B** Although individual storms or stormy seasons cannot be blamed on global warming, unusually warm water temperatures in the Gulf of Mexico in the fall of 2005 apparently contributed to Hurricane Katrina's becoming a monster storm. Recent studies suggest that the severity of Atlantic hurricanes would increase if global warming continues.

C

D

▲ **D** In much of the western United States, the freshwater supply is dependent on the melting of the winter snow pack in the Rocky Mountains. A smaller snow pack could cause streams like this one, in the Sierra Nevada, to dry up earlier in the year.

▲ **C** Hot water temperatures cause coral reefs, such as this one off the coast of the Maldives, to bleach and eventually die.

Global Locator

NATIONAL GEOGRAPHIC

A place of extremes, Death Valley in California boasts the lowest point in North America, a mostly dry lakebed called Badwater that is 86 meters below sea level (see FIGURE A). From nearby Dante's Peak, on a clear day, you can see both the lowest and the highest point (Mt. Whitney, at 4418 meters) in the continental United States. Death Valley is also notorious for its heat. The hottest temperature ever recorded in the Western Hemisphere (57°C) was measured here, at Furnace Creek, in 1913. With 4.8 centimeters of rain per year, Death Valley also ranks as one of the country's driest places. These extremes result in part from the geology of Death Valley: It is a classic rainshadow desert, lying just east of the Sierra Nevada mountain range. Its low elevation and oppressive heat come from its location in a natural basin formed by extensional stresses.

Though Death Valley is dry today, during the last ice age it was almost entirely under water. You can still see signs of Lake Manly, which once covered this area. Shoreline Butte (see FIGURE B) was once an island in the lake, and the successive levels of the water are etched into the rock.

As the lake dried up, it left large evaporite deposits, such as the salt pans at Badwater (see FIGURE C).

CHAPTER SUMMARY

1 Deserts and Drylands

1. The term **desert** refers to arid lands where annual rainfall is less than 250 millimeters. Five types of deserts have been identified: *subtropical, continental interior, rainshadow, coastal,* and *polar*. Subtropical and polar deserts result from the global wind patterns that create dry high-pressure air masses around 30° N and S and at the poles. The other three kinds of deserts result from local geologic conditions.

2. *Eolian* (wind) erosion is particularly effective in arid and *semiarid* regions. Wind moves particles through **surface creep**, **saltation**, and **suspension**. Flowing air erodes the land surface through the processes of **abrasion** and **deflation**.

3. **Dunes** are hills or ridges of sand deposited by winds. They are asymmetrical, with a gentle slope facing the wind and a steeper *slip face* on the leeward side. Common types of dunes are *barchan, transverse, star, parabolic,* and *longitudinal*. The types that will form in a given place depend on the amount of sand, the wind conditions, and the amount of vegetation.

4. Contrary to the popular image of a desert, the majority of desert lands are not covered by sand. Water erosion is an important geologic process in deserts. *Flash floods* carve deep canyons called *arroyos* and create depositional landforms such as *alluvial fans*.

5. **Desertification** involves the invasion of desert conditions into nondesert lands, which results in land degradation and loss of agricultural productivity. Desertification can be caused by natural environmental changes or by human activities, and can lead to famines affecting millions of people. Careful land use, soil management, and irrigation can prevent or mitigate many of these negative consequences.

2 Glaciers and Ice Sheets

1. The perennially frozen part of the hydrosphere is called the **cryosphere**. It includes **glaciers** on land as well as sea ice. Glaciers come in several varieties, and can form either in polar regions or at high altitudes in temperate regions. Most of the world's fresh water is locked in vast ice sheets in Antarctica and the North Pole region.

2. *Glacial ice* is formed by the compaction of grains of snow and recrystallization of small ice crystals into larger ones—processes that are quite similar to lithification and metamorphism in ordinary rocks. The mass of a glacier can change from season to season and year to year through **accumulation** and **ablation**. Accumulation (by new snowfall) predominates at the head of the glacier, and ablation (by melting) predominates at the foot.

3. Just like ordinary rocks, glaciers can move by either ductile or brittle deformation. The internal movement is ductile, and aligns the ice crystals in the direction of flow. Brittle deformation occurs when a glacier goes over a change in slope, opening up fissures called *crevasses*. At the bottom of the glacier, where *basal sliding* occurs, small fragments of rock embedded in the ice scrape away at the underlying bedrock, producing *glacial striations* and *grooves*.

4. In mountainous regions, glaciers produce a variety of distinctive erosional landforms, such as *cirques, arêtes,* and U-shaped valleys. Common periglacial features include **permafrost** and *patterned ground*, which are formed by the repeated freezing and thawing of groundwater.

5. Glacial deposits often consist of unsorted **till**. A **moraine** is a ridge or pile of debris being carried along by a glacier or deposited along its edge or terminus. When a glacier retreats, the moraine is left behind. Such deposits are useful for identifying the previous extent of glaciers that have retreated or disappeared. Other features often left behind by retreating glaciers are *kettle lakes* and elevated *eskers*.

3 Global Climate Change

1. Climate change is nothing new. Over the last 2 million years Earth has experienced repeated **glaciations** and warm *interglacial* periods. We are now in an interglacial period. The last glaciation, or ice age, peaked about 18,000 years ago. The current trend of global warming may create a more pronounced interglacial period but is unlikely to significantly delay the next ice age, which can be expected in 50,000 years or so.

2. Geologists study past climates in a variety of ways, including analysis of fossils, sediments, and ice cores. Ocean sediments contain microscopic fossils, and changes in the proportion of warm-water and cold-water plants reflect the changes in world temperatures. Ice cores preserve trapped bubbles of "fossil air" that help scientists determine the concentration of greenhouse gases in the past.

3. Climate change can come about for many reasons: tectonics, changes in ocean circulation, volcanic eruptions, and astronomical cycles. Greenhouse gases have always affected the climate. The possibility of **anthropogenic** climate change, due to the burning of fossil fuel, has arisen only in the last century.

4. The current belief of many scientists is that global warming will continue through the twenty-first century, and that its extent will depend on human actions to limit the production of greenhouse gases. The effects of the warming are so far most pronounced in the polar regions, where glaciers are retreating and ice caps are melting. Worldwide effects in the future may include a rise in sea levels, increase in the severity of weather systems, and changes in animal habitat that will cause some species to die and force others to migrate to new territory.

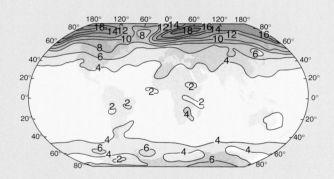 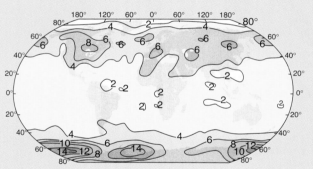

- **desert** p. 379
- **surface creep** p. 379
- **saltation** p. 379
- **suspension** p. 379
- **abrasion** p. 379

- **deflation** p. 382
- **dune** p. 382
- **desertification** p. 386
- **cryosphere** p. 389
- **glacier** p. 389

- **till** p. 397
- **moraine** p. 397
- **permafrost** p. 401
- **glaciation** p. 404
- **anthropogenic** p. 406

CRITICAL AND CREATIVE THINKING QUESTIONS

1. Look at the photograph of sandstones shown in Figure 13.5. Can you tell which way the wind was blowing when it produced these dune deposits?

2. Investigate the current status of drought and land degradation in the Sahel or elsewhere. Can you find any information about soil erosion control techniques or other methods that are being used to combat desertification?

3. Do some research on the most recent ice age. Do you live in an area that was formerly covered by ice? How thick was the ice? Is there any evidence in the landforms around you to indicate that the area was formerly glaciated?

4. Find out if your city, state, province, or country has set goals for the reduction of carbon dioxide emissions to limit its contribution to global warming. What steps have been taken to meet these goals?

5. At the height of the most recent ice age, vegetation in North America south of the ice front must have been different from the vegetation today. Do some research and find out what is known of vegetation changes in your area over the past 20,000 years.

What is happening in this picture ?

This house was built on top of permafrost in the Canadian Arctic. How might warming temperatures affect the permafrost?

Why would this cause the surface to buckle and subside?

1. _____ and _____ deserts result from the global wind patterns that create dry high-pressure air masses around 30° N and S and at the poles.

   a. Subtropical; continental interior
   b. Subtropical; polar
   c. Subtropical; rainshadow
   d. Continental interior; rainshadow
   e. Continental interior; polar

2. Deflation on a large scale takes place only where there is _____.

   a. little or no vegetation
   b. loose particles are fine enough to be picked up by the wind
   c. only in desert environments
   d. Both a and b are correct
   e. Both b and c are correct

3. A typical sand dune _____.

   a. is asymmetrical
   b. has a gentle windward slope
   c. steep leeward face
   d. All of the above statements are correct.

4. The illustration below depicts dune formation as a function of wind, sand supply, and vegetation cover. Label the ternary diagram with the following terms.

   longitudinal dunes   transverse dunes
   crescent dunes   parabolic dunes
   star dunes

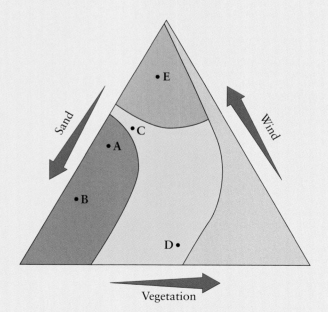

5. The major signs of desertification include all of the following except

   a. lower water tables and a reduction of surface waters.
   b. increased surface temperatures in summer months.
   c. higher levels of salt in water and topsoil.
   d. unusually high rates of soil erosion.
   e. destruction of vegetation.

6. This illustration is a block diagram showing various glaciers at Earth's surface. Label the illustration with the appropriate terms for the individual glaciers A through E.

   cirque glacier   fjord glacier
   piedmont glacier   ice cap
   valley glacier

7. In _____ meltwater and ice can exist together in equilibrium.

   a. temperate glaciers
   b. polar glaciers
   c. ice caps

8. In_____ little or no seasonal melting occurs.

   a. temperate glaciers
   b. polar glaciers
   c. ice caps

9. Glacial ice moves under the influence of gravity through
_____.

   a. basal sliding
   b. internal flow
   c. Both a and b are correct.

10. This photograph is of a glacial landscape. What is the curving ridge of sand and gravel that dominates the picture called?

   a. a drumlin
   b. an esker
   c. a kettle
   d. an end moraine

11. Study of Earth's climate record indicates that temperatures have _____ over the last two million years.

   a. fluctuated greatly
   b. remained constant
   c. been slowly decreasing
   d. been slowly increasing

12. Geologists use a number of techniques to study paleoclimate including data collected from _____.

   a. fossil pollens from peat bogs and lakes
   b. stable oxygen isotope ratios
   c. paleosols
   d. ice cores
   e. All of the above are techniques used by geologists to study paleoclimate.

13. Geologic studies have revealed a number of natural agents that can act together to cause climate change. Which of the following is not considered a possible cause of climate change?

   a. changes in the eccentricity of Earth's orbit
   b. tectonism
   c. interruptions of ocean circulation
   d. tilt and procession of Earth's rotational axis
   e. All of the above are considered possible natural causes of climate change.

14. Anthropogenic carbon dioxide in the atmosphere results from the continued burning of fossil fuels. How far back does the recent record of atmospheric carbon dioxide measurements extend?

   a. 100 years
   b. 75 years
   c. 50 years
   d. 10 years

15. What would be the effects of continued global warming over the next century?

   a. retreat of glaciers
   b. calving of large icebergs from ice shelves
   c. shrinking of some animal habitats
   d. expansion of some animal habitats
   e. All of the above are possible effects of global warming.

A Brief History of Life on Earth

The history and diversity of life, from this ferocious Tyrannosaurus Rex (on exhibit at the American Museum of Natural History) to the humblest microbe, is very much a part of the planet's geology. Throughout this book, you can see numerous examples of interactions between the biosphere and other parts of the Earth system. Plants and microorganisms accelerate the mechanical weathering of rocks and formation of soil. The skeletons of plankton sink to the bottom of the sea, where they form sediments that eventually turn into limestone. Land plants form coal, and animals leave fossils that geologists use to reconstruct the history of the planet. Biologic processes, including the transpiration of plants and respiration of animals, regulate the very air that we breathe.

This interaction also works in the opposite direction: Geology affects the course of life on this planet. Earth's atmosphere and hydrosphere provided a habitat in which life could develop and prosper. The forms that life takes are governed by the need to survive in a particular environment— the scalding waters around a mid-ocean rift, the arid land of a desert, the humid climate of a tropical rainforest, all of which result from geologic processes.

The study of geology is thus inseparable from the study of life on Earth. In this chapter, we trace the story of life from its beginnings. We examine in greater detail how life has been affected by its interactions with the atmosphere, hydrosphere, and lithosphere and how, in turn, life has shaped the Earth system.

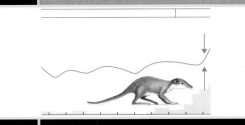
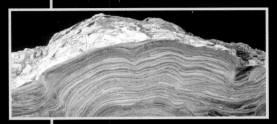

Earth's Changing Atmosphere and Hydrosphere

LEARNING OBJECTIVES

Describe how Earth's environment has changed over the last four billion years.

Explain how photosynthesis adds oxygen to the atmosphere.

Describe how the oxygen cycle affects life on Earth.

L ike other parts of the Earth system, the atmosphere has changed through time (see **FIGURE 14.1**). Compared to the Sun, which is representative of the raw materials from which Earth and the other planets of our solar system formed, Earth contains less of some volatile elements, such as nitrogen, argon, hydrogen, and helium. These elements were lost when the envelope of gases or *primary atmosphere* that surrounded early Earth was stripped away by the

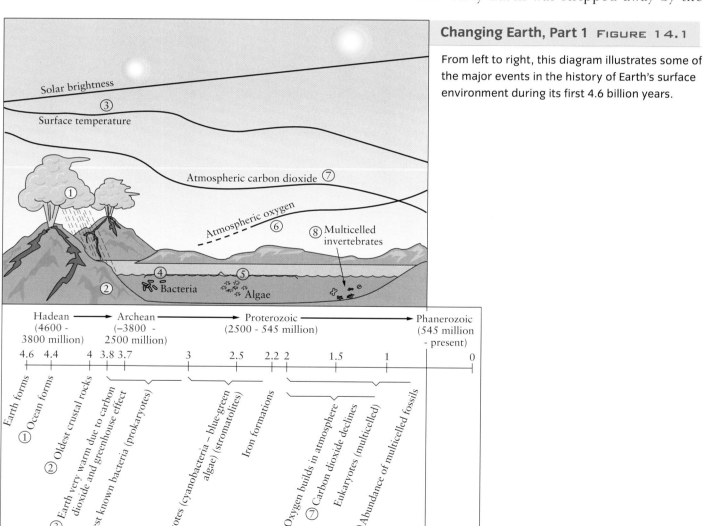

Changing Earth, Part 1 **FIGURE 14.1**

From left to right, this diagram illustrates some of the major events in the history of Earth's surface environment during its first 4.6 billion years.

A This pea plant, like all green, leafy plants, produces oxygen by the process of photosynthesis.

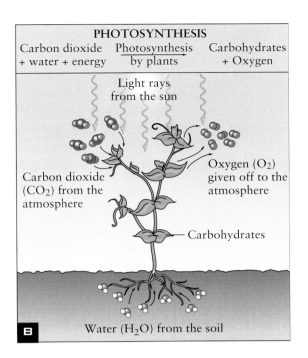

PHOTOSYNTHESIS

Carbon dioxide + water + energy	Photosynthesis by plants	Carbohydrates + Oxygen

Light rays from the sun

Carbon dioxide (CO_2) from the atmosphere

Oxygen (O_2) given off to the atmosphere

Carbohydrates

B

Water (H_2O) from the soil

B The photosynthetic reaction combines carbon dioxide (CO_2) and water (H_2O) to make carbohydrates (molecules containing C, H, and O), which the plant needs to grow. The reaction produces oxygen molecules (O_2). The plant releases the oxygen into the atmosphere through pores in its leaves. The reaction does not happen spontaneously, but requires energy from sunlight; that is why it is called *photosynthesis* (*photo-* meaning "light").

Photosynthesis FIGURE 14.2

solar wind or by meteorite impacts, or both. Little by little, the planet generated a new, *secondary atmosphere*, by volcanic outgassing of volatile materials from its interior.

Volcanic outgassing continues to be the main process by which volatile materials are released from Earth—although it is now going on at a much slower rate. The main chemical constituent of volcanic gases (as much as 97 percent by volume) is water vapor, with varying amounts of nitrogen, carbon dioxide, and other gases. In fact, the total volume of volcanic gases released over the past 4 billion years or so is believed to account for the present composition of the atmosphere, with one extremely important exception: oxygen. As you can see in Figure 14.1, Earth had virtually no oxygen in its atmosphere more than 4 billion years

ago, but the atmosphere is now approximately 21 percent oxygen.

Traces of oxygen were probably generated in the early atmosphere by the breakdown of water molecules into oxygen and hydrogen by ultraviolet light (a process called *photodissociation*). Although this is an important process, it doesn't even come close to accounting for the present high levels of oxygen in the atmosphere. Almost all of the free oxygen now in the atmosphere originated through **photosynthesis** (see FIGURE 14.2).

> **photosynthesis**
> A chemical reaction whereby plants use light energy to induce carbon dioxide to react with water, producing carbohydrates and oxygen.

Earth's Changing Atmosphere and Hydrosphere   419

Oxygen is a very reactive chemical, so at first most of the free oxygen produced by photosynthesis was combined with iron in ocean water to form iron oxide-bearing minerals. The evidence of the gradual transition from oxygen-poor to oxygen-rich ocean water is preserved in seafloor sediments. The minerals in seafloor sedimentary rocks that are more than about 2.5 billion years old contain *reduced* (oxygen-poor) iron compounds. In rocks that are less than 1.8 billion years old, *oxidized* (oxygen-rich) compounds predominate. The sediments that were precipitated during the transition contain alternating bands of red (oxidized iron) and black (reduced iron) minerals. These rocks are called *banded iron formations* (Chapter 8). Because ocean water is in constant contact with the atmosphere, and the two systems function together in a state of dynamic equilibrium, the transition from an oxygen-poor to an oxygen-rich atmosphere also must have occurred during this period.

Along with the buildup of molecular oxygen (O_2) came an eventual increase in ozone (O_3) levels in the atmosphere. Because ozone filters out harmful ultraviolet radiation, this made it possible for life to flourish in shallow water and finally on land. This critical stage in the evolution of the atmosphere was reached between 1100 and 542 million years ago. Interestingly, the fossil record shows an explosion of life forms 542 million years ago (the beginning of the Phanerozoic Eon).

Oxygen has continued to play a key role in the evolution and form of life. Over the last 200 million years, the concentration of oxygen has risen from 10 percent to as much as 25 percent of the atmosphere, before settling (probably not permanently) at its current value of 21 percent (see **FIGURE 14.3**). This increase has benefited mammals, which are voracious oxygen consumers. Not only do we require oxygen to fuel our high-energy, warm-blooded metabolism, our unique reproductive system demands even more. An expectant mother's used (venous) blood must still have enough oxygen in it to diffuse through the placenta into her unborn child's bloodstream. It would be very difficult for any mammal species to survive in an atmosphere of only 10 percent oxygen.

Changing Earth, Part 2 FIGURE 14.3

Over the last 200 million years, oxygen levels in the atmosphere have increased markedly. The rise of mammals (inset), though aided by the demise of the dinosaurs 65 million years ago, may also have resulted partly from the plentiful oxygen supply.

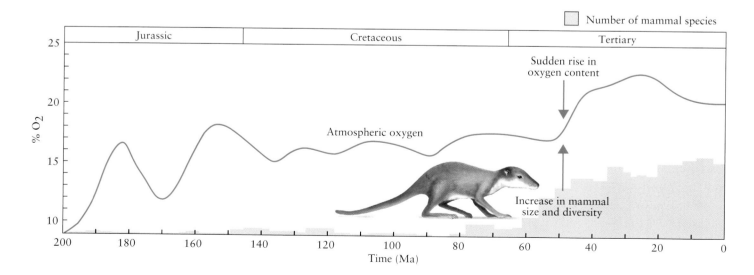

Geologists cannot yet be certain why the atmospheric oxygen levels increased, but they have a hypothesis, and it illustrates the interaction of all parts of the Earth system. First, note that photosynthesis is only one part of the oxygen cycle. The cycle is completed by decomposition, in which organic carbon combines with oxygen and forms carbon dioxide. But if organic matter is buried as sediment before it fully decomposes, its carbon is no longer available to react with the free oxygen. Thus there will be a net accumulation of carbon in sediments and of oxygen in the atmosphere.

But what could cause carbon to be buried on such a massive scale? This is where plate tectonics enter the picture. As you learned in Chapter 4, the most important tectonic event of the last 200 million years was the breakup of Pangaea and the opening of the Atlantic Ocean, which created new passive continental margins. Vast amounts of sediment accumulated on the new margins, accelerating the burial rate of organic matter. In this way the lithosphere, hydrosphere, atmosphere, and biosphere have all interacted to produce the atmosphere we breathe today.

The role of life in the chemical evolution of the atmosphere and hydrosphere does not stop with the oxygen cycle. Nitrogen, carbon, phosphorus, and sulfur also circulate through the Earth system, and their flows are also strongly affected by living organisms.

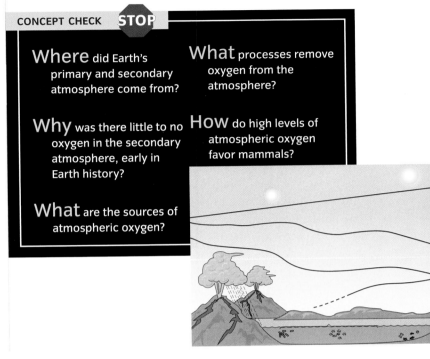

CONCEPT CHECK STOP

Where did Earth's primary and secondary atmosphere come from?

Why was there little to no oxygen in the secondary atmosphere, early in Earth history?

What are the sources of atmospheric oxygen?

What processes remove oxygen from the atmosphere?

How do high levels of atmospheric oxygen favor mammals?

Early Life

Scientists have not found conclusive evidence for life on Earth prior to the end of the Hadean Eon and the beginning of the Archaean Eon, 3.8 billion years ago. In rocks of that age, isotopic signals from graphite in metamorphosed shale suggest that the carbon in the graphite had once been part of living cells. The environment of Earth before 3.8 billion years ago was absolutely hostile to life, with a noxious atmosphere, constant volcanic eruptions, and a very high rate of bombardment by colossal-size meteorites. Yet a remarkably short time later, 3.55 billion years ago, entire colonies of bacterial life forms had developed, as shown by layered structures called stromatolites that are fossilized in rocks of this age. How microscopic life emerged and spread so quickly on a previously barren and inhospitable planet is still one of the greatest unsolved mysteries of science.

LIFE IN THREE NOT-SO-EASY STEPS

What is a living organism? There is no universal agreement on this question, but it seems clear that any living organism must have at least two abilities. First, it must have a means of *replication*, creating a more or less accurate copy of itself. Second, it must have a *metabolism*, a means of extracting energy and material sustenance from its environment.

In addition, all life on Earth (though perhaps not life on other planets) has certain other properties in common. All organisms on Earth use the same set of 20 carbon-based molecules, called *amino acids*, as "building blocks." Furthermore, all known living organisms are made of one or more **cells**, enclosed by a *membrane*. The cell membrane serves a dual purpose: maintaining a relatively constant chemical environment (or *homeostasis*) within the cell, but allowing materials and energy to pass in and out as needed.

cell The basic structural and functional unit of life; a complex grouping of chemical compounds enclosed in a porous membrane.

Carbon is essential to life on Earth because of its ability to *polymerize*, or form very long chains and complex molecules (such as proteins and enzymes), which enable a cell to function. The first step in the emergence of life was likely the creation of these chemicals out of inorganic ingredients. However, amino acids and biopolymers do not form by themselves. At minimum, they require an energy source. In the 1950s, Stanley Miller (see **FIGURE 14.4**) showed that it was possible to assemble amino acids out of a mixture of inorganic ingredients by passing an electric spark (for example, from lightning) through them. However, it is still unclear whether the earliest amino acids really formed this way.

Replication in modern life forms relies on a truly remarkable mechanism. Every cell nucleus contains a molecule called **DNA** (*deoxyribonucleic acid*). DNA is shaped like a twisted ladder (see **FIGURE 14.5**), in which each rung consists of two complementary amino acids, which fit together like a lock and key.

DNA is far too complex to have been the first self-replicating molecule. Most experts believe that a simpler, single-stranded molecule called **RNA** (*ribonucleic acid*) must have appeared first, and they have dubbed the hypothetical pre-DNA version of life the "RNA world." But even RNA is too complex and fragile to assemble by itself; it requires proteins to do the job. How would these proteins work together without any DNA or RNA to direct them? This modern-day version of the chicken-and-egg paradox has still not been resolved.

DNA deoxyribonucleic acid; a double-chain biopolymer that contains all the genetic information needed for organisms to grow and reproduce.

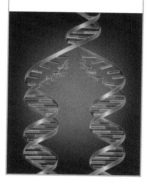

NATIONAL GEOGRAPHIC

Although present-day organisms use one method of replication exclusively (DNA), several different kinds of metabolism have evolved. Plants obtain energy via photosynthesis. Animals obtain energy from the food they eat, and oxygen is needed to release the energy—this is known as *aerobic* metabolism. Many kinds of bacteria have *anaerobic* metabolisms, for example obtaining energy through the nonoxygenated breakdown of food by the process of *fermentation*. Other one-celled organisms can extract energy directly from hydrogen, sulfur, or salt, via *chemosynthesis*. This allows them to survive in high-temperature or high-salinity environments that scientists once considered uninhabitable (see **FIGURE 14.6**). Such organisms are often called *extremophiles*, because they live in conditions that are the extreme end of the chemical and thermal circumstances in which organic molecules can remain chemically stable.

The first not-so-easy step: Making organic molecules
FIGURE 14.4

Miller's experiment used a "primordial soup" of methane, ammonia, and hydrogen, in the central flask shown here. Unfortunately, geologists now believe that he used the wrong ingredients: The primitive atmosphere that developed from Earth's outgassing probably consisted mainly of carbon dioxide, water vapor, and nitrogen. No one has come up with a convincing mechanism for producing organic molecules out of these ingredients in conditions that resemble those of early Earth.

The second not-so-easy step: Replication FIGURE 14.5

The two strands of the twisted molecule of DNA are held together by organic molecules called amino acids. Amino acids come in complementary pairs (shown as orange and brown, pink and green). When the two strands are separated, either one can act as a sort of "photographic template" to recreate the other. This allows for the duplication of genetic information, which is needed for an organism to grow or reproduce. ▶

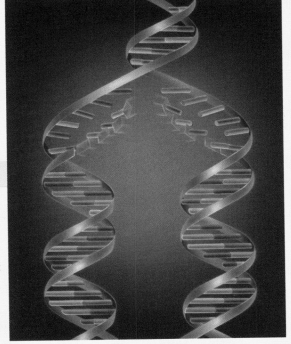

The third not-so-easy step: Metabolism FIGURE 14.6

Seafloor hydrothermal vents give us an idea of how life might have emerged in the very ▼ hot environment of early Earth.

A "Black smokers," such as this vent near the Galapagos Islands, release superheated water and a variety of minerals. Sunlight never penetrates to these depths, so organisms must rely on chemosynthesis rather than photosynthesis to obtain energy. **B** One-celled organisms like these extract energy from the chemicals produced by the vent and form the base of a food chain that includes tubeworms, crabs, and shrimp (**C**).

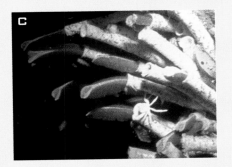

ARCHAEAN AND PROTEROZOIC LIFE

Although, as mentioned earlier in the chapter, some chemical signs of life date as far back as the beginning of the Archaean Eon, the most ancient fossils known are the remains of microscopic **prokaryotes**. Prokaryotes are still present on Earth today: All bacteria are prokaryotes, and the extremophiles shown in Figure 14.6B are part of a separate **domain** of prokaryotes called *Archaea*. Prokaryotes are unicellular (although they may form colonies, they can survive on their own), and they have rudimentary nuclei. One should not draw the tempting conclusion that bacteria are a "less advanced" life form than we are. They are well adapted to their own environments, and far more numerous than anything else on Earth. Your own body contains ten times as many bacterial cells as human cells!

> **prokaryotes**
> Single-celled organisms with no distinct nucleus—that is, no membrane separates their DNA from the rest of the cell.

For the first two billion years of life, the fossil record is very sketchy, because the only forms of life were microscopic and had no hard parts to preserve. However, we do have one important piece of indirect evidence of ancient life. *Stromatolites* are mound-like structures consisting of many thin layers of calcium carbonate. Similar structures can be found today; they are formed in seawater by the action of photosynthetic bacteria called *cyanobacteria* (see FIGURE 14.7). Fossilized stromatolites have been found in rocks up to 3.55 billion years old. Following the Principle of Uniformitarianism (Chapter 1), and assuming they were made the same way, this is strong evidence for the development of photosynthesis.

> **domain** The broadest taxonomic category of living organisms; biologists today recognize three domains: bacteria, *Archaea*, and eukaryotes.

An ancient life form FIGURE 14.7

A These odd-looking bumps in Shark's Bay, Western Australia, are stromatolites, which are formed in warm, shallow seas by photosynthetic bacteria.

B This 2.2-billion-year-old fossilized stromatolite from Michigan shows a pattern of growth rings that is identical to a cross-section of a modern-day stromatolite.

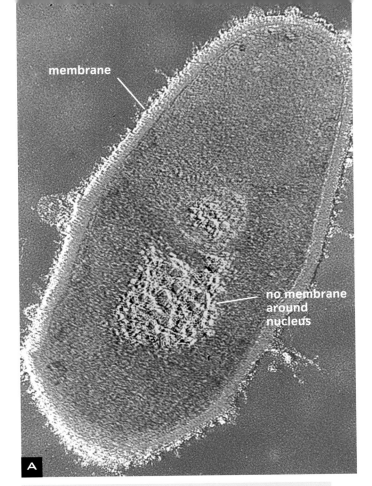

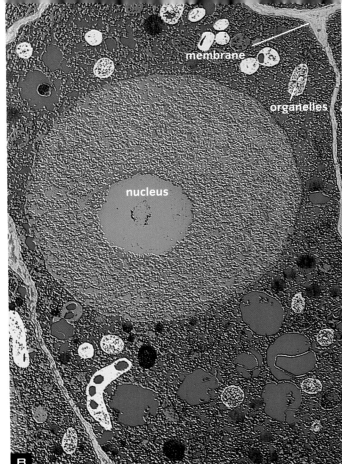

Prokaryotes and eukaryotes FIGURE 14.8

Both cells have been stained to enhance their visibility under a microscope.

A In prokaryotic cells, such as this bacterium, the nucleus is poorly defined and not contained by a membrane.

B In eukaryotic cells, such as this cell from a plant root, the nucleus and several other structures called organelles are clearly defined.

At some time in the Proterozoic Eon, a new kind of life form appeared: **eukaryotes** (from the Greek words meaning "true nucleus"). In addition to a nucleus and membrane, eukaryotes (see **FIGURE 14.8**) contain a number of smaller structures called *organelles*, some of which may originally have been prokaryotic bacteria that were engulfed by the larger eukaryotes and came to live in symbiosis with them.

Eukaryotes appeared at least 1.4 billion years ago, and perhaps as long ago as 2.7 billion years. Like most of the other dates in the early history of life, these dates are uncertain because the first eukaryotes were microscopic in size and lacked hard parts, and therefore are not well preserved as fossils. It is no accident that they emerged after the transition to an oxygenated atmosphere, because they had aerobic metabolisms, and thus were better equipped to make use of free oxygen than prokaryotic bacteria had been. The use of oxygen as a fuel had profound ramifications. Aerobic metabolisms are more efficient than anaerobic metabolisms, allowing more energy to be extracted from each molecule of food. Eukaryotic cells grew larger and became more complex than prokaryotic cells. Eukaryotes were also more tolerant of crowding than anaerobic bacteria had been. Although the first eukaryotes were still single-celled organisms, similar to modern slime molds, the stage was set for multicellular life.

eukaryotes Organisms composed of eukaryotic cells— that is, cells that have a well-defined nucleus and organelles.

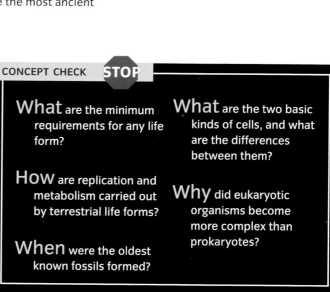

▲

A *Mawsonia spriggi* was probably a floating, disc-shaped animal like a jellyfish, 13 cm in diameter.

B *Dickinsonia costata* was a worm-like creature, 7.5 cm in diameter. ▶

The first multicelled animals FIGURE 14.9

These strangely shaped fossils, specimens of the Ediacara fauna, are the most ancient evidence of multicelled animals.

The earliest fossils of multicellular eukaryotic organisms appear just at the end of the Proterozoic Eon, in rocks about 635 million years old. These fossils, which have now been found in a number of locations, are called the *Ediacara fauna* after the site where they were first discovered, the Ediacara Hills of South Australia. The Ediacara fauna lived in quiet marine bays. They were jelly-like animals with no hard parts (see FIGURE 14.9). These organisms represent a huge jump in complexity from the first unicellular eukaryotes, which appeared at least 800 million years earlier. Scientists still do not know much about what happened during those 800 million years because fossil evidence is sparse and difficult to interpret.

CONCEPT CHECK STOP

What are the minimum requirements for any life form?

How are replication and metabolism carried out by terrestrial life forms?

When were the oldest known fossils formed?

What are the two basic kinds of cells, and what are the differences between them?

Why did eukaryotic organisms become more complex than prokaryotes?

Evolution and the Fossil Record

At the beginning of the Phanerozoic Eon, about 542 million years ago, the fossil record suddenly shows an explosion in the diversity of species. We have explored some of the reasons for this explosion: the oxygenated atmosphere, the protective layer of ozone, and the emergence of eukaryotes and then multicellular organisms. At this point it seems appropriate to pause in our history to describe *how* new species arise, and how they are preserved as fossils.

EVOLUTION AND NATURAL SELECTION

On December 27, 1831, Charles Darwin set sail from England as an unpaid naturalist and "gentleman's companion" for the captain of the *H.M.S. Beagle* (see **FIGURE 14.10**). Trained as a clergyman, Darwin at the time of his departure believed in the Biblical account of creation and the fixity of species. By the time he returned in 1836, his views had changed considerably, though he still saw the hand of God at work.

Darwin kept scrupulous notes and made paintings and sketches of the plants, animals, and fossils he saw on the voyage. In 1859, long after his return to England, Darwin published his observations and ideas in a book called *On the Origin of Species by Means of Natural Selection.* He waited so long to publish his findings because he was concerned about the uproar it might—and did—cause.

In his book, Darwin outlined the theory of **evolution**, which basically says that new **species** develop from existing ones by a gradual process of change through inheritable characteristics. All present-day organisms are descendants of different kinds of organisms that existed in the past, whose populations slowly changed in response to changing environmental conditions. Darwin was not the first to suggest evolution as an explanation for the variety and distribution of species on Earth, but he was the first to provide such

evolution The process by which inherited traits become more common through successive generations; can eventually lead to the emergence of new species.

species A population of genetically and/or morphologically similar individuals that can interbreed and produce offspring.

B Decipherer of evolution's clues

FIGURE 14.10

Though he was a retiring and unpretentious man, Charles Darwin (**A**) wrote one of the most controversial and influential books in scientific history, *On the Origin of Species* (**B**). Originally printed in a small edition of 1250 copies, his book has now been through at least 400 printings and translated into at least 29 languages.

Process Diagram

Darwin's finches FIGURE 14.11

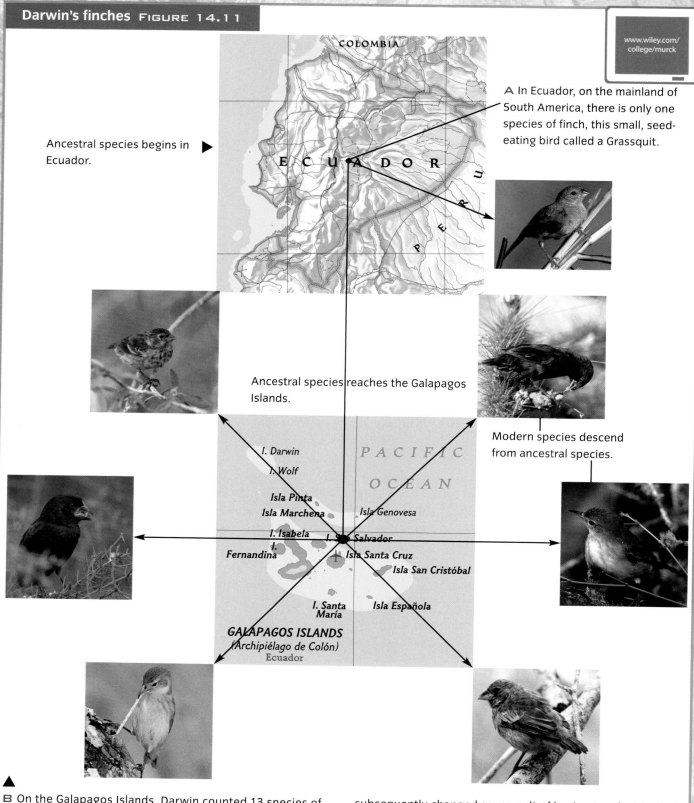

A In Ecuador, on the mainland of South America, there is only one species of finch, this small, seed-eating bird called a Grassquit.

Ancestral species begins in Ecuador. ▶

Ancestral species reaches the Galapagos Islands.

Modern species descend from ancestral species.

I. Darwin
I. Wolf
Isla Pinta
Isla Marchena
I. Isabela
I. Fernandina
I. Salvador
Isla Santa Cruz
Isla San Cristóbal
I. Santa María
Isla Española
Isla Genovesa

PACIFIC OCEAN

GALAPAGOS ISLANDS
(Archipiélago de Colón)
Ecuador

COLOMBIA

E C U A D O R

P e r u

▲

B On the Galapagos Islands, Darwin counted 13 species of apparently related birds, but with different beak shapes and different diets. What could account for such a dramatic difference in diversity between the islands and the mainland? Darwin reasoned that long ago, finches from the mainland had colonized the volcanic islands, and had subsequently changed as a result of having to adapt to their new environments. Different beak shapes had developed as the birds adapted to different diets. It was a logical idea, but it contradicted the widely believed theological doctrine that no new species had appeared on Earth since the Creation.

natural selection The process by which individuals that are well-adapted to their environment have a survival advantage, and pass on their favorable characteristics to their offspring.

a thorough discussion of the evidence, gathered during his voyage and over the following years (see FIGURE 14.11). Most importantly, Darwin was the first to propose a reasonable mechanism by which evolution could be achieved. That mechanism was **natural selection**, or "survival of the fittest."

Because the word "theory" is often misunderstood (Chapter 4), it is important to emphasize that there is no doubt among legitimate biologists that evolution does occur, has occurred throughout the history of life on this planet, and has repeatedly been shown to occur in laboratory situations. The question is *how* it occurs. The late paleontologist Stephen Jay Gould said that Darwin's two great accomplishments were establishing the *fact* of evolution, and proposing the *theory* of natural selection to explain it. There is an analogous situation in physics, where it is a *fact* (established by Galileo) that gravitation exists, but there are different *theories* (proposed by Newton and Einstein) of how it works. Darwin was the Galileo and Newton of biology, rolled into one.

Darwin was motivated to publish *On the Origin of Species* when another young naturalist, Alfred Russel Wallace, independently hit on the same idea of natural selection. According to this theory, any generation of a species will have a broad range of genetic characteristics. Individuals who are better adapted to their surroundings will have more survival success, and more reproductive success. In later generations their descendants will be more numerous than those of less well-adapted individuals. Over time, the entire population will evolve as natural selection favors the better-adapted individuals (see Figure 14.11).

Natural selection also allows for relatively rapid changes. In modern laboratories, scientists can observe evolution in progress, through the expression of specific genetic traits in organisms that reproduce rapidly, such as fruit flies. In nature, if a useful new trait emerges that provides a competitive advantage for an individual and its descendants—for example, a different beak shape, which might help a finch procure seeds—then after several generations, most individuals will have that trait. (Note that the organism does not develop this trait in response to an environmental need or constraint; the trait occurs by chance, but its occurence enhances the survival and therefore the reproductive success of the individual.) Since Darwin's time, the science of genetics has greatly strengthened his argument. Darwin did not know how organisms transmit their traits to later generations; now we know that heritable traits are encoded in *genes* on an organism's DNA. The random variation that is essential to Darwin's theory is explained by *mutations* in those genes—accidental substitutions of one amino acid for another, or deletions or transpositions.

There is still a great deal of scientific debate over the relative importance of gradual change versus rapid change in evolution. Darwin favored gradualism, whereas Gould, among others, advocated *punctuated equilibrium*, in which species persist for a very long time with few changes and undergo occasional periods of very rapid change. In recent years, scientists have begun to recognize the importance of occasional catastrophic events. The best-documented example (see the end of this chapter) is the meteorite impact that is thought to have wiped out the dinosaurs. These animals died out, in all likelihood, not because they were less well-adapted to their environment than other species, but because of a chance event that they had never experienced before and never had an opportunity to adapt to.

HOW FOSSILS FORM

The best evidence for evolution lies in the vast numbers of **fossils** chronicling the succession of species that no longer exist on Earth. Although fossilization can occur in many ways, it is always a relatively rare event. If a dead animal or

fossil Remains of an organism of a past geologic age, embedded and preserved in Earth's crust.

plant is exposed to air, running water, scavengers, or bacteria, it will decompose or be eaten, and its parts will be scattered or destroyed. Hard parts such as bones, shells, and teeth are less easily destroyed and thus more likely to be preserved than soft or delicate parts like skin, hair, feathers, or leaves. In any case, for an organism to be preserved as a fossil it must be quickly covered up by a layer of protective material—usually sand or mud, sometimes tree sap, ice, tar, or volcanic ash.

Sometimes a deceased organism is preserved with little or no alteration. For example, insects many millions of years old have been trapped in tree sap, which hardens into amber (see FIGURE 14.12A). This seals the specimen off from the elements so completely that parts of its original DNA can still be recovered. Ice and tar are also excellent preservatives. In dry climates, natural *mummification* can occur, in which the soft parts dry and harden before they have a chance to decompose.

More often, however, fossils reflect the original shape of an organism but do not contain the original materials. Bones and other hard parts are replaced by minerals carried in solution by groundwater (a process called *mineralization*). This is the process that creates *petrified wood* (see FIGURE 14.12B). The remains of plants are sometimes preserved by carbonization, which occurs when volatile material in the plant evaporates, leaving behind a thin film of carbon.

Some fossils do not contain any actual parts of the organism. For example, the organism may leave behind an imprint or *mold* in the soft sediment that covered it, as in the Ediacara fauna. Other kinds of indirect evidence of previous animal life include eggshells and trace fossils (see FIGURE 14.13). Finally, prehistoric animals left behind fecal droppings, which are called *coprolites* when preserved and fossilized. In spite of their unappealing origin, such fossils provide useful clues about the animals' characteristics, habits, and diets.

trace fossil
Fossilized evidence of an organism's life processes, such as tracks, footprints, or burrows.

CONCEPT CHECK STOP

Where did Charles Darwin find the evidence that led him to the theory of natural selection?

Why does natural selection rely upon random variations in the genetic code?

Why do hard-bodied animals (or hard parts such as teeth) appear more often as fossils than soft-bodied animals or soft parts?

How are soft-bodied animals sometimes preserved?

Not all fossils are bodies FIGURE 14.13

A This two-meter-long fossilized dinosaur (oviraptor) was found curled protectively around a nest containing at least 20 eggs. This is considered to be the first proof that dinosaurs cared for their young.

B Over 65 million years ago, hadrosaurs in what is now Argentina left their tracks in soft, red mud, which turned to rock. The formation of the Andes Mountains tilted it to such an extent that the formerly horizontal mud flat is now a vertical rock wall.

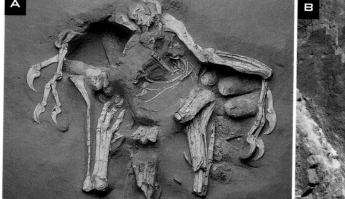
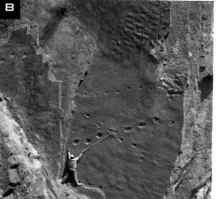

Life in the Phanerozoic Eon

LEARNING OBJECTIVES

Describe the dramatic change in Earth's biota during the Cambrian Period.

Identify the requirements for a living organism to survive on land.

Describe how plants, amphibians, and reptiles met the above requirements.

Outline the most recent evolutionary steps in the development of humans.

Define mass extinctions and identify two theories of what causes them.

During most of the history of life on Earth—3 billion years of it—the only living organisms were of microscopic size. But about 635 million years ago, with the appearance of larger multicellular animals, life forms began to diversify very rapidly into a wide variety of sizes and body types. We will take a whirlwind tour through this last phase of development of life, which continues to the present day.

THE CAMBRIAN EXPLOSION

The Phanerozoic Eon, which starts with the Cambrian Period, was a time of incredible diversification of life (see What a Geologist Sees). Why was this so? One hypothesis is that sexual reproduction, which developed with the eukaryotes, afforded a more rapid way for new "experimental" forms of life to emerge. Another hypothesis is that there was finally enough oxygen in the atmosphere to support the metabolism of larger organisms. As noted earlier, the newly developed ozone layer would have shielded Cambrian life forms from harmful ultraviolet radiation. The rising oxygen content of the atmosphere may have influenced the biochemistry of calcium phosphate and calcium carbonate in seawater, allowing animals to grow skeletons and/or shells. There are other hypotheses too; one is that predation started and this led to the preservation of creatures that had developed protective hard parts; another is that extreme climactic changes at the end of the Proterozoic Eon led to accelerated evolutionary responses of surviving organisms.

A Sample of Cambrian Life

A *Trilobites* were one of the first animals to develop a hard covering, presumably to defend against predators. They are ubiquitous among Cambrian fossils.

B This specimen of *Waptia fidensis*, a soft-bodied arthropod, is preserved in the Burgess Shale of British Columbia. This species did not have a hard shell as the trilobites did, and therefore fossils of it are much rarer.

What a Geologist Sees

488

ORDOVICIAN

⊘
444

SILURIAN

416

Green algae, fungi colonize land
First chordates . . . Jawless fish . . . Many new marine invertebrates   85%

First vascular plants . . . Ferns
Jawed fish . . . First invertebrates on land (scorpions, millipedes)

Whatever the reasons, a great many changes began to occur about 540 million years ago, in what has been called the *Cambrian explosion* or *Cambrian radiation*. Compact animals were evolving to replace the soft-bodied, jelly-like organisms of Ediacara times. These included trilobites such as the one in What a Geologist Sees, mollusks (clams and sea snails), and echinoderms (sea urchins). All of the new animals were equipped with gills, filters for feeding, efficient guts, a circulatory system, and other characteristics that have continued to serve animals well up to the present.

The Cambrian Period also saw the development of skeletons, both internal and external. Skeletons gave many organisms a selective advantage, protecting them against predators, against drying out, against being injured in turbulent water, and so on. It is not surprising that hard-shelled organisms, such as trilobites, are better preserved in the fossil record. However, a wide variety of soft-bodied creatures have been found in the Burgess Shale and even richer deposits in Chengjiang, China. From these deposits we now know that every **phylum** in the animal **kingdom** was present in the Cambrian Period, as well as a number of phyla that no longer exist. In this sense, the Cambrian marine environment had the greatest diversity of life forms in Earth's history.

kingdom The second-broadest taxonomic category. There are six recognized kingdoms, including animals and plants.

FROM SEA TO LAND

The great proliferation of life in the Cambrian explosion was confined to the sea. In order to spread to land, it was essential for living organisms to meet certain requirements. First is some means of *structural support*. Whereas aquatic organisms are buoyed up by water, land-based organisms must contend with gravity. But because water remains critical to all the chemical processes of life, any land-based organism must maintain an *internal aquatic environment*. Living in an environment surrounded by air, rather than water, the organism must develop some way of *exchanging gases* with air instead of water. Finally, all sexually reproducing organisms require a *moist environment for the reproductive system*. The first organisms to overcome these four hurdles were plants.

Plants   Evidence suggests that land plants evolved from green algae more than 600 million years ago (see **FIGURE 14.14**). Eventually, vascular plants evolved, which have structural support from stems and limbs (requirement 1) and a set of vessels through which water and dissolved elements are transferred from the roots to the leaves (requirement 2). Gas exchange (requirement 3) occurs by diffusion through adjustable openings in the leaves called *stomata*. When carbon dioxide pressure inside the leaf is high, the stomata open; when it is low, they close. The stomata also close when the plant is short of water, thereby protecting it from drying out.

The earliest life on land? FIGURE 14.14

The first plants to colonize the land may have looked like the green algae shown in this photograph.

416   DEVONIAN   359

Mosses, ferns still prevalent . . . First naked seed plants (gymnosperms) . . . First trees
Early sharks . . . Age of fishes . . . First land vertebrates (amphibians) evolve from lobe-finned fish   83%

The evolution of plant life FIGURE 14.15

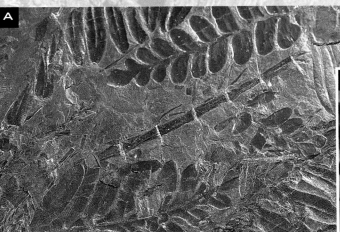

A This fossil fern, preserved in shale, is about 350 million years old.

B Compare it to a modern fern. This photograph shows the spore-producing organs, the dark spots on the underside of the frond.

C Naked-seed plants developed from the seedless plants late in the Devonian Period. These are leaves of modern and fossilized ginkgos.

www.wiley.com/college/murck

D Flowers and fruits represented an advantageous adaptation for plants, because they serve as an incentive for insects to do the work of distributing pollen. This 15-million-year-old fossil found in Idaho is shown next to its modern equivalent, the sweet gum fruit.

The means of reproduction in plants (requirement 4) has passed through several evolutionary stages. The earliest plants were seedless (see FIGURE 14.15), and some seedless plants still exist, such as mosses and ferns. These can reproduce either sexually or asexually, but for the sexual phase they require a body of water where the male and female reproductive bodies (spores) meet and fuse. Thus mosses and seedless plants in general have always depended on a rather moist environment. Seedless plants reached

359   CARBONIFEROUS   299

                                                                                              PERMIAN

Heyday of giant mosses, tree ferns...Extensive coal-forming swamps
First winged insects . . . Amphibians increase . . . Largest insects and highest oxygen levels ever . . . First reptiles

their peak in the Carboniferous Period, and their fossils created the extensive deposits of coal (carbon-rich sedimentary rocks) that gave the period its name.

In the middle Devonian Period, some plants began to evolve their own moist environment to facilitate sexual reproduction. The first plants to do this were **gymnosperms**, which include ginkgos and conifers. The female cell of a gymnosperm is attached to the vascular system and therefore has a supply of moisture. The male cell is carried in a pollen grain with a waxy coating. When the two fuse, a seed results. The seed contains moisture and a food supply that sustains the growth of the young plant until it can support itself. With the evolution of seeds, vascular plants were able to spread beyond swampy lowlands to other habitats. Gymnosperms dominated the plant world during the Triassic and Jurassic Periods—the periods we think of as the "age of dinosaurs."

■ **gymnosperm**   Naked-seed plant.
■ **angiosperm**   Flowering plant, or enclosed-seed plant.

Gymnosperms, however, have one important liability. The male cell-carrier, the pollen, is spread through the air. This is extremely inefficient. What chance does a pollen grain in the air have of finding a female cell? Eventually, **angiosperms**—flowering plants—found a more efficient solution. For a small incentive, such as nectar or a share of the pollen, insects deliver pollen directly from one flower to another, or from one part of a plant to another. After pollination, the plant develops a seed in much the same way as gymnosperms. In many cases, birds and other animals help distribute the seeds by eating the plant's seed-bearing fruits and distributing the seeds in their feces.

The last frontier for plants—the dry steppes, savannahs, and prairies—was not colonized until the Tertiary Period, when grasses evolved. This process also involved the assistance of animals, in particular the great grazing herds that lived on all continents except Antarctica. As these examples show, animals and plants have evolved in parallel, so it is time for us to go back to the Paleozoic Era and retrace the history of the animal kingdom.

Arthropods   Many of the creatures in the Cambrian seas belonged to the phylum *Arthropoda*, named after their jointed legs. Modern arthropods include crabs, spiders, centipedes, and insects—the most diverse phylum on Earth. They were the first creatures to make the transition from sea to land.

With a few exceptions, early arthropods were small and light. They were covered with a hard shell of *chitin* (a fingernail-like material). Thus, they were well adapted to life on land, in regard to the need for structural support and water conservation. The first arthropods on land were probably centipedes and millipedes, in the Silurian Period. By the Mississippian Period, insects were abundant and included dragonflies with a wing span of up to 60 centimeters (see **FIGURE 14.16**). But for all their success as land creatures, the arthropods have very primitive respiratory and vascular systems. For example, insects breathe through tiny tubes that penetrate their outer coating. This mode of respiration severely limits

Ancient insects FIGURE 14.16

Arthropods, such as this ancient dragonfly from the early Cretaceous Period (about 140 million years ago), were among the first and most successful land animals. This fossil from Brazil is about 7 cm in length.

| PERMIAN | ⊘ 251 | TRIASSIC | ⊘ 200 |

Pangaea forms . . . Conifers appear, colonize uplands
First beetles and flies . . . Mammal-like reptiles (therapsids)   95%

First dinosaurs appear . . . First true mammals   80%

B

Pioneer fish? FIGURE 14.17

A When a living coelacanth was caught in the Indian Ocean in 1938, it created a sensation because not only its species but its entire order, the crossopterygians, had been believed to be extinct. The first fish to haul themselves onto land may have been relatives of the coelacanth. Today's coelacanths, like this one photographed near the Comoro Islands, are exclusively deep-sea creatures. B Another candidate for the first fish to make the transition to land is the lungfish. Unlike the coelacanth, today's lungfish still survive for short periods on land without water.

the size of an insect and explains why almost all insects are small.

The "blood" of arthropods is simply body fluid that bathes the internal organs; it does not circulate in blood vessels. The fluid is generally kept in motion by a sluggish "heart" that is little more than a contracting tube. However, primitive does not mean poorly adapted. The arthropods have branched into more than a million species, and they are close to indestructible. Who ever heard of a cockroach having a heart attack?

Fishes and amphibians

The phylum of greatest interest to most of us, because humans belong to it, is the *chordates*. These are animals that have at least a primitive version of a spinal cord (called a *notochord*). Like all the other animal phyla, chordates can be found in fossils of the Cambrian Period, although they are relatively inconspicuous. The earliest so far discovered is *Haikouichthys*, a jawless fish akin to a hagfish or a lamprey that lived 525 million years ago—only 20 million years after the beginning of the "Cambrian explosion."

Jawed fish arrived next. With jaws, fish could lead a predatory lifestyle and grow to much larger sizes; the jawless fish had been limited to filtering food out of the water or dredging it from the sea floor. Among the first large jawed fish were sharks and ray-finned fish. The earliest known intact shark fossil is 409 million years old.

The first fish to venture onto land may have been a member of an obscure order called *Crossopterygii*, or lobe-finned fish (see FIGURE 14.17). These fish had several features that could have enabled them to make the transition to land. Their lobe-like fins contained all the elements of a quadruped limb. They had internal nostrils, characteristic of air-breathing animals. As fish, they had already developed a vascular system that was adequate for life on land.

Life in the Phanerozoic Eon   435

The first terrestrial chordates, amphibians, have never become wholly independent of aquatic environments, because they have not developed an effective method for conserving water. To this day they retain permeable skins. They have also never really met the reproductive requirement for life on land. In most amphibian species, the female lays her eggs in water and the male fertilizes them there after a courtship ritual. The young (e.g., tadpoles) are fish-like when first hatched. Just like the seedless plants, amphibians have kept one foot (figuratively speaking) on land and one foot in the water. They originated in the Devonian Period and have not diversified much since then.

Reptiles, birds, and mammals

One branch of the amphibians, however, evolved into reptiles—the first fully terrestrial animal. Reptiles were freed from the water by evolving an egg that contained amniotic fluid for the young to grow in, and by developing a watertight skin. These two evolutionary advances enabled them to occupy many terrestrial niches that the amphibians had not been able to exploit because of their need to live near water. The amniotic egg led to an explosion in reptile diversity, much as the evolution of the jaw had done for fishes. Moving out of the Mississippian and Pennsylvanian swamps, some colonized the land, some moved back into the water, and a few took to the air. Not only did reptiles greatly increase in diversity during the Jurassic Period, they also grew to tremendous size. The dinosaurs were the largest land animals that ever lived, possibly ranging up to 100 metric tons in weight and 35 meters in length.

Birds first appeared near the end of the Jurassic Period, and they are now considered to be direct

Early birds and mammals FIGURE 14.18

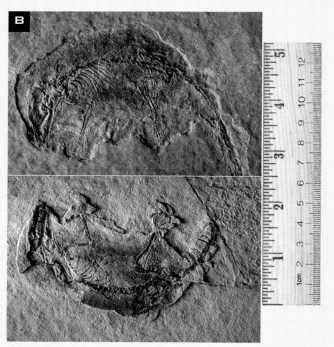

▲
A The skeletons and teeth of *Archaeopteryx* were very similar to those of dinosaurs. However, the very detailed impressions of feathers in this fossil identify *Archaeopteryx* as a bird.

▲
B Discovered in 2002 in China, this shrew-sized *Eomaia scansoria* specimen is the oldest known fossil of a placental mammal (that is, a mammal that gives live birth). It lived 125 million years ago, during the height of the dinosaur age.

145

CRETACEOUS

⊘
65

Angiosperms (flowering plants) appear and proliferate...
Dinosaurs reach largest sizes ever...Marsupials and placental mammals...Meteorite impact causes mass extinction

70%

descendants of the dinosaurs. An early bird, *Archaeopteryx* (see FIGURE 14.18A), would have been classified as a dinosaur if not for the discovery that it had feathers. Even before *Archaeopteryx*, vertebrates had made the transition to the air in the form of pterosaurs—flying reptiles with long wings and tails. Curiously, pterosaurs are *not* considered by most paleontologists to be true dinosaurs, although they were very closely related. The details of the transition from reptiles to birds remains one of the most highly controversial topics in paleontology today.

Mammals descended from the reptiles too, though not directly from the dinosaurs. Their ancestors were a class of "mammal-like reptiles" that existed as long ago as the Permian Period. The transition from reptiles to mammals is not nearly as clear-cut as the transitions from fishes to amphibians. It is difficult to pick out a single mammalian adaptation comparable to the jaws of fish, the eggs of reptiles, or the feathers of birds. During the Cretaceous Period, mammals certainly did not outcompete the dinosaurs; they survived by being small and inconspicuous (see FIGURE 14.18B–D).

Evidence suggests that an accident—a giant meteorite impact—was at least partially responsible for the environmental changes that wiped out the great reptiles at the end of the Cretaceous Period, giving mammals a chance to grow larger and to diversify. In the Cenozoic Era, mammals have benefited from the atmosphere's high oxygen level, conducive to a fast metabolism. Unlike reptiles, whose brain sizes have not grown (relative to their body size), mammals have continued to evolve toward larger brain size throughout the Cenozoic Era. This may be one key to mammalian success, because it has enabled them to diversify their lifestyles to a much greater extent than reptiles ever could have.

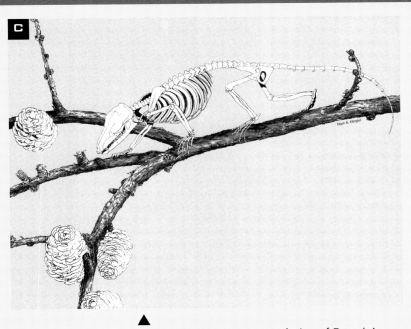

▲
C This line drawing is a rendering of *Eomaia's* skeleton.

D The shape of *Eomaia's* claws shows that it was a climber that could grasp tree branches. It is now known that small, tree-dwelling mammals began to experiment with gliding flight as early as 130 million years ago—almost as early as the first birds began flying. ▶

65   1.8   Present
←——||—————————————————————————||
            TERTIARY   QUATERNARY   ∅ Possible mass extinction
                                                                                                                 due to human impact on
        Grasses evolve . . . Mammals increase in diversity and size . . . Grazing mammals . . .   Genus *Homo* . . .   environment?
                High oxygen levels enable large brains, fast metabolism . . . First hominids . . . Ice ages   Modern Humans

A

Lucy and family FIGURE 14.19

◀ **A** This drawing by Michael Rothman depicts a mother and child of the species *Australopithecus afarensis*. These human-like individuals lived together in small groups, formed lasting bonds with mates, and looked after their children through infancy.

B This 3.3-million-year-old fossil is the skull of an *A. afarensis* baby, who probably looked much like the child in the drawing. ▶

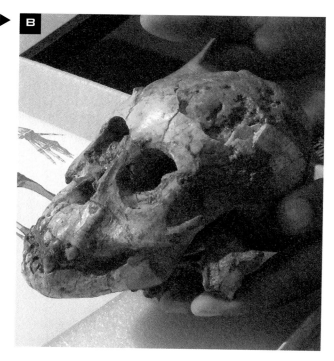

B

C

C From footprints like these, preserved in soft volcanic mud, scientists know that australopithecines walked upright on two feet. This 70-meter trail includes the footprints of two adults and possibly a child, stepping in the footprints of one of the adults. To the right are footprints of an extinct three-toed horse. ▶

The human family Charles Darwin was often accused of believing that humans are descended from the apes. In fact, the family of humans, *Hominidae*, and the family of apes, *Pongidae*, probably both descended from an earlier common ancestor. The emergence of humans is one of the most complex and controversial fields in paleontology, in part because of the sketchiness of the fossil record and the lack of transitional forms. But it is clear that **hominids**—human-like organisms—are a very recent evolutionary development. The first hominid that was clearly *bipedal* (walked upright) was *Australopithecus*, of which the famous fossil "Lucy" is an example (see FIGURE 14.19). These hominids were only about 1.2 meters in height, but had a brain capacity larger than chimpanzees. Their fossils range from about 3.9 million to 3.0 million years in age. From the shape of its pelvis and from footprints left in soft volcanic mud (see Figure 14.19B), we know that *Australopithecus* walked upright, though its skull looked more apelike than human. Lucy's descendents never spread beyond Africa, and they disappeared altogether about 1.1 million years ago.

Homo erectus, probably the first species of our own genus (*Homo*), was more widely traveled than *Australopithecus*. Fossils of *Homo erectus*, dating back about 1.8 million years, have been found in Africa, Europe, China, and Java. Even earlier, a problematic species called *Homo habilis* used stone tools. Because toolmaking is the distinguishing feature of the genus *Homo*, some experts include *habilis* in this genus; others argue that the skull of this species is more like that of the Australopithecines.

Homo erectus disappeared 300,000 years ago, and was replaced by *Homo neandertalensis* ("Neandertal man") no later than 230,000 years ago. Unfortunately, the poor fossil record between 400,000 and 100,000 years ago has made it difficult for scientists to determine how this transition occurred. We do know from burial sites that Neandertals practiced some form of religion. Because of similarities in teeth and brain size (slightly larger than our own), some experts have argued that Neandertal was part of our own species; however, recent DNA studies suggest that *Homo sapiens* is not a direct descendant of Neandertals. The Neandertals disappeared about 30,000 years ago and were replaced rather suddenly by the biologically modern Cro-Magnon people, the first indisputable members of our own species, *Homo sapiens*.

Did the Cro-Magnons evolve from the Neandertals, or were they a distinct species? What happened during the 5,000-year period when both kinds of humans were alive, and overlapped geographically in Europe? Did the Cro-Magnons kill the Neandertals? These and many other questions await answers, as paleontologists continue to look for clues in the fossil record.

MASS EXTINCTIONS

At several places in the geologic column, paleontologists have found abrupt and profound changes in the

> **mass extinction**
>
> A catastrophic episode in which a large fraction of living species become extinct within a geologically short time.

fossil assemblage—not just in one location, but worldwide. These dramatic changes record **mass extinctions**. (In the geologic time line on the previous eight pages, the boundaries marked in red correspond to mass extinctions.) For geologists, mass extinctions are a convenience, because they delineate so many of the geologic periods, but they are also a conundrum. What could possibly cause most of the species and many of the genera on Earth to become extinct in a short period of time?

The most famous mass extinction occurred 65 million years ago, at the boundary between the Cretaceous (K) and Tertiary (T) Periods. (This is also the boundary between the Mesozoic and Cenozoic Eras.) Paleontologists estimate that around 70 percent of all species died out in the K-T extinction, including the dinosaurs. Many of the species that survived must also have come perilously close to extinction.

In the last quarter-century, scientific opinion has coalesced around the hypothesis that the K-T extinction was at least partly caused by environmental changes that resulted from the impact of a large meteorite with Earth (see **FIGURE 14.20** on the following page). When it was proposed by Walter and Luis Alvarez in 1980, the meteorite-impact theory was highly controversial, but many lines of evidence now support it (see **FIGURE 14.21** on page 441). We can be reasonably certain now that the events described in Figure 14.21 did happen, although we cannot be absolutely certain that this event killed all the dinosaurs.

The biggest remaining problem with the meteorite impact theory is that the great K-T extinction was not unique, nor was it even the most dramatic of all extinctions. The most devastating mass extinction occurred 245 million years ago, at the end of the Permian Period, when as many as 96 percent of all species died out. There have been at least 5 and possibly as many as 12 mass extinctions during the last 250 million years: some of them can be linked to massive meteorite impacts, but others cannot. Several other known craters on Earth rival the size of the Chicxulub crater, yet the ages of the craters do not correlate well with the dates of other mass extinctions. On the other hand, some

Visualizing

The Effects of a Meteorite Impact

IMPACT

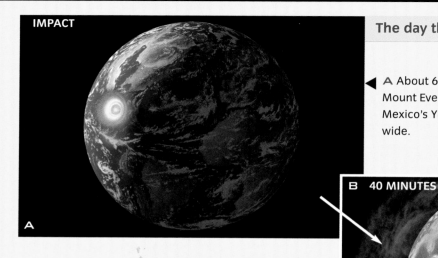

A

The day the dinosaurs died? FIGURE 14.20

◄ A About 65 million years ago, a meteorite as large as Mount Everest struck Earth at a spot just offshore from Mexico's Yucatan Peninsula. It blasted a crater 180 km wide.

B 40 MINUTES

B The blast wave itself would have killed animals and plants all over the Western Hemisphere. As broiling-hot debris from the impact rained down, forests ignited, creating continent-wide forest fires. ►

ONE WEEK

C

◄ C Soot from the fire may have remained in the atmosphere for months, or even years, blocking out the sunlight and bringing photosynthesis all over the world to an effective halt.

ONE YEAR

D A year after the impact, algae and ferns may have begun to grow again, but the forests were still bare. ►

D

Evidence of impact FIGURE 14.21

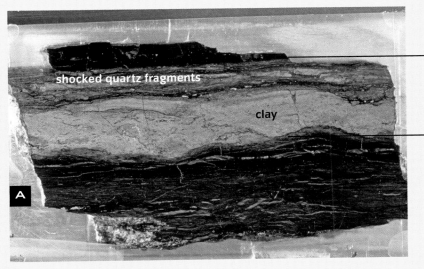

A — shocked quartz fragments

clay

impact debris

A A thin layer of clay, shown here in a sample from the Raton Basin in Colorado, can be found all over the world in rocks at the K-T boundary. Here the whitish layer of clay is about 2 centimeters thick. Above it lies a thinner layer containing shocked quartz particles, which is enriched in iridium, an element that is rare on Earth but more common in meteorites. This evidence strongly suggests that both layers were formed by a meteorite impact. In addition, the clay layer becomes thicker as one gets closer to the hypothesized impact site in Mexico. Dinosaur fossils can be found beneath the clay layer but never above, unless the sequence of strata has been disturbed.

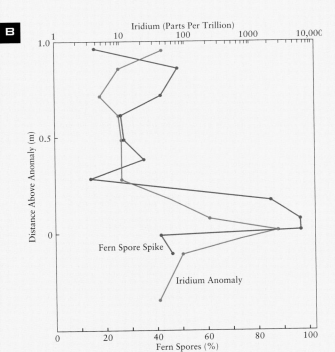

B

Iridium (Parts Per Trillion)

Distance Above Anomaly (m)

Fern Spore Spike

Iridium Anomaly

Fern Spores (%)

B In layers just above the impact, fern spores are much more prevalent than angiosperm pollen—a sign of a regenerating plant community.

C Geologists believe they have found the site where the meteorite landed. A crater lies underneath Mexico's Yucatan Peninsula. Although the crater is now completely covered by sediments, a satellite radar image shows a faint circular depression of the ground around its edge.

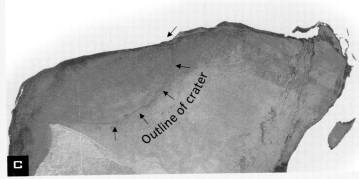

C

Outline of crater

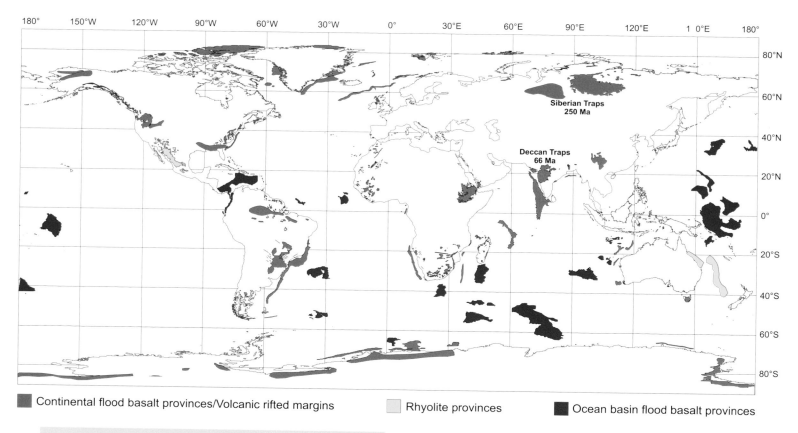

| 180° | 150°W | 120°W | 90°W | 60°W | 30°W | 0° | 30°E | 60°E | 90°E | 120°E | 1 0°E | 180° |

Siberian Traps
250 Ma

Deccan Traps
66 Ma

■ Continental flood basalt provinces/Volcanic rifted margins   □ Rhyolite provinces   ■ Ocean basin flood basalt provinces

Volcanism and mass extinctions FIGURE 14.22

Several of the world's largest flood basalt deposits formed at roughly the same time as mass extinctions. Geologists are still debating whether, and how, an enormous lava flow in one region could cause mass extinctions all over the world. (Ma = millions of years before present.)

geologists have pointed out that a strong correlation exists between the eruption of flood basalts (Chapter 6) and mass extinctions. The K-T boundary lies very close in time to the eruption of the Deccan Traps in India, and the Permian extinction occurred around the time of the largest known flood basalts, the Siberian Traps (see FIGURE 14.22).

Perhaps massive outbreaks of volcanism, producing worldwide climate change, are the "normal" agent of mass extinctions, and the K-T meteorite was merely an interloper that happened to strike Earth at the wrong time. The study of mass extinctions, and the role that impacts may have played in them, continues to be an area of active research. It is especially appropriate today, because the human footprint on planet Earth is so heavy (FIGURE 14.23 on pages 444–445) that the current rate of species extinctions now rivals the rate of extinctions during the previous die-offs.

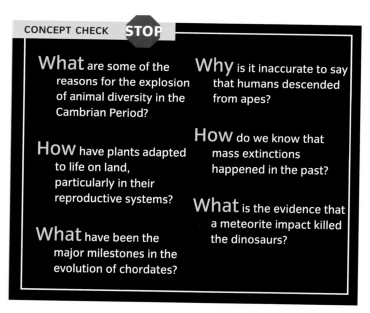

CONCEPT CHECK STOP

What are some of the reasons for the explosion of animal diversity in the Cambrian Period?

How have plants adapted to life on land, particularly in their reproductive systems?

What have been the major milestones in the evolution of chordates?

Why is it inaccurate to say that humans descended from apes?

How do we know that mass extinctions happened in the past?

What is the evidence that a meteorite impact killed the dinosaurs?

Amazing Places: The Burgess Shale

In 1909, the American paleontologist Charles Walcott discovered the world's pre-eminent site for Cambrian fossils, a rock formation called the Burgess Shale, about 90 kilometers away from Banff, British Columbia. It is a treasure trove for specimens such as this trilobite (FIGURE A), as well as more exotic organisms such as the five-eyed *Opabinia* (FIGURE B) and others. It was only in the 1970s that paleontologists realized that many of the organisms found there are unrelated to any living animals. They are not only extinct species, but members of extinct phyla. Exploring the Burgess Shale is like exploring the evolutionary paths that life could have taken—but didn't.

Many of the animals preserved in the Burgess Shale were soft-bodied. Such creatures fossilize only under very special circumstances. This is how scientists believe it happened: The Burgess fauna lived in shallow, oxygen-rich waters atop an algal reef with steep sides (FIGURE B). Periodically, mudslides driven by turbidity currents (Chapter 8) would sweep some unlucky animals off the side of the reef and deposit their bodies at the base (FIGURE C). At that time the deep waters were

still so oxygen-poor that no microorganisms existed at that depth to decompose them. The fine silt particles from the mudslide encased the bodies and hardened into shale, preserving a highly detailed imprint of the animals. The *Waptia fidensis* fossil in What a Geologist Sees (page 431) is an example.

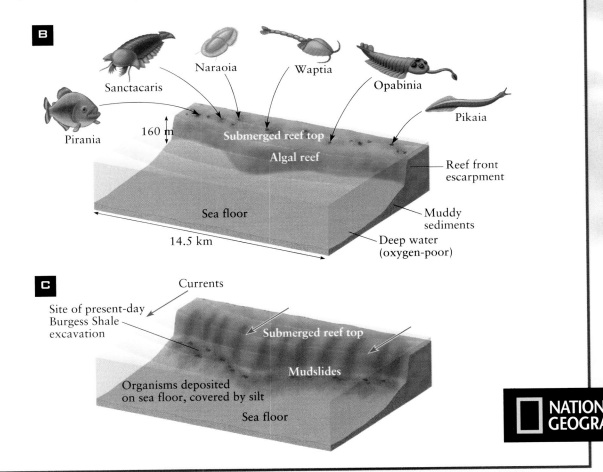

Visualizing

Human footprint on Earth's surface FIGURE 14.23

Developed by the Wildlife Conservation Society, the footprint
is a measure of population density and land use. The areas
least impacted by humans are in green.

**The human footprint
on planet Earth**

Most impacted

Least impacted

CHAPTER SUMMARY

1 Earth's Changing Atmosphere and Hydrosphere

1. The history of life is intertwined with that of the atmosphere and hydrosphere. Earth's early atmosphere consisted primarily of water vapor, carbon dioxide, and nitrogen, the products of outgassing from volcanoes.

2. Oxygen gradually accumulated over more than a billion years, primarily through the process of **photosynthesis**. Photosynthesis, a reaction by which plants convert carbon dioxide and water into carbohydrates, is by far the most important source of oxygen in our atmosphere.

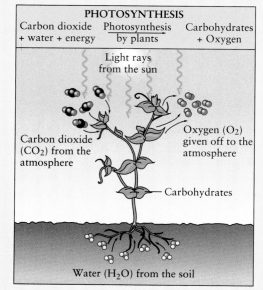

PHOTOSYNTHESIS

Carbon dioxide + water + energy — Photosynthesis by plants → Carbohydrates + Oxygen

Light rays from the sun

Carbon dioxide (CO_2) from the atmosphere

Oxygen (O_2) given off to the atmosphere

Carbohydrates

Water (H_2O) from the soil

3. The buildup in oxygen was accompanied by a buildup of ozone in the upper atmosphere. Earth's land surfaces were probably uninhabitable until the ozone layer began blocking harmful ultraviolet radiation from the Sun.

4. Living plants produce oxygen. When dead organisms decay, the bacteria that decompose them consume oxygen and release carbon dioxide. This cycle maintains the balance of oxygen in the atmosphere and hydrosphere. Nevertheless, atmospheric oxygen levels have fluctuated dramatically over geologic time. One reason this can occur is the burial of organic matter before it has completely decomposed. Plate tectonics can accelerate the rate of burial of organic matter, and thereby affect the composition of the atmosphere.

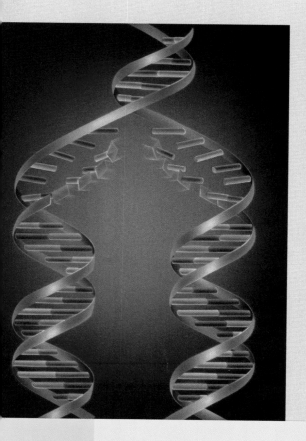

2 Early Life

1. The minimum requirements for life are a *metabolism* and a means of *replication*. All life on Earth is based on a small set of carbon-based building blocks called *amino acids*. Also, all living organisms use **DNA**, a biopolymer, to copy genetic information. However, different organisms have evolved different kinds of metabolism, and this is one reason that life is able to flourish in a great variety of environments.

2. The fundamental unit of life is the **cell**, which has two fundamental varieties: **prokaryotes** and **eukaryotes**. Eukaryotic cells enclose their DNA within a *membrane*, whereas prokaryotic cells lack a well-defined nucleus.

3. Biologists have classified all living organisms into three **domains**—bacteria, *Archaea*, and eukaryotes. The first two domains are prokaryotes, which are exclusively single-celled organisms. Prokaryotes evolved long before eukaryotes did. Eukaryotes evolved at least 1.4 billion years ago, and can be either one-celled or multicelled.

4. The most ancient fossils are about 3.5 billion years old, although chemical evidence of the presence of life is preserved in rocks considerably older than this. The best pieces of evidence for early life are fossilized *stromatolites*, thick mats of one-celled organisms that can still be found in some places today. Most other fossils of early life are microscopic, and consequently they are hard to find and difficult to interpret. The earliest known fossils of large multicellular organisms date back 635 million years.

3 Evolution and the Fossil Record

1. As a result of **evolution**, all present-day organisms are descendants of different kinds of organisms that existed in the past. The mechanism by which evolution occurs is **natural selection**, in which well-adapted individuals tend to have greater survival and more breeding success, and thus pass along their traits to later generations.

2. New **species** may emerge slowly, as the environment changes gradually over time, or quickly, when a beneficial *mutation* appears or when a population moves to a new habitat. It is not clear which process is more common.

3. Organisms can be preserved as **fossils** in many different ways. The body may be kept essentially intact, or its shape may be preserved while minerals replace its contents. Some organisms leave behind evidence, such as *molds* of footprints or **trace fossils**, without leaving behind any body parts.

4 Life in the Phanerozoic Eon

1. The Cambrian Period, at the beginning of the Phanerozoic Eon, was a time of explosive diversification of life forms. Several factors may have contributed to "the *Cambrian explosion*": the emergence of multicellular life (just before the Cambrian Period), the protective ozone layer, the rise in oxygen levels, and the ability to form shells.

2. All of the extant **phyla** in the animal **kingdom**, as well as several extinct phyla, emerged during the Cambrian Period. All of the Cambrian fauna were aquatic, and some had hard internal or external skeletons. In order to colonize land, plants and animals had to develop a means of *structural support*, an *internal aquatic environment*, and a way of *exchanging gases* with the atmosphere. For sexual reproduction, a *moist environment* was also essential.

3. The earliest land plants were seedless plants, such as ferns. The next major event in the plant kingdom was the emergence of **gymnosperms** or naked-seed plants, which flourished during the age of dinosaurs. **Angiosperms**, or flowering plants, developed last, with a much more efficient reproductive system facilitated by insect pollination.

4. The earliest land animals were arthropods (a phylum that includes insects). Fish first ventured onto land during the Devonian Period, and eventually gave rise to the amphibians. These, however, were limited in their geographic range because they still required water to spawn in.

5. Reptiles evolved a watertight skin and eggs that could be incubated outside of water. This freed them from depen-

(continued)

dence on watery environments. Both mammals and birds descended from reptiles, by different evolutionary pathways.

6. Mammals have existed since the Jurassic Period. It was apparently the disappearance of the dinosaurs, along with a rise in atmospheric oxygen, that gave mammals the opportunity to diversify and increase in size. **Hominids** are a very recent family of mammals, which appeared within the last 3.9 million years. Anatomically modern humans appeared only 30,000 years ago. Scientists are not certain whether Neandertals, which died out around the same time, were a distinct species.

7. Several geologic time periods are delineated by **mass extinctions**. The causes of mass extinctions are not thoroughly understood, but most scientists now believe that the Cretaceous-Tertiary extinction, which killed the dinosaurs, resulted from a meteorite impact. Other mass extinctions may have been caused by volcanism.

KEY TERMS

- **photosynthesis** p. 419
- **cell** p. 422
- **DNA** p. 422
- **prokaryotes** p. 424
- **domain** p. 424

- **eukaryotes** p. 425
- **evolution** p. 427
- **species** p. 427
- **natural selection** p. 429
- **fossil** p. 429

- **trace fossil** p. 430
- **kingdom** p. 432
- **gymnosperm** p. 434
- **angiosperm** p. 434
- **mass extinction** p. 439

CRITICAL AND CREATIVE THINKING QUESTIONS

1. Recall from Chapter 1 that Earth and Venus are so similar in size and overall composition that they are almost "twins." Why did these two planets evolve so differently? Why is Earth's atmosphere rich in oxygen and poor in carbon dioxide, whereas the reverse is true on Venus? What would happen to Earth's oceans if Earth were a little bit closer to the Sun?

2. One classical criticism of the theory of evolution was the paucity of transitional species. However, *Archaeopteryx* is such a species. In what ways did it resemble its dinosaur predecessors? In what ways was it like or unlike modern birds?

3. Another common criticism of Darwin's theory was its alleged inconsistency with the Biblical account of creation. Yet Darwin himself was a Christian, and many scientists since him have had no difficulty in reconciling the evidence for evolution with their Christian faith. Investigate how they have done so. Do you personally feel that evolution conflicts with your religious beliefs?

4. What do you think might have happened to mammals if the K-T extinction had not wiped out the dinosaurs?

What is happening in this picture ?

In this fossil bird and fossil frog, found in the Messel Oil Shale near Darmstadt, Germany, the animals' soft tissue, such as feathers and skin, are exceedingly well-preserved. (Note the faint imprint of the frog's skin around its bones.) Why do you think the soft body parts were preserved in this case, even though they are seldom found in most fossils?

1. Which of the following statements best describes changes in Earth's environment over the last four billion years?

   a. Solar brightness and oxygen generally increases while surface temperature and atmospheric CO_2 decreases.
   b. Solar brightness and oxygen generally decreases while surface temperature and atmospheric CO_2 increases.
   c. Solar brightness and surface temperature generally increases while oxygen and atmospheric CO_2 generally decreases.
   d. Solar brightness and surface temperature generally decreases while oxygen and atmospheric CO_2 generally increases.

2. Photosynthesis is fundamentally important to life on Earth. On this illustration, label the input and output of the photosynthetic process using the following terms:

   carbon dioxide CO_2 light rays
   water H_2O carbohydrates
   oxygen O_2

3. Although geologists cannot be certain why the atmospheric oxygen levels increased, they hypothesize that atmospheric oxygen has increased over time _____.

   a. as a continued increase in vegetation and photosynthesis
   b. because of increased rates of seafloor spreading
   c. because organic matter is buried before it decomposes, reducing the formation of carbon dioxide (CO_2)
   d. All of the above statements are correct.

4. The minimum requirements for life are _____.

   a. photosynthesis and a means of replication
   b. mobility and a means of replication
   c. metabolism and a means of replication
   d. mobility and metabolism

5. This illustration shows two cells. One is from a prokaryotic and the other a eukaryotic organism. Label structures in each of the cells with the following terms, and then label each cell either *prokaryotic* or *eukaryotic*.

   organelles   cell membrane   nucleus

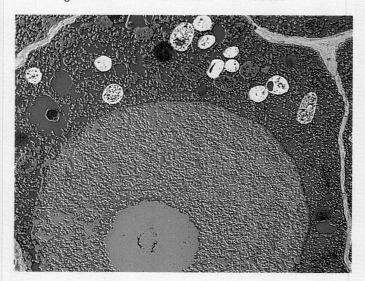

6. _____ is an example of anaerobic metabolism.

   a. Chemeosynthesis
   b. Photosynthesis
   c. Fermentation
   d. Both a and b are correct.
   e. Both a and c are correct.

7. Natural selection is the process by which _____.

   a. individuals migrate to environments better suited to survival of the species
   b. individuals adapt to their environment over time
   c. well-adapted individuals pass on their survival advantages to their offspring
   d. All of the above are correct.

8. As a mechanism for evolution, natural selection requires the passing on of specific traits from one generation to the next. These specific traits may reflect _____ passed on through an organism's _____.

   a. genetic mutations; metabolism
   b. genetic mutations; DNA
   c. genetic mutations; RNA
   d. metabolism; DNA
   e. metabolism; RNA

9. How are organisms preserved as fossils within the rock record?

   a. through the preservation of the organism in substances such as sap, tar, or ice
   b. through minerals replacing the contents of the dead organism (mineralization)
   c. by the leaving of impressions, molds or footprints
   d. through mummification in especially dry environments
   e. All of the above are methods through which organisms could be preserved as fossils.

10. Which of the following hypotheses do geologists think best explains the explosion of life forms represented by the Cambrian fossil record?

   a. Sexual reproduction began in the early Cambrian leading to a greater diversity of living organisms.
   b. An oxygen-rich atmosphere allowed for the metabolism of larger organisms.
   c. Predation began in the early Cambrian, leading to the development of many more organisms with shells and skeletons that might be better preserved in the fossil record.
   d. Extreme climatic changes at the end of the Proterozoic Eon led to accelerated evolutionary responses in surviving organisms.
   e. All of the above hypotheses are considered equally valid by geologists today.

11. Which of the following is not a requirement for a living organism to survive on land?

   a. structural support
   b. a means of locomotion
   c. an internal aquatic environment
   d. a mechanism of exchanging gases with the air
   e. if sexually reproducing, a moist environment for the reproductive system

12. Vascular plants exchange gas with the air by _____.

   a. diffusion through adjustable openings in the leaves called stomata
   b. osmosis involving oxygenated water in the root system
   c. photosynthesis
   d. All of the above.

13. Reptiles were freed from the water by _____.

   a. evolving an egg that contained amniotic fluid for the young to grow in
   b. developing a watertight skin
   c. developing a warm-blooded metabolism
   d. Both a and b are correct.
   e. Both b and c are correct.

14. _____ is probably the first species of our own genus.

   a. *Australopithecus*
   b. *Homo erectus*
   c. *Homo neandertalensis*
   d. *Homo sapiens*

15. Mass extinctions on Earth may be correlated with _____.

   a. massive eruptions of flood basalts
   b. the impact of massive meteorites
   c. glaciation
   d. Both a and b are correct.
   e. Both b and c are correct.

Understanding Earth's Resources

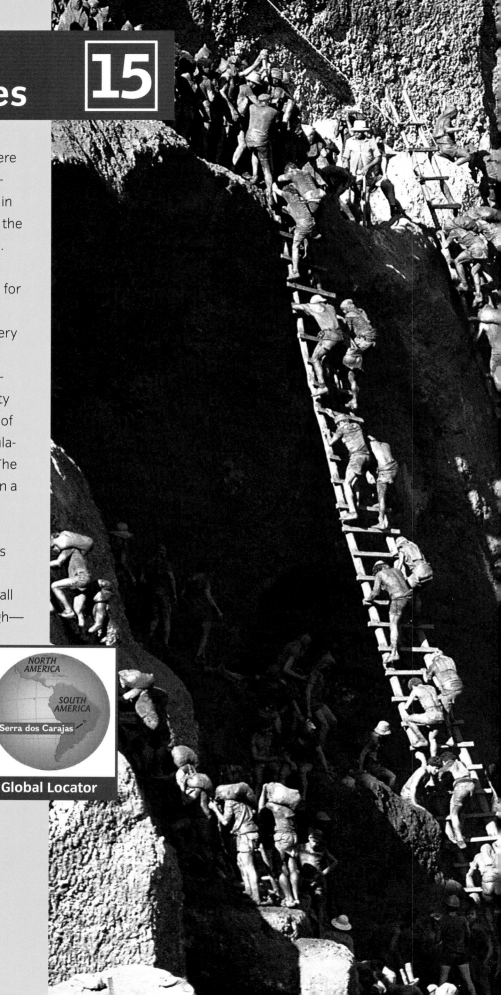

I n the 1980s, these gold miners in Brazil were each awarded a small claim by the government, as large as one of the square pedestals in the photograph. As the mine became deeper, the miners had to carry the ore out on their backs. Accidents, injury, and death were common.

Why are people willing to risk life and limb for gold? Gold is rare, and scarcity drives up its value—especially for these miners, who are very poor. It is nearly indestructible, making it an excellent store of value. It doesn't tarnish, corrode, or dissolve in common liquids. Its scarcity also ensures that owners take very good care of it. Nearly all the gold ever mined is still in circulation, making it the ultimate recyclable metal. The gold in Cleopatra's bracelets may now reside in a filling or a wedding band.

Because gold is so durable and so assiduously guarded, we can estimate how much has been mined in human history: about 150,000 tons. If all of this gold were stacked on a football field, it would make a pile only 1.75 meters high—not much of a return on 17,000 years of work!

Gold is only one of many resources that humans extract from Earth; others include copper, iron, tin, and zinc. Most are less romantic than gold, but much more essential to our daily lives. It may seem our supply of these resources is limitless, but every mineral and ore comes in finite amounts. As time goes by, the limits will become more and more apparent.

Global Locator

NORTH AMERICA

SOUTH AMERICA

Serra dos Carajas

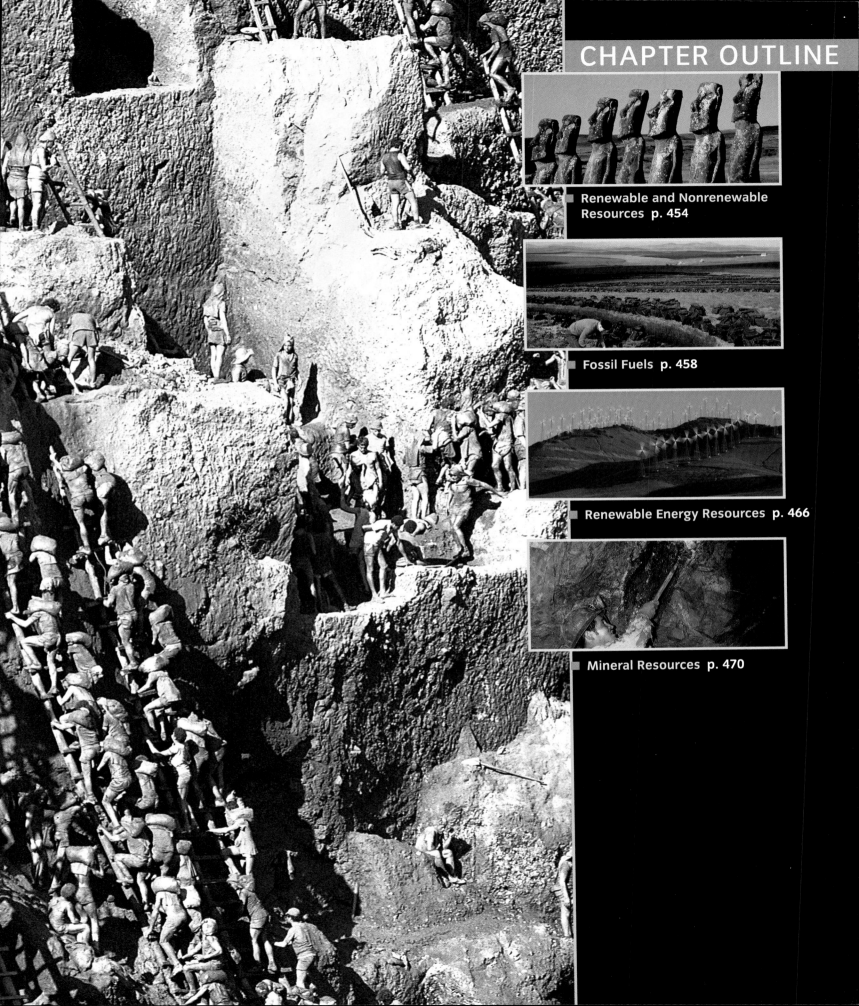

Renewable and Nonrenewable Resources

LEARNING OBJECTIVES

Describe how the loss of resources has affected past civilizations.

Explain the difference between renewable and nonrenewable resources.

Identify two kinds of nonrenewable resources that are crucial to modern society.

From the time our ancestors first picked up conveniently shaped stones and used them for hunting and skinning wild animals, to the present day, humans have relied on a bountiful Earth to supply a seemingly endless flow of useful materials. Today it is hard to imagine life without a myriad of such resources making life easier and more convenient. But, as you will learn, history suggests that there may be limits to the bounty.

NATURAL RESOURCES AND ANCIENT HISTORY

Silent and inscrutable after ten centuries, the giant stone heads of Easter Island (see **FIGURE 15.1**) are the principal remains of a civilization that once flourished on this remote outpost in the South Pacific. The Rapa Nui people transported hundreds of the megaliths, called *moai*, over hilly terrain, from the quarry where they were carved to their final locations. Archaeologists believe that the inhabitants used tree trunks and ropes to roll the stone figures across the land.

Just as interesting as the questions of how and why the Rapa Nui transported the moai is the reason they stopped. Scientists have found that palm trees grew on Easter Island until around 1500. Apparently the islanders cut down every last tree, leaving the ground bare and vulnerable to rapid erosion. In so doing, the Rapa Nui lost not only the ability to transport megaliths but also the resources necessary to sustain their civilization, which was decimated by famine and warfare over the next century.

The fall of past civilizations is an endlessly fascinating and endlessly controversial topic. No civilization's decline can be attributed to a single cause. Nevertheless, a common thread runs through the stories of many societies that flourished and then collapsed. That thread is the depletion of **natural resources**. The Easter Islanders failed to conserve their trees, a *biologic resource*, and soil, an *Earth resource*.

> **natural resources** Useful materials that are obtained from the lithosphere, atmosphere, hydrosphere, or biosphere.

Easter Island megaliths
FIGURE 15.1

The Polynesians who erected these stone heads on Easter Island were unwittingly sowing the seeds of their own culture's demise. They stripped the island of its palm trees, at least in part because they needed the trunks to roll the megaliths into place.

foggaras

Ghost town in the Sahara FIGURE 15.2

A An ancient city in the Saharan desert, Garama was the center of the Garamantian culture from 500 BC to 500 AD, and as many as 10,000 people may have lived here.

B Such a large population was supported by a vast underground network of foggara, or aqueducts, whose openings to the surface can still be seen.

Another critical Earth resource is water. Halfway around the globe, in present-day Libya, the Garamantian culture arose in the Sahara Desert around the same time as the Roman Empire, and prospered for about a thousand years in spite of the nearly complete lack of rainfall. The Garamantes survived by tapping into immense underground aquifers and transporting the groundwater through a channel of aqueducts called *foggara* (see FIGURE 15.2). Unfortunately, around 500 AD they apparently reached the end of their technological ability to retrieve groundwater. The foggara had been depleted, and their cities became ghost towns.

Like the Rapa Nui and the Garamantes, every human society depends on natural resources. Besides the most important resources—soil, water, and air—Earth's resources include building materials, metals, fertilizers, oil, coal, and gemstones. Biologic resources include crops, wild plants, and animals. Some of these resources are **renewable**. For example, even though we may consume a food crop each season, a new crop grows during the following season. A layer of soil lost to erosion will eventually be regenerated by the physical, chemical, and biologic processes of soil formation. Groundwater drawn from wells may eventually be replenished by rainwater. But what does "eventually" mean? Some resources take a very long time to regenerate—longer than humans are willing or able to wait. For example, the aquifers under the Sahara Desert formed tens of thousands of years ago, when the climate was moister than today, and they will not be replenished until the climate changes again, perhaps tens of thousands of years from now. For practical purposes, we consider these resources to be **nonrenewable**.

> **renewable resource** A resource that can be replenished or regenerated on the scale of a human lifetime.

> **nonrenewable resource** A resource that cannot be replenished or regenerated on the scale of a human lifetime.

An average American uses energy, directly or indirectly, at a rate equivalent to burning more than a hundred fifty 75-watt light bulbs every minute of the day, every day of the year. (Canadians, with their cold winters, use even more energy.)

A Here a highway sign admonishes Los Angeles commuters, "Don't be fuelish—be carpoolish." ▶

United States

Watts/person

= 75 watts of energy usage

World Average

Haiti

B At the other extreme, Haiti uses the least energy per capita of any country in the western hemisphere—about 1.5 light bulbs per inhabitant. In the town of La Victoire, there are no cars in evidence, and only one television for the whole town (right). The photographer reported, "Sometimes it even works." ▶

Resources such as coal, oil, copper, iron, gold, and fertilizers are mined from mineral deposits. Mineral deposits are known to be forming today, but the rate of formation is exceedingly slow. For example, it may take 600,000 years for a large copper deposit to be formed. From a human point of view, all mineral resources are one-crop resources, and Earth's supply of those "crops" is fixed.

Humans are the first and only species to routinely use nonrenewable resources. It does not seem to be in our nature to stop. However, history suggests that we should use such resources very judiciously and monitor how much we have left. That is what the Rapa Nui and the Garamantes failed to do. After the cultures that produced the moai and the foggara died out, their descendants had to adapt to the new conditions. The Easter Islanders developed new rituals that allocated their limited food resources. The inhabitants of the Sahara developed a nomadic lifestyle that was not dependent on the extraction of deeply buried groundwater. Like them, our descendents may have to make serious changes in their lifestyles as critical resources, such as oil, become scarce.

RESOURCES AND MODERN SOCIETY

In our world of the twenty-first century, there are now 6.5 billion human inhabitants, and the population is growing larger by millions each year. Each of us uses—directly or indirectly—a very large amount of material derived from nonrenewable *mineral resources*. Nearly every chemical element is used in one way or another, and they are extracted from more than 200 different minerals. Without them we could not build planes, cars, televisions, or computers. We could not distribute electric power or build tractors to till the fields and produce food. The metals needed to build machines for manufacturing, transportation, and communication are nonrenewable resources extracted from the ground. Without these resources, industry would collapse and living standards would deteriorate dramatically.

We are equally dependent on *energy resources.* Imagine what life would be like if we had to rely entirely on human muscle power. If a healthy adult rides an exercise bike that drives an electric generator hooked to a light bulb, the best an average person can do in a non-stop eight-hour day of pedaling is to keep a 75-watt bulb burning. In North America the same amount of electricity can be purchased from a power company for about 10 cents. Viewed in this way, we can see that human muscle power is puny. Over many thousands of years our ancestors found ways to supplement muscle power. At first, they did this by domesticating beasts of burden such as horses, oxen, camels, elephants, and llamas. They later learned to make sails to use wind power, dams to use water power, and engines to convert the heat energy of wood, oil, and coal into mechanical power.

Today we use supplementary energy in every part of our lives, from food production and transportation to housing and recreation. North Americans are among the world's biggest energy consumers (see **FIGURE 15.3**). Whereas soil and water were the most critical nonrenewable resources for earlier civilizations, our society's weakest link may be its excessive dependence on nonrenewable energy sources. How and whether we can reduce this dependence is both a political and social problem. Geology plays an important role in three ways: identifying new supplies of traditional resources, allowing us to estimate how much we have left, and evaluating the consequences of using new and unconventional resources.

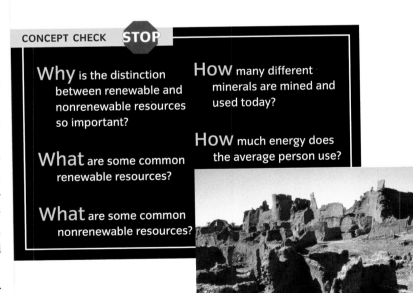

CONCEPT CHECK **STOP**

Why is the distinction between renewable and nonrenewable resources so important?

What are some common renewable resources?

What are some common nonrenewable resources?

How many different minerals are mined and used today?

How much energy does the average person use?

Fossil Fuels

People make use of renewable energy sources, such as wood, wind, solar, and water power, and nonrenewable energy sources. Everywhere in the world, even in the least developed countries, nonrenewable sources supply at least half of the energy used (see FIGURE 15.4), and the main nonrenewable source of energy is **fossil fuels**.

Fossil fuels are animal or plant remains that have been buried and trapped in sediment or sedimentary rock, and have undergone chemical and physical changes during and after burial. All fossil fuels are composed primarily of *hydrocarbon* (hydrogen + carbon) compounds. Because fossil fuels are derived from the altered remains of plants or animals, their energy content is derived originally from the Sun, via photosynthesis (Chapter 14). The principal fossil fuels are peat, coal, oil, and natural gas. The kind of sediment, the specific type of organic matter, and the post-burial changes that occur determine which type of fossil fuel is formed.

> **fossil fuel** Combustible organic matter that is trapped in sediment or sedimentary rock.

Energy sources and uses FIGURE 15.4

In the United States, fossil fuels (oil, natural gas, and coal) account for 85 percent of the energy used. The large amount of lost energy—nearly half, as shown in the upper right arrow—arises both from inefficiencies in energy use and from the fundamental physical limits on the efficiency of any heat engine.

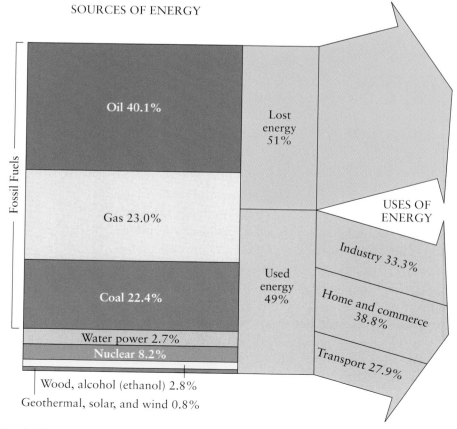

SOURCES OF ENERGY

Fossil Fuels

Oil 40.1%

Gas 23.0%

Coal 22.4%

Water power 2.7%

Nuclear 8.2%

Wood, alcohol (ethanol) 2.8%

Geothermal, solar, and wind 0.8%

Lost energy 51%

Used energy 49%

USES OF ENERGY

Industry 33.3%

Home and commerce 38.8%

Transport 27.9%

Coalification FIGURE 15.5

The conversion of plant matter to coal, or *coalification*, happens over a period of millions of years, as layers of peat are buried and compressed by overlying sediments.

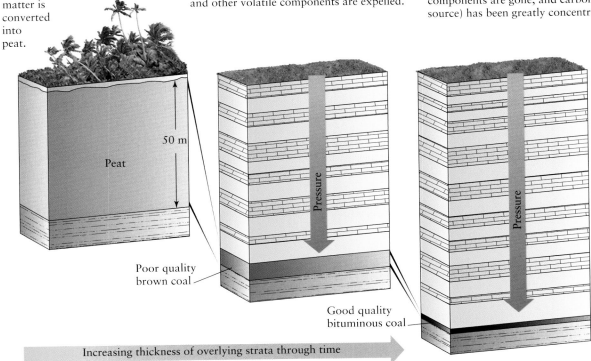

1. Swamps are thick with the organic remains of vegetation. As the organic matter decomposes, it is buried by more vegetation and sediment. The plant matter is converted into peat.

2. As the thickness of overlying sediment increases over time, causing higher pressures and temperatures on the organic layer, water and other volatile components are expelled.

3. By the time a layer of peat has been converted into coal, its thickness has been reduced by 90 percent, most of the volatile components are gone, and carbon (the heat source) has been greatly concentrated.

50 m

Peat

Pressure

Pressure

Poor quality brown coal

Good quality bituminous coal

Increasing thickness of overlying strata through time

PEAT AND COAL

Organic matter that accumulates on land comes from trees, bushes, and grasses. These land plants are rich in organic compounds that tend to remain solid after burial. In water-saturated places such as swamps and bogs, the remains accumulate to form **peat**.

Peat is the initial stage in the formation of **coal** (see FIGURE 15.5). As layers of peat are compressed and heated by being buried under more and more sediment, water and gaseous compounds such as carbon dioxide (CO_2) and methane (CH_4) escape. The loss of these constituents leaves a higher proportion of carbon in the residue, and carbon is what we want for burning—the higher the carbon content, the greater the *rank* of the coal. The lowest rank of coal is called *lignite*, whereas *anthracite* is the name for the highest rank of coal. Much of the world's coal is *bituminous* coal, intermediate between lignite and anthracite. Anthracite

peat A biogenic sediment formed from the accumulation and compaction of plant remains from bogs and swamps, with a carbon content of about 60 percent.

coal A combustible rock formed by the compression, heating, and lithification of plant-rich sediment (peat).

is so changed from its original form that it is considered a metamorphic rock.

Coal occurs in strata (miners call them *seams*) along with other sedimentary rocks, mainly shales and sandstones. Anthracite occurs with slates—low-grade metamorphic rocks. Coal seams tend to occur in groups. For example, 60 seams of bituminous coal have been found in western Pennsylvania. This abundant coal powered Pittsburgh's steel mills and made it into the "Steel City."

Peat (see FIGURE 15.6) and coal have been forming more or less continuously since vascular plants first appeared on Earth about 450 million years ago. However, the size of peat swamps has varied greatly throughout Earth's history, and therefore the amount of coal formed has also varied. By far the largest amounts of peat swamp formation occurred on Pangaea during the warm Carboniferous and Permian Periods, 360 to 245 million years ago. The great coal seams of Europe and eastern North America formed from peat deposited at that time, when the swamp plants were different from today and included giant ferns and scale trees. A second great period of peat formation peaked during the Cretaceous Period, 144 to 66.4 million years ago. During this period the plants of peat swamps were flowering plants, much like those found in swamps today.

The luxuriant growth needed to form thick and extensive coal seams is most likely to occur in a tropical or semitropical climate. Because of this, we can deduce that either the global climate was warmer in the past, or the lands where these seams occur were once in the tropics. Probably both conditions were involved. Coal deposits that were formed in warm low-latitude environments but are now in frigid polar lands provide compelling evidence for plate tectonics. The many places where coal is found in the United States today provide an excellent example (see FIGURE 15.7). Today peat formation is occurring in wetlands such as the Okefenokee Swamp in Florida and the Great Dismal Swamp in Virginia and North Carolina (see Figure 15.7B).

PETROLEUM

In the ocean, microscopic phytoplankton (tiny floating plants) and bacteria are the principal sources of organic matter in sediment. However, the chemicals found in these organisms differ from those in land plants. When buried and compressed, under the right conditions they are slowly transformed into **petroleum**.

Oil first came into widespread use in about 1847, when a merchant in Pittsburgh started bottling and selling oil from natural seeps. Five years later a Canadian chemist discovered that heating and distilling oil yields *kerosene*, a liquid that could be used in lamps. Kerosene lamps quickly replaced whale oil lamps and greatly reduced the use of candles in household lighting. Soon oil wells were being dug by hand near Oil Springs, Ontario. In 1859, the first commercial oil well was drilled in the United States (see FIGURE 15.8), and a modern industry was born.

petroleum Naturally occurring gaseous, liquid, and semisolid substances that consist chiefly of hydrocarbon compounds.

oil The liquid form of petroleum.

NATIONAL GEOGRAPHIC

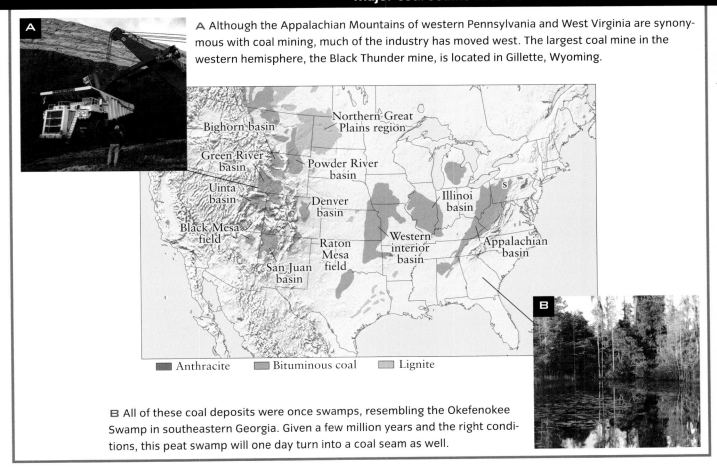

A Although the Appalachian Mountains of western Pennsylvania and West Virginia are synonymous with coal mining, much of the industry has moved west. The largest coal mine in the western hemisphere, the Black Thunder mine, is located in Gillette, Wyoming.

Bighorn basin

Northern Great Plains region

Green River basin

Powder River basin

Uinta basin

Denver basin

Illinoi basin

Black Mesa field

Raton Mesa field

Western interior basin

Appalachian basin

San Juan basin

■ Anthracite   ■ Bituminous coal   ■ Lignite

B All of these coal deposits were once swamps, resembling the Okefenokee Swamp in southeastern Georgia. Given a few million years and the right conditions, this peat swamp will one day turn into a coal seam as well.

---

natural gas The gaseous form of petroleum.

The modern use of **natural gas** started with an accidental discovery at Fredonia, New York, in 1821, where an ordinary water well began to produce bubbles of a mysterious gas. The gas was accidentally ignited and produced a spectacular flame. Wooden pipes were installed to carry the gas to a nearby hotel, where it was used in gaslights. Though rarely used today, gaslights preceded the electric light bulb and were considered by many people to be superior to early bulbs.

Today, petroleum products are used for a wide variety of purposes in addition to their main use as fuel for heating and operating vehicles and machines. Different components of petroleum are used in fertilizers,

The first oil well FIGURE 15.8

On August 27, 1859, Edwin Drake's drill "struck oil" in a deposit 21 meters underground in Titusville, Pennsylvania. This shed at Drake's Well Museum is a reconstruction of the world's first commercially productive oil well.

lubricants, asphalt, and the most ubiquitous material in contemporary life, plastic (see **FIGURE 15.9**).

Deposits of petroleum are nearly always found in association with sedimentary rocks that formed in a marine environment. When marine microorganisms die, their remains settle to the bottom and collect in the fine seafloor mud, where they start to decay. The decay process quickly uses up any oxygen that is present. The remaining organic material is thereby preserved without decaying and covered with more layers of mud and decaying organisms. The increasing heat and pressure associated with burial initiate a series of complex physical and chemical changes, called *maturation*, which break down the solid organic material and turn it into liquid and gaseous hydrocarbons. Meanwhile, the sediment itself is turned into rock, such as shale or limestone. A rock in which organic material has been converted into oil and natural gas is called a *source rock*. Oil and gas occupy more volume than solid organic matter, so the conversion process creates internal stresses that cause small fractures to form in the source rock and the fractures allow oil and gas to escape into adjacent, more porous rocks.

Being light, oil and gas slowly migrate toward the surface. If nothing happens to stop this slow percolation, they will eventually reach the surface and evaporate into the atmosphere. It is estimated that 99.9 percent of the petroleum in the world escapes this way.

On occasion, however, the migrating oil and gas will encounter an obstacle—a layer of impermeable rock that prevents it from reaching the surface. Such a formation is called a *cap rock*, and it most commonly consists of shale. A geologic situation that includes a source rock to contribute organic material, a porous *reservoir rock* in which the oil accumulates, and a cap rock to stop the migration is referred to as a petroleum trap (or a hydrocarbon trap). Although this may seem like a rather rare combination of circumstances, several types of petroleum traps are known (see **FIGURE 15.10**). Recognizing and finding these formations is the job of a petroleum geologist.

One raw material, many uses FIGURE 15.9

Crude oil consists mostly of hydrocarbons—molecules containing carbon and hydrogen but no oxygen. The molecules come in many different sizes. At an oil refinery, the crude oil is distilled into heavier and lighter components. Each type of hydrocarbon has different uses. The ones with fewer carbon atoms generally have a lower boiling point and are more useful as fuels. (The "octane" in gasoline, for instance, has eight carbon atoms.)

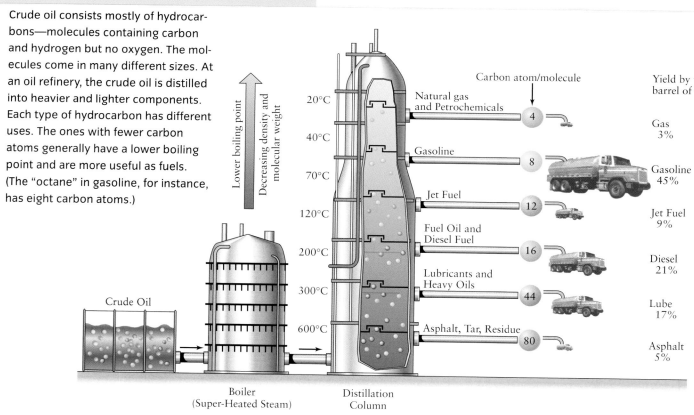

Oil and gas traps FIGURE 15.10

This figure illustrates six geologic circumstances in which oil can become trapped. Each kind of trap requires a source rock, a reservoir rock to contain the oil or gas, and a cap rock. Note that oil does not form an underground pool; it is trapped in the pores of the reservoir rock. Oil is usually found together with natural gas and water. The gas lies on top, because it is the least dense, and the water lies underneath, because it is densest.

www.wiley.com/college/murck

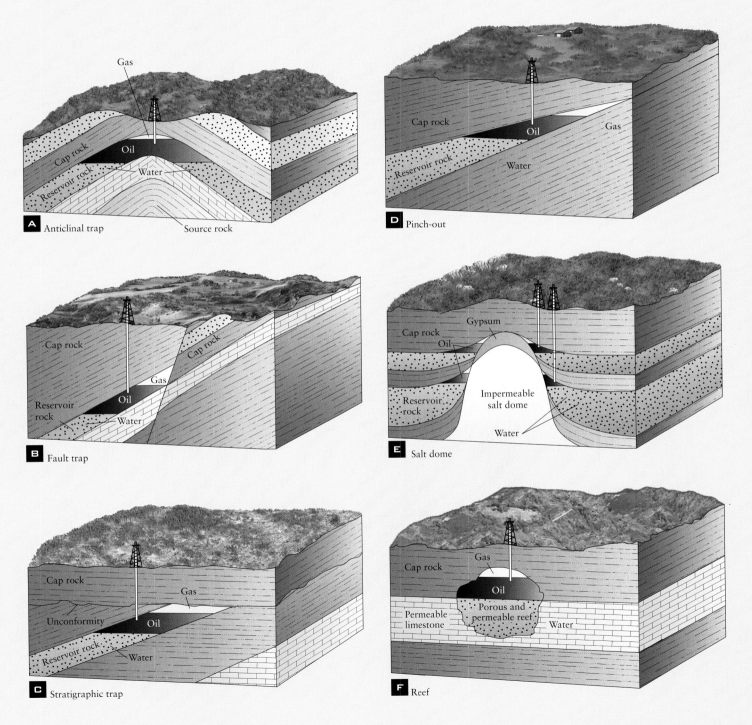

A Anticlinal trap

D Pinch-out

B Fault trap

E Salt dome

C Stratigraphic trap

F Reef

TAR SANDS AND OIL SHALES

Oil that is too viscous (thick) to be readily pumped is called *tar*. **Tar sands**, in which such deposits occur, are a promising and so far underexploited source of oil. The world's largest known deposit of tar sands is in Alberta, Canada (see FIGURE 15.11A). Geologists estimate that it contains as much as 600 billion barrels of tar. (For comparison, the world's annual oil production is about 30 billion barrels.) The sandstones in this region are cemented together by tar. To soften the tar and extract it from the sand, workers cook the sandstone with hot water and steam inside a rotating drum.

> **tar sand** A sediment or sedimentary rock in which the pores are filled by dense, viscous, asphalt-like oil.

Another potential source of petroleum is a waxlike organic substance called *kerogen*, which is found in certain very fine-grained sedimentary rocks such as shales. If burial temperatures are not high enough to form oil and natural gas, kerogen will be formed instead. If the kerogen is mined and heated, it breaks down and forms oil and gas. To be worth extracting, the kerogen in **oil shales** must yield more energy than the amount needed to mine and heat it. The world's largest deposit of rich oil shale is found in Colorado, Wyoming, and Utah (see FIGURE 15.11B). The U.S. Geological Survey estimates that these shale deposits, if mined and processed, could yield about 2000 billion barrels of oil. Like tar sands, the

> **oil shale** A fine-grained sedimentary rock with a high content of kerogen.

Nontraditional sources of oil FIGURE 15.11

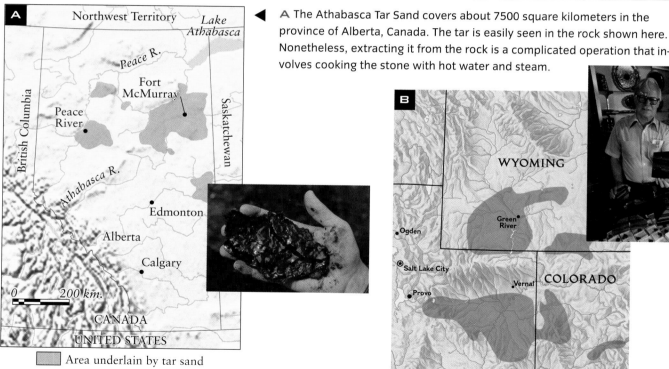

A The Athabasca Tar Sand covers about 7500 square kilometers in the province of Alberta, Canada. The tar is easily seen in the rock shown here. Nonetheless, extracting it from the rock is a complicated operation that involves cooking the stone with hot water and steam.

Area underlain by tar sand

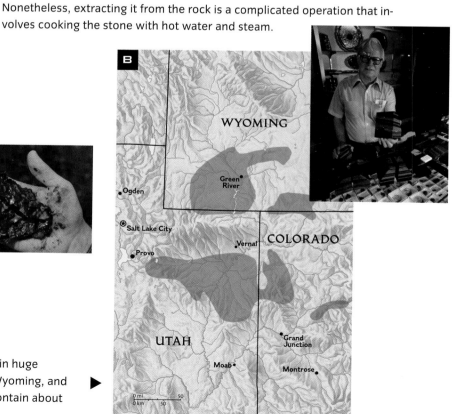

B Another hydrocarbon-rich rock is oil shale, found in huge amounts in the Green River formation in Colorado, Wyoming, and Utah. The shale bookends held by this gem dealer contain about half a pint of oil.

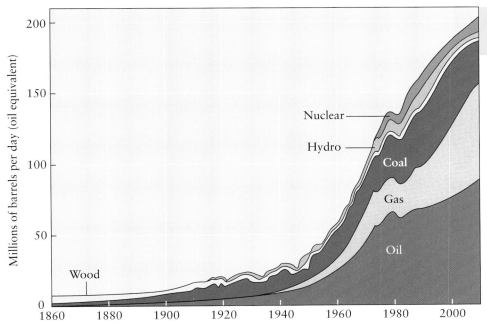

The history of world energy consumption shows oil's rise to prominence in the current energy mix. The brief decline in energy use in the late 1970s and early 1980s was a result of the "energy crises" of 1973 and 1979, when countries in the Middle East cut production and industrialized countries began to take conservation more seriously.

oil shale would have to be "cooked" to recover the crude oil.

RELYING ON FOSSIL FUELS

Because most of us grew up in an oil-powered economy, it is easy to forget just how recent our dependence on oil is. FIGURE 15.12 shows the history of energy consumption and the changing character of the world's energy "mix." Until 1860 or so, the world ran mainly on energy provided by wood; from then until World War II, coal was the dominant energy source. The huge increase in demand for energy after World War II has mostly been filled by oil and natural gas. The ability to discover new fossil fuel deposits to meet these increasing demands was greatly improved by the advent of new technologies for exploration, including seismic studies, satellite imagery, computer modeling, and digital recording methods.

Despite our improved ability to locate and retrieve fossil fuels, we cannot ignore the fact that they are nonrenewable resources. Most projections indicate that world production of oil will peak and begin to decrease before the year 2020. What will happen then? One possibility is that, as in the 1970s, increased efficiency and conservation measures will reduce the demand for energy. There may also be an increased contribution from so-called *unconventional* oil and gas reserves, including tar sands and oil shales. Coal is still abundant in the United States, but it is unlikely to replace oil, because it creates more severe air pollution. Some researchers are working on new technologies for cleaner-burning coal. However, to fulfill the worldwide demand for energy in a reliable, affordable, and environmentally acceptable manner, it will be necessary to place greater emphasis on other sources of energy besides fossil fuels.

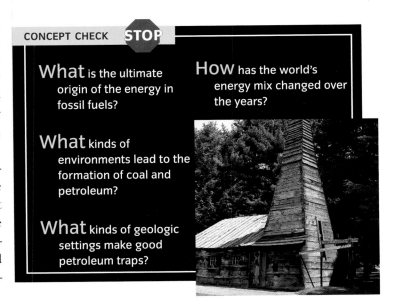

CONCEPT CHECK STOP

What is the ultimate origin of the energy in fossil fuels?

How has the world's energy mix changed over the years?

What kinds of environments lead to the formation of coal and petroleum?

What kinds of geologic settings make good petroleum traps?

Renewable Energy Resources

LEARNING OBJECTIVES

Identify the major alternatives to fossil fuels.

Explain why biomass, wind, and wave energy are indirect forms of solar energy.

Explain why solar energy is only a minor part of our energy mix.

Describe how nuclear power works, and its advantages and disadvantages.

 e noted earlier that each of the world's 6.5 billion people use energy at an average rate of approximately 2370 watts, for a total rate of 15.2×10^{12}. However, this is only a minute fraction of the amount of energy that enters the Earth system, primarily from the Sun (see **FIGURE 15.13**). It is clear, then, that we are not about to run out of energy in an absolute sense. The question is whether we are clever enough to learn how to use alternative sources of energy rather than relying so heavily on nonrenewable fossil fuels.

The potential alternatives to fossil fuels come from solar, biomass (fuel wood and animal waste), wind, wave, tidal, hydroelectric, nuclear, and geothermal sources. Most of these are renewable. Nuclear power is technically nonrenewable, but the amount of energy available is so vast that it is, for all practical purposes, inexhaustible. (The real issues for nuclear power are safety and waste disposal.) Sources that derive their energy directly or indirectly from the Sun will not be exhausted as long as the Sun continues to shine. Tidal energy too will continue as long as Earth rotates faster than the Moon revolves around it.

Earth's energy budget FIGURE 15.13

Earth's primary energy sources are the Sun, geothermal energy, and the tides. The incoming energy from the Sun—173,000 terawatts—dwarfs the other two sources. Most of our other energy sources are, in effect, converted solar energy. For instance, the energy in fossil fuels comes from photosynthesis by plants, and hydro power comes from the energy stored in evaporated water that has been precipitated on land.

www.wiley.com/college/murck

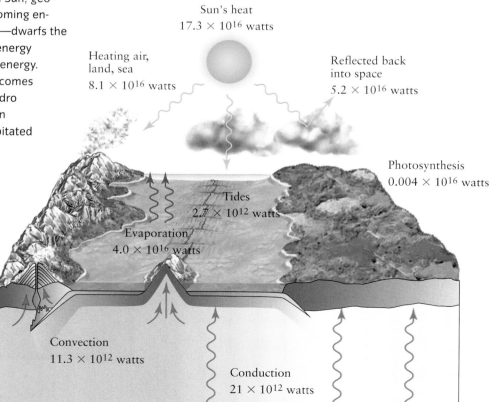

Solar energy FIGURE 15.14

The solar panels here were installed on the roof of San Francisco's Moscone Convention Center in 2004. With 5400 solar panels generating 825,000 kilowatt hours of electricity per year—about enough electricity to power 100 homes—it is the largest city-owned solar installation in the United States.

POWER FROM SUN, WIND, AND WATER

Solar energy reaches Earth from the Sun at a rate more than 10,000 times greater than the sum of all human energy demands. The hyper-abundance of solar energy makes it a very logical candidate for the power source of the future. However, the reality is that the solar future is still far away. Solar energy is best used for direct heating at or below the boiling point of water, as in greenhouses and solar water heaters. But it will make a significant dent in our energy budget only if we can convert it directly into electricity. Unfortunately, the best available solar (or *photovoltaic*) cells (see FIGURE 15.14) are still too costly and too inefficient for most uses. Photovoltaic technology is constantly improving, however, and the cost of energy generated in this manner is decreasing.

Plants are very efficient converters of solar energy. Our oil-based economy runs on the energy stored by plants millions of years ago. However, living plant matter also contains stored solar energy—and, unlike fossil fuels, it is renewable. **Biomass energy** derived from fuel wood was the dominant source of energy until the end of the nineteenth century, when it was displaced by coal. Biomass fuels are still widely used throughout the world, particularly in less developed countries, where the cost of fossil fuel is very high in relation to income.

> **biomass energy** Any form of energy that is derived more or less directly from plant life, including fuel wood, peat, animal dung, and agricultural wastes.

Wind is another indirect form of solar energy, because the heating of land and oceans creates the giant convection cells that drive the winds (Chapter 12). For thousands of years wind has been used as a source of power for ships and windmills. Today, huge windmill "farms" are being erected in particularly windy places (see FIGURE 15.15). In Denmark, about 3000 wind turbines supply electricity throughout the breezy coastal country. Although windmills are still expensive, it seems likely that their cost

Wind energy FIGURE 15.15

The windmills at this wind farm at Tehachapi Pass, California, generate "clean" (pollution-free) electricity by harnessing the energy of the wind. Each fan's rotary motion turns an electric generator. Wind farms can operate only where steady surface winds prevail year-round; consequently, wind energy is likely to remain a niche energy source.

Hydropower FIGURE 15.16

The Hoover Dam on the Colorado River in Arizona and Nevada looms in the background of this photograph. Behind it lies Lake Mead, an artificial lake created by the construction of the dam. Water from the lake drains through the hydroelectric power station (foreground). The ultimate sources of hydroelectric energy are the Sun, which powers the processes of the hydrologic cycle, and gravity, which causes the water behind the dam to fall and release its stored potential energy.

> **hydroelectric energy** Electricity generated by running water.

will soon be competitive with coal-burning electric power plants.

Several methods have been proposed to harness the power of waves, tides, or sea currents. However, **hydroelectric energy** is at present the only form of water-derived power that meets a significant portion of the world's energy needs. In order to convert the power of flowing water into electricity, it is necessary to build dams (see FIGURE 15.16). The flowing water is used to run turbines, which in turn convert the energy into electricity. The total recoverable energy from the water flowing in all of the world's streams is estimated to be equivalent to the energy obtained by burning 15 billion barrels of oil per year. Thus, even if all the potential hydropower in the world were developed, it could not meet all of today's energy needs. Another problem is that reservoirs eventually fill up with silt, so even though the source is inexhaustible, dams and reservoirs have limited lifetimes.

An energy source that will undoubtedly play a larger role in the future is hydrogen, which can be used to power *fuel cells*. Like batteries, fuel cells depend on chemical energy. But unlike batteries, they are easily replenished with new fuel. The only waste product produced by a hydrogen fuel cell is water (H_2O), so it is very clean compared with fossil fuels. However, the downside to fuel cells is that there is no place to "mine"

hydrogen. It can be extracted by separating water into hydrogen and oxygen, but the separation process uses up just as much energy as it produces. Hence hydrogen cannot really be thought of as a fuel in its own right, but as a way of storing energy that is produced in some other way (perhaps with solar or nuclear power). Even so, there is a great deal of excitement these days about the "hydrogen economy," and future developments will be worth watching.

NUCLEAR AND GEOTHERMAL POWER

Of the alternative power sources we mentioned earlier, *nuclear energy* is probably the only one with the potential to completely replace fossil fuel energy. However, it comes with serious drawbacks that have so far prevented it from reaching that potential.

Present-day nuclear reactors exploit the energy-producing process called *radioactive decay* (Chapter 3). Two isotopes of uranium, a radioactive element, occur naturally in certain minerals, and can be mined and used as fuel for nuclear reactors. The natural radioactivity of uranium-235 is augmented by purifying and concentrating it, and by bombarding it with neutrons to stimulate the process of fission. Under precisely controlled conditions, this gives rise to a *chain reaction* that sustains itself as long as there is enough uranium fuel present (see FIGURE 15.17A). An uncontrolled chain reaction would cause a nuclear explosion. But if

A

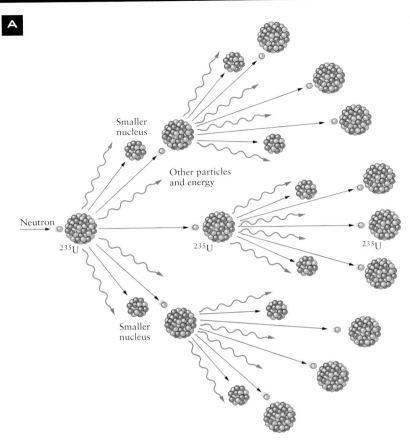

Neutron

²³⁵U

Smaller nucleus

Other particles and energy

²³⁵U

²³⁵U

Smaller nucleus

◀ A tremendous source of energy?

A In a chain reaction, a neutron strikes a uranium nucleus and causes it to split into smaller nuclei. The process releases two more neutrons, which can go on to collide with more uranium nuclei. Each time a uranium nucleus splits, it releases heat. An atomic power plant uses that heat to generate steam, which turns turbines that generate electricity.

...Or a tremendous source of toxic waste?

B Nuclear reactors produce radioactive waste that must be isolated from human contact for thousands of years. Here, a technician uses a radiation detector at a waste storage site in France where reprocessed waste from ten reactors is stored for five years before final disposal.

B ▶

C

◀ C In the United States, nuclear waste is presently stored at 125 temporary sites in 39 states. No permanent storage site has been built yet. In 2002, the U.S. government finally designated Yucca Mountain in Nevada, shown here, as a permanent storage site. However, the selection remains highly controversial.

the reaction proceeds under control, a great deal of useful energy can be generated. The fissioning of just 1 gram of uranium-235 produces as much heat as the burning of 2 million grams (or 13.7 barrels) of oil. This vast difference in energy content means that nuclear energy is essentially limitless, even though it does come from a nonrenewable resource.

Approximately 17 percent of the world's electricity is derived from nuclear power plants. In several European countries and in Japan, the production of nuclear power is rising rapidly because these countries have very limited fossil fuel supplies. In the United States, however, the last new nuclear power plant was ordered in 1978. The industry has been dogged by concerns about safety, especially after an accident in a reactor in Chernobyl, Ukraine, in 1986 killed about 50 people and released radioactivity over a wide area. Another still unresolved problem is the disposal of radioactive waste (see FIGURE 15.17B).

Another possibility is *geothermal energy*, which comes from Earth's interior, produced by the radioactive decay of uranium, thorium, and other radioactive isotopes, as well as the slow escape of heat from Earth's core. As we explain in Chapters 4 and 6, this is the energy source that powers plate tectonics and volcanic eruptions. It has been utilized for heat and power generation in locations where it is easily tapped, including New Zealand, Italy, Iceland, and the United States.

For the most part, the geothermal energy used by humans comes from *hydrothermal reservoirs*, underground systems of hot water or steam that circulate in fractured or porous rocks. To be used efficiently, hydrothermal reservoirs should be 200°C or hotter, and this temperature must be reached within 3 kilometers of the surface. Therefore, most of the world's hydrothermal reservoirs are close to the margins of tectonic plates, where hot rocks or magma can be found close to the surface. The Geysers geothermal installation in northern California—the largest producer of geothermal power in the world—is a hydrothermal reservoir system.

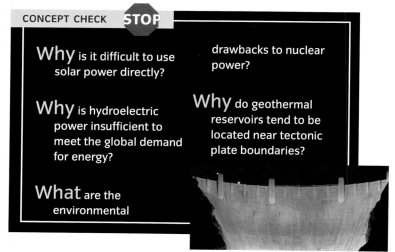

CONCEPT CHECK STOP

Why is it difficult to use solar power directly?

Why is hydroelectric power insufficient to meet the global demand for energy?

What are the environmental drawbacks to nuclear power?

Why do geothermal reservoirs tend to be located near tectonic plate boundaries?

Mineral Resources

LEARNING OBJECTIVES

Define economic geology.

Explain what distinguishes ores from other mineral deposits.

Identify six ways in which minerals become concentrated.

Describe the connection between plate tectonics and the location of mineral deposits.

The number and diversity of minerals and rocks that provide materials used by humans is so great as to defy any attempts to organize them conveniently. Nearly every kind of common rock and mineral can be used for something (see TABLE 15.1), but the economic reality is that we mine about 200 different minerals and about a dozen types of rock. Those mineral resources that are most valuable also tend to be rare. There are two broad groups of mineral resources. Metallic minerals are mined specifically for the metals that can be extracted from them. Examples are zinc, a valuable plat-

Some mineral products used in everyday life TABLE 15.1

Sand, gravel, stone, brick (clay), cement, steel, tar (asphalt)

Iron and steel, copper, lead, tin, cement, asbestos, glass, tile (clay), plastic

Mineral pigments (e.g., iron, zinc, titanium), and fillers (e.g., talc, mica, asbestos)

Iron, copper, many rare metals

Grown with mineral fertilizers; processed, packaged, and delivered by machines made of metal

ing metal refined from the mineral sphalerite (ZnS), and lead, used in automobile batteries, which we refine from the mineral galena (PbS). Nonmetallic minerals are mined for their properties as minerals, not for the metals they contain. Examples are salt, gypsum, and clay.

The branch of geology that is concerned with discovering new supplies of useful minerals is *economic geology*. Economic geologists seek mineral deposits from which the desired minerals can be recovered least expensively. To distinguish between profitable and unprofitable mineral deposits, we use the word **ore**. The difference depends on the extraction cost of the mineral and how much people are prepared to pay for it. It is quite common for a formerly unprofitable deposit to become a profitable ore, if extraction technology improves, if new uses are identified for the mineral, or if the mineral becomes scarce enough that the market price increases.

> **ore** A deposit from which one or more minerals can be extracted profitably.

FINDING AND ASSESSING MINERAL RESOURCES

Mineral resources are nonrenewable. Therefore, they can be exhausted by mining. Moreover, economically exploitable minerals are localized in distinct areas within Earth's crust, due to the way the crust has been modified by plate tectonics (see **FIGURE 15.18** on the following page). The uneven distribution of ex-

ploitable deposits means that no nation is self-sufficient in mineral supplies. For example, the United States has little or no aluminum, manganese, nickel, or chromium and must import most of what is used.

Because minerals are a nonrenewable resource, countries that can meet their needs for a given mineral today may not be able to in the future. A typical example is England, one of the first industrialized nations. A little more than a century ago, England was a great mining nation, producing and exporting such materials as tin, copper, tungsten, lead, and iron. Today, the known deposits of those minerals have been exhausted. The pattern followed by England—intensive mining followed by depletion of the resource, declining production and exports, and increasing dependence on imports—has been repeated often enough that geologists can estimate the remaining effective lifetime of a given mine or mineral resource. However, such projections are necessarily tentative; it is especially difficult to anticipate whether new deposits will be discovered.

Over the past few decades there has been a slow but steady shift of mineral exploration and production away from the industrialized nations and toward the less developed parts of the world. This shift will undoubtedly continue, where economic and social conditions permit it; as you can see from Figure 15.18, there are many parts of the world whose mineral resources are relatively unexploited. This doesn't mean that there are no more mineral deposits to be found in the industrialized countries; it just means that geologists will have to develop new exploration techniques and learn to look harder for mineral deposits in unconventional locations.

Visualizing

World map of mineral resources FIGURE 15.18

Although it is far from complete, this map hints at the complexity of the distribution of mineral resources around the world. Even a mineral-rich nation like the United States must depend on tiny Jamaica for its aluminum.

Major mines

- Al Aluminum
- Sb Antimony
- Bi Bismuth
- Cr Chromium
- Co Cobalt
- Cu Copper
- Au Gold
- Fe Iron ore
- Pb Lead
- Mn Manganese
- Mo Molybdenum
- Ni Nickel
- Pt Platinum
- Ag Silver
- Sn Tin
- Ti Titanium
- W Tungsten
- U Uranium
- Zn Zinc

Industry and mining

- Diamonds
- Phosphate
- Potash
- Al Processing plant
- Rare earth elements
- Steel Steel manufacturing

Annual crude steel production (millions of metric tons)

- Greater than 40
- 11-40
- 2.3-10
- 0-2.2

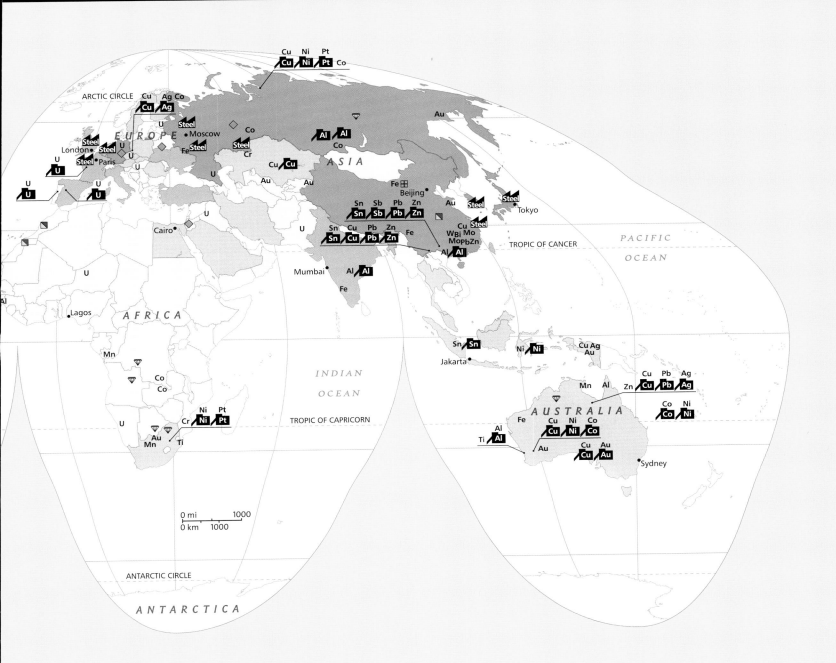

Cu Ni Pt
Cu **Ni** **Pt** Co

ARCTIC CIRCLE Cu Ag Co
Cu **Ag**
U
Steel
• Moscow
EUROPE   Co
London • **Steel** U   Co
Steel • Paris   **Steel**   **Steel**   *ASIA*
U **Steel**   Fe   **Steel**
U   U   Cr
U   **Cu** Cu
U   Au   Au   Fe ⊞
U   Beijing •
U   Cairo ◇   Au
U   Au   **Steel**   **Steel**
Sn Sb Pb Zn   Tokyo
Sn **Sb** **Pb** **Zn**   Au
U   Sn Cu Pb Zn Fe   Cu **Steel**
Al   **Sn** **Cu** **Pb** **Zn**   W Bi Mo
Mumbai •   Al **Al**   Mo Pb Zn
Al **Al**
Fe

Al
Lagos •   *AFRICA*

Mn   *INDIAN*   Sn **Sn**   Ni **Ni**   Cu Ag
▽   *OCEAN*   Au
Co   Jakarta •
▽ Co

TROPIC OF CAPRICORN   Cu Pb Ag
Ni Pt   Mn Al   Zn **Cu** **Pb** **Ag**
U Cr **Ni** **Pt**   *AUSTRALIA*   Co Ni
Au ▽ ▽   ▽   **Co** **Ni**
Mn   Ti   Fe Cu Ni Co
Al **Al**   **Cu** **Ni** **Co**
Ti **Al**   Au   Cu Au   • Sydney
Au   **Cu** **Au**

PACIFIC
OCEAN
TROPIC OF CANCER

0 mi   1000
0 km   1000

ANTARCTIC CIRCLE

ANTARCTICA

There are many ways mineral deposits can form. They are classified by the processes by which the ore minerals are concentrated.

Hydrothermal: A miner in Potosi, Bolivia, points to a rich vein containing chalcopyrite (a copper mineral), sphalerite (a zinc mineral), and galena (a lead mineral). The vein formed when volcanically heated water forced its way up into a fracture in the surrounding andesite.

Metamorphic: Hot fluids can also alter existing rocks through the process of contact metamorphism. The water often concentrates the minerals into distinct zones. This rock is from the Tem-Piute, Arizona, a metamorphic mineral deposit. The ore minerals present are sphalerite (brown, lower left), pyrite (gold-colored), and scheelite (pale grayish brown, lower right). Scheelite is a tungsten mineral.

Magmatic: Fractional crystallization in magmas can concentrate minerals. Chromite, the main ore mineral of chromium, sinks to the bottom of a basaltic magma and accumulates in an almost pure layer. In this unusually fine outcrop at Dwars River in South Africa, layers of pure chromite (black) are sandwiched by layers of plagioclase.

HOW MINERAL DEPOSITS ARE FORMED

For a mineral deposit to form, one or more processes must locally enrich the rock in one or more minerals. Minerals can become concentrated as a result of six general types of processes, which we illustrate in FIG-URE 15.19.

Hydrothermal deposits are formed when minerals precipitate out of hot water solutions. These solutions frequently deposit their mineral loads in existing

Sedimentary: Evaporite deposits, such as this salt pan in Death Valley, California, form when lake or sea water evaporates and leaves its dissolved mineral behind. Household products like baking soda and borax come from lake deposits. Death Valley itself was briefly mined for borax from 1883 to 1888.

Placer: The world's richest known gold deposit, in Witwatersrand, South Africa, is an ancient placer deposit. The layer of conglomerate that lies under the hammer head was once a loose layer of pebbles at the bottom of a shallow sea. The pebbles, along with the gold ore they contain, were eroded and washed downstream into the sea from a source vein that has never been found. The deposit was formed about 2.7 billion years ago.

Residual: Chemical weathering of rock can concentrate insoluble minerals by removing the more soluble ones. The most important mineral formed this way is bauxite, an aluminum ore. In this bauxite sample from Queensland, Australia, rounded masses of aluminum hydroxide (gibbsite) are imbedded in a matrix of iron and aluminum hydroxides.

www.wiley.com/
college/murck

cracks in the rock, creating mineral veins like the one pictured in Figure 15.19A. Such mineral deposits usually contain many different metal ores. For example, 90 percent of the silver produced in the United States is a byproduct of copper mined from hydrothermal deposits. In many cases the ore mineral is not visible to the naked eye. Remember that the size of a crystal depends on the amount of time it has to grow. Near the

Hydrothermal gold deposit FIGURE 15.20

Delicate sheets of native gold were deposited in the center of a quartz vein, Burgin Hill Mine, California.

surface, hydrothermal solutions cool very rapidly and thus do not have time to grow large crystals (but see FIGURE 15.20 for a beautiful exception).

In recent years, geologists have gotten a chance to see hydrothermal deposits forming right before their eyes in three distinct locations: a desert valley, a shallow sea, and a mid-oceanic rift zone (see What a Geologist Sees).

Many of the processes discussed elsewhere in this book play important roles in concentrating minerals and creating ore deposits. For instance, contact metamorphism and regional metamorphism (Chapter 10) give rise to *metamorphic mineral deposits* of the kind shown in Figure 15.19B. Fractional crystallization (Chapter 6) produces *magmatic mineral deposits* (see FIGURES 15.19C and 15.21). Commercially valuable ores that arise in magmatic deposits include

chromium, which is used to make steel tougher, and titanium, a corrosion-resistant metal that is lighter and stronger than steel.

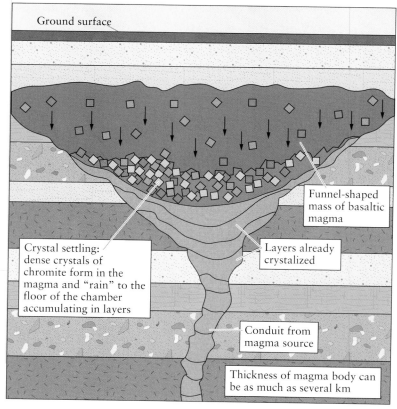

Ground surface

Funnel-shaped mass of basaltic magma

Crystal settling: dense crystals of chromite form in the magma and "rain" to the floor of the chamber accumulating in layers

Layers already crystalized

Conduit from magma source

Thickness of magma body can be as much as several km

Magmatic mineral deposits FIGURE 15.21

Chromite, the main ore mineral of chromium, crystallizes from a basaltic magma. Because they are denser than the magma, chromite crystals sink to the bottom and accumulate in an almost pure layer.

Mineral deposits in progress

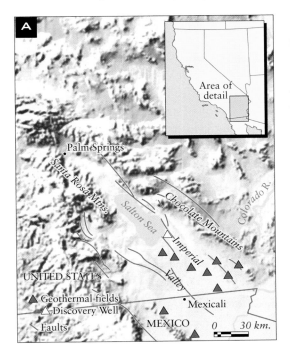

For many years, geologists could only speculate about how hydrothermal mineral deposits might have formed. Then three discoveries changed everyone's ideas about mineral deposits. First, in 1962, oil prospectors in the Imperial Valley of southern California (FIGURE A) unexpectedly tapped a superheated (320°C) brine at a depth of 1.5 kilometers.

As the brine flowed upward, it cooled and deposited minerals it had been carrying in solution. Over a period of three months, more than 1.5 tons of copper and half a ton of silver were deposited in this manner. Clearly, the same process could happen naturally under the right circumstances.

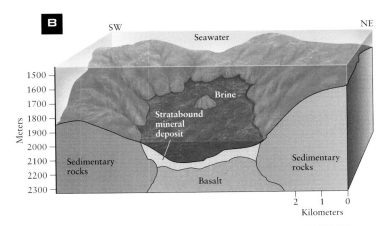

Two years later, oceanographers found a series of hot, dense brine pools at the bottom of the Red Sea (FIGURE B). Like the Imperial Valley brine, this water is trapped in a graben, formed in this case over the spreading center between the African and Arabian plates. Even more surprisingly, the oceanographers found ore minerals such as chalcopyrite, galena, and sphalerite at the bottom of the brine pools. In other words, they had found a sedimentary mineral deposit in the process of formation.

Water temperature—320°C
Black color from suspended iron sulfide

"Chimney" is made of chalcopyrite and pyrite deposited by hydrothermal solution.

Finally, in 1978, scientists in deep-diving submarines discovered 300°C hot springs emerging from the ocean floor at a mid-ocean ridge in the Pacific. Around the hot springs lay a blanket of sulfide minerals (FIGURE C). Again, they could see a modern hydrothermal ore deposit forming before their eyes.

Placer deposits always contain minerals that originated somewhere else and were transported away from their original site (sometimes not very far) by water. A careful look at Chapter 11 will show you where to look for placer deposits: They accumulate in places such as sand bars, where the water slows down and loses its carrying power (see **FIGURES** 15.19E and 15.22). Placer deposits are associated with minerals that have a high specific gravity, such as gold and platinum. Our last category, *residual deposits*, as shown in Figure 15.19F, are the minerals that are left behind by chemical weathering of a host rock (Chapter 7). In summary, nearly everything you have learned about the Earth system in this book has some relevance to economic geology.

PLATE TECTONIC CONTROLS ON ORE DEPOSITS

To a certain extent, the distribution of mineral resources can be linked to specific tectonic environments. This permits economic geologists to identify locations where certain types of mineral deposits are most likely to be found (see **FIGURE** 15.23). However, the correlation is not nearly as precise or as reliable as we would like, which is why we still need exploration geologists to go out to the sites.

Many kinds of mineral deposits occur in groups, forming what exploration geologists call *metallogenic provinces*. These are regions of the crust within which mineral deposits occur in unusually large numbers. Metallogenic provinces form as a result of either climatic (as in the formation of bauxite deposits in the tropics) or plate tectonic processes. For example, both magmatic and hydrothermal deposits typically form near present or past plate boundaries. This is hardly surprising, as the deposits are related directly or indirectly to igneous activity, and most igneous activity is related to plate tectonics.

Likewise, some energy resources are concentrated in particular tectonic environments. Geothermal energy is closely related to active plate boundaries and other locations where heat flow is high. Fossil fuels occur in environments where organic-rich sediments accumulated long ago—terrestrial swamp environments for peat and coal, and marine sedimentary basins for oil and natural gas.

WILL WE RUN OUT?

Are there enough mineral and energy resources available on this planet to raise the living standards of all people to the levels they desire? The answer to this

Placer deposits FIGURE 15.22

Dense ore minerals, such as gold, platinum, and diamond, tend to sink and accumulate at the bottom of sediment layers deposited by moving water, such as streams or longshore currents. Such concentrations are called *placers*. More than half of the gold recovered in human history has come from placers. This diagram shows three typical geologic locations where placer concentrations can be found.

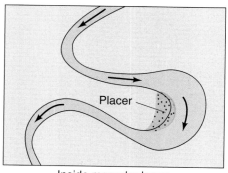

Inside meander loops

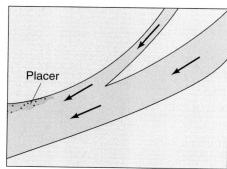

Downstream from a tributary

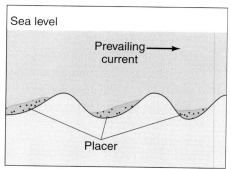

Behind undulations on ocean floor

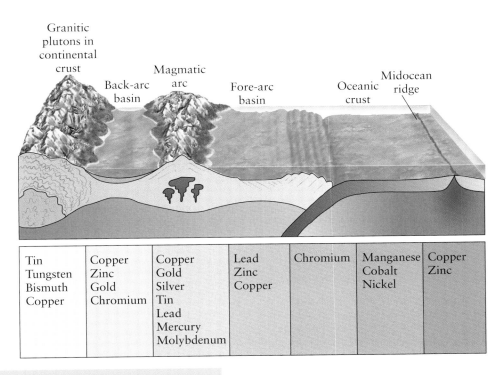

Tin Tungsten Bismuth Copper	Copper Zinc Gold Chromium	Copper Gold Silver Tin Lead Mercury Molybdenum	Lead Zinc Copper	Chromium	Manganese Cobalt Nickel	Copper Zinc

Plate tectonics and minerals FIGURE 15.23

Many materials are concentrated in particular plate-tectonic settings. For instance, tungsten is found in the metamorphic rocks adjacent to granite intrusions. Manganese is found on the deep seafloor in great abundance in "manganese nodules," which may be formed partly through chemical precipitation and partly by biochemical reactions.

question is not clear. Many geologists are concerned that shortages of nonrenewable resources will eventually hamper development. Unfortunately, there is no sure way to know exactly how much of any mineral or energy resource remains to be found. At present, it appears that shortages of energy resources are more likely to affect us in the short run than shortages of metallic minerals.

Throughout the history of energy and mineral exploration, geologists have continually refined and improved the techniques used to locate new deposits. Much ingenuity has been expended in bringing mineral and energy resource production to its present state. By a combination of new discoveries by geologists, new extraction and processing technologies, and careful use and conservation practices, it is likely that Earth's mineral and energy resources can be extended far into the twenty-first century. However, in the deep future—by the end of the twenty-first century and beyond—our descendants may face some very hard choices, as the Easter Islanders did in the 1500s. At the very least, they will have to manage Earth's resources very differently from how we and our predecessors have managed them.

CONCEPT CHECK STOP

Why are some mineral deposits not considered to be ores?

How do placer deposits form?

What kind of mineral deposit do aluminum cans come from?

What is a natural tectonic setting for magmatic mineral deposits?

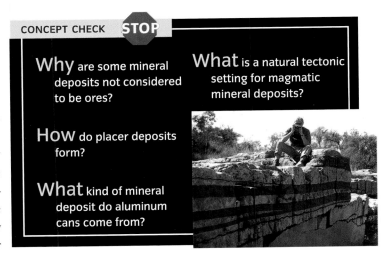

Amazing Places: Yesterday's and today's copper mines

◄ The last of our Amazing Places in this book is actually two places, which vividly illustrate the changes in mining over the last 150 years. In FIGURE A, tourists view the interior of the Delaware Copper Mine on the Keweenaw Peninsula in Michigan, the site of one of America's early large mining operations. The first mines in this area opened in 1843, and the mine shown here operated from 1847 to 1887.

Much of the ore was pure "native copper," such as the crystal shown in FIGURE B. The copper was brought to the surface from underground mine shafts. After the 1920s, though, the Michigan mines became increasingly less competitive with the much larger open-pit mines in the West. ►

◄ The Bingham Canyon Copper Mine in Utah, shown in FIGURE C, is the largest human excavation on Earth. Located near Salt Lake City, it is open to the public. It is currently 4 kilometers wide and nearly a full kilometer deep.

To fully appreciate its immensity, notice that each of the "stairsteps" in the photo is higher than a telephone pole and wide enough to comfortably fit a train (FIGURE D). In contrast to the Michigan copper deposits, the ores here are only 0.6 percent copper, and therefore must undergo extensive concentration, refining, and smelting. Nevertheless, the Kennecott Utah Copper Corp. profitably extracted 264,000 tons of copper here in 2004, along with 8500 kilograms of gold and 85,000 kilograms of silver. ►

NATIONAL GEOGRAPHIC

1 Renewable and Nonrenewable Resources

1. **Natural resources** can be either **renewable** or **nonrenewable**. Biologic resources are mostly renewable, whereas Earth's resources, particularly minerals and fossil fuels, are mostly nonrenewable. A resource such as soil or groundwater is considered nonrenewable if it cannot be replenished on the scale of a human lifetime.

2. Past civilizations have gone through major upheavals when they ran out of a nonrenewable resource. The most likely shortages to face our civilization in the near future are shortages of energy resources.

2 Fossil Fuels

1. Currently, most of the world's energy needs are met by **fossil fuels**, a nonrenewable resource. These resources form over the course of millions of years by the burial and chemical alteration of animal and plant remains.

2. The major types of fossil fuel are **peat**, **coal**, **petroleum**, and **natural gas**. Peat and coal are formed by the compression and heating of land plants in swampy or boggy conditions, with higher grades of coal being formed by deeper burial. Although peat swamps continue to exist today, the most prolific periods for the production of peat were the Carboniferous and Permian Periods, when warm and moist conditions prevailed over most of Europe and North America.

3. Petroleum and natural gas are formed by the compression and heating of the remains of marine organisms. Petroleum consists of a variety of hydrocarbon molecules, which contain hydrogen and carbon but no oxygen. The molecules with more carbon atoms tend to be heavier, less volatile, and more useful as lubricants, while the molecules with fewer carbon atoms (including natural gas) are lighter and more useful as fuel.

4. The **oil** component of petroleum can be refined into gasoline or kerosene, and it is also the raw material for plastics. Nonconventional sources of oil include **tar sand**, which is already economically important in Canada, and **oil shale**. These require more intensive processing to yield their oil, but they may become economically important in the future.

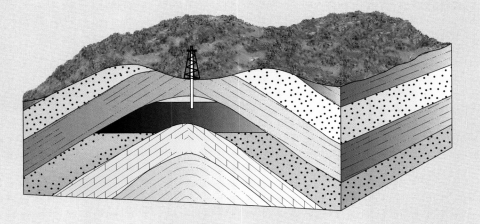

3 Renewable Energy Resources

1. By far the greatest source of energy available at Earth's surface comes from solar radiation. However, humans are so far able to capture only a minuscule fraction of that energy. Direct *solar energy* remains too expensive for electric generation in most places, although it is well suited for heating.

2. *Wind power* and **biomass energy** can be considered indirect versions of solar energy. They are renewable, but at present they provide only a small fraction of our energy. Biomass (primarily wood) was the world's leading source of energy until the late 1800s, when it was replaced by coal.

3. **Hydroelectric energy**, derived from running water, is a proven energy source in North America. It is not capable of significant further growth. Hydroelectric energy is generated when water flows over a dam and drives the turbines of a power station.

4. *Nuclear energy*, provided by the radioactive decay of uranium, is virtually limitless and probably the only resource that could in theory replace fossil fuels. In Japan and certain European countries, it is already a major and growing source of energy. However, in North America it faces a serious waste-disposal problem and public fears about its safety.

5. *Geothermal energy*, powered by the heat radiating from magma chambers and recently erupted volcanic rocks, is plentiful in certain locations near tectonic plate boundaries.

4 Mineral Resources

1. *Economic geologists* search for new mineral deposits and assess how long they are likely to last. Not all mineral deposits can be mined profitably; those that can are called **ores**. About 200 minerals are mined in the world today.

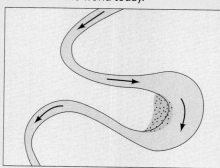

2. Mineral deposits form when natural processes concentrate certain types of material in one location. *Hydrothermal* deposits occur when superheated water cools rapidly and dissolved minerals precipitate out of it. *Metamorphic* deposits occur as a result of contact metamorphism, and *magmatic* deposits can result from crystallization. *Sedimentary* deposits include the salts left behind by evaporating water from a lake or shallow sea. *Placer* deposits form when mineral-bearing rocks are eroded and transported by water. They are usually found where the water slows down and drops its sediment load. Finally, *residual* deposits occur in the rock that is left behind after chemical weathering. Examples of ores that are formed by these six processes are (respectively) copper, zinc, chromium, gypsum, gold, and aluminum.

3. The location of several of these processes is controlled by plate tectonics, so that an understanding of an area's geologic history can guide economic geologists to promising locations. For example, magmatic mineral deposits are most likely to be found near present or former plate boundaries. Oil and gas are typically found in former marine sedimentary basins.

4. New discoveries and new technologies can prolong to some extent our supply of nonrenewable resources. However, in the long run, society must confront the challenge of managing Earth's finite resources.

KEY TERMS

- natural resource p. 454
- renewable resource p. 455
- nonrenewable resource p. 455
- fossil fuel p. 458
- peat p. 459

- coal p. 459
- petroleum p. 460
- oil p. 460
- natural gas p. 461
- tar sand p. 464

- oil shale p. 464
- biomass energy p. 467
- hydroelectric energy p. 468
- ore p. 471

CRITICAL AND CREATIVE THINKING QUESTIONS

1. Oil production in the United States satisfies only half of the country's needs; the rest is imported. If imports were cut off, what changes would you expect to occur in your lifestyle?

2. If you are interested in controversies, investigate how geophysicist M. King Hubbert predicted in 1956 that annual U.S. oil production would peak in the early 1970s (the actual peak occurred in 1971). Also, investigate the debate over whether the same predictive technique can be applied to world oil supplies. When do you think world oil production will reach its peak—or has it done so already?

3. Many hydrothermal mineral deposits of copper, gold, silver, and other metals have been found in the countries bordering on the Pacific Ocean. Can you offer an explanation for this remarkable concentration? If you were part of a team of exploration geolo-

gists looking for large copper deposits, where would you focus your search?

4. Given that we are now dependent on nonrenewable resources of energy and minerals, and that the world's population continues to increase, how do you think human societies will adjust in the future? Do we have a resource problem or a population problem (or both)?

5. Some people think that sustainable development is not a useful concept, because it may be impossible to implement—or even to define—in the case of nonrenewable resources. Others think that it is an extremely important concept, if only because it makes us think about the needs of future generations in planning resource management. What do you think?

What is happening in this picture ?

An abandoned salt mine in Germany, which lies one kilometer underground, has become a repository for the casks of spent nuclear fuel shown in this picture. What are some of the characteristics that might make this a good place to store nuclear waste?

SELF-TEST

1. Which one of the following statements is incorrect?

   a. The depletion of natural resources apparently led to the collapse of some ancient civilizations.
   b. Nonrenewable resources are never replenished.
   c. Renewable resources must be managed so they are not used at a rate that is greater than the rate of renewal or replenishment of the resource.
   d. Groundwater is, in principle, a renewable resource, but once depleted it may take a very long time to be replenished.

2. Which of the following is an example of a nonrenewable resource?

   a. petroleum
   b. wind power
   c. tidal power
   d. solar power

3. Which of the following is an example of a renewable resource?

   a. hydroelectric power
   b. nuclear power
   c. coal
   d. natural gas

4. This illustration shows the proportion of energy resources used in the world today. The sources of energy have been listed for you. Label the right side of the graph showing lost and used energy and the percentages of each. Also label the distribution of used energy to industry, home and commerce, and transport showing the percentages of each.

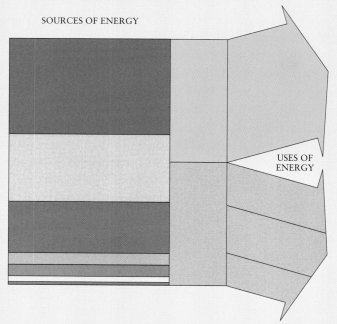

SOURCES OF ENERGY

USES OF ENERGY

5. _____ forms by the compression heating and lithification of plant-rich sediment.

   a. Petroleum
   b. Coal
   c. Biomass energy
   d. All of the above answers are correct.

6. _____ forms from the decomposition of ancient plankton buried in marine sediment.

   a. Petroleum
   b. Coal
   c. Biomass energy

7. The illustration below is a graph showing world energy consumption over time. Complete the graph by labeling the fields showing the proper distribution of types of energy consumed.

   oil   wood
   nuclear   hydroelectric
   coal   gas

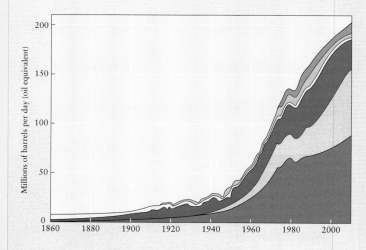

8. _____ is technically a nonrenewable resource, but the amount of energy available is so vast that it is, for all practical purposes, inexhaustible.

   a. Geothermal energy
   b. Nuclear power
   c. Biomass energy
   d. Solar power
   e. Wind energy

9. Which of the following renewable resources derives its energy from the Sun?

   a. wind energy

   b. wave energy

   c. biomass energy

   d. All of the above resources derive their energy from the Sun.

10. Why doesn't solar energy play a more significant role in electricity production?

   a. Given current photovoltaic technologies, it is very costly to produce electricity.

   b. There is not enough energy in sunlight to help with our current levels of consumption.

   c. Current photovoltaic technologies do not allow for electricity production at high enough voltages.

   d. It is too difficult to install solar technologies on existing buildings.

11. Which of the following statements about nuclear power is (are) true?

   a. Nuclear power is one of the few energy sources with the potential to completely replace fossil fuels.

   b. Nuclear reactors generate waste that must be isolated from human contact for thousands of years.

   c. Approximately 17 percent of the worlds electricity is currently generated by nuclear power.

   d. Nuclear power is generated through a controlled reaction of fission decay.

   e. All of the above statements are true.

12. _____ is the branch of geology that is concerned with discovering new supplies of useful minerals.

   a. Geophysics

   b. Environmental geology

   c. Physical geology

   d. Economic geology

   e. Mineral geology

13. What is the difference between an ore and a mineral deposit?

   a. There is no difference.

   b. Ore deposits are mineral deposits that contain metals.

   c. Ore deposits are mineral deposits that can be profitably mined.

   d. Ore deposits are mineral deposits that have formed under high temperature.

14. _____ deposits are formed when minerals precipitate out of hot water solutions.

   a. Hydrothermal

   b. Metamorphic

   c. Placer

   d. Sedimentary

   e. Magmatic

15. This illustration shows a cross-section through Earth's crust across a number of tectonic settings. Using the list below as a guide, indicate which metallic resources are associated with each tectonic setting.

tin	tungsten	copper	gold
zinc	manganese	chromium	cobalt
bismuth	lead	mercury	molybdenum
nickel	silver		

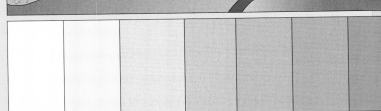

Appendix A
Units and Their Conversions

Commonly Used Units of Measure

Length

Metric Measure

1 kilometer (km)	= 1000 meters (m)
1 meter (m)	= 100 centimeters (cm)
1 centimeter (cm)	= 10 millimeters (mm)
1 millimeter (mm)	= 1000 micrometers (μm) (formerly called microns)
1 micrometer (μm)	= 0.001 millimeter (mm)
1 angstrom (Å)	= 10^{-8} centimeters (cm)

Nonmetric Measure

1 mile (mi)	= 5280 feet (ft) = 1760 yards (yd)
1 yard (yd)	= 3 feet (ft)
1 fathom (fath)	= 6 feet (ft)

Conversions

1 kilometer (km)	= 0.6214 mile (mi)
1 meter (m)	= 1.094 yards (yd)
	= 3.281 feet (ft)
1 centimeter (cm)	= 0.3937 inch (in)
1 millimeter (mm)	= 0.0394 inch (in)
1 mile (mi)	= 1.609 kilometers (km)
1 yard (yd)	= 0.9144 meter (m)
1 foot (ft)	= 0.3048 meter (m)
1 inch (in)	= 2.54 centimeters (cm)
1 inch (in)	= 25.4 millimeters (mm)
1 fathon (fath)	= 1.8288 meters (m)

Area

Metric Measure

1 square kilometer (km^2)	= 1,000,000 square meters (m^2)
	= 100 hectares (ha)
1 square meter (m^2)	= 10,000 square centimeters (cm^2)
1 hectare (ha)	= 10,000 square meters (m^2)

Nonmetric Measure

1 square mile (mi^2)	= 640 acres (ac)
1 acre (ac)	= 4840 square yards (yd^2)
1 square foot (ft^2)	= 144 square inches (in^2)

Conversions

1 square kilometer (km^2)	= 0.386 square mile (mi^2)
1 hectare (ha)	= 2.471 acres (ac)
1 square meter (m^2)	= 1.196 square yards (yd^2)
	= 10.764 square feet (ft^2)
1 square centimeter (cm^2)	= 0.155 square inch (in^2)
1 square mile (mi^2)	= 2.59 square kilometers (km^2)
1 acre (ac)	= 0.4047 hectare (ha)
1 square yard (yd^2)	= 0.836 square meter (m^2)
1 square foot (ft^2)	= 0.0929 square meter (m^2)
1 square inch (in^2)	= 6.4516 square centimeter (cm^2)

Volume

Metric Measure

1 cubic meter (m^3)	= 1,000,000 cubic centimeters (cm^3)
1 liter (l)	= 1000 milliliters (ml)
	= 0.001 cubic meter (m^3)
1 centiliter (cl)	= 10 milliliters (ml)
1 milliliter (ml)	= 1 cubic centimeter (cm^3)

Nonmetric Measure

1 cubic yard (yd^3)	= 27 cubic feet (ft^3)
1 cubic foot (ft^3)	= 1728 cubic inches (in^3)
1 barrel (oil) (bbl)	= 42 gallons (U.S.) (gal)

Conversions

1 cubic kilometer (km^3)	= 0.24 cubic miles (mi^3)
1 cubic meter (m^3)	= 264.2 gallons (U.S.) (gal)
	= 35.314 cubic feet (ft^3)
1 liter (l)	= 1.057 quarts (U.S.) (qt)
	= 33.815 ounces (U.S. fluid) (fl. oz.)
1 cubic centimeter (cm^3)	= 0.0610 cubic inch (in^3)
1 cubic mile (mi^3)	= 4.168 cubic kilometers (km^3)
1 acre-foot (ac-ft)	= 1233.46 cubic meters (m^3)
1 cubic yard (yd^3)	= 0.7646 cubic meter (m^3)
1 cubic foot (ft^3)	= 0.0283 cubic meter (m^3)
1 cubic inch (in^3)	= 16.39 cubic centimeters (cm^3)
1 gallon (gal)	= 3.784 liters (l)

Mass

Metric Measure

1000 kilograms (kg)	= 1 metric ton (also called a tonne) (m.t)
1 kilogram (kg)	= 1000 grams (g)

Nonmetric Measure

1 short ton (sh.t)	= 2000 pounds (lb)
1 long ton (l.t)	= 2240 pounds (lb)
1 pound (avoirdupois) (lb)	= 16 ounces (avoirdupois) (oz) = 7000 grains (gr)
1 ounce (avoirdupois) (oz)	= 437.5 grains (gr)
1 pound (Troy) (Tr. lb)	= 12 ounces (Troy) (Tr. oz)
1 ounce (Troy) (Tr. oz)	= 20 pennyweight (dwt)

Conversions

1 metric ton (m.t)	= 2205 pounds (avoirdupois) (lb)
1 kilogram (kg)	= 2.205 pounds (avoirdupois) (lb)
1 gram (g)	= 0.03527 ounce (avoirdupois) (oz) = 0.03215 ounce (Troy) (Tr. oz) = 15,432 grains (gr)
1 pound (lb)	= 0.4536 kilogram (kg)
1 ounce (avoirdupois) (oz)	= 28.35 grams (g)
1 ounce (avoirdupois) (oz)	= 1.097 ounces (Troy) (Tr. oz)

Pressure

Metric Measure

1 pascal (Pa)	= 1 newton/square meter (N/m^2)
1 kilogram-force square centimeter (kg/cm^2 or kgf/cm^2)	= 1 technical atmosphere (at) = 98,067 Pa = 0.98067 bar
1 bar	= 10^5 pascals (Pa) = 1.02 kilogram-force/square centimeter (kgf/cm^2 or at)

Nonmetric Measure

1 atmosphere (atm)	= 1.01325 bar = 14.696 lb/in^2 (psi)
1 pound per square inch (lb/in^2 or psi)	= 68.046 × 10^{-3} atmospheres (atm)

Conversions

1 kilogram-force/ square centimeter (kgf/cm^2 or at)	= 0.96784 atmosphere (atm) = 14.2233 pounds/square inch (lb/in^2 or psi) = 0.98067 bar = 98,067 Pa
1 bar	= 0.98692 atmosphere (atm) = 10^5 pascals (Pa) = 1.02 kilogram-force/square centimeter (kgf/cm^2)
1 Pa = 10^{-5} bar	= 10.197 × 10^{-6} kgf/cm^2 (or at) = 9.8692 × 10^{-6} atm = 145.04 × 10^{-6} lb/in^2 (or psi)
1 atm	= 101,325 Pa = 1.01325 bar

Temperature

Metric Measure

0 degrees Celsius (°C)	= freezing point of water at sea level
100 degrees Celcius (°C)	= boiling point of water at sea level
0 degrees Kelvin (K)	= −273.15°C = absolute zero
1°C	= 1K (temperature increments)
273.15K	= 0.0° C

Nonmetric Measure

Fahrenheit (°F)	= (K · 9/5) − 459.67
Fahrenheit (°F)	= (°C · 9/5) + 32

Conversions

degrees Kelvin (K)	= °C + 273.1
degrees Celcius (°C)	= K − 273.15
degrees Fahrenheit (°F)	= (°C · 9/5) + 32
degrees Celcius(°C)	= (°F − 32) · 5/9

Appendix B
Tables of the Chemical Elements and Naturally Occurring Isotopes

Alphabetical list of the elements

Element	Symbol	Atomic Number	Crustal Abundance, Weight Percent	Element	Symbol	Atomic Number	Crustal Abundance, Weight Percent
Aluminum	Al	13	8.00	Lithium	Li	3	0.0020
Antimony	Sb	51	0.00002	Lutetium	Lu	71	0.000080
Argon	Ar	18	Not known	Magnesium	Mg	12	2.77
Arsenic	As	33	0.00020	Manganese	Mn	25	0.100
Barium	Ba	56	0.0380	Mercury	Hg	80	0.000002
Beryllium	Be	4	0.00020	Molybdenum	Mo	42	0.00012
Bismuth	Bi	83	0.0000004	Neodymium	Nd	60	0.0044
Boron	B	5	0.0007	Neon	Ne	10	Not known
Bromine	Br	35	0.00040	Nickel	Ni	28	0.0072
Cadmium	Cd	48	0.000018	Niobium	Nb	41	0.0020
Calcium	Ca	20	5.06	Nitrogen	N	7	0.0020
Carbon[a]	C	6	0.02	Osmium	Os	76	0.00000002
Cerium	Ce	58	0.0083	Oxygen[b]	O	8	45.2
Cesium	Cs	55	0.00016	Palladium	Pd	46	0.0000003
Chlorine	Cl	17	0.0190	Phosphorus	P	15	0.1010
Chromium	Cr	24	0.0096	Platinum	Pt	78	0.0000005
Cobalt	Co	27	0.0028	Polonium	Po	84	Footnote[d]
Copper	Cu	29	0.0058	Potassium	K	19	1.68
Dysprosium	Dy	66	0.00085	Praseodymium	Pr	59	0.0013
Erbium	Er	68	0.00036	Protactinium	Pa	91	Footnote[d]
Europium	Eu	63	0.00022	Radium	Ra	88	Footnote[d]
Fluorine	F	9	0.0460	Radon	Rn	86	Footnote[d]
Gadolinium	Gd	64	0.00063	Rhenium	Re	75	0.00000004
Gallium	Ga	31	0.0017	Rhodium[c]	Rh	45	0.00000001
Germanium	Ge	32	0.00013	Rubidium	Rb	37	0.0070
Gold	Au	79	0.0000002	Ruthenium[c]	Ru	44	0.00000001
Hafnium	Hf	72	0.0004	Samarium	Sm	62	0.00077
Helium	He	2	Not known	Scandium	Sc	21	0.0022
Holmium	Ho	67	0.00016	Selenium	Se	34	0.000005
Hydrogen[b]	H	1	0.14	Silicon	Si	14	27.20
Indium	In	49	0.00002	Silver	Ag	47	0.000008
Iodine	I	53	0.00005	Sodium	Na	11	2.32
Iridium	Ir	77	0.00000002	Srontium	Sr	38	0.0450
Iron	Fe	26	5.80	Sulfur	S	16	0.030
Krypton	Kr	36	Not known	Tantalum	Ta	73	0.00024
Lanthanum	La	57	0.0050	Tellurium[c]	Te	52	0.000001
Lead	Pb	82	0.0010	Terbium	Tb	65	0.00010

(CONTINUED)

Element	Symbol	Atomic Number	Crustal Abundance, Weight Percent	Element	Symbol	Atomic Number	Crustal Abundance, Weight Percent
Thallium	Tl	81	0.000047	Vanadium	V	23	0.0170
Thorium	Th	90	0.00058	Xenon	Xe	54	Not known
Thulium	Tm	69	0.000052	Ytterbium	Yb	70	0.00034
Tin	Sn	50	0.00015	Yttrium	Y	39	0.635
Titanium	Ti	22	0.86	Zinc	Zn	30	0.008-2
Tungsten	W	74	0.00010	Zirconium	Zr	40	0.0140
Uranium	U	92	0.00016				

Source: After K. K. Turekian, 1969.

[a]Estimate from S. R. Taylor (1964).

[b]Analyses of cruscal rocks do not usually include separate determinations for hydrogen and oxygen. Both combine in essentially constant proportions with other elements, so abundances can be calculated.

[c]Estimates are uncertain and have a very low reliability.

[d]Elements formed by decay of uranium and thorium. The daughter products are radioactive with such short half-lives that crustal accumulations are too low to be measured accurately.

Appendix C
Tables of the Properties of Selected Common Minerals

Properties of the common minerals with metallic luster TABLE C.1

Mineral	Chemical Composition	Form and Habit	Cleavage	Hardness / Specific Gravity		Other Properties	Most Distinctive Properties
Chalcopyrite	$CuFeS_2$	Massive or granular.	None. Uneven fracture.	3.5–4	4.2	Golden yellow to brassy yellow. Dark green to black streak.	Streak. Hardness distinguishes from pyrite.
Copper	Cu	Massive, twisted leaves and wires.	None. Can be cut with a knife.	2.5–3	9	Copper color but commonly stained green.	Color, specific gravity, malleable.
Galena	PbS	Cubic crystals, coarse or fine-grained granular masses.	Perfect in three directions at right angles.	2.5	7.6	Lead-gray color. Gray to gray-black streak.	Cleavage and streak.
Hematite	Fe_2O_3	Massive, granular, micaceous.	Uneven fracture.	5–6	5	Reddish-brown, gray to black. Reddish-brown streak.	Streak, hardness.
Limonite (*Goethite* is most common.)	A complex mixture of minerals, mainly hydrous iron oxides.	Massive, coatings, botryoidal crusts, earthy masses.	None.	1–5.5	3.5–4	Yellow, brown, black, yellowish-brown streak.	Streak.
Magnetite	Fe_3O_4	Massive, granular. Crystals have octahedral shape.	None. Uneven fracture.	5.5–6.5	5	Black. Black streak. Strongly attracted to a magnet.	Streak, magnetism.

(continued)

TABLE C.1 (CONTINUED)

Mineral	Chemical Composition	Form and Habit	Cleavage	Hardness	Specific Gravity	Other Properties	Most Distinctive Properties
Pyrite ("Fool's gold")	FeS_2	Cubic crystals with striated faces. Massive.	None. Uneven fracture.	6–6.5	5.2	Pale brass-yellow, darker if tarnisned. Greenish-black streak.	Streak. Hardness distinguishes from chalcopyrite. Not malleable, which distinguishes from gold.
Sphalerite	ZnS	Fine to coarse granular masses. Tetrahedron shaped crystals.	Perfect in six directions.	3.5–4	4	Yellowish-brown to black. White to yellowish-brown streak. Resinous luster.	Cleavage, hardness, luster.

Properties of rock-forming minerals with nonmetallic luster TABLE C.2

Mineral	Chemical Composition	Form and Habit	Cleavage	Hardness	Specific Gravity	Other Properties	Most Distinctive Properties
Amphiboles. (A complex family of minerals, *Hornblende* is most common.)	$X_2Y_5Si_8O_{22}(OH)_2$ where X = Ca, Na; Y = Mg, Fe, Al.	Long, six-sided crystals; also fibers and irregular grains.	Two; intersecting at 56° and 124°.	5–6	2.9–3.8	Common in metamorphic and igneous rocks. *Hornblende* is dark green to black; *actinolite*, green; *tremolite*, white.	Cleavage, habit.
Apatite	$Ca_5(PO_4)_3(F, OH, Cl)$	Granular masses. Perfect six-sided crystals.	Poor. One direction.	5	3.2	Green, brown, blue, or white. Common in many kinds of rocks in small amounts.	Hardness, form.
Aragonite	$CaCO_3$	Massive, or slender, needle-like crystals.	Poor. Two directions.	3.5	2.9	Colorless or white. Effervesces with dilute HCl.	Effervescence with acid. Poor cleavage distinguishes from calcite.

TABLE C.2 (CONTINUED)

Mineral	Chemical Composition	Form and Habit	Cleavage	Hardness	Specific Gravity	Other Properties	Most Distinctive Properties
Calcite	$CaCO_3$	Tapering crystals and granular masses.	Three perfect; at oblique angles to give a rhomb-shaped fragment.	3	2.7	Colorless or white. Effervesces with dilute HCl.	Cleavage, effervescence with acid.
Dolomite	$CaMg(CO_3)_2$	Crystals with rhomb-shaped faces. Granular masses.	Perfect in three directions as in calcite.	3.5	2.8	White or gray. Does not effervesce in cold, dilute HCl unless powdered. Pearly luster.	Cleavage. Lack of effervescence with acid.
Feldspars: Potassium feldspar (*orthoclase* is a common variety)	$KAlSi_3O_8$	Prism-shaped crystals, granular masses.	Two perfect, at right angles.	6	2.6	Common mineral. Pink, white, or gray in color.	Color, cleavage.
Garnets	$X_3Y_2(SiO_4)_3$; X = Ca, Mg, Fe, Mn; Y = Al, Fe, Ti, Cr.	Perfect crystals with 12 or 24 sides. Granular masses.	None. Uneven fracture.	6.5–7.5	3.5–4.3	Common in metamorphic rocks. Red, brown, yellowish-green, black.	Crystals, hardness, no cleavage.
Graphite	C	Scaly masses.	One, perfect. Forms slippery flakes.	1–2	2.2	Metamorphic rocks. Black with metallic to dull luster.	Cleavage, color. Marks paper.
Gypsum	$CaSO_4.2H_2O$	Elongate or tabular crystals. Fibrous and earthy masses.	One, perfect. Flakes bend but are not elastic.	2	2.3	Vitreous to pearly luster. Colorless.	Hardness, cleavage.
Halite	NaCl	Cubic crystals.	Perfect to give cubes.	2.5	2.2	Tastes salty. Colorless, blue.	Taste, cleavage.

(continued)

TABLE C.2 (CONTINUED)

Mineral	Chemical Composition	Form and Habit	Cleavage	Hardness	Specific Gravity	Other Properties	Most Distinctive Properties
Kaolinite	$Al_2Si_2O_5$ $(OH)_4$	Soft, earthy masses. Submicroscopic crystals.	One, perfect.	2–2.5	2.6	White, yellowish. Plastic when wet; emits clayey odor. Dull luster.	Feel, plasticity, odor.
Mica: *Biotite*	$K(Mg, Fe)_3$ $Alsi_3O_{10}$ $(OH)_2$	Irregular masses of flakes.	One, perfect.	2.5–3	2.8–3.2	Common in igneous and metamorphic rocks. Black, brown, dark green.	Cleavage, color. Flakes are elastic.
Mica: *Muscovite*	$KAl_3Si_3O_{10}$ $(OH)_2$	Thin flakes.	One, perfect.	2–2.5	2.7	Common in igneous, and metamorphic rocks. Colorless, pale-green or brown.	Cleavage, color. Flakes are elastic.
Olivine	$(Mg,Fe)_2$ SiO_4	Small grains. granular masses.	None Conchoidal fracture.	6.5–7	3.2–4.3	Igneous rocks. Olive green to yellowish-green.	Color, fracture, habit.
Pyroxene (A complex family of minerals. *Augite* is most common.)	$XY(SiO_3)_2$ $X = Y = Ca$, Mg, Fe	8-sided stubby crystals. Granular masses.	Two, perfect, nearly at right angles.	5–6	3.2–3.9	Igneous and metamorphic rocks. *Augite*, dark green to black; other varieties white to green.	Cleavage
Quartz	SiO_2	6-sided crystals, granular masses.	None. Conchoidal fracture.	7	2.6	Colorless, white, gray, but may have any color, depending on impurities Vitreous to greasy luster.	Form, fracture, striation across crystal faces at right angles to long dimension.

abrasion   Wind erosion in which airborne particles chip small fragments off rocks that protrude above the surface.

air   A mixture of 78 percent nitrogen, 21 percent oxygen, and trace amounts of other gases, found in Earth's atmosphere.

alluvium   Unconsolidated sediment deposited in a recent geologic time by a stream.

angiosperm   Flowering plant, or enclosed-seed plant.

anthropogenic   Produced by human activities.

anticline   A fold in the form of an arch, with the rock strata convex upward and the older rocks in the core.

aquiclude   A layer of impermeable rock.

aquifer   A body of rock or regolith that is water-saturated, porous, and permeable.

asthenosphere   A layer of weak, ductile rock in the mantle that is close to melting but not actually molten.

Anticline

atmosphere   The envelope of gases that surrounds Earth.

atom   The smallest individual particle that retains the distinctive chemical properties of an element.

banded iron formation   A type of chemical sedimentary rock rich in iron minerals and silica.

barrier island   A long, narrow, sandy island lying offshore and parallel to a lowland coast.

batholith   A large, irregularly shaped pluton that cuts across the layering of the rock into which it intrudes.

beach   Wave-washed sediment along a coast.

beach drift   The movement of particles along a beach as they are driven up and down the beach slope by wave action.

bed load   Sediment that is moved along the bottom of a stream.

bedding   The layered arrangement of strata in a body of sediment or sedimentary rock.

bedding surface   The top or bottom surface of a rock stratum or bed.

biogenic sediment   Sediment that is primarily composed of plant and animal remains, or precipitates as a result of biologic processes.

biomass energy   Any form of energy that is derived more or less directly from plant life, including fuel wood, peat, animal dung, and agricultural wastes.

biosphere   The system consisting of all living and recently dead organisms on Earth.

body wave   A seismic wave that travels through Earth's interior.

bond   The force that holds the atoms together in a chemical compound.

brittle deformation   A permanent change in shape or volume, in which a material breaks or cracks.

burial metamorphism   Metamorphism that occurs after diagenesis, as a result of the burial of sediments in deep sedimentary basins.

cave   Underground open space; a cavern is a system of connected caves.

cell   The basic structural and functional unit of life; a complex grouping of chemical compounds enclosed in a porous membrane.

cementation   The process in which substances dissolved in pore water precipitate out and form a matrix in which grains of sediment are joined together.

channel   The clearly defined natural passageway through which a stream flows.

chemical sediment   Sediment formed by the precipitation of minerals dissolved in lake, river, or sea water.

chemical weathering   The decomposition of rocks and minerals by chemical and biochemical reactions.

clastic sediment   Sediment formed from fragmented rock and mineral debris produced by weathering and erosion.

clay   A family of hydrous aluminosilicate minerals. The term is also used for tiny mineral particles of any kind that have physical properties like the clay minerals.

coal   A combustible rock formed by the compression, heating, and lithification of plant-rich sediment (peat).

compaction   Reduction of pore space in a sediment as a result of the weight of overlying sediments.

compound   A combination of atoms of one or more elements in a a specific ratio.

compression   A stress that acts in a direction perpendicular to and *toward* a surface.

compressional wave   A seismic body wave consisting of alternating pulses of compression and expansion in the direction of wave travel; P wave or primary wave.

condensation   The process by which water changes from a vapor into a liquid or a solid.

Brittle deformation

conglomerate A clastic sedimentary rock with large fragments in a finer-grained matrix.

contact metamorphism Metamorphism that occurs when rocks are heated and chemically changed adjacent to an intruded body of hot magma.

continental crust The older, thicker, and less dense part of Earth's crust; the bulk of Earth's land masses.

continental drift The slow, lateral movement of continents across Earth's surface.

convection A form of heat transfer in which hot material circulates from hotter to colder regions, loses its heat, and then repeats the cycle.

convergent margin A boundary along which two plates come together.

core Earth's innermost compositional layer, where the magnetic field is generated and much geothermal energy resides.

core Earth's innermost compositional layer.

Coriolis effect An effect due to Earth's rotation that causes a freely moving body to veer from a straight path.

correlation A method of equating the ages of strata that come from two or more different places.

craton A region of continental crust that has remained technically stable for a very long time.

creep The imperceptibly slow downslope granular flow of regolith.

crust The outermost compositional layer of the solid Earth, part of the lithosphere.

cryosphere The perennially frozen part of the hydrosphere.

crystal structure An arrangement of atoms or molecules into a regular geometric lattice. Materials that possess a crystal structure are said to be crystalline.

crystallization The process whereby mineral grains form and grow in a cooling magma (or lava).

deflation Wind erosion in which loose particles of sand and dust are removed by the wind, leaving coarser particles behind.

delta A sedimentary deposit, commonly triangle-shaped, that forms where a stream enters a standing body of water.

Delta

deposition The laying down of sediment.

desert An arid land that receives less than 250 millimeters of rainfall or snow equivalent per year, and is sparsely vegetated unless it is irrigated.

desertification Invasion of desert conditions into nondesert areas.

dip The angle between a tilted surface and a horizontal plane.

discharge (1) The amount of water passing by a point on the channel's bank during a unit of time.

discharge (2) The process by which subsurface water leaves the saturated zone and becomes surface water.

dissolution The separation of a material into ions in solution by a solvent, such as water or acid.

divergent margin A boundary along which two plates move apart from one another.

divide A topographic high that separates adjacent drainage basins.

DNA deoxyribonucleic acid; a double-chain biopolymer that contains all the genetic information needed for organisms to grow and reproduce.

domain The broadest taxonomic category of living organisms; biologists today recognize three domains: bacteria, *Archaea*, and eukaryotes.

drainage basin The total area from which water flows into a stream.

ductile deformation A permanent but gradual change in shape or volume of a material, caused by flowing or bending.

Ductile deformation

dune A hill or ridge of sand deposited by winds.

elastic deformation A temporary change in shape or volume from which the material rebounds after the deforming stress is removed.

elastic rebound theory The theory that continuing stress along a fault results in a buildup of elastic energy in the rocks, which is abruptly released when an earthquake occurs.

element The most fundamental substance into which matter can be separated by chemical means.

eolian sediment Sediment that is carried and deposited by the wind.

epicenter The point on Earth's surface directly above an earthquake's focus.

erosion The wearing away of bedrock and transport of loosened particles by a fluid, such as water.

Erosion

estuary   A semi-enclosed body of coastal water, in which fresh water mixes with seawater.

eukaryotes   Organisms composed of eukaryotic cells—that is, cells that have a well-defined nucleus and organelles.

evaporation   The process by which water changes from a liquid into a vapor.

evaporite   A rock formed by the evaporation of lake water or seawater, followed by lithification of the resulting salt deposit.

evolution   The process by which inherited traits become more common through successive generations; can eventually lead to the emergence of new species.

fault   A fracture in Earth's crust along which movement has occurred.

flood   An event in which a water body overflows its banks.

floodplain   The relatively flat valley floor adjacent to a stream channel, which is inundated when the stream overflows its banks.

flow   Any mass-wasting process that involves a flowing motion of regolith containing water and/or air within its pores.

focus   The location where rupture commences and an earthquake's energy is first released.

fold   A bend or warp in a layered rock.

foliation   A planar arrangement of textural features in a metamorphic rock, which give the rock a layered or finely banded appearance.

fossil   Remains of an organism of a past geologic age, embedded and preserved in Earth's crust.

fossil fuel   Combustible organic matter that is trapped in sediment or sedimentary rock.

fractional crystallization   Separation of crystals from liquids during crystallization.

fractional melt   A mixture of molten and solid rock.

fractionation   Separation of melted materials from the remaining solid material during the course of melting.

Geologic Column   The succession of all known strata, fitted together in relative chronological order.

geologic cross section   A diagram that shows geologic features that occur underground.

geologic map   A map that shows the locations, kinds, and orientations of rock units, and structural features such as faults and folds.

geology   The scientific study of Earth.

glaciation   A period during which the average temperature drops by several degrees, for a long enough time that ice sheets expand significantly.

glacier   A semi-permanent or perenially frozen body of ice, consisting largely of recrystallized snow, that moves under the pull of gravity.

Glacier

gneiss   A coarse-grained, high-grade, strongly foliated metamorphic rock with separation of dark and light minerals into bands.

gradient   The steepness of a stream channel.

greenhouse effect   The process through which long-wavelength (infrared) heat energy is absorbed by gases in the atmosphere, thereby warming Earth's surface.

groundwater   Subsurface water contained in pore spaces in regolith and bedrock.

gymnosperm   Naked-seed plant.

habit   The distinctive shape of a particular mineral.

half-life   The time needed for half of the parent atoms of a radioactive substance to decay into daughter atoms.

hardness   A mineral's resistance to scratching.

high-grade   Rocks that are metamorphosed under temperature and pressure conditions higher than about 400°C and 400 MPa.

humus   Partially decayed organic matter in soil.

hydroelectric energy   Electricity generated by running water.

hydrologic cycle   A model that describes the movement of water through the reservoirs of the Earth system; the water cycle.

hydrosphere   The system comprising all of Earth's bodies of water and ice, both on the surface and underground.

igneous rocks   Rocks that form by cooling and solidification of molten rock.

infiltration   The process by which water works its way into the ground through small openings in the soil.

isostasy   The flotational balance of the lithosphere on the asthenosphere.

isotopes   Atoms with the same atomic number but different mass numbers.

Lava

joint   A fracture in a rock, along which no appreciable movement has occurred.

kingdom   The second-broadest taxonomic category. There are six recognized kingdoms, including animals and plants.

lava   Molten rock that reaches Earth's surface.

limestone   A sedimentary rock that consists primarily of the mineral calcite.

lithification   The group of processes by which loose sediment is transformed into sedimentary rock.

lithosphere   Earth's rocky, outermost layer, comprising the crust and the uppermost part of the mantle.

load   The suspended and dissolved sediment carried by a stream.

longshore current   A current within the surf zone that flows parallel to the coast.

low-grade   Rocks that are metamorphosed under temperature and pressure conditions up to 400°C and 400 MPa.

luster   The quality and intensity of light that reflects from a mineral.

magma Molten rock, which may include fragments of rock, volcanic glass and ash, or gas.

magnetic reversal A period of time in which Earth's magnetic polarity reverses itself.

mantle The middle compositional layer of Earth, between the core and the crust.

marble The product of metamorphism formed by recrystallization of limestone.

mass extinction A catastrophic episode in which a large fraction of living species become extinct within a geologically short time.

mass wasting The downslope movement of regolith and/or bedrock masses due to the pull of gravity.

mechanical weathering The breakdown of rock into solid fragments by physical processes that do not change the rock's chemical composition.

Mechanical weathering

metamorphic facies The set of metamorphic mineral assemblages that form in rocks of different compositions under similar temperature and stress conditions.

metamorphic rocks Rocks that have been altered by exposure to high temperature, high pressure, or both.

metamorphism The mineralogical, textural, chemical, and structural changes that occur in rocks as a result of exposure to elevated temperatures and/or pressures.

metasomatism The process whereby the chemical composition of a rock is altered by the addition or removal of material by solution in fluids.

meteorite A fragment of extraterrestrial material that falls to Earth.

mineral A naturally formed, solid, inorganic substance with a characteristic crystal structure and a specific chemical composition.

molecule The smallest chemical unit that has all the properties of a particular compound.

moment magnitude A measure of earthquake strength that is based on the rupture size, rock properties, and amount of ground displacement on the fault surface.

moraine A ridge or pile of debris that has been, or is being, transported by a glacier.

mudstone A very fine-grained sedimentary rock of the same composition as shale but without fissility.

natural gas The gaseous form of petroleum.

natural resources Useful materials that are obtained from the lithosphere, atmosphere, hydrosphere, or biosphere.

natural selection The process by which individuals that are well-adapted to their environment have a survival advantage, and pass on their favorable characteristics to their offspring.

nonrenewable resource A resource that cannot be replenished or regenerated on the scale of a human lifetime.

normal fault A fault in which the block of rock above the fault surface moves downward relative to the block below.

numerical age The age of a rock or geological feature in years before the present.

oceanic crust The thinner, denser, and younger part of Earth's crust, underlying the ocean basins.

Normal fault

oil shale A fine-grained sedimentary rock with a high content of kerogen.

oil The liquid form of petroleum.

ore A deposit from which one or more minerals can be extracted profitably.

orogen An elongated region of crust that has been deformed and metamorphosed by a continental collision.

ozone layer A zone in the stratosphere where ozone is concentrated.

paleomagnetism The study of rock magnetism in order to determine the intensity and direction of Earth's magnetic field in the geologic past.

paleontology The study of fossils and the record of ancient life on Earth; the use of fossils for the determination of relative ages.

paleoseismology The study of prehistoric earthquakes.

peat A biogenic sediment formed from the accumulation and compaction of plant remains from bogs and swamps, with a carbon content of about 60 percent.

percolation The process by which groundwater seeps downward and flows under the influence of gravity.

permafrost Ground that is perennially below the freezing point of water.

permeability A measure of how easily a solid allows fluids to pass through it.

petroleum Naturally occurring gaseous, liquid, and semi-solid substances that consist chiefly of hydrocarbon compounds.

Moraine

photosynthesis A chemical reaction whereby plants use light energy to induce carbon dioxide to react with water, producing carbohydrates and oxygen.

phyllite A fine-grained metamorphic rock with pronounced foliation, produced by further metamorphism of slate.

plate tectonics The movement and interactions of large fragments of Earth's lithosphere, called plates.

pluton Any body of intrusive igneous rock, regardless of size or shape.

plutonic rock An igneous rock formed underground from magma.

porosity The percentage of the total volume of a body of rock or regolith that consists of open spaces (pores).

precipitation The process by which water that has condensed in the atmosphere falls back to Earth's surface as rain, snow, or hail.

pressure A particular kind of stress in which the forces acting on a body are the same in all directions.

prokaryotes Single-celled organisms with no distinct nucleus—that is, no membrane separates their DNA from the rest of the cell.

pyroclastic flow Hot volcanic fragments (tephra) that are buoyed by heat and volcanic gases, and flow very rapidly.

quartzite The product of metamorphism formed by recrystallization of sandstone.

radioactivity A process in which an element spontaneously transforms into another isotope of the same element, or into a different element.

radiometric dating The use of naturally occurring radioactive isotopes to determine the numerical age of minerals, rocks, or fossils.

recharge Replenishment of groundwater.

recrystallization The formation of new crystalline mineral grains from old ones.

reef A hard structure on a shallow ocean floor, usually but not always built by coral.

reflection The bouncing back of a wave from an interface between two different materials.

refraction The bending of a wave as it passes from one material into another material, through which it travels at a different speed.

regional metamorphism Metamorphism of an extensive area of the crust, associated with plate convergence, collision, and subduction.

regolith A loose layer of broken rock and mineral fragments that covers most of Earth's surface.

relative age The age of a rock, fossil, or other geologic feature relative to another feature.

renewable resource A resource that can be replenished or regenerated on the scale of a human lifetime.

reverse fault A fault in which the block on top of the fault surface moves up and over the block on the bottom.

Richter magnitude scale A scale of earthquake intensity based on the recorded heights, or amplitudes, of the seismic waves recorded on a seismograph.

rift valley A linear, fault-bounded valley along a divergent plate boundary or spreading center.

rock A naturally formed, coherent aggregate of minerals and possibly other nonmineral matter.

rock cycle The set of crustal processes that form new rock, modify it, transport it, and break it down.

salinity A measure of the salt content of a solution.

saltation Sediment transport in which particles move forward in a series of short jumps along arc-shaped paths.

sand A sediment made of relatively coarse mineral grains.

sandstone A medium-grained clastic sedimentary rock in which the clasts are typically, but not necessarily, dominated by quartz grains.

schist A high-grade metamorphic rock with pronounced schistosity, in which individual mineral grains are visible.

schistosity Foliation in coarse-grained metamorphic rocks.

seafloor spreading The processes through which the seafloor splits and moves apart along a midocean ridge, and new oceanic crust forms along the ridge.

sedimentary rocks Rocks that form under conditions of low pressure and low temperature near the surface.

Sedimentary rocks

seismic discontinuity   A boundary inside Earth where the velocities of seismic waves change abruptly.

seismic wave   An elastic shock wave that travels outward in all directions from an earthquake's source.

seismogram   The record made by a seismograph.

seismograph   An instrument that detects and measures vibrations of Earth's surface.

seismology   The scientific study of earthquakes and seismic waves.

Seismogram

shale   A very fine-grained fissile or laminated sedimentary rock, consisting primarily of clay-sized particles.

shear   A stress that acts in a direction *parallel* to a surface.

shear wave   A seismic body wave in which rock is subjected to side-to-side or up-and-down forces, perpendicular to the wave's direction of travel; S wave or secondary wave.

shield volcano   A broad, flat volcano with gently sloping sides, built of successive lava flows.

slate   A very fine-grained, low-grade metamorphic rock with slaty cleavage; the metamorphic product of shale.

slaty cleavage   Foliation in low-grade metamorphic rocks that causes such rocks to break into flat, plate-like fragments.

slope failure   The falling, slumping, or sliding of relatively coherent masses of rock.

soil   The uppermost layer of regolith, which can support rooted plants.

soil horizon   One of a succession of zones or layers within a soil profile, each with distinct physical, chemical, and biologic characteristics.

soil profile   The sequence of soil horizons from the surface down to the underlying bedrock.

species   A population of genetically and/or morphologically similar individuals that can interbreed and produce offspring.

spring   Where the water table intersects the land surface, allowing groundwater to flow out.

strain   A change in shape or volume of a rock in response to stress.

stratigraphy   The science of rock layers and the process by which strata are formed.

Stratigraphy

stratovolcano   A volcano composed of solidified lava flows interlayered with pyroclastic material. Such volcanoes usually have steep sides that curve upward.

Stratovolcano

streak   A thin layer of powdered mineral made by rubbing a specimen on an unglazed fragment of porcelain.

stream   A body of water that flows downslope along a clearly defined natural passageway.

stress   The force acting on a surface, per unit area, which may be greater in certain directions than in others.

strike   The compass orientation of the line of intersection between a horizontal plane and a planar feature, such as a rock layer or fault.

strike-slip fault   A fault in which the direction of the movement is mostly horizontal and parallel to the strike of the fault.

structural geology   The study of stress and strain, the processes that cause them, and the deformation and rock structures that result from them.

subduction zone   A boundary along which one lithospheric plate descends into the mantle beneath another plate.

surf   The "broken," turbulent water found between a line of breakers and the shore.

surface creep   Sediment transport in which the wind causes particles to roll along the ground.

surface runoff   Precipitation that drains over the land or in stream channels.

surface wave   A seismic wave that travels along Earth's surface.

suspended load   Sediment that is carried in suspension by a flowing stream of water or wind.

suspension   Sediment transport in which the wind carries very fine particles over long distances and periods of time.

syncline   A fold in the form of a trough, with the rock strata concave upward and the younger rocks in the core.

system   A portion of the universe that can be separated from the rest of the universe for the purpose of observing changes that happen in it.

tar sand   A sediment or sedimentary rock in which the pores are filled by dense, viscous, asphalt-like oil.

tectonic cycle   Movements and interactions in the lithosphere and the internal Earth processes that drive them.

tension   A stress that acts in a direction perpendicular to and *away from* a surface.

thrust fault   A reverse fault with a shallow angle of dip.

tides   A regular, daily cycle of rising and falling sea level that results from the gravitational action of the Moon, the Sun, and Earth.

till   A heterogeneous mixture of crushed rock, sand, pebbles, cobbles, and boulders deposited by a glacier.

topographic map   A map that shows the shape of the ground surface, as well as the location and elevation of surface features, usually by means of contour lines.

trace fossil   Fossilized evidence of an organism's life processes, such as tracks, footprints, or burrows.

transform fault   A fracture in the lithosphere where two plates slide past each other.

transpiration   The process by which water taken up by plants passes directly into the atmosphere.

turbidity current   A turbulent, gravity-driven flow consisting of a mixture of sediment and water, which conveys sediment from the continental shelf to the deep sea.

Transform fault

unconformity   A substantial gap in a stratigraphic sequence that marks the absence of part of the rock record.

uniformitarianism   The concept that the processes governing the Earth system today have operated in a similar manner throughout geologic time.

viscosity   The degree to which a substance resists flow; a less viscous liquid is runny, whereas a more viscous liquid is thick.

volcanic rock   An igneous rock formed from lava.

volcano   A vent through which lava, solid rock debris, volcanic ash, and gases erupt from Earth's crust to its surface.

water table   The top surface of the saturated zone.

wave-cut cliff   A coastal cliff cut by wave action at the base of a rocky coast.

weathering   The chemical and physical breakdown of rock exposed to air, moisture, and living organisms.

Wave-cut cliff

Chapter 14

Pages 416–17: ©Louis Psihoyos/Matrix International, Inc.; page 419: (left) ©Royalty-Free/Corbis; page 423: (top left) O. Louis Mazzatenta/NG Image Collection; page 423: (center right) Kirk Moldoff/NG Image Collection; page 423: (bottom left) ©WHOI, D. Foster/Visuals Unlimited; page 423: (bottom center) Photo courtesy of Colleen M. Cavanaugh, Department of Organismic and Evolutionary Biology, Harvard University; page 423: (bottom right) Woods Hole Oceanographic Institution; page 424: (left) O. Louis Mazzatenta/NG Image Collection; page 424: (right) Francois Gohier/Photo Researchers, Inc.; page 425: (left) (c)Dr. Jeremy Burgess/ SPL/Photo Researchers, Inc.; page 425: (right) ©CNRI/SPL/Photo Researchers, Inc.; page 426: (left) O. Louis Mazzatenta/NG Image Collection; page 426: (right) O. Louis Mazzatenta/NG Image Collection; page 427: (left) Robert Clark/NG Image Collection; page 427: (right) Mary Evans Picture Library/Photo Researchers, Inc.; page 428: (center left) Ralph Lee Hopkins/NG Image Collection; page 428: (top right) ©J. Dunning/VIREO; page 428: (top left) ©Gerald & Buff Corsi/Visuals Unlimited; page 428: (center right) ©Fritz Polking/Visuals Unlimited; page 428: (center right) ©Joe & Mary Ann McDonald/Visuals Unlimited; page 428: (bottom left) Tierbild Okapia/Photo Researchers, Inc.; page 428: (bottom right) Eric Hosking/Photo Researchers, Inc.; page 430: (top left) Paul Zahl/NG Image Collection; page 430: (top right) Stephen St. John/NG Image Collection; page 430: (bottom left) ©Louis Psihoyos/Matrix International, Inc.; page 430: (bottom right) Louie Psihoyos/NG Image Collection; page 431: (left) O. Louis Mazzatenta/NG Image Collection; page 431: (right) Robert Clark/NG Image Collection; page 432: Raymond Gehman/NG Image Collection; page 433: (top left) ©Edward R. Degginger/Bruce Coleman, Inc.; page 433: (top right) Chris Johns/NG Image Collection; page 433: (bottom left) ©Breck P. Kent/Animals Animals/Earth Scenes; page 433: (bottom right) ©Theodore Clutter/Photo Researchers, Inc.; page 434: ©John Cancalosi/DRK Photo; page 435: (left) Hans Fricke/NG Image Collection; page 435: (right) Tom McHugh/Photo Researchers, Inc.; page 436: (left) ©Francois Gohier/Photo Researchers, Inc.; page 436: (right) Carnegie Museum of Natural History; page 437: (left) Mark A. Klingler/Carnegie Museum of Natural History; page 437: (right) Mark A. Klingler/Carnegie Museum of Natural History; page 438: (left) ©Michael Rothman; page 438: (right) ©John Reader/ Science Photo Library/Photo Researchers, Inc.; page 440: ©Chris Butler; page 441: (top) Jonathan Blair/NG Image Collection; page 441: (center right) Virgil L. Sharpton, University of Alaska, Fairbanks; page 441: (bottom) Courtesy NASA; page 442: Courtesy Dr. Richard Ernst; page 443: (top) Michael Melford/NG Image Collection; page 449: Courtesy NASA

Chapter 15

Pages 452–53: ©Stephanie Maze/Woodfin Camp & Associates; page 454: James L. Amos/NG Image Collection; page 455: (left) Toby Savage/Saudi Aramco World/PADIA; page 455: (right) ©Nick Brooks; page 455: (right inset) ©Nick Brooks; page 456: (top) Michael Nichols/NG Image Collection; page 456: (bottom) James P. Blair/NG Image Collection; page 460: Robert Sisson/NG Image Collection; page 461: (top left) Richard Olsenius/NG Image Collection; page 461: (center right) Grant Heilman Photography; page 461: (bottom) ©Runk Shoenberger/Grant Heilman Photography; page 464: (top right) Sarah Leen/NG Image Collection; page 464: (center left) Jonathan P. Blair/NG Image Collection; page 467: (top) Sarah Leen/NG Image Collection; page 467: (bottom) Marc Moritsch/NG Image Collection; page 468: Jim Richardson/NG Image Collection; page 469: (right) Emory Kristof/NG Image Collection; page 469: (bottom left) Emory Kristof/NG Image Collection; page 474: (top) Brian J. Skinner; page 474: (center) William Sacco; page 474: (bottom) Brian J. Skinner; page 475: (top) ©Martin Miller; page 475: (center) Brian J. Skinner; page 475: (bottom) ©William E. Ferguson; page 476: (top) ©Ken Lucas/Visuals Unlimited; page 477: (bottom) Woods Hole Oceanographic Institution; page 480: (top left) Phil Schermeister/NG Image Collection; page 480: (top right) Photo copyright Wendell E. Wilson; page 480: (bottom left) James L. Amos/NG Image Collection; page 480: (bottom right) James L. Amos/NG Image Collection; page 483: ©Roger Rossmeyer/Corbis

TABLE AND LINE ART CREDITS

Chapter 2

What A Geologist Sees, page 50, top: Adapted from Mineral Information Institute. "How Many Minerals and Metals Does It Take to Make a Light Bulb?" Retrieved July 5 2006 from www.mii.org/lightbulb.html. Reprinted by permission of The Mineral Information Institute, www. mii.org.

Chapter 4

Figure 4.1 and Figure 4.9: From Skinner, Brian J. and Stephen C. Porter, *The Dynamic Earth: An Introduction to Physical Geology*. Copyright 2003 John Wiley & Sons, Inc. Reprinted with permission of John Wiley & Sons, Inc. *Amazing Places*, page 111, Figure C: Adapted from Rubin, Ken. "Loihi Volcano—The Bathymetric Map of Loihi Seamount." Hawaii Center for Volcanology website. 1997. Retrieved October 6, 2006, from http://www.soest.hawaii.edu/GG/HCV/loihi.html. Reprinted by permission of the Hawaii Center for Volcanology, Ken Rubin.

Chapter 5

Case Study, page 122, Figure A: Satake, Kenji, "2004 Sumatra Earthquake." Figure S7. December 2004. Retrieved October 3, 3006 from http://staff.aist.go.jp/kenji.satake/animation.gif. Reprinted by permission of K. Satake, Geological Survey of Japan, AIST.

Chapter 6

Figure 6.9: Adapted from Myers and Brantley, "United States Geological Survey Open File Report 95-231," 1995. Figure 6.11B: Adapted from Sigurdsson, Haraldur, "Volcanic pollution and climate: The 1783 Laki Eruption," *EOS*, vol. 63, pages 601-602, 1982. Copyright 1982, American Geophysical Union. Modified by permission of the American Geophysical Union.

Chapter 9

Amazing Places, page 268, Figure C: "Figure 18.32," adapted from EARTH STRUCTURE, SECOND EDITION by Ben A. van der Pluijm and Stephen Marshak. Copyright © 2004 by W. W. Norton & Company, Inc. Used by permission of W. W. Norton & Company, Inc. This selection may not be reproduced, stored in a retrieval system, or transmitted in any form or by any means without the prior written permission from the publisher.

Chapter 11

Figure 11.8: From The New York Times Graphics, "Coastal Defenses Are Disappearing." August 30, 2005. Used by permission of The New York Times Agency. NOTE: Any future revisions, editions thereof in print and any other format require clearance by The New York Times Agency. *What a Geologist Sees*, page 313, Figure B: Adapted from Revenga, C., S. Murray, J. Abramovitz, and A. Hammond. "Watersheds of the World: Ecological Value and Vulnerability." Washington, DC: World Resources Institute. 1998. Modified by permission of World Resources Institute. 2006. Earth Trends: Environmental Information. Available at http://earthtrends.wri.org. Washington DC: World Resources Institute.

Chapter 12

Figure 12.1A: We have made every effort to secure proper permission for reproducing this image. However, it has not been possible to determine the current copyright owner. If you have information indicating who the copyright owner may be, please contact us at rflahive@wiley.com. Figure 12.10: A H. J. de Blij and Peter O. Muller, *Physical Geography and the Global Environment*, copyright © 1996 John Wiley & Sons, Inc. Reprinted by permission of John Wiley & Sons, Inc. Figure 12.17B: From Skinner, Brian J. and Stephen C. Porter, *The Dynamic Earth: An Introduction to Physical Geology*. Copyright 2003 John Wiley & Sons, Inc. Reprinted with permission of John Wiley & Sons, Inc.

Chapter 13

Geomorphology, Figure 13.5A: Adapted from K. Pye and L. Tsoar, *Aeolian Sand and San Dunes*, Figure 7.1, Chapman & Hall, 1990. Reprinted with kind permission of Springer Science and Business media. Figure 13.6, diagram: Adapted from John T. Hack, *The Geographical Review*, vol. 31, fig. 19, page 260 by permission of the American Geographical Society: *What a Geologist Sees*, page 400, Figures A and B: Adapted from A. H. Lachenbruch in Rhodes W. Fairbridge, ed., *The Encyclopedia of* Van Nostrand Reinhold, New York. Reprinted with kind permission of Springer Science and Business media. Figure 13.21: After Nigel Calder, *The Weather Machine*, BBC Publications, London 1974. Used by permission of Mr. Nigel Calder. Figure 13.23: From Skinner, Brian J. and Stephen C. Porter, *The Dynamic Earth: An Introduction to Physical Geology*. Copyright 2003 John Wiley & Sons, Inc. Reprinted with permission of John Wiley & Sons, Inc.

Chapter 14

Figure 14.1: MacKenzie, Fred. T. OUR CHANGING PLANET AN INTRODUCTION TO EARTH SYSTEM SCIENCE AND GLOBAL ENVIRONMENTAL CHANGE, 3rd Edition, © 2003, p. 205. Reprinted by permission of Pearson Education, Inc., Upper Saddle River, NJ. Figure 14.3: From Falkowski, Paul G. et al, "The Rise of Oxygen over the Past 205 Million Years and the Evolution of Large Placental Mammals." *Science* 30 Sep. 2005, Vol. 309, no. 5744, pages 2202–2204. Reprinted with permission of AAAS. Figure 14.21B: From Nichols, D.J, D. M. Jarzen, C. J. Orth, and P. Q. Oliver. "Palynological and Iridium Anomalies at Cretaceous-Tertiary Boundary, South-Central Saskatchewan". *Science* 14 February 1986, Vol. 231, no. 4739, pages 714–717. Reprinted with permission of AAAS.

Chapter 15

Figure 15.12: Adapted from Bookout, J. F., in *Episodes*, vol. 12 (4), 1989, figure 1. Used with the permission of the International Union of Geological Sciences. Figure 15.22: From Skinner, Brian J. and Stephen C. Porter, *The Dynamic Earth: An Introduction to Physical Geology*. Copyright 2003 John Wiley & Sons, Inc. Reprinted with permission of John Wiley & Sons, Inc.

Line drawings in the following figures have been adapted from Murck, Barbara W. and Brian J. Skinner, *Geology Today: Understanding Our Planet*. Copyright 1999 John Wiley & Sons, Inc. Reprinted with permission of John Wiley & Sons, Inc.

Chapter 1: 1.4; 1.5; *What a Geologist Sees*, page 8, diagram; 1.7; 1.8; 1.9; 1.10. **Chapter 2:** 2.1, left; 2.2; 2.3 row 2, column 2; 2.3 row 4, column 2; 2.5; Table 2.1; 2.13; 2.14; 2.15. **Chapter 3:** 3.3; 3.6; 3.8 3.13; 3.14 3.15; 3.16; *Case Study*, page 80, diagram.: **Chapter 4:** 4.2A and B; 4.3A, B, and C; 4.4B; 4.5;4.6; 4.7; 4.12; 4.13; 4.14; 4.15; *Amazing Places*, page 111, Figure D. **Chapter 5:** 5.3A; 5.6; 5.8B; 5.9; 5.10A and B; 5.11; 5.12 right; *What a Geologist Sees*, page 138, Figure C; 5.14A; 5.15; 5.16. **Chapter 6:** *Case Study*, page 159, Figures A and B; 6.10; 6.14; 6.15; 6.16; 6.20; 6.21; 6.22. **Chapter 7:** 7.1; 7.5; 7.8; 7.11; 7.12A; 7.13A; 7.17. **Chapter 8:** 8.1, top row; 8.2A and B; 8.5, top and middle; 8.12A ; 8.15; 8.16. **Chapter 9:** 9.2; 9.8; 9.9; 9.11; 9.12; 9.13A; 9.14; 9.15; 9.16; *What a Geologist Sees*, page 259, Figure B; 9.17B;

MAP CREDITS

Chapter 1
Page 9 (left): NG Map Collection; page 20–21: NG Map Collection

Chapter 2
Page 50: NG Map Collection

Chapter 4
Pages 102-103: NG Map Collection

Chapter 5
Page 118: NG Map Collection; page 122: NG Map Collection; page 144: NG Map Collection

Chapter 7
Page 205: NG Map Collection

Chapter 8
Page 220: NG Map Collection

Chapter 11
Page 323: NG Map Collection. Graph: Water by Volume) Peter Gleick, The Pacific Institute; (Map: Primary Watersheds) World Resources Institute; Peter Gleick, The Pacific Institute; (Artwork: "Hydrologic Cycle") Don Foley; (Map: Access to Fresh Water) World Health Organization; (Graph: Water Withdrawals) AQUASTAT-FAO. Water Consultant AARON WOLF, Oregon State University

Chapter 13
Pages 402-403: NG Map Collection

Chapter 14
Page 428: NG Map Collection; pages 444–45: NG Map Collection. (Map: World) Human Footprint Project © Wildlife Conservation Society (WCS) and Center for International Earth Science Information Network (CIESIN) 2006; Project leads: Eric Sanderson, Kent Redford, WCS; Marc Levy, CIESIN; Funding:Center for Environmental Research and Conservation (CERC) at Columbia University, ESRI Conservation Program, Prospect Hill Foundation

Chapter 15
Page 464 (right): NG Map Collection; page 472-73: NG Map Collection

INDEX

Note to reader: *italicized* entries refer to figures, locators with a "t" refer to tables.